Phosphors

Phosphors

Synthesis and Applications

edited by

Sanjay J. Dhoble
B. Deva Prasad Raju
Vijay Singh

PAN STANFORD PUBLISHING

Published by

Pan Stanford Publishing Pte. Ltd.
Penthouse Level, Suntec Tower 3
8 Temasek Boulevard
Singapore 038988

Email: editorial@panstanford.com
Web: www.panstanford.com

British Library Cataloguing-in-Publication Data
A catalogue record for this book is available from the British Library.

Phosphors: Synthesis and Applications

ISBN 978-981-4774-49-9 (Hardcover)
ISBN 978-042-9460-99-9 (eBook)

Contents

Preface

This book compiles and details cutting-edge research on luminescence. Rare earth–doped luminescent materials play an integral role in modern life, due to their tremendous applications ranging from scintillators, color displays, fluorescent lamps, and intensifying screens to dosimetry of ionizing radiations. The chemical composition, degree of structural disorder, defects, and the presence of dopants/impurities have a notable influence on the electronic and optical properties of luminescent materials. In recent decades, light-emitting diodes, more commonly referred to as LEDs, have brought about a major revolution in general illumination. LEDs are winning the global market with their worthy applications in newer fields, in addition to already existing ones.

This book focuses on research on luminescence materials for use in diverse fields, demonstrates the underlying physics based on luminescence, and lists its applications. It is illustrated throughout with excellent figures and references accompanying each chapter. Edited by prominent luminescence researchers, this book will appeal to anyone involved in luminescence research and its applications, especially advanced undergraduate-, graduate-, and postgraduate-level students of spectroscopy, solid state physics, luminescence, material synthesis, and optical properties and researchers working on the synthesis of optical materials, the characterization of luminescence materials, solid state lighting, radiation dosimetry luminescence, and phosphor applications.

Sanjay J. Dhoble

B. Deva Prasad Raju

Vijay Singh

2018

Chapter 1

Hydrodynamic Aspects on Sonoluminescence

Ho-Young Kwak
Mechanical Engineering Department, Chung-Ang University, Seoul 06974, Korea
kwakhy@cau.ac.kr

1.1 Introduction

The discovery of fogging on photographic plates by ultrasound waves by Marinesco and Trillat [1], and the fogging accompanied by a faint luminescence [2], led to obtaining sonoluminescence (SL) in various liquids [3]. Photomultiplier tubes (PMTs) enable us to find that the sonoluminescent flashes occur at the end of collapse [4] and occur periodically matched to the frequency of the ultrasound [5]. Fifty-three years after the discovery of SL, Gaitan's success [6, 7] in trapping a single bubble in a standing ultrasonic wave enabled systematic experiments on SL, which unveiled various exotic phenomena on SL [8–10].

SL is the light emission associated with the catastrophic bubble collapse of a gas bubble oscillating under an ultrasonic field [7]. SL is characterized by a few tens to hundreds of picosecond flashes of

Phosphors: Synthesis and Applications
Edited by Sanjay J. Dhoble, B. Deva Prasad Raju, and Vijay Singh

ISBN 978-981-4774-49-9 (Hardcover), 978-042-9460-99-9 (eBook)
www.panstanford.com

a continuous spectrum with no major peaks [8, 9], and the jitter between flashes is less than 50 ps [10].

In this article, how single-bubble sonoluminescence (SBSL) in water and sulfuric acid solutions and from a cloud of bubbles (multibubble sonoluminescence [MBSL]) in water can be explained hydrodynamically will be discussed. Also, the radiation mechanism from a sonoluminescing gas bubble and a bubble cloud will be discussed with respect to the gas temperature and pressure obtained from the hydrodynamic calculations. As an application of SL phenomena, synthesis of specialty nanomaterials at the MBSL condition will also be discussed.

1.2 Sonoluminescence from a Single Bubble

1.2.1 Hydrodynamics of a Single-Bubble Motion under Ultrasound

Consider a trapped bubble that oscillates synchronous to the ultrasonic wave. As is well known, one cannot obtain the instantaneous properties such as density, pressure, and temperature for the gas inside the bubble using the ideal gas law with a polytropic relation [11]. Instead, one should solve the mass, momentum, and energy equations to understand the gas behavior inside a bubble. In spherical symmetry, the mass and momentum equations for the gas inside the bubble are given by

$$\frac{\partial \rho_g}{\partial t} + \frac{1}{r^2}\frac{\partial}{\partial r}\left(\rho_g u_g r^2\right) = 0, \tag{1.1}$$

$$\frac{\partial}{\partial t}(\rho_g u_g) + \frac{1}{r^2}\frac{\partial}{\partial r}\left(\rho_g u_g^2 r^2\right) + \frac{\partial P_b}{\partial r} = 0, \tag{1.2}$$

where r is the distance from center, ρ_g is the gas density, and u_g is the gas velocity, which obeys $u_g(R_b,t) = U_b$. The solutions satisfying Eqs. 1.1 and 1.2 were obtained by Kwak and Yang [12]. These are given with time-derivative notation as

$$\rho_g = \rho_0 + \rho_r, \tag{1.3a}$$

$$u_g \frac{\dot{R}_b}{R_b} r, \tag{1.3b}$$

$$P_b = P_{b0} - \frac{1}{2}\left(\rho_0 + \frac{1}{2}\rho_r\right)\frac{\ddot{R}_b}{R_b}r^2, \tag{1.3c}$$

where $\rho_o(R_b)^3$ = const. and $\rho_r = ar^2/(R_b)^5$. The constant a is related to the gas mass inside a bubble by $a/m = 5(1 - \text{NBC})/(4\pi)$, with NBC = $(P_{bo}(R_b)^3/T_{bo})/(P_\infty(R_e)^3/T_\infty)$ where T_b and T_∞ are gas and ambient temperatures, respectively; R_e is the equilibrium bubble radius; and the subscript "o" denotes the properties at the bubble center. The linear velocity profile showing the spatial inhomogeneity inside the bubble is a crucial ansatz for the homologous motion of spherical objects [13], which is encountered in the gravitational collapse [14], and the quadratic pressure profile given in Eq. 1.3c, was verified by comparisons with direct numerical simulations [15].

Assuming that the internal energy for the gas inside a bubble is a function of gas temperature only as $de = C_{v,b}dT_b$, where $C_{v,b}$ is a constant volume specific heat, the energy equation is as follows:

$$\rho_g C_{v,b}\frac{DT_b}{Dt} = -\frac{P_b}{r^2}\frac{d}{dr}(r^2u_g) - \frac{1}{r^2}\frac{d}{dr}(r^2q_r), \tag{1.4}$$

where q_r is the radial component of heat flux inside a bubble. The viscous dissipation term in the internal energy equation also vanishes because of the linear velocity profile. Since the solutions given in Eq. 1.3 also satisfy the kinetic energy equation, only Eq. 1.4 needs to be solved. Wu and Roberts [16] and Moss et al. [17] tried to solve numerically the total energy equation without the heat transfer term along with the mass and momentum equations given in Eqs. 1.1 and 1.2, respectively. However, their calculations without considering the heat transfer inside the bubble and at the bubble wall overestimated the gas temperature inside the bubble at the collapse.

With some manipulation of the energy equation [12] one may obtain the following heat transport equation for the gas inside the bubble:

$$\frac{(\gamma-1)d}{r^2}\frac{d}{dr}(r^2q_r) = -\frac{\gamma P_b}{r^2}\frac{d}{dr}(r^2u_g) - \frac{DP_b}{Dt}, \tag{1.5}$$

where γ is the specific heat ratio of the gas. One can solve Eq. 1.5 by using the Fourier law with the solutions for ρ_g, u_g, and P_b that are given in Eqs. 1.3a, 1.3b, and 1.3c, respectively. The final solution of the heat transport equation can be represented by the superposition of the temperature distributions caused by the uniform pressure and by the radial pressure variation induced by the rapid change of the bubble wall acceleration, as can be seen in Eq. 1.3c, that is,

$$T(r) = T_b(r) + T_b'(r). \tag{1.6}$$

The temperature profile due to the uniform pressure distributions is well known [12] and is perfectly valid for a nonsonoluminescing gas bubble:

$$T_h(r) = \frac{B}{A}\cdot\left[-1+\sqrt{\left(1+\frac{A}{B}T_{b0}\right)^2 - 2\eta\frac{A}{B}(T_{bl}-T_\infty)\left(\frac{r}{R_b}\right)^2}\right], \tag{1.7}$$

where A and B are the coefficients in the temperature-dependent gas conductivity having a form such as $k_g = AT + B$, $\eta = (R_b/\delta)/(B/k_l)$, and k_l is the thermal conductivity of the liquid. The thickness of the thermal boundary layer, δ, for thermal conduction may be determined from the mass, momentum, and energy equations for liquid with the assumption of the quadratic temperature profile in that layer [18].

Abrupt temperature rise and subsequent quenching due to the bubble wall acceleration and the increase and decrease in the acceleration may be treated in another timescale [19], different from that of the bubble motion. This temperature distribution $T_b'(r)$, which considers no temperature gradient at the bubble center is given by [20]

$$T_b'(r) = -\frac{1}{40(\gamma-1)k_g'}\left(\rho_0+\frac{5}{21}\rho_r\right)\cdot\left[(3\gamma-2)\frac{\dot{R}_b\ddot{R}_b}{R_b^2}+\frac{\dddot{R}_b}{R_b}\right]r^4 + C. \tag{1.8}$$

The coefficient C may be determined from a boundary condition $k_g' dT_b/dr = k_l dT_b/dr$ at the wall, where T_l is the temperature distribution in the thermal boundary layer with different thickness δ'. That is,

$$C = \frac{1}{20(\gamma-1)}\left[(3\gamma-2)\dot{R}_b\ddot{R}_b R_b + \dddot{R}_b R_b^2\right] \cdot\left[\frac{\delta'}{k_l}\left(\rho_0+\frac{5}{14}\rho_{r=R_b}\right)+\frac{R_b}{2k_g'}\left(\rho_0+\frac{5}{21}\rho_{r=R_b}\right)\right]. \tag{1.9}$$

The temperature distribution inside a bubble may be estimated from Eqs. 1.7 and 1.8 with appropriate values of δ' and k_g'. The temperature distribution from Eq. 1.7 can be regarded as a background one. The gas conductivity at ultrahigh temperatures may be obtained from collision integrals [21]. The value of δ' is chosen so that proper bouncing motion results after the collapse, which is about 0.1 μm. Without heat loss, that is, $\delta' \to \infty$ or $k'_g \to 0$, the temperature of the gas inside a bubble goes up to 10^7 K, as confirmed in numerical calculations [16, 17].

The instantaneous radius of a bubble is obtained by the equation from the Keller and Miksis (KM) equation [22], which may be obtained from the mass and momentum conservation for the liquid surrounding the bubble as follows:

$$\left(1-\frac{U_\mathrm{b}}{C_\mathrm{B}}\right)R_\mathrm{b}\frac{dU_\mathrm{b}}{dt}+\frac{3}{2}U_\mathrm{b}^2\left(1-\frac{U_\mathrm{b}}{3C_\mathrm{B}}\right)=\frac{1}{\rho_\infty}\left(1+\frac{U_\mathrm{b}}{C_\mathrm{B}}+\frac{R_\mathrm{b}}{C_\mathrm{B}}\frac{d}{dt}\right)$$
$$\cdot\left[P_\mathrm{B}-P_\mathrm{s}\left(t+\frac{R_\mathrm{b}}{C_\mathrm{B}}\right)-P_\infty\right], \qquad (1.10)$$

where R_b is the bubble radius, U_b is the bubble wall velocity, C_B is the sound speed in liquid at the bubble wall, and ρ_∞ is the medium density. The liquid pressure on the external side of the bubble wall, P_B, is related to the pressure inside the bubble wall P_b by $P_\mathrm{B} = P_\mathrm{b} - 2\sigma/R_\mathrm{b} - 4\mu U_\mathrm{b}/R_\mathrm{b}$, where σ and μ are the surface tension and dynamic viscosity of the liquid, respectively. The pressure of the deriving sound field, P_s, may be represented by a sinusoidal function such as $P_\mathrm{s} = P_\mathrm{A}\sin\omega t$, where P_A is the deriving sound amplitude $\omega = 2\pi f_\mathrm{d}$ and f_d is the driving frequency. At the limit of infinite sound velocity, the KM equation reduces to the well-known Rayleigh–Plesset (RP) equation.

The mass and energy equation for the liquid provides a time-dependent first-order equation for the thermal boundary layer thickness δ with a quadratic in temperature profile, which is given by [23]

$$\left[1+\frac{\delta}{R_\mathrm{b}}+\frac{3}{10}\left(\frac{\delta}{R_\mathrm{b}}\right)^2\right]\frac{d\delta}{dt}=\frac{6\alpha}{\delta}-\left[2\frac{\delta}{R_\mathrm{b}}+\frac{1}{2}\left(\frac{\delta}{R_\mathrm{b}}\right)^2\right]\frac{dR_\mathrm{b}}{dt}$$
$$-\delta\left[1+\frac{1}{2}\frac{\delta}{R_\mathrm{b}}+\frac{1}{10}\left(\frac{\delta}{R_\mathrm{b}}\right)^2\right]\frac{1}{T_\mathrm{bl}-T_\infty}\frac{dT_\mathrm{bl}}{dt}, \qquad (1.11)$$

where α is the thermal diffusivity of liquid. The above equation determines the heat flow rate through the bubble wall. The values for instantaneous bubble radius, bubble wall velocity, and acceleration and the thermal boundary layer thickness obtained from Eqs. 1.10 and 1.11 are used to calculate the density by Eq. 1.3a, velocity by Eq. 1.3b, pressure by Eq. 1.3c, and temperature profile for the gas inside the bubble by Eq. 1.6, with Eqs. 1.7 and 1.8 without any further assumptions.

1.2.2 Numerical Integration of Equations for Single-Bubble Motion

Usually, it is not easy to integrate Eq. 1.10 numerically without normalization using appropriate physical variables. It is natural to choose normalization variables, which can characterize the bubble motion. For instance, normalization of the governing equation may be done as follows [12, 20]. The radius is compared to the equilibrium radius R_0, and the velocity and pressure are related to constants, $u_0 = (P_\infty/\rho_\infty)^{1/2}$ and $P_0 = P_\infty$, respectively. The constants for normalizing other physical quantities such as time, dynamic viscosity, and surface tension of the liquid were obtained from the condition that the KM equation becomes homogeneous after the normalization procedure. The governing equation may be normalized by the timescale of $t_0 = R_0/u_0$, the characteristic time of the bubble motion.

Also, the time t and the driving frequency f_d in the sinusoidal term in Eq. 1.10 were normalized by t_0 and $1/t_0$, respectively [24, 25], so that the nondimensional form of the sinusoidal term becomes $\cos(2\pi ft)$, where $f = f_d t_0$ is the nondimensional frequency (single-timescale normalization [STN] method). However, the STN method yields a bubble behavior moving in phase with the driving ultrasound, which might cause an artificial resonance [26]. The reduction in the nondimensional frequency in the sinusoidal term with STN yields quite different results for bubble motion in numerical evaluation. Note also that the expansion ratio R_{max}/R_0 [24] and consequently the gas temperature and pressure and the bubble wall velocity at the collapse point increase considerably by the reduction in the nondimensional frequency $f_d t_0$ with the STN method. Certainly, the STN method for numerical evaluation of Eq. 1.10 is correct in the mathematical sense. However, the method does not yield the correct value for the expansion ratio [26]. For a sonoluminescing gas bubble with an equilibrium radius of R_0 = 5.0 μm at the driving frequency of 12.926 kHz and the driving amplitude of 1.45 atm, the maximum bubble radius estimated using the RP equation or the KM equation with STN is 86 μm, while the observed maximum radius is about 48.8 μm [27].

In general, the characteristic frequency of the driving force, f_0, differs from the natural frequency of bubble oscillation, t_0. Because the bubble wall motion described by Eq. 1.10 is also normalized with this normalization method, but the nondimensional time is recovered by $1/f_0$ in the numerical procedure (two-timescale normalization [TTN] method), there is a lag time of the bubble motion with respect to the characteristic time of the applied ultrasound, f_0, which is defined as [26]

$$\tau = \frac{1}{f_0} - t_0. \tag{1.12}$$

The bubble behavior under ultrasound can be described correctly with the concept of such lag time [26, 28]. It has also been found experimentally that a single bubble SL at a low driving frequency cannot be up-scaled significantly because of the lagging motion of the bubble to the ultrasound [27]. The TTN method introduced above is a physical approximation that represents the lagging behavior of bubble motion (a nonlinear oscillator) under ultrasound. The calculated radius–time curves for the bubble with an equilibrium radius of 8.5 μm by the KMNS method (KMNS = Keller–Miksis equation and the solutions for the Navier–Stokes equation for the gas inside the bubble) with TTN, driven by the ultrasonic field with a frequency of 26.5 kHz and amplitude of 1.075 atm is shown in Fig. 1.1. The case shown in Fig. 1.1 is certainly below the SL threshold. The calculated values of the maximum radius and the period for each bouncing motion are in good agreement with the observed one [29]. However, the bubble radius–time curve obtained by the RP method with STN shows 10 number of bouncing motions rather than 7. Also the magnitude of the maximum bubble radius at the first bounce is significantly less than the observed one.

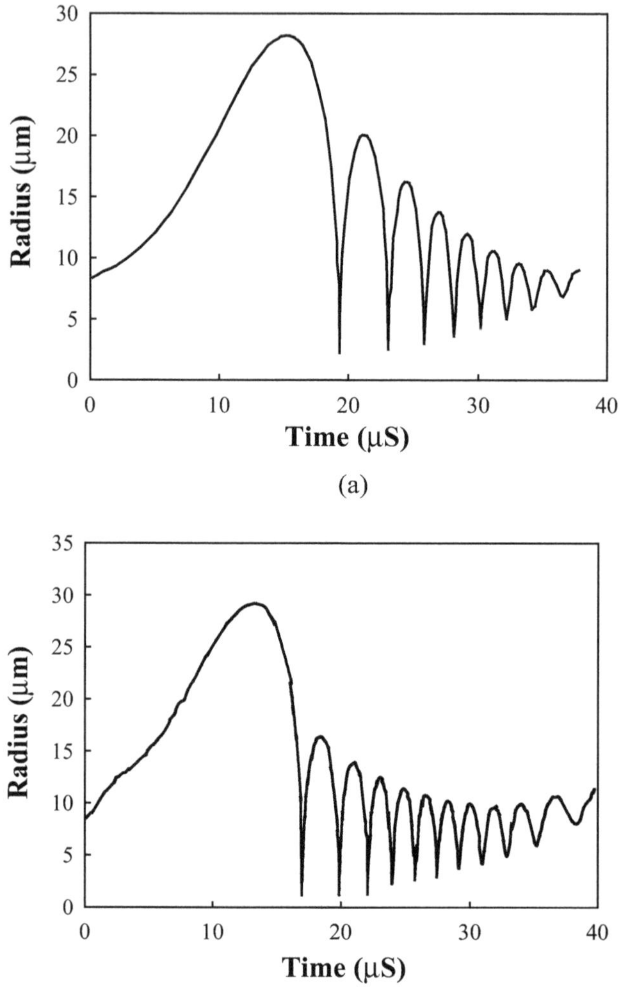

(a)

(b)

Figure 1.1 Theoretical radius–time curves for an air bubble of R_0 = 8.5 μm at P_A = 1.075 atm and f_d = 26.5 kHz in water by the KMNS method with TTN (a) and by the RP method with STN (b) [12].

The change in the driving frequency due to the introduction of STN gives quite different results in the bubble wall velocity, which was unexpected; the bubble wall velocity at the collapse point exceeds 2000 m/s [25], which is an invalid result by the KM equation that employs acoustic approximation, because the observed value was well below 1000 m/s [30]. KMNS with TTN provides a bubble wall velocity of about 600 m/s at the collapse point, a velocity that is closer to the observed one [30]. The calculated bubble wall velocity by the RP method with $\tau = 0$ for a case of $R_0 = 5.4$ μm at $P_A = 1.37$ atm and $f_d = 28.84$ kHz well exceeds the sound velocity of water 1481 m/s at collapse. However, the calculated maximum velocity by the KMNS method with $\tau = 0.26$ μs is about 900 m/s. Considering the influence of Mie lobe clusters and the changes in the refractive indices at the end phase of the bubble collapse, Gompf and Pecha [31] obtained a bubble wall velocity of 1 ns before the SL pulse of about 950 m/s, which is a much lower value than that predicted by the RP method with STN.

1.2.2.1 SBSL in water

The temperature distribution for the gas inside the bubble is crucially dependent on the boundary condition at the bubble wall and the thermal conductivity for the gas inside the bubble. Consider a bubble with an equilibrium radius of $R_0 = 4.1$ μm driven at $f_d = 40$ kHz and $P_A = 1.45$ atm [32] as an example case. However, the adjusted parameter of $R_0 = 5.5$ μm was used in this study so that the calculated radius–time curve around the collapse point is well fit to the observed one. As shown in Fig. 1.2, it is very hard to obtain the observed radius–time curve around the collapse point with a bubble of $R_0 = 4.1$ μm, driven at $P_A = 1.45$ atm and $f_d = 40$ kHz by the RP method with STN [32]. Even with this small equilibrium radius, the minimum bubble radius predicted by the RP method with STN is much smaller than the observed value. Further deviations from the bubble radius–time curve predicted by the RP method with STN are apparent, which yield over-prediction in the bubble wall velocity and correspondingly in the acceleration. The bubble wall acceleration calculated by the KMNS method with TTN of a relaxation time of 0.31 μs is about 3.46×10^{12} m/s^2, which is close to the observed value of 3.5×10^{12} m/s^2, while the RP method with STN yields a bubble wall acceleration of 1.6×10^{15} m/s^2 [33]. The calculation results discussed in this article are from the RP method with STN and the KMNS method with TTN generally.

Consider the maximum temperature achieved for the gas inside the bubble due to compression either by the adiabatic or by the polytropic process. Assume that the conductivity for the gas is linearly dependent on the gas temperature, such as $k_g = AT + B$. For air, $A = 5.528 \times 10^{-5}$ W/mK^2 and $B = 1.165 \times 10^{-2}$ W/mK give a good fit to the thermal conductivity in the range of 200 K < T < 3000 K [11]. With this conductivity of air, the gas temperatures at the bubble center and bubble wall by Eq. 1.7 with the KM equation are 8043 K and 5004 K, respectively, as shown in Fig. 1.3. An unreasonably small value of the heat conductivity for the gas inside the bubble yields the highest temperature of 12,303 K at the center and the heat bath boundary condition of T_{bl} = 304 K at the bubble wall, which is a limiting case of no heat transport inside the bubble.

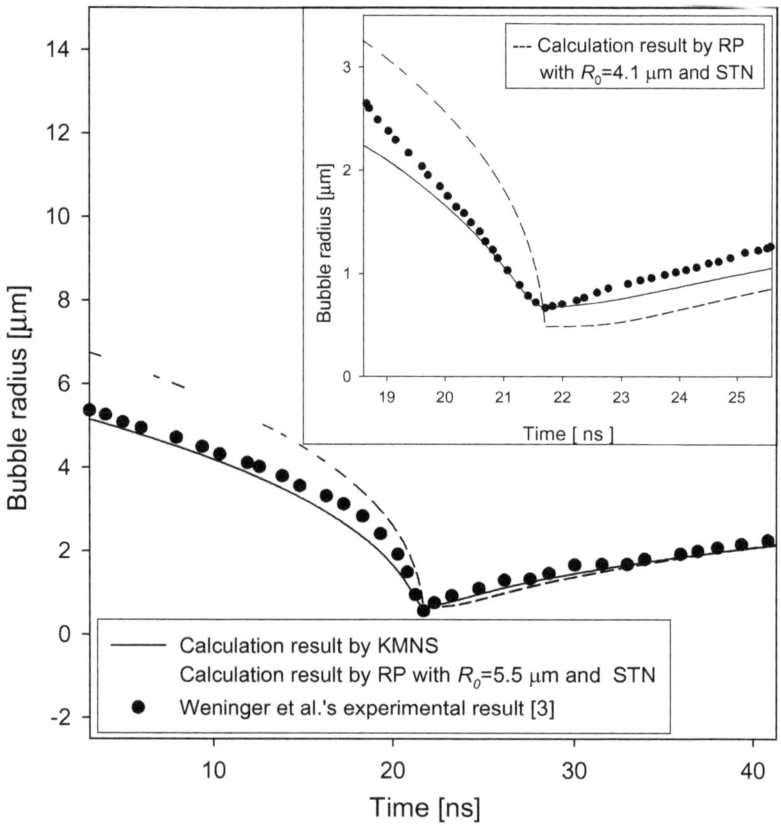

Figure 1.2 Bubble radius–time curve around the collapse point for a bubble of R_0 = 5.5 μm driven at f_d = 40 kHz with P_A = 1.45 atm. The solid line indicates calculation results by the KMNS method with TTN, and the dashed line shows calculation results by the RP method with STN [33]. Full circles denote experimental results by Wininger et al. [32].

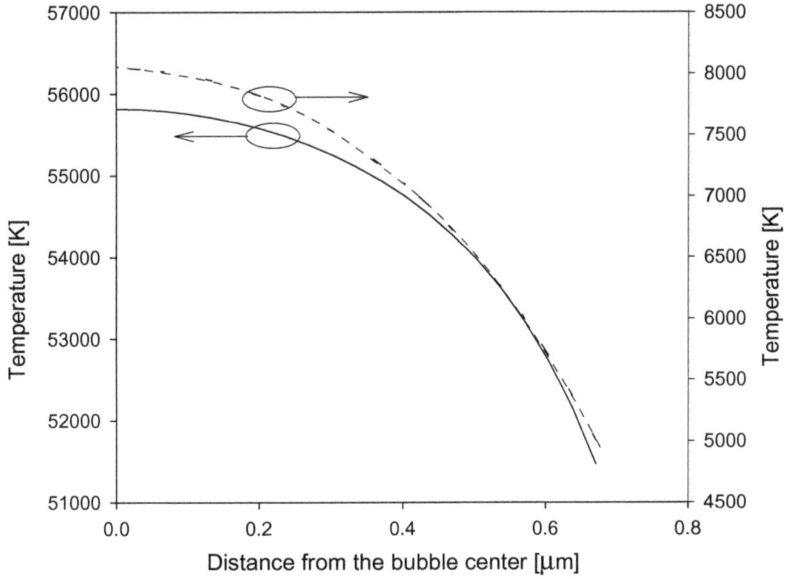

Figure 1.3 Temperature distribution at the collapse point for a bubble of R_0 = 5.5 μm driven at f_d = 40 kHz and P_A = 1.45 atm for the case calculated with Eq. 1.7 (-) and with Eq. 1.6 (—) [33].

Of course, finite heat transfer at the bubble wall may be considered by using a polytropic relation of the form, TV^{n-1} = const., where n is the polytropic index. One can obtain the gas temperature of 8000 K at the collapse point [34] with the RP method and the polytropic index of 1.30. However, the use of a polytropic relation in the study of the sonoluminescing gas bubble is not adequate because the bubble behavior cannot be described by either isothermal or adiabatic processes and the temperature distribution inside the bubble at the collapse is no longer uniform since the relaxation time of the vibrational motion (10^{-6}–10^{-9} s) is longer than the characteristic time at that moment [35, 36].

The gas conductivity obtained from the collision integral [21] was taken to be 10.0 W/mK because of very dense gas density at the collapse point [37]. The temperature distribution at the collapse point may be obtained from Eq. 1.6 with the boundary condition of the continuity of heat flux at the wall, such as $k'_g dT_b/dr = k_l dT_l/dr$. The conductivity of liquid k_l also affects the gas temperature at the bubble center significantly. The conductivity of liquid around the bubble wall was taken to be 1.0 W/mK in this study. The calculated temperature at the center and the wall by the KMNS method with TTN are 55,819 K and 51,893 K, respectively, as can be seen in Fig. 1.3. On the other hand, with the lower values of the thermal conductivity of $k_g' = 0.07$ W/mK and $k_l = 0.01$ W/mK, the uniform temperature obtained is as much as 3,061,589 K. Certainly the non-uniformity in the pressure and temperature gives significant deviations from the simple uniform model.

It is not a surprising result that Ruuth et al. [38] obtained a maximum temperature of 500,000 K at the center by using a molecular dynamic simulation with the heat bath boundary condition.

1.2.2.2 SBSL in sulfuric acid solutions

Various nonlinear behaviors for the SL bubbles in a sulfuric acid solution were observed, which are also captured by the KMNS method. The calculated radius–time curve for a bubble with R_0 = 15 μm, driven by the ultrasonic field with a frequency 37.8 kHz and amplitude of 1.5 atm, in an aqueous solution of sulfuric acid is shown in Fig. 1.4a. The calculated radius–time curve that exactly mimics the alternating pattern of the observed result shows two different states of bubble motion: a light-emitting cycle after a non-light-emitting cycle is repeated. This happens due to the heat transfer across the bubble wall: more heat transfer in one cycle induces a low minimum radius at the collapse point, which, in turn, produces a larger maximum bubble radius in another cycle. Furthermore, the added mass due to the increase in medium density and the heat transfer through the bubble wall reduce the expansion ratio considerably. The calculated minimum bubble radius for the light-emitting cycles, 4.6 μm, is close to the observed value of 3.7 μm [39]. As shown in Fig. 1.4b, the calculated radius–time curve obtained by using the RP method with a polytropic relation does not show the alternating pattern [40].

Figure 1.5a shows the time-dependent bubble wall velocity and the variation of the bubble wall acceleration around the collapse point for the bubble shown in Fig. 1.4a. The calculated magnitude of the minimum velocity at the collapse point for the light emitting cycles is about 115 m/s, which is close to the observed velocity of 120 m/s, whereas the maximum bubble wall velocity for non-light-emitting cycle is about 88 m/s, which is also close to the observed results of 80 m/s [39]. However, the magnitude of the minimum velocity, which is about 900 m/s, calculated by the RP method with a polytropic relation, is much higher than the observed value. The maximum bubble wall acceleration is about 10^{10} m/s^2. This value is smaller than the case of the sonoluminescing gas bubble in water by 2 orders of magnitude, so the gas pressure inside the bubble is almost uniform and the temperature increase due to the bubble wall acceleration is as small as 300°C.

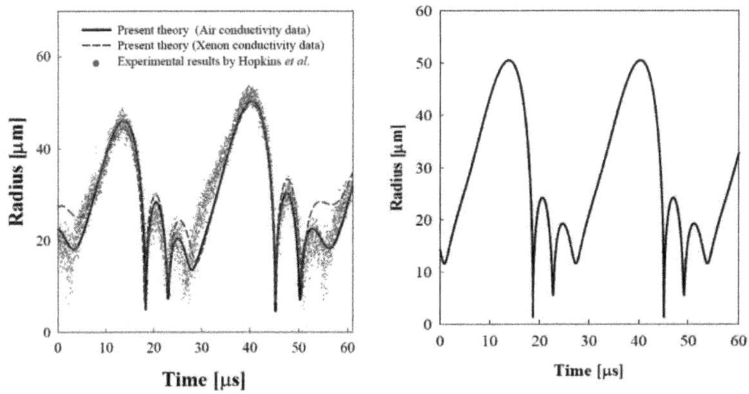

(a) (b)

Figure 1.4 Theoretical radius–time curve obtained by the KMNS method with TTN (a) along with the observed one by Hopkins et al. [39] for a xenon bubble of R_0 = 15.0 μm at P_A = 1.50 atm and f_d = 37.8 kHz in a sulfuric acid solution, and the curve calculated by the RP method with a polytropic relation (b) [40].

Figure 1.5b shows the calculated time-dependent bubble center temperature and the temporal emission die to electron–atom collisions from the bubble. The peak temperature calculated at the bubble center is about 8200 K, which is in excellent agreement with the observed value of 7000 K. In fact, the effective temperature of the light emission is about 7000 K because a considerable temperature drop occurs at the bubble wall. Such a low gas temperature of 8000 K and the steep gradient in temperature inside the bubble are due to extensive heat transfer through the bubble wall, which is possible when the thermal conductivity of the gas has similar value as the one of the medium. Our calculated gas pressure at the collapse point for the argon bubble with $R_0 = 13$ μm driven at $P_A = 1.4$ bar and $f_d = 28.5$ kHz in a sulfuric acid solution is about 2800 atm, which is also close to the lower bound value of the observed result, 1600 atm [41].

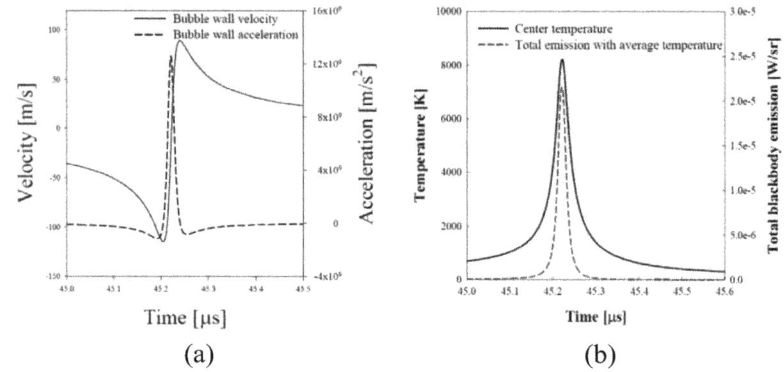

Figure 1.5 Calculated bubble wall velocity and acceleration near the collapse point by the KMNS method with TTN (a) for the bubble shown in Fig. 1.4a. Gas temperature at the bubble center and the corresponding total emission due to electron–atom collisions with the average temperature around the collapse point [40].

1.3 Radiation Mechanism for a Sonoluminescing Gas Bubble

1.3.1 Theory

Many attempts have been made to model the spectrum of SL. A continuous spectrum with no major peaks may be obtained from black-body radiation or *bremsstrahlung* and/or possibly from recombination radiation [42]. The source for the SL spectrum was considered as *bremsstrahlung* in partially or fully ionized gases produced due to the ultrahigh temperature. In such a high-temperature environment, even the total radiation power whose full width at half maximum (FWHM) is less than 500 ps, the gas molecules may dissociate and/or ionize, and hence, elastic collisions between ions and electrons can be imagined.

The classical approach with consideration of the Maxwell distribution of electrons and population of ions [43] may provide the emission coefficient j_ω due to *bremsstrahlung* from the gas for one polarization. However, a quantum theory calculation provides a more viable result, which is good at low as well as high frequencies [44]. That is,

$$j_\omega = n_i n_e Z^2 V_b \left(\frac{e^2}{4\pi\varepsilon_0}\right)^3 \left(\frac{m_e}{2\pi K_B T_e}\right)^{1/2} \cdot \frac{1}{m_e^2 c^3} \frac{4}{3\pi} \left[\frac{\pi}{\sqrt{3}} \bar{G}(T,\omega)\right], \quad (1.13)$$

where n_i and n_e are the number densities of ions and electrons, respectively, in the bubble volume V_b. The electron temperature T_e is obtained by assuming a Maxwellian electron velocity distribution, which characterizes the isotropicity of the spectrum. In a high-frequency regime (Born approximation), the Gaunt factor obtained by summation over the electron velocity distribution [44] is

$$\bar{G}(T,\omega) = \sqrt{\frac{3}{\pi}} \exp\left(-\frac{\hbar\omega}{k_B T_e}\right) \Big/ \sqrt{\hbar\omega / k_B T_e}. \quad (1.14)$$

The corresponding spectral radiance turns out in terms of wavelength λ to be

$$j_\lambda = n_i n_e Z^2 V_b \cdot \left[\frac{4\pi^3}{3} \left(\frac{8 k_B T_e}{\pi m_e}\right)^{1/2} \left(\frac{e^2}{4\pi\varepsilon_0}\right)^3 \frac{1}{m_e c^3 h} \frac{\pi}{\sqrt{3}} \bar{G}(T,\omega)\right]$$
$$\cdot \left[\sqrt{\frac{\pi}{3}} G(T,\lambda)\right] \frac{hc}{k_B T_e \lambda} \cdot \frac{1}{\lambda} \cdot \frac{1}{2\sqrt{3}\pi^2}. \quad (1.15)$$

The spectral radiance is maximum when $dj_\lambda/d\lambda = 0$ [20, 35], which results in

$$\lambda_{\max} = \frac{2hc}{3 k_B T_e}. \quad (1.16)$$

The emission intensity given in Eq. 1.15 varies as $1/\lambda^{2.5}$ as observed in experiments [8]. On the other hand, the maximum occurs at $5126/T_b$ (μm) for black-body radiation. The total emission of *bremsstrahlung* is obtained by integrating j_ω over all frequencies. The result is found to be applicable, which is given by [43]

$$j = \int j_\omega d\lambda = n_i n_e Z^2 V_b \left[\frac{4\pi^3}{3} \left(\frac{8 k_B T_e}{\pi m_e}\right)^{1/2} \left(\frac{e^2}{4\pi\varepsilon_0}\right)^3 \frac{1}{m_e c^3 h}\right]. \quad (1.17)$$

The Rosseland mean free path l_R for the case of the fully ionized gas with absorption by electron–ion *bremsstrahlung* [45], which corresponds to the emission coefficient given in Eq. 1.13 is given by

$$l_R = \frac{1}{\kappa_\lambda} = \left[\frac{4}{3}\pi^3\left(\frac{8k_B T_e}{\pi m_e}\right)^{1/2}\frac{\lambda^3}{hm_e c^4}\left(\frac{e^2}{4\pi\varepsilon_0}\right)^3\right]^{-1}\frac{2\sqrt{3}\pi^2 k_B T_e}{n_i n_e Z^2}. \quad (1.18)$$

The hemispherical spectral radiance from a light source in any medium may be described as [45–47]

$$j_\lambda^\Delta = 4\pi R_b^2 e_{b\lambda} \times \left[1 + \frac{\exp(-2\kappa_\lambda R_b)}{\kappa_\lambda R_b} + \frac{\exp(-2\kappa_\lambda R_b) - 1}{2\kappa_\lambda^2 R_h^2}\right], \quad (1.19)$$

where κ_λ is the absorption coefficient of the photon and $e_{b\lambda}$ is the hemispherical emissive power from a black-body source, which is given by

$$e_{b\lambda} = \frac{2\pi hc^2}{\lambda^5 (e^{hc/\lambda k_B T} - 1)}. \quad (1.20)$$

For the case of small absorption, Eq. 1.19 may be written within the limit of $\kappa_\lambda R_b < 1$ as

$$j_\lambda^\Delta = \frac{4}{3}\kappa_\lambda R_b \cdot 4\pi R_b^2 e_{b\lambda}. \quad (1.21)$$

The above equation, which represents the case of light emission with finite absorption or optical thickness of $4\kappa_\lambda R_b/3$, yields the same spectral radiance for *bremsstrahlung* due to ion–electron collision with the Rosseland free path by taking one polarization except $1/\sqrt{hc/\lambda k_B T_e}$ in the Gaunt factor.

The mean absorption coefficient in a weakly ionized medium may be calculated using the following equation with an induced emission correction [45]:

$$\kappa_\nu = n_a n_e a_{\nu,\text{class}} = \left(\frac{e^2}{4\pi\varepsilon_0}\right)\frac{n_e \nu_{\text{eff}}}{\pi mc\nu^2} = \frac{1}{l_A} \quad (1.22)$$

Here $\upsilon_{\text{eff}} = n_a \upsilon \sigma_{\text{tr}}$, where υ_{eff} is the effective frequency of electron–atom collisions, v is particle velocity, σ_{tr} is the transport scattering cross-section, and n_a is the number density of atoms. With this absorption coefficient, the spectral radiation due to electron–atom collisions for one polarization may be written as

$$j_\lambda = n_i n_e V_b \left[\left(\frac{8k_B T_e}{\pi m_e} \right)^{1/2} \left(\frac{e^2}{4\pi\varepsilon_0} \right) \frac{h\sigma_{tr}}{\pi m_e c} \right] \frac{1}{\lambda^3} \Big/ \left[\exp\left(-\frac{hc}{k_B T_e \lambda} \right) - 1 \right]. \tag{1.23}$$

The corresponding total emission for one polarization is given by

$$j = n_i n_e V_b \left[\frac{\pi^2}{6} \left(\frac{8k_B T_e}{\pi m_e} \right)^{1/2} \left(\frac{e^2}{4\pi\varepsilon_0} \right) \frac{h\sigma_{tr}}{\pi m_e c} \right] \left(\frac{k_B T_e}{hc} \right)^2. \tag{1.24}$$

The number density of electrons, n_e, was obtained using the following the law of mass action for the simple-ionization reaction, which is pressure dependent [43]:

$$\frac{\phi^2}{1-\phi^2} = \frac{II}{P_b} \left(\frac{2\pi m_e}{h^2} \right)^{3/2} (k_B T_b)^{5/2} \exp\left(-\frac{\theta_i}{T_b} \right), \tag{1.25}$$

where ϕ is the degree of ionization and θ_i is the characteristic temperature for ionization. If electron excitation above the ground level is ignored, the value of and the product of θ_i, which is related to the internal partition function due to the electron excitation, becomes nearly constant. It is noted that departure from the Saha equation for ionization should be recognized in such transient event [43] like SL.

The emission by electron–atom scattering, given in Eq. 1.23 becomes comparable [47] to that by electron–ion scattering, given in Eq. 1.15, if the degree of ionization is less than 0.01, which corresponds to an electron temperature of approximately 20,000 K [48]. For the light from the MBSL condition where the gas temperature is approximately 10,000 K, the spectral radiation due to electron–atom collisions can be used. Once the time-dependent temperature of the gas inside the bubble around the collapse point is known, one can calculate the total radiance using either Eq. 1.17 or Eq. 1.24 and obtain the FWHM for the light emission.

Assuming that the light emission is due to black-body radiation, the time-dependent temperature change and the corresponding radiance during the collapsing phase for the bubble considered in Fig. 1.2 are shown in Fig. 1.6. The data in Fig. 1.6 are calculated by using the RP equation with a polytropic relation and with STN, a conventional calculation method in the SL study [24, 25]. As clearly seen in Fig. 1.6, the light pulse duration, which may be considered as the time duration of the gas temperature above 2000 K is about 1.6 ns, even though the FWHM of the spectrum is about 230 ps. Furthermore, the rising half of the pulse calculated is not Gaussian and the pulse shape is rather lopsided, which is quite different from the observed shape. The temporal change of the gas temperature and the corresponding radiance with the electron–ion *bremsstrahlung* assumption for the bubble shown in Fig. 1.2 are shown in Fig. 1.7. The calculated FWHM of the radiance is about 166 ps, which is in close agreement with the observed value of 150 ps. As clearly seen in Fig. 1.7 (inset), the ionization rate is positive until the collapse point so that the recombination reaction hardly occurs.

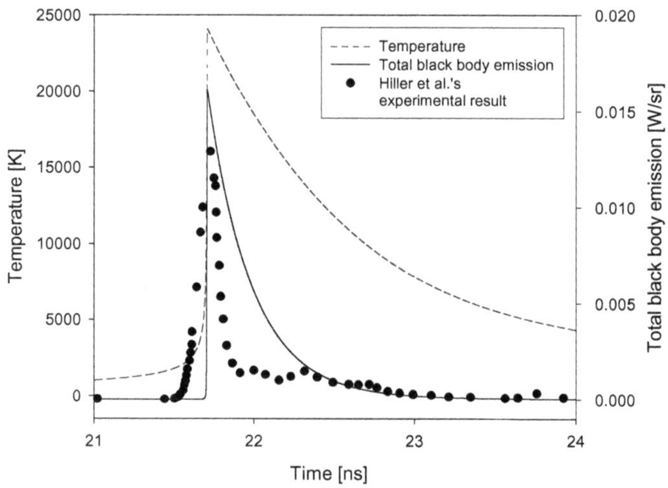

Figure 1.6 Total black-body emission and the gas temperature depending on time for the bubble shown in Fig. 1.2. The calculation was done by the RP method with STN [33]. Full circles denote experimental results by Weninger et al. [32].

Recombination radiation, which broadens the radiance, is possible and was considered during the quenching after the collapse point. The overall shape of the SL pulse is Gaussian, as shown in Fig. 1.7. However, Moran and Sweider [49] observed a small after pulse at about 250 ps, which might be due to a brief heating of the gas inside the bubble associated with reflected pressure waves [50, 51]. Also note that the Rosseland mean free path l_R given in Eq. 1.18 at a temperature range of 50,000 K, which is about 2.5 mm, and the photon–matter mean free path l_A given in Eq. 1.22, at a temperature of 13,000 K, are much greater than the bubble radius at the collapse point, so black-body radiation hardly occurs. Furthermore, the upper limit of the size of the light-emitting region, which is about 0.2 μm [52], is much smaller than the photon–matter mean free path calculated at a temperature range between 15,000 K and 50,000 K.

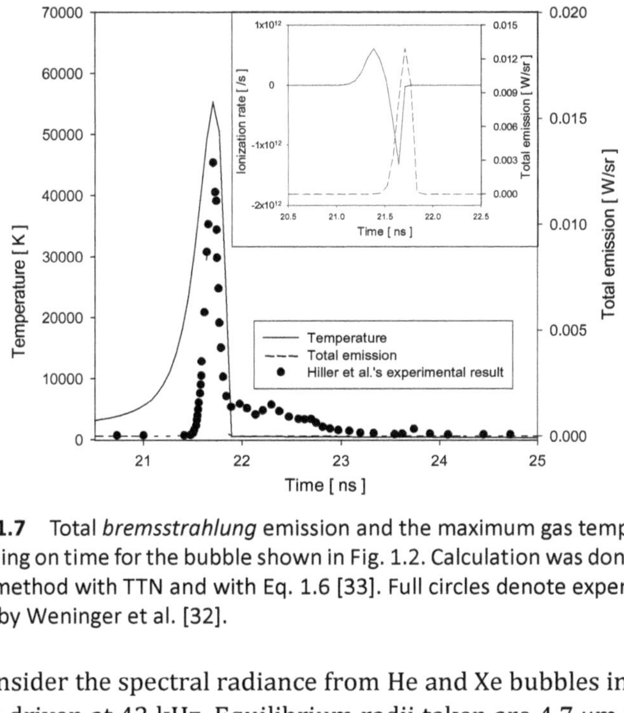

Figure 1.7 Total *bremsstrahlung* emission and the maximum gas temperature depending on time for the bubble shown in Fig. 1.2. Calculation was done by the KMNS method with TTN and with Eq. 1.6 [33]. Full circles denote experimental results by Weninger et al. [32].

Consider the spectral radiance from He and Xe bubbles in water (23°C) driven at 42 kHz. Equilibrium radii taken are 4.7 μm for the He bubble and 5.5 μm for the Xe bubble, which are the measured values by light-scattering technique. The driving amplitudes chosen in the calculation are 1.45 atm for the He bubble and 1.34 atm for the Xe bubble. The spectral radiances calculated with these parameters at the peak temperature at the collapse point as well as the observed ones are shown in Fig. 1.8. The calculation results suggest that the gas temperature, which shapes the spectral radiance, is determined by the hydrodynamic parameters such as equilibrium radius and driving amplitude. The two bubbles should have different temperatures so that the observed spectrum from the He bubble is comparable to the one from the Xe bubble in its magnitude. Further the two bubbles have different emission origin, for example, electron–atom *bremsstrahlung* for the Xe bubble having temperature of 12500 K, and electron–ion *bremsstrahlung* for the He bubble having temperature of 48000 K, which are shown in Fig. 1.8.

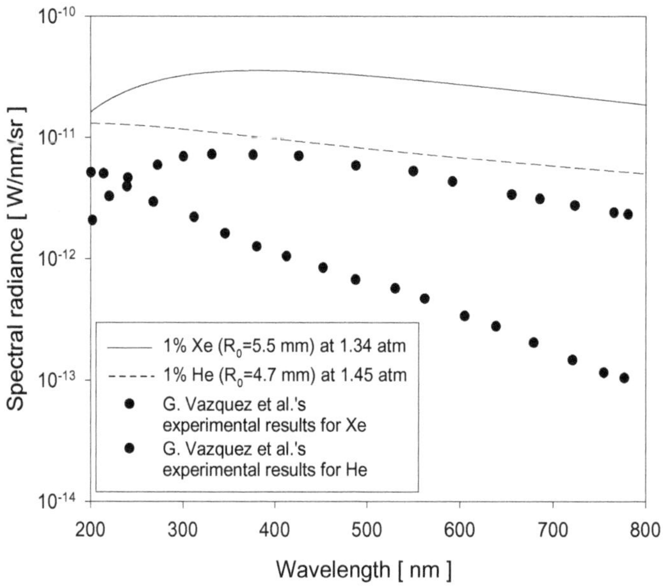

Figure 1.8 Spectral radiance from a bubble of He (150 Torr) and of Xe (3 Torr) in water driven at 40 kHz [33]. Observed data are taken from Ref. [34].

Camara et al.'s argument [53] on the black-body emission from a sonoluminescing gas bubble driven at 30 ~ 40 kHz is as follows. Because the core region of the gas bubbles at the collapse point is so hot that the radius light-emitting region R_e is greater than the photon–matter mean free path l_R defined in Eq. 1.18, light emits at the bubble's surface whose temperature is about 8000 K. However, Camera et al.'s argument is not true as discussed in the previous paragraph.

1.3.2 Spectrum Measurements

A schematic of the experimental setup to measure the spectral intensity depending on wavelength [42] is illustrated in Fig. 1.9. A microbubble is acoustically levitated in a quartz cylindrical cell with equal dimension of diameter and height of 3.8 cm. The bubble is driven by a cylindrical lead zirconate titanate (PZT) transducer (Channel Inc. USA) attached to the square glass plate, which is cemented on the top of the test cell. The resonant frequency is approximately 38.6 kHz at room temperature in water. The transducer has dimensions of 44.5 mm diameter and 25.4 mm height. A nichrome wire for bubble generation was installed inside the resonator. The temperature of water in the test cell was monitored by a T-type thermocouple.

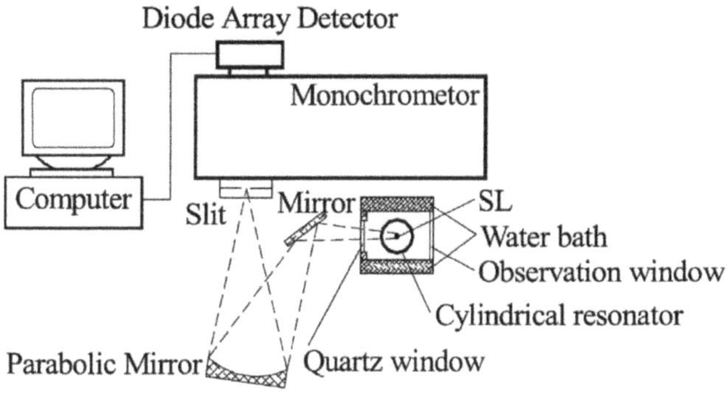

Figure 1.9 Experimental apparatus for spectrum measurement of sonoluminescence [42].

To control the temperature of liquid in the test cell, it was placed in a rectangular box in which the temperature of air inside was controlled by circulating water. The box was constructed with an aluminum frame and was covered by foam material, or insulation. The internal volume is 105 mm by 120 mm by 150 mm tall. A rectangular quartz window was made on one side to allow passage of light out of the box to the mirror. At the opposite side, a plain glass window was made to monitor the light. The system allows the resonator temperature to be controlled for about 5°C to room temperature.

As shown in Fig. 1.9, the light from a sonoluminescing gas bubble, which was collected by a parabolic mirror with a focal length of 250 mm was made to be focused at the slit of the monochrometer (Spectra Pro 275). The 50 mm D flat mirror, which is 85 mm from the bubble, was employed to put the bubble at the focal point of the parabolic mirror. About 1/200 of the light for the SL bubble can be reasonably collected by such an arrangement without serious chromatic aberration (the solid angle related to the projected area of the flat mirror from the bubble center is about 0.068). A grating blazed at 300 nm was employed in these measurements. The entrance and exit slits were open to 3 mm so that the resolution or bandpass was about 10 nm. Micrometer positioners were used to align the bubble position to the entrance slit of the monochrometer.

An image intensified 512 diode array (Princeton instruments IRY-512N) was used to collect data at about a 40 nm range per scans so that 16 scans can cover the entire spectral range measured from 200 nm to 800 nm. The Spectra Pro utilized direct digital scanning which provides precise linear scanning with respect to wavelength. The time required to take a single spectral scan was about 2 min. The spectral interval measured with this monochrometer was about 0.0772 nm. Data at a spectral component were obtained by averaging the data in the range of 15 nm to reduce random noises.

Figures 1.10a and 1.10b show the spectral intensity obtained from an air bubble in water at 10°C and at 20°C, respectively. In addition to our experimental results, the calibrated data for the spectral intensity measured by Hiller et al. [8] are shown. For the wavelength range of 200 to 400 nm, a 30 W deuterium and for the range of 250 to 950 nm, a 45 W quartz-tungsten-halogen (QTH) incandescent lamp were used in the calibration of grating and the PMT (Hamamatsu R2027) efficiency. Because the data by Hiller et al. are normalized ones, the maximum intensity is closely fitted to the one of our data. Although our data are slightly larger above 350 nm, both experimental results yield similar spectral behavior remarkably. The calculated SL spectra from an air bubble of equilibrium radius of 4.5 μm driven by an ultrasound with a frequency 38.6 kHz and amplitude of 1.44 atm are also shown in the figures. The peak temperature inside the bubble, calculated from Eq. 1.6, is about 34,000 K, so the spectral intensity has a broad peak at 280 nm.

The calculated and measured spectrums by us and by Hiller et al. [8] show a common shape in the visible region with power-law dependence on the wavelength, with an exponent of −2.5, as shown in Fig. 1.10. Such power law dependence in the visible region was also observed for the Xe and He bubbles [54]. Taking into account the fact that square root of the light transmittance through quartz and water at the 200 nm is about 0.5 [55] the calculated spectrum is in good agreement with the measured data, even below 300 nm. In fact, the measured spectrum by Hiller et al. is in good agreement with the calculated spectra even in ultraviolet (UV) region as shown in Fig. 1.10b. In this respect, the graphs in Fig. 1.10a show the basic properties of SL spectra; the spectrum that is continuous from the UV cut-off of water at 200 nm to the infrared has a broad maximum with no line or band absorptions.

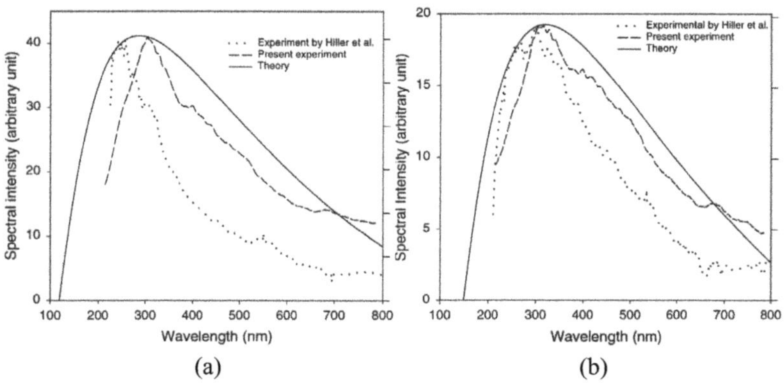

Figure 1.10 Calculated and observed data for the spectral intensity of SL for air bubble in water at 10°C (a) and at 20°C (b). The experimental data observed by Hiller et al. [8] are shown as a dotted line [42].

Considerable absorption below 250 nm [56] yields quite different shapes in the spectral intensity between theory and our measurements as can be seen in Fig. 1.10 because of light absorption in water. The calculated spectrum whose broad peak occurs at 320 nm is closely fitted to the spectrum intensities observed at 20°C, as shown in Fig. 1.10b. Again the observed and calculated spectral intensities in the visible region show the power dependence on a wavelength as $\lambda^{-2.5}$. If the calculated spectrum in the visible region is examined closely, the broad peak obtained at water temperature of 10°C should be shifted to the left, which, in turn, yields a higher maximum temperature inside the bubble. As shown in Figs. 1.10a and 1.10b, the maximum spectral intensity at 10°C is two times greater than that observed at 20°C. Slight change in water temperature reduces the spectrum intensity considerably.

1.3.3 Pulse Width Measurements

Figure 1.11 shows a schematic diagram of a light scattering-and time-correlated single-photon-counting (TC-SPC) experimental system [57]. The scattering intensity of the bubble illuminated by the filtered light from a 10 mW He-Ne laser was received by a PMT (Hamamatsu R2027). An 80° scattering angle was chosen such that there is a one-to-one relationship between the scattered intensity and the bubble radius [58]. Photons from an SL gas bubble were detected by two fast microchannel-plate photomultiplier tubes (MCP-PMTs; Hamamatsu R3809U-50) with a sensitive spectral response in the range of 160 to 850 nm. The Hamamatsu R3809U-50 MCP-PMT has a transit time spread of less than 25 ps. Multiphotons detected by one (#1 in Fig. 1.11) of the two MCP-PMTs generates a start time signal to the time-to-amplitude converter (TAC) at each SBSL flash. On the other hand, a single photon from the same pulse detected by another MCP-PMT (#2 in Fig. 1.11) generates a signal to stop the TAC. Because the start signal is a trigger that corresponds to the emission time of each SBSL flash, the measured distribution represents an average of the SBSL pulse shape [59]. It is necessary to carefully adjust the sensitivity of the two MCP-PMTs differently to obtain the actual SBSL pulse shape. If the start signal is taken from a single-photon detector, as done by Gompf et al. [9], the temporal distribution becomes an autocorrelation between the flashes. This method is hardly applicable to a moving bubble at the velocity node. The true SBSL pulse shape can be extracted from the measured data by deconvolution of the TC-SPC instrument response function (IRF). When the pulse width of SL is stable for at least several minutes, the stop signal, which is stored and accumulated in memory over time, can produce the SBSL pulse shape. However, it is very difficult to obtain a stable oscillating bubble at the velocity node in a sulfuric acid solution.

In this study, we use the ORTEC model 9327 consisting of a 1 GHz amplifier and a timing discriminator that is optimized for use with the millivolt signals produced by an MCP-PMT. The timing discriminator is the only component that is different from the TC-SPC systems used by Gompf et al. [9] and Moran and Sweider [49]. The zero-crossing technique employed in the timing discriminator results in minimal timing jitter in the SL pulse measurements. The stop signal delayed before entering the TAC by an ORTEC nanosecond delay (Model 425A). The delay time was chosen to be 32 ns by trial and error. A standard time-to-amplitude converter (ORTEC 566) and a multichannel analyzer (ORTEC 926-M32-USB) convert the time interval between the instants of the start and stop inputs to their corresponding amplitude and send the digitalized data to a personal computer that is interfaced to utilize the data acquisition.

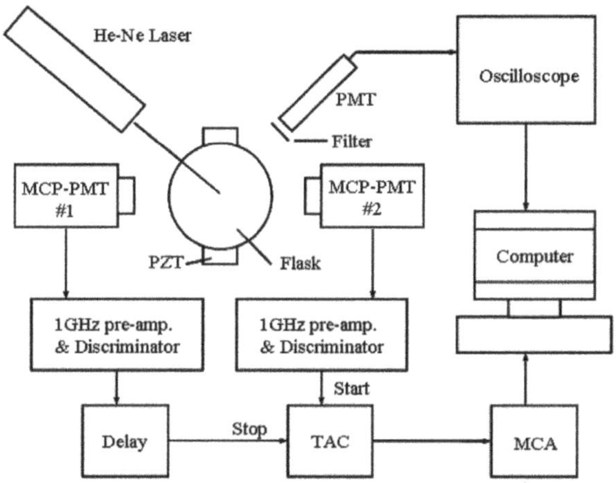

Figure 1.11 Experimental setup for time-correlated single-photon counting (TC-SPC) and light scattering [57].

The bubble radius–time curve for an air bubble in a 100 torr sulfuric acid solution with an ultrasound frequency of 28.5 kHz is shown in Fig. 1.12a. No bubble was trapped at the velocity node below this pressure head. The theoretical fit parameters derived for the bubble motion [12, 20] are R_0 = 6 μm and P_A = 1.45 atm. The estimated bubble wall velocity and bubble wall acceleration at the collapse point are 492 m/s and 10^{12} m/s^2, respectively; these values are similar to those obtained for the SL air bubble in water. The calculated radius of the bubble at the collapse point is approximately 0.7 μm. Figure 1.12b shows the pulse shape for the air bubble in a sulfuric acid solution. The FWHM obtained is about 165.6 ps, which is almost identical to the value obtained for the air bubble in water [9].

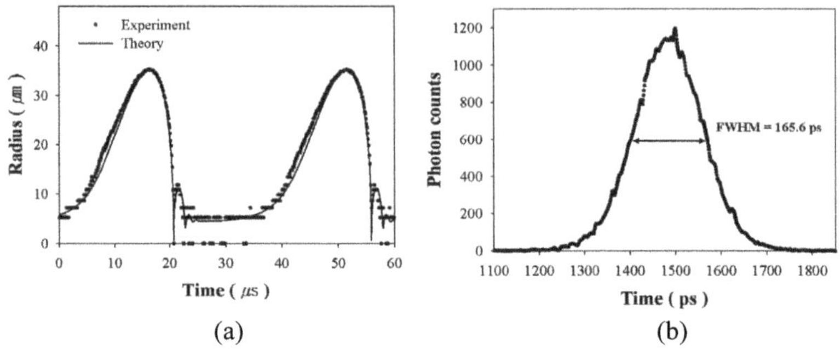

Figure 1.12 Measured bubble radius–time curve (a) and pulse width (b) from an air bubble of R_o = 6 μm in a sulfuric acid solution with an ultrasound frequency of 28.5 kHz and an amplitude of 1.45 atm [57].

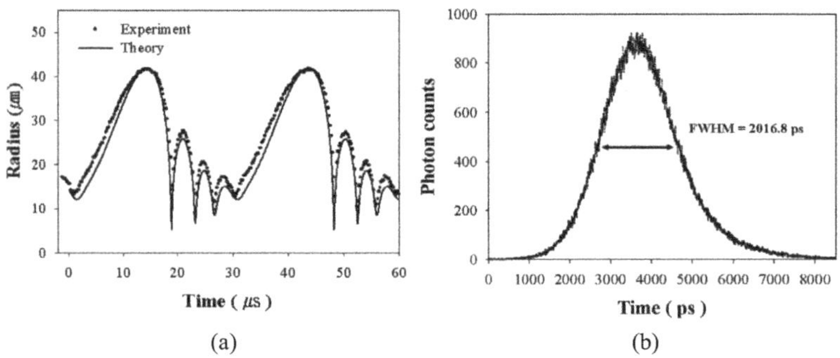

Figure 1.13 Measured bubble radius–time curve (a) and pulse width (b) from a xenon bubble of R_o = 12 μm in a sulfuric acid solution with an ultrasound frequency of 27.4 kHz and an amplitude of 1.35 atm [57].

The bubble radius–time curve for xenon bubble formed in a 50 torr sulfuric acid solution with different ultrasound frequencies of 27.4 kHz is shown in Fig. 1.13. The theoretical fit parameters are R_o = 12 μm and P_A = 1.35 atm for the curve. The estimated bubble wall velocity and bubble wall acceleration at the collapse point are 67 m/s and 0.5×10^{10} m/s^2, respectively, for the curve. The bubble radius at the collapse point is 5.12 μm, which is much larger than the radius obtained for the air bubble in water.

The pulse shape measured for the xenon bubble in a 50 torr sulfuric acid solution for the case in Fig. 1.13a is shown in Fig. 1.13b. The measured FWHM for the bubble is 2.02 ns. This value is approximately 14 times longer than the pulse width for air bubble in water. In this case, one may measure the pulse width directly without any significant error by TC-SPC because the pulse width is considerably longer than the IRF.

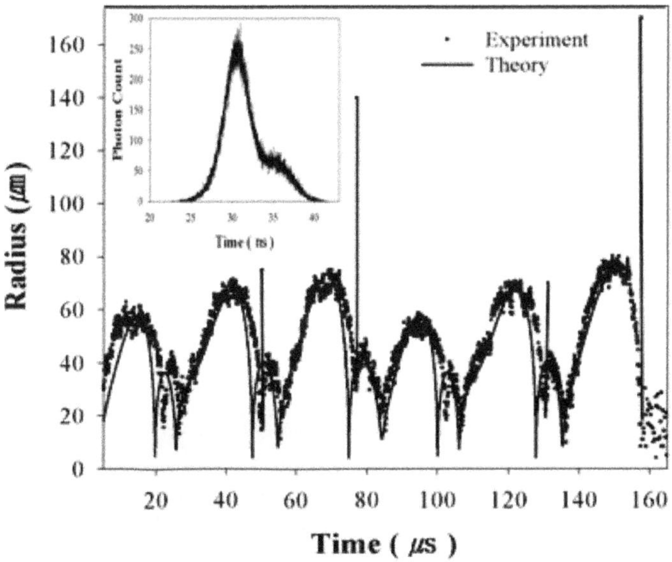

Figure 1.14 Scattered intensity from an sonoluminescence bubble with an ultrasound frequency of 37.4 kHz and an amplitude of approximately 1.7 atm. The theoretical bubble radius–time curve for R_0 = 14 μm is also shown as a solid line. The sharp peaks represent the sonoluminescence intensities of the bubble. The pulse shape is shown in the inset [57].

At the ultrasound frequency of 37.4 kHz, which was facilitated by a spherical flask of 44 mm diameter that was attached to a cylindrical PZT, various nonlinear behaviors of the SL bubble in a sulfuric acid solution were observed by causing slight changes in the ultrasound amplitude. One of them is shown in Fig. 1.14, which shows the scattered data of the He-Ne laser that was obtained from the SL bubble. A cycle of alternating strong and weak light intensities, represented by sharp peaks, appears repeatedly, as shown in Fig. 1.14. In the inset, the measured pulse shape of the bubble is shown. The occurrence of another pulse in addition to the major one is due to the chaotic behavior of the SL bubble. For this particular bubble, a theoretical fit with an equilibrium radius of 14 μm, an ultrasound frequency of 37.4 kHz, and an amplitude of 1.7 atm is found to be close to the observed behavior of the bubble, as shown in Fig. 1.14.

1.4 Multibubble Sonoluminescence

1.4.1 Introduction

It is well known that in an MBSL condition, several thousands of microbubbles are generated and collapsed synchronously with an applied ultrasound [60–62]. Measurements on the pulse width of a cloud of bubbles subjected to ultrasound using a time-correlated single-photon-counting technique indicated that the bubbles in a cloud collapse simultaneously emit a light that is synchronous with the applied ultrasound [63].

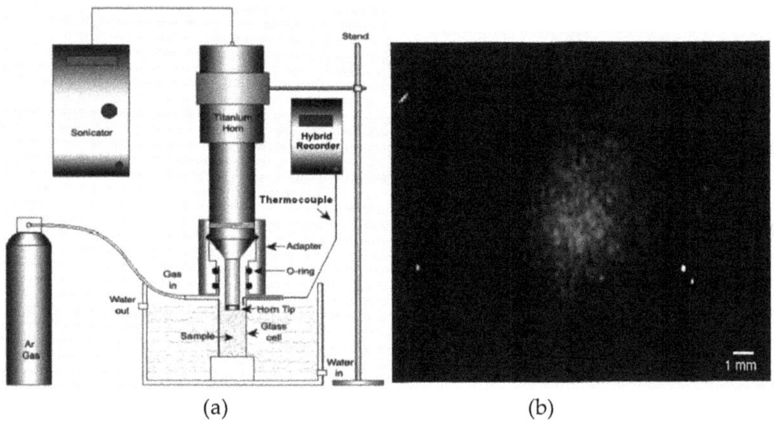

Figure 1.15 Typical experimental apparatus for multibubble sonoluminescence (a) and light from a bubble cloud (b) in an applied ultrasound field with a frequency of 20 kHz and a power of 165 W [64].

An experimental apparatus [64] for MBSL consists of a cylindrical quartz cell into which a 5 mm diameter titanium horn (Mixonix XL2010, USA) is inserted, as shown in Fig. 1.15a. The system was operates at 20 kHz and a power input of 165 W, which is about 30% of the maximum power of 500 W. An adapter was used to connect the titanium horn and the cylindrical cell whose diameter and height are 25 and 93 mm, respectively. Three O-rings provide a seal between the horn and the cell and environment. The cell has two ports: one is for a temperature sensor and the other is for the input of argon gas. The solution in the test cell is kept at constant temperature in an argon atmosphere. Continuous circulation of water to the bath in which the cell is immersed keeps the temperature of the solution inside the cell around 20°C. The MBSL condition can be found by trial and error at a proper ultrasound intensity, liquid temperature, and distance between the horn tip and the bottom of the cell. The emission of blue light concentrated around the horn tip by the argon bubble cloud is shown in Fig. 1.15b. This picture was captured in a dark room by a 35mm camera with a 30 min exposure time.

A spherical bubble cloud subjected to harmonic far-field pressure excitation was investigated by Omta [65] and D'Agostino and Brennen [66]. They revealed that the natural frequency of the bubble cluster is always less than the natural frequency of the individual bubbles. Oguz and Prosperetti [67] investigated the interaction of two 100 μm bubbles subjected to an ultrasound with moderate amplitude. Mettin et al. [68] considered the mutual interaction between two microbubbles ($R_0 < 10$ μm) in a strong acoustic field ($P_A > 1$ bar, $f_d =$ 20 kHz). They found that the strength and direction of the secondary Bjerknes forces due to the radiation generated by other bubbles differed from the forces predicted by linear theory. Yasui et al. [69] performed numerical simulations on a system of two bubbles and considered the interactions between n number of bubbles. They found that the expansion of a bubble during the rarefaction phase of ultrasound was strongly reduced by the presence of other bubbles in the cluster. They also obtained the pressure field of the center of a cloud of similarly sized, homogeneously distributed bubbles that pulsated together with an applied ultrasound. An [70] obtained the radiation sound pressure from the other bubbles acting on a particular bubble in a cluster. They investigated the collective motion of similarly sized microbubbles in a cluster and found that the radiation pressure term added in the KM equation considerably suppresses bubble motion. Recently, Dzaharudin et al. [71] performed numerical simulations of a cluster of encapsulated microbubbles by adding the interaction term in the KM equation. They found that the oscillation amplitude of microbubbles that are close together was reduced for a given applied ultrasound power.

MBSL was studied hydrodynamically to obtain the velocity profile and radiation pressure field by solving the continuity and momentum equations for a spherical cluster containing numerous microbubbles in this article [72]. The calculated pulse width and spectral radiance for a bubble with the radiation pressure added in the KM equation are compared with the measured pulse width and spectral radiance values of the MBSL.

1.4.2 Hydrodynamics for a Bubble Cluster

Consider a spherical bubble cluster containing bubbles under ultrasound. Assume that the bubbles in the cluster are homogeneously distributed and pulsate together synchronously with the applied ultrasound, which was considered by Yasui et al. [69] and An [70]. The density of the bubble-liquid mixture of the cluster is approximately given by [73]

$$\bar{\rho} = \alpha_l \rho_l + \alpha_g \rho_g \cong \rho_l (1 - \alpha_g), \tag{1.26}$$

where α_l and α_g are the liquid and gas volume fraction, respectively, and ρ_l is the liquid density. The gas volume, which depends on the instantaneous radius of the bubble, R_b, in the cluster, is given by

$$\alpha_g = 4\pi R_b^3 N_b / 3, \tag{1.27}$$

where N_b is the number density of bubbles in the cluster. Assuming that the mass per unit volume in the cluster does not change with time [73], we have

$$\bar{\rho}/N_b = \bar{\rho}_o / N_{bo}, \tag{1.28}$$

where the subscript "o" denotes the initial state. Using Eqs. 1.26, 1.27, and 1.28, we have the relationship between α_g and α_{go} and the relationship between N_b and N_{bo} from Eq. 1.28 as

$$\alpha_g = \alpha_{go} (R_b/R_o)^3 / \left\{ 1 - \alpha_{go} \left[1 - (R_b/R_o)^3 \right] \right\}, \tag{1.29}$$

$$N_b = N_{bo} / \left\{ 1 - \alpha_{go} \left[1 - (R_b/R_o)^3 \right] \right\}. \tag{1.30}$$

The continuity equation for the cluster with the average density given in Eq. 1.26 is as follows:

$$\frac{D\bar{\rho}}{Dt} + \frac{\bar{\rho}}{r^2} \frac{\partial}{\partial r} (r^2 u) = 0, \tag{1.31}$$

where D/Dt is the material derivative. Using Eq. 1.28 the above continuity equation can be rewritten in terms of the number density of bubbles as

$$\frac{DN_b}{Dt}+\frac{N_b}{r^2}\frac{\partial}{\partial r}(r^2u)=0. \tag{1.32}$$

Using Eqs. 1.29 and 1.30, one can obtain the velocity in the cluster from the continuity equation, Eq. 1.32. The velocity field in the cluster is given by

$$u=\frac{1}{3}\left(-\frac{1}{\alpha_g}\frac{d\alpha_g}{dt}+\frac{3}{R_b}\frac{dR_b}{dt}\right)r. \tag{1.33}$$

With the help of Eq. 1.29, the velocity profile in the cluster simply reduced to

$$u=\alpha_g\frac{\dot{R}_b}{R_b}r. \tag{1.34}$$

The velocity profile given in Eq. 1.34 indicates that the cluster behaves in the same way that the bubbles in the cluster do. However, the degree of the coupled behavior is expected to be weak because the value of α_g is much smaller than unity, which is drastically different from the gas motion inside the spherical bubble oscillating under ultrasound, described by Eq. 1.3b.

To calculate the radiation pressure due to the pulsating bubble inside the cluster, one should solve the following momentum equation [66], which is valid when α_g is much less than unity for bubble-liquid mixture:

$$\bar{\rho}\frac{Du}{Dt}=-\frac{\partial p}{\partial r} \tag{1.35}$$

Using the velocity profile given in Eq. 1.34, one can explicitly obtain the quadratic radiation pressure field inside the cluster from the momentum equation (Eq. 1.35):

$$p_{\text{rad}}(r)=p_{\text{cl}}+\frac{\bar{\rho}\alpha_g}{2}\left(\frac{r_{\text{cl}}}{R_b}\right)^2\left[R_b\ddot{R}_b+2(1-\alpha_g)\dot{R}_b^2\right]\left[1-\left(\frac{r}{r_{\text{cl}}}\right)^2\right]. \tag{1.36}$$

Using Eq. 1.27, the radiation pressure field can be approximated as

$$p_{\mathrm{rad}}(r)=p_{\mathrm{cl}}+\frac{\bar{\rho}n}{2}\left(\frac{R_{\mathrm{b}}}{r_{\mathrm{cl}}}\right)\left[R_{\mathrm{b}}\ddot{R}_{\mathrm{b}}+2(1-\alpha_{\mathrm{g}})\dot{R}_{\mathrm{b}}^2\right]\left[1-\left(\frac{r}{r_{\mathrm{cl}}}\right)^2\right], \quad (1.37)$$

where p_{cl} is the pressure at the cluster wall. In the above equation, the term $4\pi(r_{\mathrm{cl}})^3N_{\mathrm{b}}/3$ is replaced by the number of bubbles in the cluster, n. Taking β as the ratio of the radiation pressure at the cluster wall to the radiation pressure at the center, the average pressure field inside the cluster can be obtained as

$$p_{\mathrm{rad}}(r)=\frac{1}{2(1-\beta)}\bar{\rho}n\left(\frac{R_{\mathrm{b}}}{r_{\mathrm{cl}}}\right)\left[R_{\mathrm{b}}\ddot{R}_{\mathrm{b}}+2(1-\alpha_{\mathrm{g}})\dot{R}_{\mathrm{b}}^2\right]\left[1-(1-\beta)\left(\frac{r}{r_{\mathrm{cl}}}\right)^2\right]. \quad (1.38)$$

From Eq. 1.38, the average pressure filed inside the cluster can be approximated as

$$\begin{aligned}\bar{p}_{\mathrm{rad}}&=\frac{(2+3\beta)}{10(1-\beta)}\bar{\rho}n\left(\frac{R_{\mathrm{b}}}{r_{\mathrm{cl}}}\right)\left[R_{\mathrm{b}}\ddot{R}_{\mathrm{b}}+2(1-\alpha_{\mathrm{g}})\dot{R}_{\mathrm{b}}^2\right]\\&=\xi\rho_\infty\left[R_{\mathrm{b}}\ddot{R}_{\mathrm{b}}+2(1-\alpha_{\mathrm{g}})\dot{R}_{\mathrm{b}}^2\right], \quad (1.39)\end{aligned}$$

where $\xi=[(2+3\beta)/(1-\beta)/10](1-\alpha_{\mathrm{g}})nR_{\mathrm{b}}/r_{\mathrm{cl}}$.

The motion of a bubble inside a cluster is affected by the radiation pressure field generated by the surrounding bubbles pulsating under an applied ultrasound. If the radiation pressure field is considered an imposed pressure, the KM equation may be written as

$$\begin{aligned}&\left(1-\frac{U_{\mathrm{b}}}{C_{\mathrm{b}}}\right)R_{\mathrm{b}}\frac{dU_{\mathrm{b}}}{dt}+\frac{3}{2}\left(1-\frac{U_{\mathrm{b}}}{3C_{\mathrm{b}}}\right)U_{\mathrm{b}}^2=\frac{1}{\rho_\infty}\left(1+\frac{U_{\mathrm{b}}}{C_{\mathrm{b}}}+\frac{R_{\mathrm{b}}}{C_{\mathrm{b}}}\frac{d}{dt}\right)\\&\cdot\left[P_{\mathrm{B}}-P_{\mathrm{s}}\left(t+\frac{R_{\mathrm{b}}}{C_{\mathrm{b}}}\right)-p_\infty-\bar{p}_{\mathrm{rad}}\right]. \quad (1.40)\end{aligned}$$

Using Eq. 1.39, Eq. 1.40 can be written more explicitly as

$$\begin{aligned}&\left[1-\frac{U_{\mathrm{b}}}{C_{\mathrm{b}}}+\xi\left(1+\frac{U_{\mathrm{b}}}{C_{\mathrm{b}}}\right)\right]R_{\mathrm{b}}\frac{dU_{\mathrm{b}}}{dt}+\frac{3}{2}\left[1-\frac{U_{\mathrm{b}}}{3C_{\mathrm{b}}}+\frac{4}{3}\xi(1-\alpha_{\mathrm{g}})\left(1+\frac{U_{\mathrm{b}}}{C_{\mathrm{b}}}\right)\right]U_{\mathrm{b}}^2\\&=\frac{1}{\rho_\infty}\left(1+\frac{U_{\mathrm{b}}}{C_{\mathrm{b}}}+\frac{R_{\mathrm{b}}}{C_{\mathrm{b}}}\frac{d}{dt}\right)\left[P_{\mathrm{B}}-P_{\mathrm{s}}\left(t+\frac{R_{\mathrm{b}}}{C_{\mathrm{b}}}\right)-p_\infty\right]. \quad (1.41)\end{aligned}$$

The behavior of a bubble inside a bubble cluster may be estimated using Eq. 1.11 and the modified KM equation given in Eq. 1.41. The added terms due to radiation pressure in the KM equation may be considered as the change in the effective mass for the bubble [74] because the time rate change of the kinetic energy with the effective mass yields the key terms in the left-hand side of Eq. 1.41:

$$\frac{d}{dt}\left(\frac{1}{2}M_{\mathrm{eff}}\dot{R}_{\mathrm{b}}^{2}\right)=2M_{\mathrm{eff}}\frac{\dot{R}_{\mathrm{b}}}{R_{\mathrm{b}}}\left(R_{\mathrm{b}}\ddot{R}_{\mathrm{b}}+\frac{3}{2}\dot{R}_{\mathrm{b}}^{2}\right), \tag{1.42}$$

where $$M_{\mathrm{eff}}=4\pi\rho_{\infty}R_{\mathrm{b}}^{3}.$$

The number of bubbles, which was estimated from the degradation rate of methylene blue in water under the MBSL condition, was approximately 3000 per cycle [64]. The size distribution of the bubble at the MBSL condition was measured by the phase-Doppler technique in which He-Ne laser (633 nm) and two avalanche photodiodes were used as receiving optics [63]. The measured volume that was focused on an expected MBSL region is approximately 10^{-3} cm^{3}. The mean diameter was calculated from the acquisition of around 3000–6000 individual bubbles. The average and Sauter mean diameters detected during the period of the bubble expansion phase including the maximum size of argon bubbles under the MBSL condition were approximately 36.4 μm and 54.8 μm, respectively [63], so the average equilibrium radius in the MBSL condition was estimated to be approximately 5 μm. On the basis of these experimental data, a spherical bubble cluster containing 3000 microbubbles with a radius of 3.5 mm was considered in this study. The equilibrium radius and ultrasound amplitude were 5 μm and 1.4 atm, respectively, resulting in a maximum bubble radius of 25 μm in the MBSL environment.

Figure 1.16 shows the bubble radius–time curves for a single bubble and a bubble in a cluster of 3000 bubbles in the same ultrasound field with a frequency of 20 kHz and an amplitude of 1.4 atm. The relaxation time of the bubble motion relative to the ultrasound filed was taken as 0.5 μs in this case. In this study, the radiation pressure at the cluster wall was taken as zero or β was taken as zero. The maximum bubble radius of a bubble in the cluster was considerably smaller than that of a single bubble in the same applied ultrasonic field. In fact, the maximum bubble radius of a bubble in the cluster decreased as the number of bubbles in the cluster increased. Consequently, the minimum radius of a bubble in the cluster increased, and the gas pressure at the collapse point as well as the maximum temperature decreased as the number of bubbles in the cluster increased. This may be due to an increase in the effective mass of the bubble [74] by the radiation pressure field retarding the compression process by the ultrasound field, causing the amount of time from equilibrium point to the collapse point to increase.

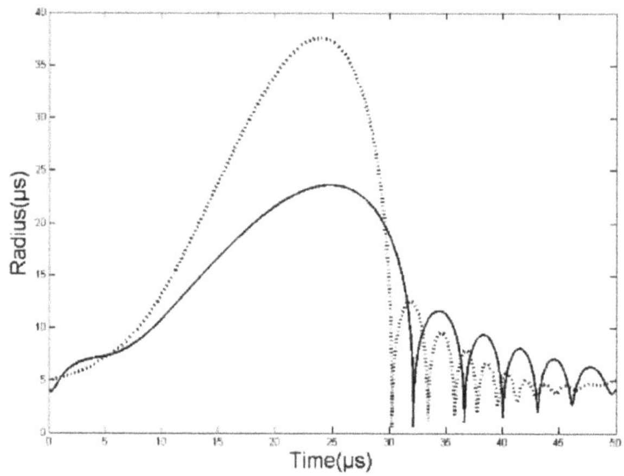

Figure 1.16 Time-dependent bubble radius–time curve for a single bubble (dotted line) and a bubble (solid line) in a cluster of 3000 bubbles in a water solution subjected to ultrasound with a frequency of 20 kHz and an amplitude of 1.4 atm [72].

No light emission can be expected when the number density of bubble in the cluster exceeds 28/mm^3, which corresponds to the case of a 5030 bubbles in a cluster of 3.5 mm in radius. Above the limit, with the number density of bubble, for example, 29/mm^3, the average gas temperature reached around the collapse point is approximately below 4000 K so that the maximum spectral radiance due to the *bremsstrahlung* cannot be measured.

The temporal change of the gas temperature and the corresponding total emission with the electron–atom *bremsstrahlung* radiation for the bubble shown in Fig. 1.16 are shown in Fig. 1.17. Figure 1.17 shows the maximum temperature calculated by Eq. 1.6, corresponding to light emission, occurs 400 ps prior to the collapse point where the bubble radius is at its minimum and the gas pressure is at its maximum as can be seen clearly in Fig. 1.17. On the other hand, the temperature calculated by Eq. 1.7 with the uniform pressure approximation inside the bubble has its maximum at the collapse point as can be seen from the slash-dot line in Fig. 1.17. Equation 1.6, which includes the acceleration and deceleration of the bubble wall, produces a time of the light pulse occurrence that agrees with the experimental results, indicating that the light pulse bursts nanoseconds prior to the bubble collapse [75]. Note that the total emission was calculated by multiplying the number of bubbles in the cluster with the value estimated by Eq. 1.24.

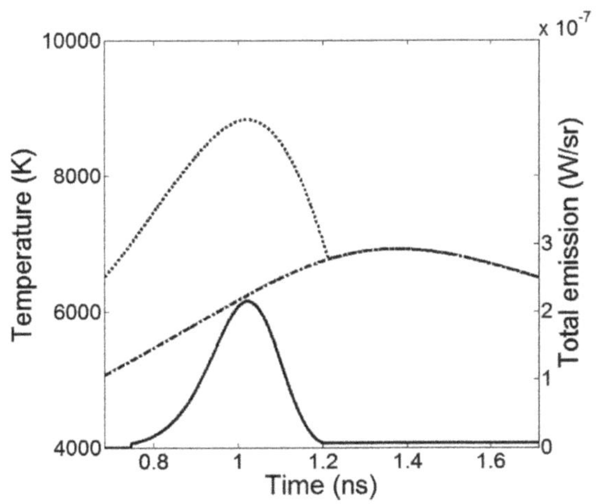

Figure 1.17 Temporal temperature change obtained from Eq. 1.6 (dotted line) and from Eq. 1.7 (slash-dot line) and total emission due to electron–atom collisions (solid line) for the case shown in Fig. 1.2 [72].

The spectral radiance estimated from Eq. 1.23 by multiplying the number of bubbles inside the cluster at a temperature of 8000 K, which is the case of 3000 bubbles in a cluster, has broad peak near 600 nm as shown in Fig. 1.18. If the gas temperature inside the bubble increases, the broad peak shifts to a lower wavelength, the peak occurs at 430 nm at a gas temperature of 12,000 K for a bubble in a cluster of 1000 bubbles under the same ultrasound filed. On the other hand, the peak occurs at a higher wavelength of 750 nm when the gas temperature is 6800 K, which is the case of 4000 bubbles in a cluster, as can be seen in the inset in Fig. 1.18.

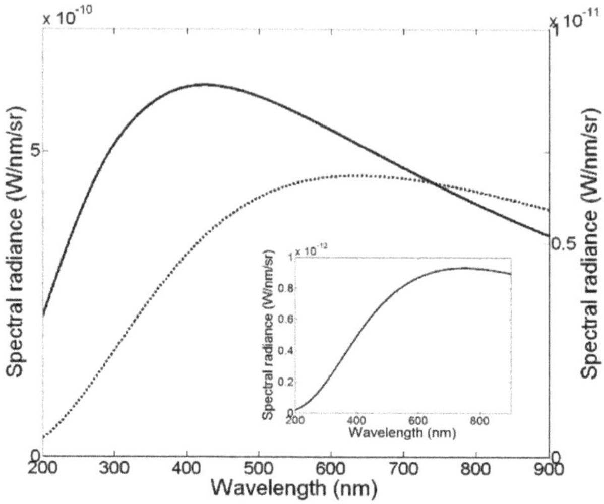

Figure 1.18 Spectral radiance from argon bubble clouds containing 1000 bubbles (solid line: left) and 3000 bubbles (dotted line: right) evaluated at 12,000 K and 8000 K, respectively. The inset shows the spectral radiance from an argon bubble cloud containing 4000 bubbles evaluated at 6800 K [72].

Figure 1.19 shows the SL pulse width from a bubble cloud of argon bubbles under MBSL conditions in pure water. Surprisingly, the measured pulse width of 251.9 ps from a bubble cloud is comparable to the pulse width from a single sonoluminescing bubble, as shown in Fig. 1.12b. Differences in the data from the SBSL were partly due to the photon flashes being detected 1.5 ns before and 1.5 ns after the point of maximum intensity, as can be seen from a raw data shown in the inset. This may indicate that several thousand bubbles synchronously collapsed and emitted light simultaneously in small, scattered events. This observation is quite different from that of Giri and Arakeri [76], who argued that long SL flash durations (1–10 ns) obtained in their experiment are related to the multibubble environment due to the convolution or superposition of the light signals from a large number of SL events. The calculated FWHM of the radiance is 210 ps, as shown in Fig. 1.17, which is smaller than the observed value of 251.9 ps for a cluster of argon bubbles [63]. Another difference in the pulse shapes between SBSL and MBSL could be attributed to the lower number of photon counts at maximum intensity in the case of MBSL. This is because the light intensity under MBSL conditions is weak, and the longest possible test duration for the MBSL is about 1 hr.

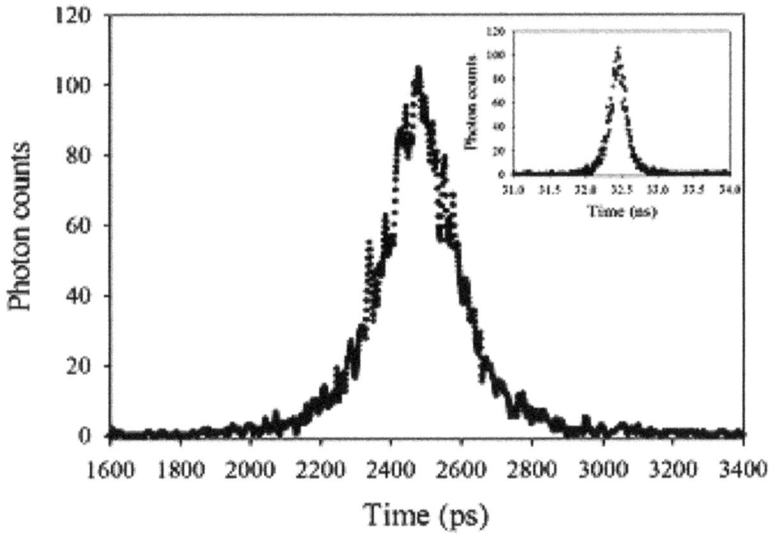

Figure 1.19 Measured pulse from an argon bubble cloud in water at the MBSL condition at an ultrasound frequency of 20 kHz. In the inset, raw data for the pulse is shown.

Spectral characteristics with a broad peak around 450 nm for MBSL in a water solution were obtained by Didenko and Pugach [77], Ashok Kumar and Grieser [78], and Wall et al. [79]. Recently, a similar radiation spectrum with a broad peak around 500 nm was observed for MBSL in Hg [80]. On the other hand, Didenko and Gordeychuk [81] observed a broad spectrum with a peak around 300 nm and with a prominent emission of the excited OH radical at 310 nm in a water solution saturated with noble gases. Experimental results that show different locations of maximum peaks in spectral radiation distributions may be due to clusters containing different numbers of bubbles.

The spectrum of MBSL contains molecular emission bands, for example, carbon bands in organic liquid [82] such as dodecane and the spectral bands of the OH radical in water [81], which are not discussed in this article.

1.4.3 Applications

Sonochemistry involves an application of SL. The intense local heating and high pressure inside the bubbles and liquid adjacent bubble wall from such collapse can give rise to unusual effects in chemical reactions. The estimated temperature and pressure in the liquid zone around the collapsing bubble with equilibrium radius 5 μm, an average radius of bubbles generated in a sonochemical reactor at a driving frequency of 20 kHz with an input power of 179 W, is about 1000°C and 500 atm, respectively. The MBSL facilitates the transient supercritical state [83] in the liquid layer where rapid chemical reactions can take place. In fact, methylene blue, which is one of a number of typical textile dyes, degraded fast at the MBSL condition, while methylene blue does not degrade under simple ultrasonic irradiation [64].

The MBSL condition, which is shown in Fig. 1.15b, has proven to be a useful technique to make novel materials with unusual properties. In our study, various metal oxides such as ZnO powder [84], used as a primary reinforcing filler for elastomer, homogeneous $Li_4Ti_5O_{12}$ nanoparticles [85], used for electrode materials, and core/shell nanoparticles such as CdS coating on TiO_2 nanoparticles [86] and ZnS coating on TiO_2 nanoparticles [87], which are very likely to be useful for the development of inorganic dye-sensitized solar cells, were synthesized through a one-pot reaction under the MBSL condition.

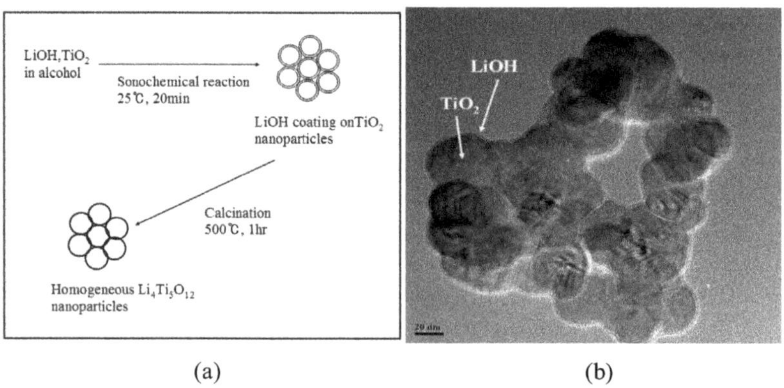

(a) (b)

Figure 1.20 (a) Two-step method for synthesizing $Li_4Ti_5O_{12}$ nanoparticles at the MBSL condition and (b) TEM image of LiOH-coated TiO_2 nanoparticles prepared [85].

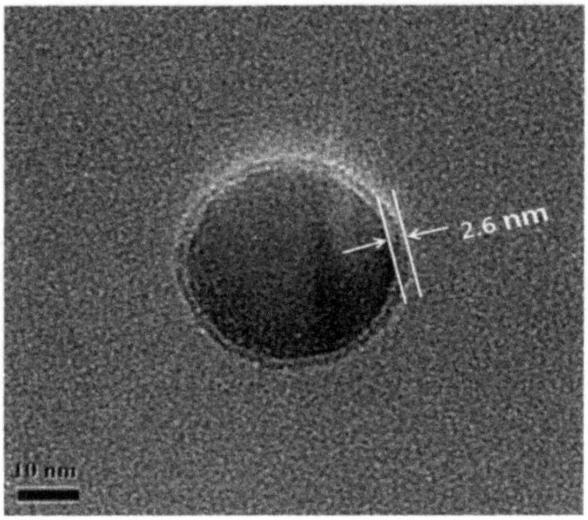

(a)

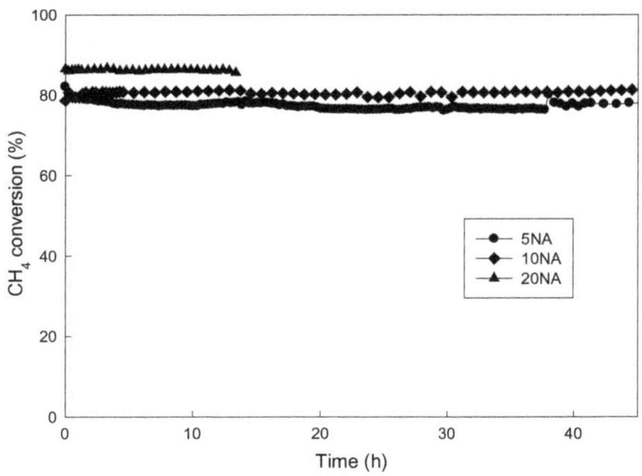

(b)

Figure 1.21 (a) TEM image of Ni/Al_2O_3 nanopowder with Ni loading of 10 wt% and (b) CH_4 conversion with time for a stream over prepared supported Ni-based catalysts with Ni loadings of 5 wt% (5NA), 10 wt% (10NA), and 20 wt% (20NA) [88].

A simple sonochemical method for producing homogeneous LTO nanoparticles is shown schematically in Fig. 1.20a. First, LiOH and TiO_2 nanoparticles were used to prepare LiOH-coated TiO_2 nanoparticles, as shown in Fig. 1.20a, at the MBSL condition.

Second, the resulting nanoparticles were thermally treated at 500°C for 1 hr to prepare LTO nanoparticles. Figure 1.20b shows a high-resolution transmission electron microscopy (TEM) image of LTO nanoparticles having an average grain size of 30–40 nm. All nanoparticle synthesized at the MBSL condition are very pure in phase and quite homogeneous in their size and shape.

Recently we succeeded in synthesizing a supported nickel catalysts with core/shell structure such as Ni/Al_2O_3, $Ni/MgO-Al_2O_3$, and Ni/CeO_2 at the MBSL condition, which turned out to be effective for various methane reforming reactions [88–90]. Figure 1.21a shows TEM images of supported Ni catalysts on Al_2O_3. The thickness layer is approximately 2.6 nm with 10% Ni loading. Figure 1.21b shows the methane conversion efficiency for the supported Ni catalysts with various Ni loadings with partial oxidation reforming at a reaction temperature of 700°C [88].

Sonication alone enables the rapid dispersion of solids, the decomposition of organics, and the formation of porous materials and nanostructures [91]. Consequence of such action of sonication, sonochemistry provides a new way to more rapidly and efficiently synthesize many specialty nanoparticles with a low level of waste [92].

This clean technology enables the preparation of new materials such as nanostructured amorphous iron and alloys, nanostructured Fe on silica, nanocolloids of Fe, nanostructured Mo_2C and MoS_2, and biomaterials [93] and various alloys [94].

1.5 Conclusions

Twenty-five years ago, the energy-focusing phenomenon of converting sound energy into light, SL, was unbelievable to everyone. After Gaitan's success [6, 7] in trapping a single bubble in a standing ultrasonic wave, the phenomenon has been unveiled by careful and ingenious experiments by many researchers in this field. At present, SL characterized by hundreds of picosecond to nanosecond flashes of a continuous spectrum with a broad maximum could be understood by homologous solutions for the Navier–Stokes equations of gases inside the bubble and solutions of the liquid outside the bubble wall for the sonoluminescing bubble under an ultrasonic field. Heat transfer across the bubble–liquid interface plays a crucial role in the evolution of the sonoluminescing gas bubble, which was confirmed by molecular dynamics simulation [95].

References

1. Marinesco, N., and Trillat, J. J. (1933). Chimie physique-action des ultrasons sur les plaques photographiques, *C. R. Acad. Sci. Paris*, **196**, p. 858.
2. Frenszel, J., and Schultes, H. (1934). Luminescenz im ultraschallbeschickten Wasser, Kurze Mitteilung, *Zeit für Phys. Chem.*, **B27**, pp. 421–424.
3. Chambers, L. A. (1937). The emission of visible light from cavitated liquids, *J. Chem. Phys.*, **5**, pp. 290–292.
4. Meyer, E., and Kuttruff, H. (1959). Zur Phasenbezielung zwischen Sonolumineszenz und Kavitationsvorang bei periodischer Anregung, *Zeit Angew. Phys.*, **9**, p. 325.
5. Günther, P., Zeil, W., Grisar, U., and Heim, E. (1957). Versuche über die Sonolumineszenz wässriger Loesungen, *Zeit für Electrochem.*, **61**, p. 188.
6. Gaitan, D. F. (1990). An experimental investigation of acoustic cavitation in gaseous liquids (PhD Dissertation, The University of Mississippi).
7. Gaitan, D. F., Crum, L. A., Church, C. C., and Roy, R. A. (1992). Sonoluminescence and bubble dynamics for a single, stable, cavitation bubble, *J. Acoust. Soc. Am.*, **91**, pp. 3166–3183.
8. Hiller, R., Putterman, S. J., and Barber, B. P. (1992). Spectrum of synchronous picosecond sonoluminescence, *Phys. Rev. Lett.*, **69**, pp. 1182–1184.
9. Gompf, B., Guenter, R., Nick, G., Pecha, R., and Eisenmenger, W. (1997). Resolving sonoluminescence pulse width with time-correlated single photon counting, *Phys. Rev. Lett.*, **79**, pp. 1405–1408.
10. Hiller, R., Putterman, S. J., and Weninger, K. R. (1998). Time resolved spectra of sonoluminescence, *Phys. Rev. Lett.*, **80**, pp. 1090–1093.
11. Prosperetti, A., Crum, L. A., and Commander, K. W. (1988). Nonlinear bubble dynamics, *J. Acoust. Soc. Am.*, **83**, pp. 502–514.
12. Kwak, H., and Yang, H. (1995). An aspect of sonoluminescence from hydrodynamic theory, *J. Phys. Soc. Jpn.*, **64**, pp. 1980–1992.

13. Padmanabhan, T. (2001). *Theoretical Astrophysics* (Cambridge University Press, Oxford).

14. Jun, J., and Kwak, H. (2000). Gravitational collapse of Newtonian stars, *Int. J. Mod. Phys. D*, **9**, pp. 35–42.

15. Lin, H., Storey, B. D., and Szeri, A. J. (2002). Inertially driven inhomogeneities in violently collapsing bubbles: the validity of the Rayleigh-Plesset equation, *J. Fluid Mech.*, **452**, pp. 145–162.

16. Wu, C. C., and Roberts, P. H. (1993). Shock-wave propagation in a sonoluminescing gas bubble, *Phys. Rev. Lett.*, **70**, pp. 3424–3427.

17. Moss, W. C., Clarke, D. B., White, J. W., and Young, D. A. (1994). Hydrodynamic simulations of bubble collapse and picosecond sonoluminescence, *Phys. Fluids*, **6**, pp. 2979–2985.

18. Theofanous, T., Biasi, L., and Isbin, H. S. (1969). A theoretical study on bubble growth in constant and time-dependant pressure fields, *Chem. Eng. Sci.*, **24**, pp. 885–897.

19. Davidson, R. C. (1972). *Method in Nonlinear Plasma Theory* (Academic Press, New York).

20. Kwak, H., and Na, J. (1996). Hydrodynamic solutions for a sonoluminescing gas bubble, *Phys. Rev. Lett.*, **77**, pp. 4454–4457.

21. Boulos, M. I., Fauchais, P., and Pfender, E. (1994). *Thermal Plasma*, Vol. 1, (Plenum Press, New York).

22. Keller, J. B., and Miksis, M. (1980). Bubble oscillations of large amplitude, *J. Acoust. Soc. Am.*, **68**, pp. 628–633.

23. Kwak, H., Oh, S., and Park, C. (1995). Bubble dynamics on the evolving bubble formed from the droplet at the superheat limit, *Int. J. Heat Mass Trans.*, **38**, pp. 1709–1718.

24. Hilgenfeldt, S., and Lohse, D. (1999). Predictions for upscaling sonolumi-nescence, *Phys. Rev. Lett.*, **82**, pp. 1036–1039.

25. Prosperetti, A., and Hao, Y. (1999). Modeling of spherical gas bubble oscillations and sonoluminescence, *Phil. Trans. R. Soc. Lond. A*, **357**, pp. 203–223.

26. Kwak, H., Lee, J., and Karng, S. W. (2001). Bubble dynamics for single bubble sonoluminescence, *J. Phys. Soc. Jpn.*, **70**, pp. 2909–2917.

27. Jeon, J., Lee, J., and Kwak, H. (2003). Possibility of upscaling for single bubble sonoluminescence at a low driving frequency, *J. Phys. Soc. Jpn.*, **72**, pp. 509–515.

28. Karng, S. W., and Kwak, H. (2006). Relaxation behavior of microbubbles in ultrasonic field, *Jpn. J. Appl. Phys.*, **45**, pp. 317–322.

29. Lőfstedt, R., Barber, B. P., and Putterman, S. J. (1993). Toward a hydrodynamic theory of sonoluminescence, *Phys. Fluids A*, **5**, pp. 2911–2928.

30. Delgadino, G. A., and Bonetto, F. J. (1997). Velocity interferometry technique used to measure the expansion and compression phases of a sonoluminescing bubble, *Phys. Rev. E*, **56**, pp. R6248–R6251.

31. Gompf, B., and Pecha, R. (2000). Mie scatteing from a sonoluminescing bubble in high spatial and temporal resolution, *Phys. Rev. E*, **61**, pp. 5253–5256.

32. Weninger, K. R., Evans, P. G., and Putterman, S. J. (2000). Time correlated single photon Mie scattering from a sonoluminescing bubble, *Phys. Rev. E*, **61**, pp. R1020–R1023.

33. Byun, K.-T., Kim, K. Y., and Kwak, H. (2005). Sonoluminescence characteristics from micron and submicron bubbles, *J. Kor. Phys. Soc.*, **47**, pp. 1010–1022.

34. Vazquez, G., Camara, C., Putterman, S. J., and Weninger, K. (2001). Sonoluminescence: nature's smallest blackbody, *Opt. Lett.*, **26**, pp. 575–577.

35. Kwak, H., and Na, J. (1997). Physical processes for single bubble sonoluminescence, *J. Phys. Soc. Jpn.*, **66**, pp. 3074–3083.

36. Vincenti, W. G., and Kruger, C. H. (1977). *Introduction to Physical Gas Dynamics* (Robert E. Krieger Publishing Co., New York).

37. Hirschfelder, J. O., Curtis, C. F., and Bird, R. B. (1954). *Molecular Theory of Gases and Liquids* (John Wiley & Sons, New York).

38. Ruuth, S. J., Putterman, S., and Merriman, B. (2002). Molecular dynamics simulation of the response of a gas to a spherical piston: Implications for sonoluminescence, *Phys. Rev. E*, **66**, paper #036310.

39. Hopkins, S. D., Putterman, S. J., Kappus, B. A., Suslick, K. S., and Camara, C. G. (2005). Dynamics of a sonoluminescing bubble in sulfuric acid, *Phys. Rev. Lett.*, **95**, paper #254301.

40. Kim, K. Y., Byun, K.-T., and Kwak, H. (2006). Characteristics of sonoluminescing bubbles in aqueous solutions of sulfuric acid, *J. Phys. Soc. Jpn.*, **75**, paper #114705.

41. Flannigan, D. J., Hopkins, S. D., Camara, C. G. Putterman, S. J., and Suslick, K. S. (2006). Measurement of pressure and density inside a single sonoluminescing bubble, *Phys. Rev. Lett.*, **96**, paper #204301.

42. Jeon, J., Yang, I., Na, J., and Kwak, H. (2000). Radiation mechanism from a single bubble sonoluminescence, *J. Phys. Soc. Jpn.*, **69**, pp. 112–119.

43. Mitchner, M., and Kruger, C. H. (1973). *Partially Ionized Gases* (John Wiley and Sons, New York).
44. Bekefi, G. (1966). *Radiation Processes in Plasmas* (John Wiley and Sons, New York).
45. Zeldovich, Ya. B., and Raizer, Yu. P. (1966). *Physics of Shock Waves and High-Temperature Hydrodynamics Phenomena* (Academic Press, New York).
46. Siegel, R., and Howell, J. R. (1981). *Thermal Radiation Heat Transfer* (Hemisphere Publishing Corporation, New York).
47. Hilgenfeldt, S., Grossman, S., and Loshe, D. (1999). A simple explanation of light emission in sonoluminescence, *Nature*, **398**, pp. 402–405.
48. Hilgenfeldt, S., Grossman, S., and Loshe, D. (1999). Sonoluminescence light emission, *Phys. Fluids*, **11**, pp. 1318–1330.
49. Moran, M. J., and Sweider, D. (1998). Measurements of sonoluminescence temporal pulse shape, *Phys. Rev. Lett.*, **80**, pp. 4987–4990.
50. Vuong, V. Q., and Szeri, A. J. (1996). Sonoluminescence and diffusive transport, *Phys. Fluids*, **8**, pp. 2354–2364.
51. Kwak, H., Lee, Y. P., and Karng, S. W. (1999). Pressure wave propagation inside a sonoluminescing gas bubble, *J. Phys. Soc. Jpn.*, **68**, pp. 705–708.
52. Dam, J. S., and Levinsen, M. T. (2004). Size of light-emitting region in a sonoluminescing bubble, *Phys. Rev. Lett.*, **92**, paper #144301.
53. Camara, C., Putterman, S., and Kirilov, E. (2004). Sonoluminescence from a single bubble driven a 1 Megahertz, *Phys. Rev. Lett.*, **92**, paper #124301.
54. Barber, B. P. (1992). Synchronous picosecond sonoluminescence (PhD Dissertation, UC at Los Angeles).
55. Griem, H. R. (1983). *Basic Plasma Physics I*, eds. Galeev, A. A., and Sudan, R. N. (North Holland, Amsterdam).
56. Hiller, R. (1995). The spectrum of single bubble sonoluminescence (PhD Dissertation, UC at Los Angeles).
57. Jeon, J.-S., Lim, C., and Kwak, H. (2008). Measurement of pulse width of sonoluminescing gas bubble in sulfuric acid solution, *J. Phys. Soc. Jpn.*, **77**, paper #033703.
58. Hansen, G. M. (1984). Mie scattering as a technique for the sizing of air bubble (PhD Dissertation, The University of Mississippi).
59. Becker, W. (2005). *Advance Time-Correlated Single-Photon Counting Technique* (Springer, Berlin).

60. Tsochatzidis, N. A., Guiraud, P., Wilhelm, A. M., and Delmas, H. (2001). Determination of velocity, size and concentration of ultrasonic cavitation bubbles by the phase-Doppler technique, *Chem. Eng. Sci.*, **56**, pp. 1831–1840.

61. Servan, G., Laborde, J. L., Hita, A., Caltagirone, J. P., and Gerard, A. (2003). On the interaction between ultrasound waves and bubble in mono- and dual- frequency sonoreactors, *Ultrason. Sonochem.*, **10**, pp. 347–355.

62. Birkin, B., Offin, D. G., Vian, C. J. B., and Leigton, T. G. (2011). Multiple observation of cavitation cluster dynamics close to an ultrasonic tip, *J. Acoust. Soc. Am.*, **130**, pp. 3379–3388.

63. Ko, I., and Kwak, H. (2010). Measurement of pulse width from a bubble cloud under multibubble sonoluminescence condition, *J. Phys. Soc. Jpn.*, **79**, paper #124401.

64. Byun, K., and Kwak, H. (2005). Degradation of methylene blue under multibubble sonoluminescence condition, *J. Photochem. Photobiol. A*, **175**, pp. 45–50.

65. Omta, R. (1987). Oscillations of a cloud of bubbles of small and not so small amplitude, *J. Acoust. Soc. Am.*, **82**, pp. 1018–1033.

66. Dagostino, L., and Brennen, C. E. (1989). Linearized dynamics of spherical bubble clouds, *J. Fluid Mech.*, **199**, pp. 155–176.

67. Oguz, H. N., and Prosperetti, A. (1990). A generalization of the impulse and virial theorems with an application to bubble oscillation, *J. Fluid Mech.*, **218**, pp. 143–162.

68. Mettin, R., Akhatov, I., Parlitz, U., Ohl, C. D., and Lauterborn, W. (1997). Bjerknes forces betwwen small cavitation bubbles in a strong acoustic field, *Phys. Rev. E*, **56**, pp. 2924–2931.

69. Yasui, K., Iida, Y., Tuziuti, T., Kozuka, T., and Towata, A. (2008). Strongly interacting bubbles under ultrasonic horn, *Phys. Rev. E*, **77**, paper #016609.

70. An, Y. (2011). Formulation of multibubble cavitation, *Phys. Rev. E*, **83**, paper #066313.

71. Dzaharudin, F., Suslov, S. A., Manasseh, R., and Ooi, A. (2013). Effects of coupling, bubble size, and spatial arrangement of microbubble cluster in ultrasonic fields, *J. Acoust. Soc. Am.*, **134**, pp. 3425–3434.

72. Mahmood, S., Yoo, Y., Oh, J., and Kwak, H. (2014). Hydrodynamic approach to multibubble sonoluminescence, *Ultrason. Sonochem.*, **21**, pp. 1512–1518.

73. Lahey Jr., R. T., Taleyarkhan, R. P., Nigmatulin, R. I., and Akhatov, I. S. (2006). Sonolumineacence and the search for sonofusion, *Adv. Heat Transfer*, **39**, pp. 1–168

74. Apfel, E. (1981). Acoustic cavitation, in *Methods of Experimental Physics*, Vol. 19, *Ultrasonics*, ed. Edmonds, P. D. (Academic Press Inc., New York), p. 373.

75. Baber, S. J., and Putterman, S. J. (1992). Light scattering measurements of the repetitive supersonic implosion of a sonoluminescing bubble, *Phys. Rev. Lett.*, **69**, pp. 3839–3842.

76. Giri, A., and Arakeri, V. H. (1998). Measured pulse width of sonoluminescence flashes, *Phys. Rev. E*, **58**, pp. R2713–R2716.

77. Didenko, Y. T., and Pugach, S. P. (1994). Spectra water sonoluminescence, *J. Phys. Chem.*, **98**, pp. 9742–9749.

78. Ashok Kumar, M., and Greiser, F. (1999). Sonophotoluminescence from aqueous and non-aqueous solutions, *Ultrason. Sonochem.*, **6**, pp. 1–5.

79. Wall, M., Ashok Kumar, M., Tronson, R., and Grieser, F. (1999). Multibubble sonoluminescence in aqueous salt solutions, *Ultrason. Sucochem.*, **6**, pp. 7–14.

80. Troia, A., and Ripa, D. M. (2013). Sonoluminescence in liquid metals, *J. Phys. Chem. C*, **117**, pp. 5578–5583.

81. Didenko, Y. T., and Gordeychuk, T. V. (2000). Multibubble sonoluminescence spectra of water which resemble single bubble sonoluminescence, *Phys. Rev. Lett.*, **84**, pp. 5640–5643.

82. Flint, E. B., and Suslick, K. S. (1991). The temperature of cavitation, *Science*, **253**, pp. 1397–1399.

83. Hua, I., Hochemer, R. H., and Hoffmann, M. R. (1995). Sonolytic hydrolysis of p-nitrophenyl acetate: the role of supercritical water, *J. Phys. Chem.*, **99**, pp. 2335–2342.

84. Byun, K., Seo, K. W., Shim, I. W., and Kwak, H. (2008). Syntheses of ZnO and ZnO-coated TiO_2 nanoparticles in various alcohol solutions at multibubble sonoluminescence condition, *Chem. Eng. J.*, **135**, pp. 168–173.

85. Lee, S. S., Byun, K., Park, J. P., Kim, S. K., Kwak, H., and Shim, I. W. (2007). Preparation of $Li_4Ti_5O_{12}$ nanoparticles by a simple sonochemical method, *Dalton Trans.*, pp. 4182–4184.

86. Lee, S. S., Seo, K. W., Yoon, S. H., Shim, I. W., Byun, K., and Kwak, H. (2005). CdS coating on TiO_2 nanoparticles under multibubble sonoluminescence condition, *Bull. Korean Chem. Soc.*, **126**, pp. 1579–1581.

87. Lee, S. S., Byun, K., Park, J. P., Shim, I. W., and Kwak, H. (2008). Homogeneous ZnS coating onto TiO_2 nanoparticles by a simple one pot sonochemical method, *Chem. Eng. J.*, **139**, pp. 194–197.

88. Kim, H. W., Kang, K. M., and Kwak, H. (2009). Preparation of supported Ni catalysts with a core/shell structure and their catalytic tests of partial oxidation of methane, *Int. J. Hydrogen. Energy*, **34**, pp. 3351–3359.

89. Kim, H. W., Kang, K. M., Kwak, H., and Kim, J. H. (2011). Preparation of supported catalysts on various metal oxides with core/shell structures and their tests for the steam reforming of methane, *Chem. Eng. J.*, **168**, pp. 775–783.

90. Kang, K. M., Kim, H. W., Shim, I.-W., and Kwak, H. (2011). Catalytic test of Ni catalysts with core/shell structure for dry reforming of methane, *Fuel Process. Technol.*, **92**, pp. 1236–1243.

91. Suslick, K. S. (1990). Sonochemistry, *Science*, **347**, pp. 1439–1445.

92. Cintas, P., and Luche, J-L. (1999). Green chemistry: the sonochemical approach, *Green Chem.*, **1**, pp. 115–125.

93. Suslick, K. S., Fang, M. M., Hyeon, T., and Mdleleni, M. M. (1999). Application of sonochemistry to material synthesis, in *Sonochemistry and Sonoluminescence*, eds. Crum, L. A., et al. (Kluwer Academic Publishers, Netheland), pp. 291–320.

94. Manoharan, S. S., and Rao, H. L. (2004). Sonochemical synthesis of nanomaterials, in *Encyclopedia of Surface and Colloid Science*, Vol. 10, ed. Nalwa, H. S. (CRC Press), pp. 67–82.

95. Kim, K. Y., Lim, C., Kwak, H., and Kim, J. H. (2008). Validation of molecular dynamics simulation for a collapsing process of sonoluminescing gas bubble, *Mol. Phys.*, **106**, pp. 967–975.

Chapter 2

Persistent Luminescence: Cerium-Doped Phosphors

Suchinder K. Sharma
Institute of Applied Physics, TU Bergakademie Freiberg, Leipziger Str. 23, 09596 Freiberg, Germany
suchindersharma@gmail.com

2.1 Introduction

Luminescent materials are found in a broad range of everyday applications. The existence of luminous organisms such as bacteria in the sea and in decaying organic matters, glowworms, and fireflies have mystified and thrilled man since time immemorial. The phenomenon of certain kinds of substances emitting light on absorbing various kinds of energies without heat generation is called *luminescence*. While in the seventies and eighties, the field of luminescent materials seemed to be fairly well covered, research in the nineties has been revitalized both in industry and academia. Improvements over the last three decades have led to the phosphor materials that operate close to their physical limits. In the present chapter, an interesting subclass of luminescent materials, long

Phosphors: Synthesis and Applications
Edited by Sanjay J. Dhoble, B. Deva Prasad Raju, and Vijay Singh

ISBN 978-981-4774-49-9 (Hardcover), 978-042-9460-99-9 (eBook)
www.panstanford.com

persistent materials, is discussed. Various aspects of persistent materials, which include their historical perspective, the important parameters for good persistent phosphors, a review of different mechanisms, and their applications, especially bioimaging, are discussed. In the later sections, cerium (Ce^{3+})-doped luminescent materials and the influences of symmetry and coordination on luminescence output of Ce^{3+}-doped materials are discussed. Ce^{3+}-doped persistent materials and an interesting case of codoped materials exhibiting white, long persistent behavior have been discussed. The future scope of long persistent materials with Ce^{3+} as the dopant has also been included to motivate the potential reader.

2.2 Historical Perspective

Luminescence phenomena exist in many different forms and have been observed from the earliest times. The *Chinese Book of Odes* (the Shih Ching) from the period of 1500–1000 BC as well as the ancient Indian holy scriptures (the Vedas) refer to light emission from fireflies and glowworms [1–3]. In the Vedas, the word "khadyota," which means "glowworm" in Sanskrit, has been mentioned frequently [2]. Luminescence from bacteria, fungi, and decaying fish was documented by Aristotle (384–322 BC) in his *De Coloribus* (or "About Colors"): "Some bodies, though they are not fire, nor participate in any way of the nature of fire, yet seem to produce light" [1–3]. Many stories of light coming from living and inanimate objects were told over the centuries. Some stories include quotes from Strabo (58 BC–ca. AD 25) of luminous fish living in the Nile of ancient Egypt and Pliny the Elder (AD 23–79) told of glowworms, lampyrides, luminous mollusks, and jellyfish in his *Natural History* [2]. Herodotus, the father of history, was among the first to describe mineral luminescence when he spoke of a temple in Tyre where a smaragdine column, which is presumed to have been made of fluorspar, or false emerald, shone in the evening [2]. As the superstition and belief in magic led to the development of science in many civilizations, observations led to further examination and actual studies [4]. This transition evolved over nearly 2000 years, and in 1565, a Spaniard, Nicolas Monardes, observed intense blue emission from an aqueous wood extract (*Lignum nephriticum*).

Scientists such as Athanasius Kircher of Germany, Francesco Grimaldi of Italy, and Robert Boyle and Isaac Newton of England studied this solution almost 90 years later and reported that upon illumination with white light, blue light appeared in the solution by reflection and yellow by transmission [3]. But it wasn't until much later (1852) that George Stokes of England identified this phenomenon as luminescence emission [3]. More discoveries of luminescence from inorganic materials were made in the 17th and 18th centuries [1]. Alchemists were the first to actually synthesize luminescent materials, even though this was mainly by accident in their attempts to transmute metal to gold [2, 3]. In 1603, Vincenzo Cascariolo, a Bolognese cobbler and alchemist, heated barium sulphate powder with coal, creating a porous cake that glowed bluish purple at night after having been exposed to sunlight during the day [2, 3–5]. The stone was given the name *Bolognian stone* and later named *lapis solaris*, or "solar stone," as well as *spingiasolis*, or "sun sponge" [2, 3]. Samples of this stone were presented to Scipio Bagatello (famous alchemist), Giovanni Antonio Magini (math and astronomy professor at the University of Bologna), Galileo Galilei, and finally Giulio CesaeLa Galla (professor at Collegio Romano). It was La Galla who wrote in 1612 the first publication on the first man-made luminescent material, entitled *De Phoenomenis in Orbe Lunae Novi Telescopii Usu* [2, 3]. Another important publication is a monograph on the Bolognian stone written by Fortvnii Licetientitled in 1640, *Litheosphorus Sive de Lapide Bononiensi*, for "stony phosphorus," where "phosphorus" means "light bearer," as shown in Fig. 2.1 [3]. Thus, the term "phosphor" was later coined to mean any "microcrystalline solid luminescent material" and to distinguish it from the elemental phosphorous later discovered in 1669 by Hennig Brand [3]. Since the Bolognian stone exhibited a long-lasting glow, long-lived luminescence became known as *phosphorescence* [3]. Eilhard Wiedemann (German physicist) later introduced the term "luminescence" in 1888 to include all light emission not caused solely by a temperature rise and included both fluorescence (short-lived luminescence) and phosphorescence (long-lived luminescence) [3]. The 19th century gave birth to the categorization of various types of luminescence. The categorization was developed to differentiate between the various luminescent excitation methods [3].

Figure 2.1 The book titled *Litheosphorvs Sive de Lapide Bononiensi* by author Fortvnii Liceti in 1640. Reproduced with permission from Ref. [3].

The history of long persistent phosphors is much shorter than the phosphors' history, although some of the phosphors found in the early times might have exhibited an afterglow for a short period of time. When a phosphor has defects with a certain trapping depth, which can be effectively activated at room temperature, it will show persistent phosphorescence. The term "persistent phosphors" has been applied to materials exhibiting phosphorescence lasting from minutes to hours. Long-lasting phosphor (LLP) materials are characterized by their phosphorescence duration and total light output. Long persistent phosphors are also known as LLPs, long-duration phosphors, long-lived phosphors, and long-afterglow phosphors. LLP materials can be compared to an optical battery. At first, the material is charged with a light for few minutes. When the excitation is stopped and material is taken in a dark room, one can observe the emission of light coming out, easily observable by the naked eye during several hours, in the most favorable case. Emergency signage used in case of electricity failures is one of the main application of LLPs. Other applications were also proposed, which include watch dials, decorative objects, toys [6], and more recently energy [7] and outdoor applications [8]{NOTE:http://www.studioroosegaarde.net/}. In most cases, taking into account the average human eye sensitivity and color perception, the emission needs to be persistent in the green range [9].

The physics behind the persistent-luminescence phenomenon is, however, not that simple, and intensive research has been pursued on such materials in the past 20 years. Although the basic principles of persistent luminescence are well understood by now, many details and mechanisms are still a subject of scientific debate. In LLP materials, two kinds of active centers are involved: traps and emitter centers. In some cases, the emitting center can also play the role of a trap center. Trap levels can be due to lattice defects, impurities [10, 11], or various codopants, which are introduced to enhance traps within the material in the energy bandgap mainly just below the conduction band (CB) in the case of electron traps or just above the valence band (VB) in the case of hole traps [12–14]. Under light, excited levels are generated and the excitation instead of being radiatively relaxed can be nonradiatively captured in the traps. Traps usually do not emit radiation but store the excitation energy for some time. Hence, this phenomenon can be related to an optical battery, where the filling of traps corresponds to the charging process. Then, in a second step corresponding to the discharge, the energy can be released by a thermal or optical source [15, 16] or by other physical stimulations (mechanostimulation for, instance [17]), resulting in stimulated emissions from the active centers. Thus, emitters are the centers that emit in the wavelength range of interest.

2.3 Properties of Persistent Materials

A good LLP material possesses the following properties:

- The first property is the initial afterglow intensity and the afterglow duration. While the initial intensity is an important factor for safety signage applications, the afterglow duration can be defined as the duration below which the photopic intensity decreases to an eye-perceivable intensity value of 0.32 mcd/m^2.
- The activation energy of traps that store charges (electrons/holes) must be between 0.5 and 1.0 eV, which can be emptied continuously at room temperature. The activation energy can be measured using the thermoluminescence (TL) technique [18]. The traps with activation energy greater than 1.0 eV are suitable for storage devices and for luminescence dating. The traps require additional energy for charge release, which can

be supplied by either thermal energy or optical energy [18, 19]. TL glow curves responsible for persistent luminescence and the effective technique of phototransfer are shown in Fig. 2.2.

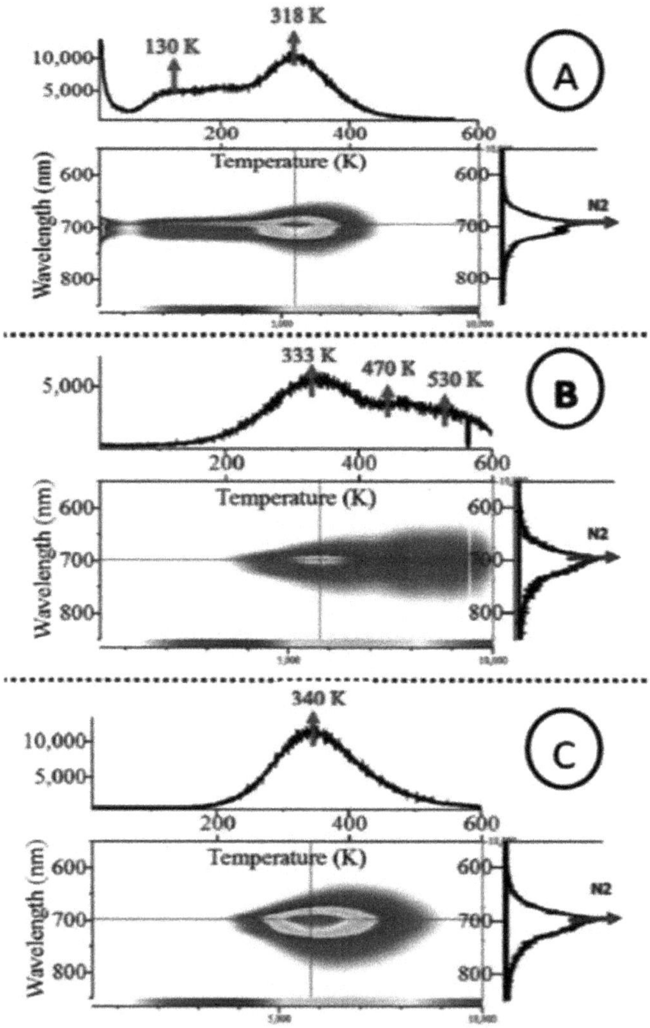

Figure 2.2 Thermoluminescence contour plots in $ZnGa_2O_4:Cr^{3+}$ under (A) 254 nm excitation, (B) 580 nm excitation, and (C) 580 nm excitation, followed by photostimulation using 977 nm laser. A clear transfer of charges from deep traps to shallow traps can be observed after optical stimulation. Reprinted from Ref. [20], Copyright (2017), with permission from Elsevier.

- The frequency factor or an attempt to escape frequency (s, in sec^{-1}) is another parameter for good LLP materials. The delay in phosphorescence is generally due to the time spent by the charges (e.g., electrons) in the electron traps. The mean time spent in a trap at temperature T is given by Eq. 2.1

$$\tau = s^{-1} \exp\left(\frac{E}{kT}\right), \tag{2.1}$$

where s is the frequency factor, E is the activation energy of traps or trap depth, and k is the Boltzmann constant. The charges once trapped are released slowly from traps at room temperature, delaying the overall recombination process. LLP materials have two basic processes: excitation and recombination. Initially, when the traps are empty and can store charges, the excitation process leads to charge trapping, and it takes some time for charges to saturate. After the initial few seconds of excitation, there is a competition between the electrons that are getting trapped and those that are keen to come out of the traps to recombine at the luminescent center. During this process, some charges again get trapped at the traps, and a different phenomenon, *retrapping*, can happen. If a single process of charge detrapping from a single trap is dominant, a single exponential behavior of the decay curve is observed. However, due to competition between *trapping-retrapping* processes, a multiexponential or hyperbolic decay curve is obtained. The frequency factor (s) is an important parameter, and its value depends upon the competition between these competitive processes. The typical value of s is between 10^6 and 10^{14} sec^{-1} [18]. In the literature, $s = 10^{11}$ s^{-1} is used, which underestimates the overall phenomenon. Hence, due care should be taken while evaluating the relevant parameters and processes in LLP materials.

2.4 The Modern Era

Hoogenstraaten was first to report the interesting properties of ZnS and Cu and Co phosphors in 1953 with prominent emission at 530 nm [21]. The use of ZnS:Cu, Co phosphors was limited for practical applications due to shorter decay time and stability in a humid environment. The study was followed by other reports in

1971 on the optical and electrical properties of $SrAl_2O_4:Eu^{2+}$ [22]. A few decades later in 1996, Matsuzawa et al. published an article on the long phosphorescent properties of aluminate-family-based $SrAl_2O_4:Eu^{2+},Dy^{3+}$ phosphors with prominent emission in the green wavelength region (520 nm) [23]. The article received over 1500 citations (data from Google Scholar). Takasaki et al. complemented these results and compared them with traditional ZnS:Cu, Co phosphors [24]. Later, Eu^{2+}-activated aluminate hosts were studied extensively by many [25–27]. Apart from aluminate hosts, the Dy^{3+}-codoped silicate phosphor ($Sr_2MgSi_2O_7:Eu^{2+},Dy^{3+}$) was established as an efficient LLP phosphor in 2001 by Lin et al. [28]. Eu^{2+} was favored over other activators to prepare LLP materials, with a comprehensive review written by Eeckhout et al. in 2010 [29]. The previously considered dormant topic in research is now a vibrant topic, which involves fields like biology, chemistry, physics, and materials science. The typical applications of persistent materials are decoration and safety signage, watch dials, solar panels, and medical imaging. The persistent materials can be found in ancient Chinese paintings to distinguish colors in daylight and at night. Toys with persistent materials can be easily found in the market. The radium in watch dials has been replaced by $SrAl_2O_4$:Eu,Dy for their long decay time. Some LLP materials used for decoration are shown in Fig. 2.3.

Figure 2.3 LLP materials for decoration. Reproduced from Ref. [30] with permission of The Royal Society of Chemistry (RSC) on behalf of the European Society for Photobiology, the European Photochemistry Association and the RSC.

2.5 Mechanisms

The mechanisms of persistent luminescence are most controversial of all. The mechanisms can be categorized into two categories: delocalized and localized mechanisms. The delocalized mechanism can be understood in a simple way as the one in which the charge trapping-detrapping is through the CB (for electrons) and through participation of the VB (for holes). Upon excitation, the charges move to the CB, where they are free to move and get stored at defects with a trap depth of 0.5–1.0 eV. But due to thermal energy at room temperature, the electrons are detrapped and again move to the CB. Out of all these electrons, some recombine at luminescence center, with others retrapped at the same traps. The localized mechanism does not involve charge trapping-detrapping through the CB and will be discussed separately in a later section.

The delocalized mechanism for persistent luminescence has been studied extensively. There are two active centers for charge trapping, electrons and holes, with defects formed during synthesis. The general charge-trapping processes for electron and hole trapping are shown in Figs. 2.4a and 2.4b, respectively. The most controversial delocalized mechanism is for $SrAl_2O_4:Eu^{2+},Dy^{3+}$. Matsuzawa (1996) first explained it using photoconductivity measurements by assuming holes as the main charge carriers [23], as shown in Fig. 2.5a. The holes are due to Sr^{2+} vacancies, which acted as traps. The incident photon excites the Eu^{2+}, and a hole escapes to the VB, leaving behind the Eu^{+} ion, which is captured by the Dy^{3+} ion, forming Dy^{4+}. The thermal energy provided by room temperature detraps these holes into the VB. From the VB, these holes moves back to Eu^{+} ions, forming Eu^{2+} with the release of photons. The second model was proposed by Aitasalo in 2003 for $CaAl_2O_4:Eu^{2+}$ [13], as shown in Fig. 2.5b.

The electrons are directly trapped at defects, and the holes are trapped at calcium vacancies (V_{Ca}''). The room temperature removes electrons from traps, which recombine at oxygen vacancies via energy transfer to the europium ion. The Aitasalo model rejected the Matsuzawa model as it failed to explain the LLP behavior of non-Dy-codoped $SrAl_2O_4:Eu$, which also shows persistence. Also, the possibility of Eu^{+} and Dy^{4+} was minimal in these hosts due their instability in these hosts. As a common feature of both models, holes were considered as charge carriers and the charge trapping-detrapping involves the VB.

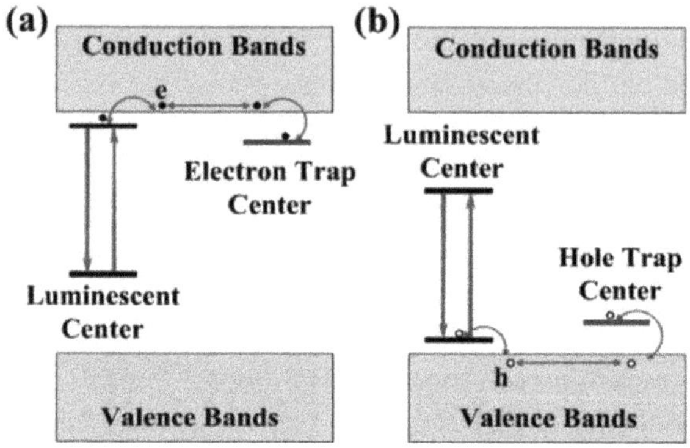

Figure 2.4 Persistent-luminescence mechanism due to electron and hole traps. Reprinted with permission from Ref. [31]. Copyright (2016) American Chemical Society.

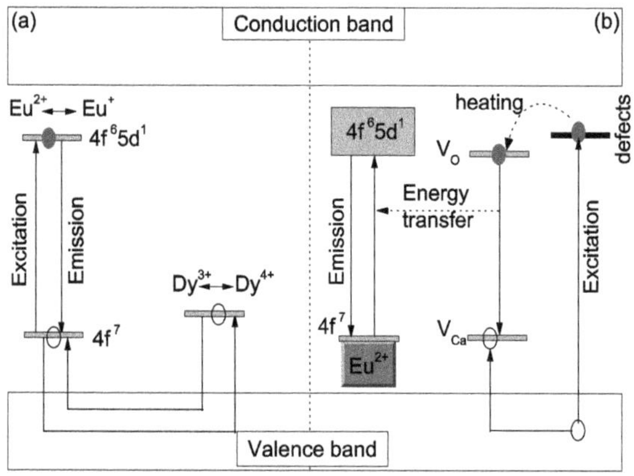

Figure 2.5 Persistent-luminescence mechanism, as explained by (a) Matsuzawa and (b) Aitasalo using hole trapping via the valence band.

In 2005, two new models were proposed by Dorenbos and Clabauto to explain the LLP mechanism, as shown in Figs. 2.6a and 2.6b, respectively [32, 33]. Dorenbos observed that the 5d level of Eu^{2+} was within the CB such that the electrons, as explained in the Matsuzawa model, were within the CB and trapped by Dy^{3+}, forming Dy^{2+}. The traps in both cases were found to be 0.9 eV. The recombination occurs at the recombination center. On the other hand, Clabau used electron spin resonance (ESR) spectroscopy to conclude that there was a decrease in the concentration of Eu^{2+} ions upon excitation, followed by an increase in their concentration upon the complete detrapping process. It was in contradiction to the Aitasalo model. There was no migration of electrons through the CB, and the nature of traps remains the same upon Dy^{3+} doping. So the influence of Dy^{3+} was minimal, as explained by Dorenbos, and the main charge-trapping vacancies were oxygen vacancies. In both these models, electrons were considered as the charge carriers. Similarly, there have been several efforts to understand the type and nature of defects, the luminescence center, and the charge trapping-detrapping mechanism [34–36].

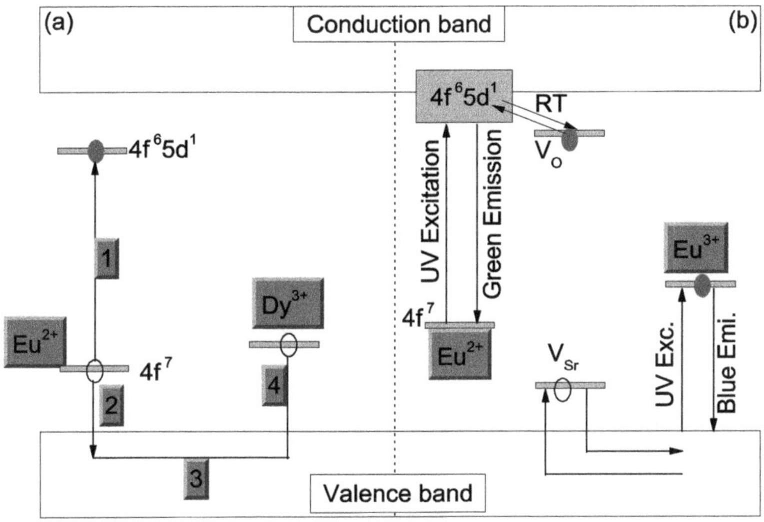

Figure 2.6 Persistent-luminescence mechanism proposed by (a) Dorenbos and (b) Clabau.

2.6 Bioimaging

Bioimaging has been most interesting application of persistent materials. As a pre-requisite, LLP materials must emit in the "biological optical window," where the tissues are most transparent, that is, between 600 nm and 1100 nm [37, 38]. In this method, specially designed LLP nanoparticles, which are able to circulate in the blood vessels of small animals, are used as luminescent biomarkers. Making use of red persistent luminescence for bioimaging presents numerous advantages over conventional fluorescence techniques. The biomarkers are irradiated before injection into the animal body (ex vivo), making the exposure of the animal to high-energy radiation unnecessary. Since the nanoparticles are irradiated outside the animal body and the emission is in the biological optical window (650 nm to 900 nm), autofluorescence of living tissues can be prevented, improving the signal-to-noise ratio. Red LLP materials with good properties are difficult to produce for the following two reasons: (i) It is not straightforward to predict properties of Eu^{2+}-doped oxide materials, and in most oxides, the 5d state of Eu^{2+} lies within the CB, and (ii) the human eye sensitivity to red light is much lower when compared to the green or blue region (also known as the Purkinje effect) [39].

The first material ever reported with the red LLP features was $MgSiO_3:Eu^{2+},Dy^{3+},Mn^{2+}$, reported by Wang et al. [40]. The 660 nm LLP emission of Mn^{2+} was observed due to efficient energy transfer from rare earth ions to Mn^{2+} ions (for ~4 hs). However, the first demonstration of the bioimaging technique was demonstrated in 2007 using silicate nanoparticles $Ca_{0.2}Zn_{0.9}Mg_{0.9}Si_2O_6:Eu^{2+},Dy^{3+},Mn^{2+}$, with Mn^{2+} as a luminescent ion [37]. In this method, the specially designed LLP nanoparticles, which are able to circulate in the blood vessels of small animals, are used as luminescent biomarkers. Though intended for diagnosis applications in living animals, first-generation LLP probes (silicates and phosphates) suffered from several limitations. In particular, the first generation of persistent-luminescence nanoparticles had to be excited ex vivo by ultraviolet (UV) light prior to its systematic administration, producing a persistent-luminescence signal only detectable during ~1 hr after the injection and preventing long-term imaging in living animals. Thus, this first-generation probe does not allow observing of, for example, in vivo tumor-homing strategies such as those relying on slow accumulation of stealth nanocarriers within malignant stroma by the enhanced permeability and retention effect [41, 42]. Indeed, depending on the nanoparticle characteristics, such retention process usually requires from 2 to 24 hr [43–45]. This is far too long in relation to the emission from persistent-luminescence nanoparticles, which hardly exceeds 1 hr in vivo.

An important red LLP material was prepared in 2011, when a novel compound $ZnGa_2O_4:Cr^{3+}$ with enhanced persistent time was reported [38]. Later, as a major improvement in the technique, a second-generation material allowing visible-light excitation of LLP was tested and verified [47]. $ZnGa_2O_4$ crystallizes in the normal spinel structure (space group O_h^7 (Fd-3m)), with Zn^{2+} ions occupying tetrahedral sites and Ga^{3+} ions occupying octahedral sites, as shown in Fig. 2.7a. Cr^{3+} ions substitute octahedrally coordinated Ga^{3+} ions in the host lattice. Upon excitation, Cr^{3+} emits via its ^2E (^2G) $\rightarrow$ $^4A_2$ (^4F) 3d–3d transition with a near-infrared (NIR) luminescence peak at 696 nm, a wavelength suitable for in vivo imaging. Zinc gallate proved to be an exciting candidate for the application, owing to the fact that its red LLP can be triggered with the visible-light excitation, enabling the re-excitation of nanoparticles inside the animal body [47].

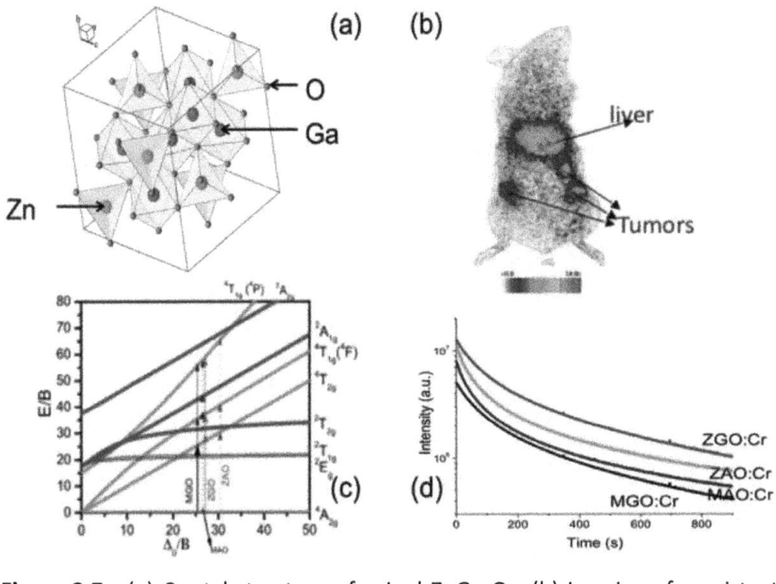

Figure 2.7 (a) Crystal structure of spinel $ZnGa_2O_4$; (b) imaging of persistent luminescent nanoparticles in tumor-bearing mouse; (c) comparison of crystal field splitting of Cr^{3+} in different spinels; and (d) comparison of persistent luminescence decay curves in different spinels. (a) Reprinted with permission from Ref. [46]. Copyright (2014) American Chemical Society. (b–d) Reprinted from Ref. [47], Copyright (2014), with permission from Elsevier.

With the second-generation materials, prior ex vivo excitation remains possible but was not necessary as was the case with first-generation silicate probes. The nanoparticles injected were allowed to circulate into the blood vessels of a small animal for some time. After the end of the excitation, the carriers were progressively detrapped by thermal activation to feed luminescent centers and are detected using a suitable imaging camera. With such nanoprobes, in vivo excitation was possible many times and long-term imaging could be envisioned. For instance, Fig. 2.7b shows the persistent-luminescence image of a tumor-bearing mouse [47]. CT26 tumor fragments of about 3 mm^3 were implanted in both flanks of the mouse. The image shows the localization of the nanoparticles in the tumors, probably through the enhanced permeability and retention (EPR) effect. This result constitutes proof of in vivo tumor passive targeting with stealth persistent-luminescence nanoparticles and constitutes considerable progress compared to the first generation of nanoprobes. Other in vivo applications of these nanoprobes are reported in Ref. [47]. A comparison of crystal field splitting among different spinels is shown in Fig. 2.7c based on the Tanabe–Sugano diagram. The persistent luminescence could be seen for different spinel hosts, whose decay curves have been compared in Fig. 2.7d.

Cr^{3+}-doped zinc gallogermanates were also shown to possess very long persistent luminescence, which could be even triggered by sunlight [48, 49]. Another compound, Cr^{3+}-doped $MgGa_2O_4$, was reported for optical imaging, showing broader red-shifted persistent luminescence as compared to $ZnGa_2O_4:Cr^{3+}$ [50]. Several other hosts were also studied and in vivo tests were performed [51–55]. Third-generation persistent luminescent nanoprobes were developed recently. In this strategy, the photostimulation-based luminescence phenomenon was used to probe small animals in vivo. The work in this direction was reported later by many other research groups [53, 55–57]. The use of LLP materials for multifunctionalities was also proposed.

2.7 Localized Mechanism of Charge Recombination

There are two excitation paths for persistent-luminescence compounds, (i) one more efficient with high energy through the CB and (ii) a more local effect using less energy. The origin of persistent luminescence in the localized mechanism was due to the presence of neighboring antisite defects that are close to Cr^{3+} substituting the Ga^{3+} ion in $ZnGa_2O_4$. Antisite defects were the defects resulting from the exchange in site positions of zinc and Ga ions, forming $Ga_{Zn}^{\bullet}$ and $Zn_{Ga}^{'}$. The reader is referred to Refs. [58–60] for the EPR and PL spectroscopy correlation and to Refs. [58, 61–63] for the optical spectroscopy of chromium. An important point in this mechanism was that the Cr^{3+} ion liberates an electron–hole pair during excitation and does not change its oxidation state during excitation [59]. The first step was the excitation of Cr_{N2} ions, as shown in Fig. 2.8a, and then excitation was dissociated by the local electric field into an electron and a hole that get trapped at $Ga_{Zn}^{\bullet}$ and $Zn_{Ga}^{'}$ neighboring antisite defects, respectively, as shown in Fig. 2.8b. The excitation is thus trapped in the vicinity of chromium in the form of a pair of neutral defects Ga_{Zn}^{x} and Zn_{Ga}^{x}, while Cr^{3+} returns to its $^4A_2$ ground state (see Fig. 2.8c). Electrons and holes can then migrate far from the Cr^{3+} ion, so this storage mechanism can proceed many times with the same Cr_{N2} ion. The electron (or hole) trapping by $Ga_{Zn}^{\bullet}$ ($Zn_{Ga}^{'}$) was strictly equivalent to an optically induced donor–acceptor (D-A) transfer from $Zn_{Ga}^{'}$ (D) to $Ga_{Zn}^{\bullet}$ (A) [58, 59]. The reverse reaction (electron–hole release and capture by Cr^{3+}) was thermally activated, with the activation energy responsible for the main TL peak. All this indicates that the mechanism of charge trapping-detrapping was initiated and localized in the vicinity of Cr^{3+} ions, while chromium $^4T_2$ and ^2E manifolds are well below the CB edge (see Fig. 2.8d) [46, 47].

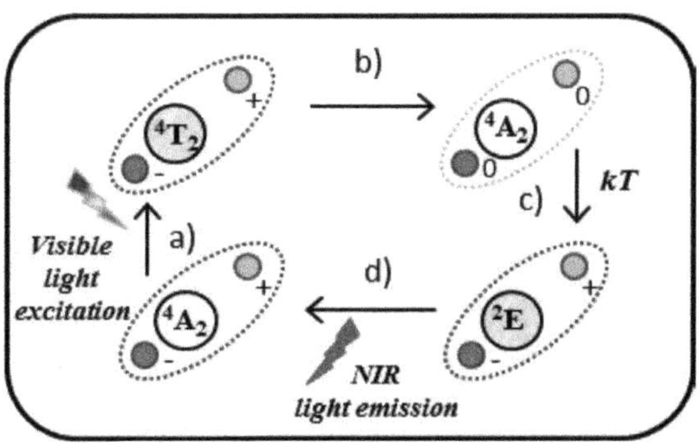

Figure 2.8 The localized persistent-luminescence mechanism. Reprinted with permission from Ref. [46]. Copyright (2014) American Chemical Society.

2.8 Cerium-Doped Luminescent Materials

The lanthanides, or rare earth ions, or elements with the 4f configuration, are positioned in the fourth period in the periodic table with the electronic configuration $[Xe]4f^n6s^2$, except for lanthanides like lanthanum (La), cerium (Ce), gadolinium (Gd), and lutetium (Lu), which have an electronic configuration of $[Xe]4f^{n-1}5d^16s^2$. Most elements are expressed by coupling of the spin and angular momenta. However, lanthanides possess an additional quantum number known as the Russell–Saunders quantum number or the spin–orbit coupling expressed by the symbol "J" with $^{2S+1}L_J$ as the common formalism. The lanthanides that are characterized by a partially occupied inner shell have Ce as the first member. The fundamental state is $^2F_{5/2}$, as described by Hund [64]. Later, Lang proved Hund's predictions using X-ray spectra, stating that the lowest Ce state was a doublet $^2F_{5/2}$ -$^2F_{7/2}$ separated among each other by 2253 cm^{-1} and 50,000 cm^{-1} from the higher 5d level [65–67]. Ce^{3+} possesses one electron in the $4f^1$ ground state, with shielding from filled $5s^2$ and $5p^6$ orbitals. The $4f^1$ state cannot be perturbed by the type of compound/host. However, on excitation to the 5d orbital a strong interaction exists, which should be taken care of while interpreting the luminescence spectra. Depending upon the site symmetry, at the most five distinct 4f $\rightarrow$ 5d transitions can be observed in the absorption spectra. Since the electron–phonon coupling spectra of the 4f and 5d electrons are different, the parity allowed electron–dipole transitions between the 4f and 5d states exhibit broad absorption and emission bands. Due to interaction with the crystal field, the average position of 5d bands decreases when compared to the position of free ions, which is also called *centroid shift* or *barycentre*. The combined effect of spin–orbit interactions and crystal field leads to a red shift (denoted by D) of the first 4f $\rightarrow$ 5d transitions. Hence, the total crystal field splitting (ε_{cfs}) can be defined as the difference between the lowest and highest 5d levels. The relation between different parameters can be defined using Eq. 2.2 as

$$D(A) = \varepsilon_C(A) + \frac{\varepsilon_{cfs}(A)}{r(A)} - 1890\,\text{cm}^{-1}, \qquad (2.2)$$

where A is the host lattice and $r(A)$ is the fraction of the total crystal field splitting that contributes to the red shift.

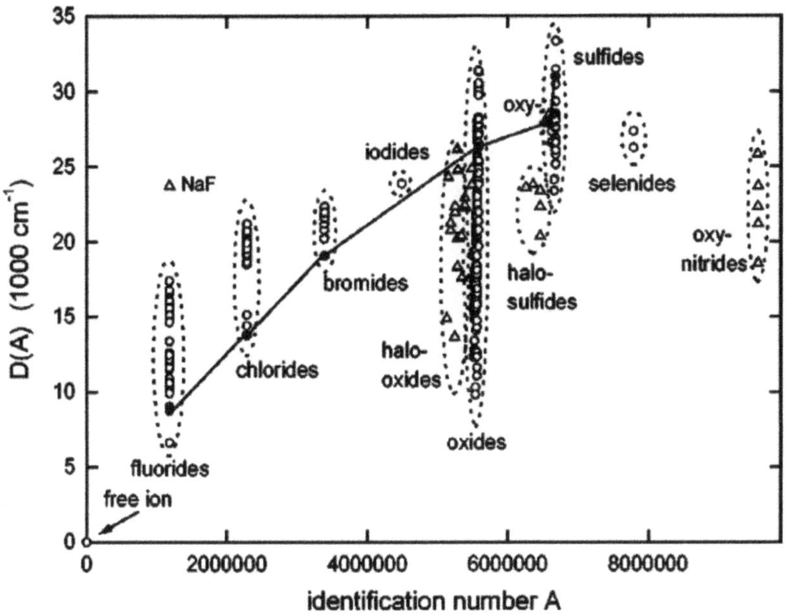

Figure 2.9 The red shift in inorganic materials. Reprinted with permission from Ref. [70]. Copyright (2000) by the American Physical Society.

The spectroscopic red shift of the trivalent lanthanides in different hosts is shown in Fig. 2.9. The centroid shift in fluorides varies between 5130 and 18,700 cm^{-1} for different hosts. One can expect that if the ionic radius of the cation is small and possesses a larger valency, it is more attracted to the fluoride ligands. The attractive forces on the fluoride change by cations other than Ce^{3+} along with the Ce^{3+}-to-fluoride-ion distance, and the anion coordination also plays a crucial role in the centroid shift. The maximum variation in the red shift can be observed for oxide hosts with a range between 9200 and 33500 cm^{-1}, as observed in Fig. 2.9. The shift can be maximized for anions with high polarizability, while the average cation electronegativity should be minimized [68, 69].

2.9 Influence of Symmetry and Coordination

The optical transitions of Ce^{3+} in any host can be determined by the symmetry and coordination of Ce^{3+} in the host lattice. The point charge electrostatic model (PCEM) explains the effect of the crystal field on luminescence features. For cubic and octahedral symmetries, eightfold or sixfold coordination exists to the ligands. For Ce^{3+} ions coordinated eightfold to O^{2-} ligands, the polyhedron is approximately cubic, with one plane containing four O^{2-} ligands at 45° around the *c* axis with respect to another plane. The absorption bands contain orbital doublet (E_g) and triplet (T_{2g}) levels, respectively.

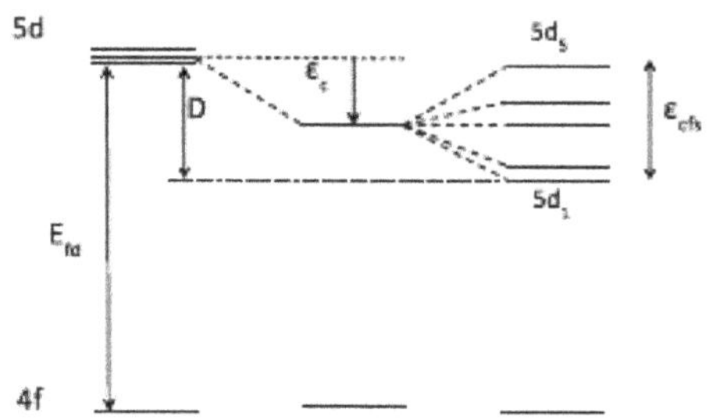

Figure 2.10 Influence of crystal field splitting and centroid shift on the Ce^{3+} 5d level. A splitting into five different levels is presented. Reproduced from Ref. [73] with permission from The Electrochemical Society.

Similarly, for sixfold coordination, the higher excited state (^{2}D) splits to E_g and T_{2g} orbitals in O_h symmetry, as shown in Fig. 2.10. The energy levels of the lowest excited state of Ce^{3+} are lower in energy in cubic symmetry than in octahedral symmetry because the shift of the E_g state from the center of gravity of the ^{2}D degenerate state is 1.5 times larger than that of the T_{2g} state. The reduction may be caused by the covalency or the expansion of the 5d wave function toward the ligands (*nephelauxetic effect*) [71]. The maximum crystal field splitting can be observed for octahedral coordination, followed by cubic, dodecahedral (ddh), and smallest for a tricapped trigonal prism (3ctp) and cuboctahedron (cubo) coordination, as explained by Dorenbos [72].

2.10 Cerium-Doped Long Persistent Materials

Ce^{3+} is a good candidate for studying the behavior of 5d electrons with the possibility of tuning their luminescence from the red to the UV region due to strong crystal field dependence of its 5d–4f transition energy [74]. The lifetime of Ce^{3+} 5d–4f transitions is about

3 orders of magnitude shorter than their 4f–4f transition lifetimes in different hosts. The Ce^{3+} persistent luminescence can be observed in different hosts, such as oxides, sulfides, silicates, and garnets. The different hosts doped with only Ce^{3+} that exhibit persistent luminescence are tabulated in Table 2.1.

Table 2.1 LLP emission for different hosts doped with Ce^{3+} only

LLP emission maximum (nm)	Host	Decay time (min)	Reference
385	$SrAl_2O_4$	600	[75, 76]
400	$Sr_2Al_2SiO_7$	2	[77]
400	$CaAl_2O_4$	600	[78–80]
402	$SrMgAl_2SiO_7$	2	[77]
400–417	$Ca_2Al_2SiO_7$	60	[81–84]
400–430	Lu_2SiO_5	180	[85, 86]
402–450	$BaAl_2O_4$	600	[87]
412	$Ca_4Al_6Si_3O_{19}$	60	[88]
420	$CaAl_4O_7$	600	[75]
425	$CaYAl_3O_7$	60	[89]
435	$Sr_3Al_2O_5Cl_2$	180	[90]
472–511	$Sr_4Al_{14}O_{25}$	10	[91]
505	$Y_3Al_{5-x}O_{12}$ (x=2.5-3.5)	180–1200	[92]
508–568	CaS	5	[93]
525	$Y_3Al_5O_{12}$	2	[94, 95]

The hosts that contain Ce^{3+} as a dopant and other lanthanide/transition metal ions as a codopant, which helps in producing/increasing persistent luminescence, are presented in Table 2.2. For only Ce^{3+}-doped phosphor hosts, a clear variation in the persistent luminescence emission maximum from 385 nm to 525 nm can be observed. $SrAl_2O_4$:Ce^{3+} possesses LLP emission at 385 nm with an afterglow of over 10 hr [75, 76]. The garnet $Y_3Al_{5-x}O_{12}$:Ce^{3+} possesses an LLP emission maximum at 505 nm, which lasts for 3–20 hr, depending on the Ce^{3+} concentration. However, most hosts possess blue LLP emission (400–435 nm) upon Ce^{3+} doping. The large variation in LLP emission (386–525 nm) can be attributed to the large crystal field effect on 5d levels and the location of Ce^{3+} in the host lattice. Ce^{3+}-doped alkaline earth aluminates MAl_2O_4(M = Ca, Sr, Ba) are most studied and possess persistent luminescence for over 10 hr.

Table 2.2 LLP emission for different hosts doped-codoped with Ce^{3+} and other lanthanides/transition metal ions

LLP emission maximum (nm)	Host	Codopant	Decay time	Reference
350–435 nm	$Ba_5(PO_4)_3Cl$	Eu^{2+}	05 min	[96]
375, 515	$SrAl_2O_4$	Mn^{2+}	600 min	[76]
386–591	$Ca_{0.5}Sr_{1.5}Al_2SiO_7$	Tb^{3+}	01 min	[97]
408–573	$Sr_2Al_2SiO_7$	Dy^{3+}	60 min	[98]
410–588	$Sr_2Al_2SiO_7$	Tb^{3+}	01 min	[99]
408–680	$BaMg_2Si_2O_7$	Mn^{2+}	120 min	[100]
435–620	$Sr_3Al_2O_5Cl_2$	Eu^{3+}	180 min	[101]
505, 700	$Y_3Al_2Ga_3O_{12}$	Cr^{3+}	min	[102]
525	$CaAl_2O_4$	Mn^{2+}	600 min	[103]
540	$Gd_3Al_2Ga_3O_{12}$	Cr^{3+}	120 min	[104, 105]
543	$CaAl_2O_4$	Tb^{3+}	600 min	[106, 107]
550	$Ca_2Al_2SiO_7$	Mn^{2+}	600 min	[103]
808–1335	$Y_3Al_2Ga_3O_{12}$	Nd^{3+}	600 min	[108]

To exhibit a persistence behavior, the mobility of the electron or its delocalization is an important factor. From the delocalized band, there must be a trapping state to offer relocalization for the charge carriers. The delocalized band wave function may be continuous or discrete at the dopant site, depending upon the position of energetically excited states of the dopant. From earlier discussions for Ce^{3+}, it was anticipated that the T_{2g} state can be populated via a charge transfer process from the ligand orbitals (composing the host VB). The relaxation of this charge transfer state back to the ligand orbital was an important point of due consideration. The relaxation from the T_{2g} (5d) level to the $^2F_{(7/2,\ 5/2)}$ ground states of Ce^{3+} was clearly required for Ce^{3+} emission, though. The relative position of

Ce^{3+} ground states with respect to the 5d excited states (E_g, T_{2g}) can be determined from the host excitation data and the excitation-emission spectra using vacuum-referred binding energy (VRBE) diagrams.

A comparison of LLP emission spectra for different Ce^{3+}-doped and Ce^{3+}-codoped hosts is presented in Fig. 2.11.

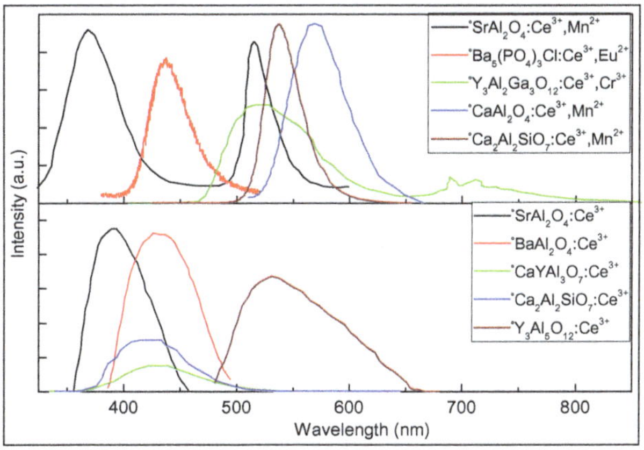

Figure 2.11 LLP emission spectra for different Ce^{3+}-doped and Ce^{3+}-codoped samples.

The variation in LLP emission from the near-UV to the deep-red region of the electromagnetic spectrum could be seen. The LLP emission in only a Ce^{3+}-doped sample can be attributed to typical 5d $\rightarrow$ 4f emission, which can be deconvoluted into two bands corresponding to $^2F_{7/2}$ and $^2F_{5/2}$ levels, respectively. Looking on to the role of trapping centers for photoenergy storage in long persistent photostimulable and thermostimulable phosphors, important information about the interaction between the conduction carriers and rare earth ions such as Ce^{3+} is required. The Ce^{3+} lowest 5d level ($5d_1$) has to be close to the CB to produce persistent luminescence. Some measurements relative to photoconductivity are already reported in this direction [108]. Utilizing the host bandgap data and knowing the energetic excitation that causes a maximum photocurrent, the information about the lowest T_{2g} (for a cubic environment) can be determined. Afterglow emissions of few seconds to maximum of 1200 min are reported. The afterglow depends on delocalization of 5d electrons of Ce^{3+}, leading to photoionization of Ce^{3+}. Electron traps situated within the host bandgap can thus trap the electrons from the delocalization band. The slow release of electrons from these trapping levels leads to room-temperature phosphorescence. Codoping different Ce^{3+}-doped phosphors with other suitable codopant ions (e.g.. Mn^{2+}, $Eu^{2+/3+}$, Tb^{3+}) increases the persistence time of such phosphor, and a new series of materials using the energy transfer phenomenon can be prepared with prior knowledge of electronic structure or VRBE calculations [108].

2.11 Cerium-Doped White, Long Persistent Materials

An interesting observation of some Ce^{3+}-doped LLP materials is the ability to exhibit white LLP upon codoping with some other codopant [97–100]. As the LLP materials can be used for various safety- and emergency-related applications, the research on white LLP materials can be considered promising. The emission spectra for different phosphors exhibiting white LLP are presented in Fig. 2.12. The mechanism in white LLP materials was dominated by the energy transfer process between different ions. For example, in the case of $Sr_2Al_2SiO_7$:Ce^{3+},Dy^{3+}, energy transfer between Ce^{3+} and Dy^{3+} leads to an emission from Ce^{3+} + Dy^{3+} in the LLP spectra, as shown in Fig. 2.13. For an energy transfer mechanism, Dexter's theory can be considered more appropriate, which explains that the spectral overlap between the donor emission and the acceptor excitation is required for efficient energy transfer [109]. In $Sr_2Al_2SiO_7$:Ce^{3+}, there is an emission band at 408 nm (upon 335 nm excitation), which has two components at 400 and 433 nm because of different 5d $\rightarrow$ 4f transitions.

For a $Sr_2Al_2SiO_7$:Dy^{3+}-doped sample, the emission bands were observed at 491 and 573 nm for different excitation bands between 230 and 400 nm [98]. Due to the overlap between the emission bands of Ce^{3+}-doped and excitation bands of Dy^{3+}-doped samples, efficient transfer was possible for excitation at 350 nm. The TL glow peaks showed maxima at 338, 381, and 469 K, where the 338 K peak was responsible for the LLP. The mechanism of LLP in $Sr_2Al_2SiO_7$:Ce^{3+},Dy^{3+} was considered to be a four-step process [98]:

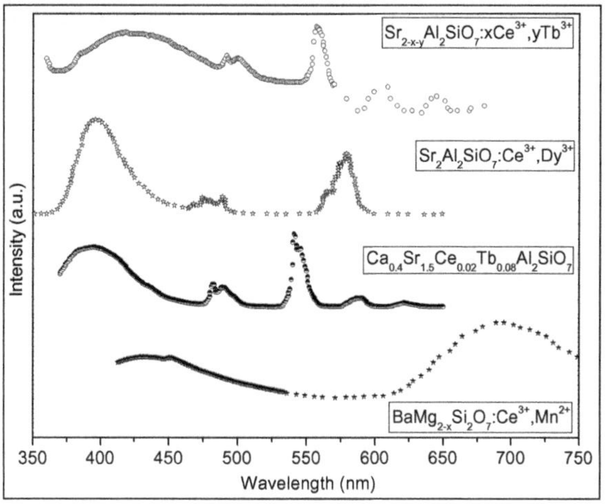

Figure 2.12 White LLP emission spectra for some phosphors.

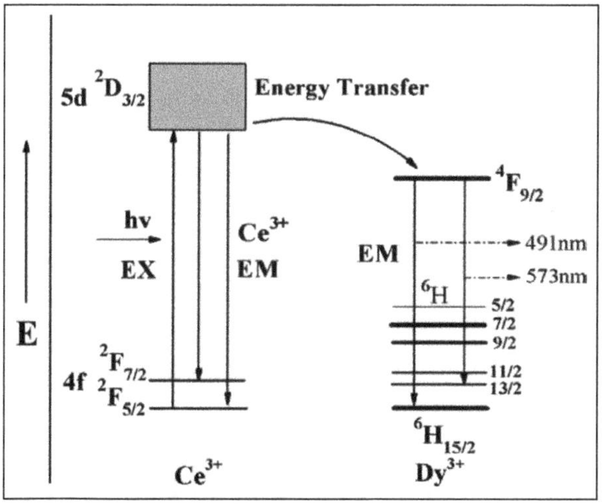

Figure 2.13 Energy transfer process for charge transfer in $Sr_2Al_2SiO_7$. Reproduced from Ref. [98] with permission from The Electrochemical Society.

1. Upon excitation, the electron moves from the Ce^{3+} 4f state to the Ce^{3+} 5d state close to the CB.
2. The electron at room temperature moves to the CB, where it gets trapped by the electron traps, with the hole trapped at the Dy^{3+} ground state.

3. Upon detrapping, a part of electrons coming back to the Ce^{3+} ground state transfer energy to Dy^{3+}.
4. At room temperature, the trapping-detrapping in the shallow trap continues and the detrapped electrons recombine at Ce^{3+} and Dy^{3+}, producing white LLP.

2.12 Conclusions and Future Direction

The number of publications in the field of persistent materials has increased drastically in the past 20 years, as presented in Fig. 2.14. However, research in the field of Ce^{3+}-doped persistent materials is limited. The lack of information could be due to the reason that the Ce^{3+} emission in different phosphors is usually in UV. The possibility of new persistent materials emitting in UV is still unexplored (UV-LLP). The UV-persistent materials are expected to find use in bioapplications, especially in medical skin treatment, crime scene inspections, document and forgery analysis, currency, high-security markers, etc.

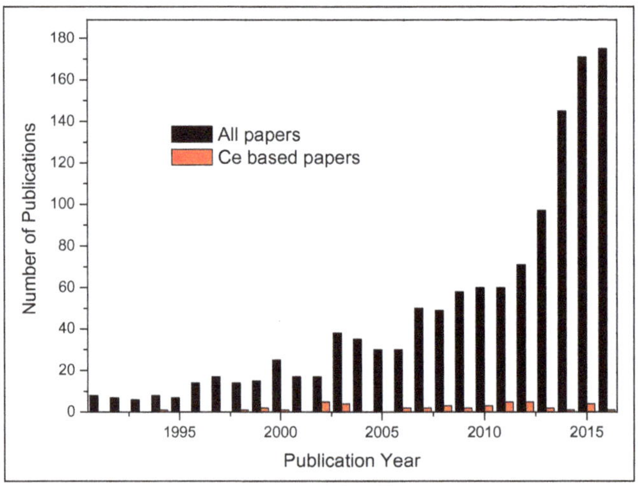

Figure 2.14 Year-wise publications on persistent materials (black bars) and on Ce-based persistent materials (red bars).

The phenomenon of persistent luminescence is governed by the presence of shallow defects with activation energy between 0.6 and

1.0 eV. However, in most cases, some other defects with activation energy > 1.0 eV are also present. These are also termed as "deep defects" in the literature. The charges stored in these deep defects have a long lifetime, typically between a few weeks to several million years. Technically, it is possible to use these deep defects for long-term monitoring. By way of thermal stimulation or optical activation, the charges stored in these deep defects can be transferred to shallow defects, where persistent luminescence can be observed at room temperature. Optical stimulation or phototransfer has gained fresh impetus in recent years with applications in bioimaging. The particles stored in small animals can be excited using a 977 nm laser to produce persistent luminescence, which can be used for long-term in vivo imaging [20]. The trapping capability of Cr^{3+}-doped compounds and a combination of transition-metal-lanthanide-doped compounds is tremendous and can be successfully explored to prepare materials emitting in the red or the deep-red region.

Many compounds with excellent persistent luminescence properties have been reported. However, challenges like particles with a small grain size, desired morphology, desired emission window, and high efficiency are still open. It is known that persistent luminescence decreases with decreasing grain size of particles. The ways to solve this problem are still open. Another area of immense interest for researchers working in LLP materials is to prepare good efficient phosphors for security and surveillance applications.

Acknowledgments

The author would like to thank Prof. Didier Gourier, Dr. Bruno Viana, and Dr. Cyrille Richards for their discussions on persistent materials and for their continous encoragement during the author's stay at Chimie Paristech, Paris.

References

1. Harvey, E. N. (1957). *A History of Luminescence from the Earliest Times until 1900* (J. H. Furst Company, Baltimore, Maryland).
2. Arpiarian, N. (1966). The centenary of the discovery of luminescent zinc sulphide, *Proc. Int. Conf. Luminescence, Budapest*, pp. 903–906.

3. Leverenz, H. W. (1968). *An Introduction to Luminescence of Solids* (Dover, New York).
4. Durant, W. (1963). *Our Oriental Heritage: The Story of Civilization* (MJF Books, New York).
5. Yocom, P. N. (1996). Future requirements of display phosphors from an historical perspective, *J. SID*, **4**(3), pp. 149–152.
6. http://www.glotechint.com/
7. Yu, F., Yang, Y., Su, X., Mi, C., and Seo, H. J. (2015). Novel long persistent luminescence phosphors: Yb^{2+} codoped MAl_2O_4 (M = Ba, Sr), *Opt. Mater. Express*, **5**, pp. 585–595.
8. http://www.studioroosegaarde.net/
9. Poelman, D., and Smet, P. F. (2010). Photometry in the dark: time dependent visibility of low intensity light sources, *Opt. Express*, **18**, pp. 26293–26299.
10. Casillas-Trujillo, L., Andersson, D. A., Dorado, B., Nikl, M., Sickafus, K. E., McClellan, K. J., and Stanek, C. R. (2014). Intrinsic defects, nonstoichiometry, and aliovalent doping of $A^{2+}B^{4+}O_3$ perovskite scintillators, *Phys. Status Solidi B*, **251**, pp. 2279–2286.
11. Holsa, J., Aitasalo, T., Jungner, H., Lastusaari, M., Niittykoski, J., and Spano, G. (2004). Role of defect states in persistent luminescence materials, *J. Alloys Compd.*, **374**, pp. 56–59.
12. Bessière, A., Lecointre, A., Benhamou, R. A., Suard, E., Wallez, G., and Viana, B. (2013). How to induce red persistent luminescence in biocompatible $Ca_3(PO_4)_2$, *J. Mater. Chem. C*, **1**, pp. 1252–1259.
13. Aitasalo, T., Deren, P., Holsa, J., Jungner, H., Krupa, J. C., Lastusaari, M., Legendziewicz, J., Niittykoski, J., and Strek, W. (2003). Persistent luminescence phenomena in materials doped with rare earth ions, *J. Solid State Chem.*, **171**, pp. 114–122.
14. Bos, A. J. J., Dorenbos, P., Bessière, A., and Viana, B. (2008). Lanthanide energy levels in YPO_4, *Radiat. Meas.*, **43**, pp. 222–226.
15. Chuang, Y.-J., Zhen, Z., Zhang, F., Liu, F., Mishra, J. P., Tang, W., Chen, H., Huang, X., Wang, L., Chen, X., Xie, J., and Pan, Z. (2014). Photostimulable near-infrared persistent luminescent nanoprobes for ultrasensitive and longitudinal deep-tissue bio-imaging, *Theranostics*, **4**, pp. 1112–1122.
16. Eeckhout, K. Van den, Bos, A. J. J., Poelman, D., and Smet, P. F. (2013). Revealing trap depth distributions in persistent phosphors, *Phys. Rev. B*, **87**, pp. 045126-30.

17. Botterman, J., Eeckhout, K. Van den, De Baere, I., Poelman, D., and Smet, P. F. (2012). Mechanoluminescence in $BaSi_2O_2N_2$:Eu, *Acta Mater.*, **60**, pp. 5494–5500.

18. Furetta, C. (2003). *Handbook of Thermo-Luminescence* (World Scientific, Singapore).

19. Bøtter-Jensen, L., McKeever, S. W. S., and Wintle, A. G. (2003). *Optically Stimulated Luminescence Dosimetry* (Elsevier, Amsterdam).

20. Sharma, S. K., Gourier, D., Viana, B., Maldiney, T., Teston, E., Scherman, D., and Richard, C. (2017). Persistent luminescence induced by near infra-red photostimulation in chromium-doped zinc gallate for in vivo optical imaging, *Opt. Mater.*, **63**, pp. 51–58.

21. Hoogenstraaten, W., and Klasens, H. A. (1953). Some properties of zinc sulfide activated with copper and cobalt, *J. Electrochem. Soc.*, **100**, pp. 366–375.

22. Abbruscato, V. (1971). Optical and electrical properties of $SrAl_2O_4$:Eu^{2+}, *J. Electrochem. Soc.*, **118**, pp. 930–933.

23. Matsuzawa, T., Aoki, Y., Takeuchi, N., and Murayama, Y. (1996). A new long phosphorescent phosphor with high brightness, $SrAl_2O_4$:Eu^{2+},Dy^{3+}, *J. Electrochem. Soc.*, **143**, pp. 2670–2673.

24. Takasaki, H., Tanabe, S., and Hanada, T. (1996). Long-lasting afterglow characteristics of Eu, Dy codoped SrO-Al_2O_3 phosphor, *J. Ceram. Soc. Jpn.*, **104**, pp. 322–326.

25. Katsumata, T., Nabae, T., Sasajima, K., Komuro, S., and Morikawa, T. (1997). Effects of composition on the long phosphorescent $SrAl_2O_4$:Eu^{2+},Dy^{3+} phosphor crystals, *J. Electrochem. Soc.*, **144**, pp. L243–L245.

26. Katsumata, T., Nabae, T., Sasajima, K., and Matsuzawa, T. (1998). Growth and characteristics of long persistent $SrAl_2O_4$- and $CaAl_2O_4$-based phosphor crystals by a floating zone technique, *J. Cryst. Growth*, **183**, pp. 361–365.

27. Hölsä, J., Jungner, H., Lastusaari, M., and Niittykoski, J. (2001). Persistent luminescence of Eu^{2+} doped alkaline earth aluminates, MAl_2O_4:Eu^{2+}, *J. Alloys Compd.*, **323–324**, pp. 326–330.

28. Lin, Y., Tang, Z., Zhang, Z., Wang, X., and Zhang, J. (2001). Preparation of a new long afterglow blue-emitting $Sr_2MgSi_2O_7$-based photoluminescent phosphor, *J. Mater. Sci. Lett.*, **20**, pp. 1505–1506.

29. Eeckhout, K. Van den, Smet, P. F., and Poelman, D. (2010). Persistent luminescence in Eu^{2+}-doped compounds: a review, *Materials*, **3**, pp. 2536–2566.

30. Kumar, A., Kedawat, G., Kumar, P., Dwivedi, J., Gupta, B.P. (2015). Sunlight-activated Eu^{2+}/Dy^{3+} doped $SrAl_2O_4$ water resistant phosphorescent layer for optical displays and defence applications, *New J. Chem.*, **39**, pp. 3380–3387.

31. Qu, B., Zhang, B., Wang, L., Zhou, R., Zeng, X. C., and Li, L. (2016). Persistent luminescence hole-type materials by design: transition-metal-doped carbon allotrope and carbides, *ACS Appl. Mater. Interfaces*, **8**, pp. 5439–5444.

32. Dorenbos, P. (2005). Mechanism of persistent luminescence in $Sr_2MgSi_2O_7$:Eu^{2+}; Dy^{3+}, *Phys. Status Solidi B*, **242**, pp. R7–R9.

33. Clabau, F., Rocquefelte, X., Jobic, S., Deniard, P., Whangbo, M. H., Garcia, A., and Le Mercier, T. (2005). Mechanism of Phosphorescence Appropriate for the Long-Lasting Phosphors Eu^{2+}-Doped $SrAl_2O_4$ with Codopants Dy^{3+} and B^{3+}, *Chem. Mater.*, **17**, pp. 3904–3912.

34. Bos, A. J. J., Dorenbos, P., Bessière, A., and Viana, B. (2008). Lanthanide energy levels in YPO_4, *Radiat. Meas.*, **43**, pp. 222–226.

35. Carlson, S., Hölsä, J., Laamanen, T., Lastusaari, M., Malkamäki, M., Niittykoski, J., and Valtonen, R. (2009). X-ray absorption study of rare earth ions in $Sr_2MgSi_2O_7$:Eu^{2+}, R^{3+} persistent luminescence materials, *Opt. Mater.*, **31**, pp. 1877–1879.

36. Hölsä, J., Laamanen, T., Lastusaari, M., Malkamäki, M., Welter, E., and Zajac, D. A. (2010). Valence and environment of rare earth ions in $CaAl_2O_4$:Eu^{2+},R^{3+} persistent luminescence materials, *Spectrochim. Acta, Part B*, **65**, pp. 301–305.

37. le Masne de Chermont, Q., Chanéac, C., Seguin, J., Pellé, F., Maîtrejean, S., Jolivet, J. P., Gourier, D., Bessodes, M., and Scherman, D. (2007). Nanoprobes with near-infrared persistent luminescence for in vivo imaging, *Proc. Natl. Acad. Sci. U. S. A.*, **104**, pp. 9266–9271.

38. Bessière, A., Jacquart, S., Priolkar, K., Lecointre, A., Viana, B., and Gourier, D. (2011). $ZnGa_2O_4$:Cr^{3+}: a new red long-lasting phosphor with high brightness, *Opt. Express*, **19**, pp. 10131–10137.

39. Poelman, D., and Smet, P. F. (2011). Photometry in the dark: time dependent visibility of low intensity light sources, *Opt. Express*, **19**, pp. 18808–18809.

40. Wang, X. J., Jia, D., and Yen, W. M. (2003). Mn^{2+} activated green, yellow, and red long persistent phosphors, *J. Lumin.*, **103**, pp. 34–37.

41. Jain, R. K., and Stylianopoulos, T. (2010). Delivering nanomedicine to solid tumors, *Nat. Rev. Clin. Oncol.*, **7**, pp. 653–664.

42. Danhier, F., Feron, O., and Préat, V. (2010). To exploit the tumor microenvironment: passive and active tumor targeting of nanocarriers for anti-cancer drug delivery, *J. Control. Release*, **148**, pp. 135–146.

43. Al-Jamal, W. T., Al-Jamal, K. T., Tian, B., Cakebread, A., Halket, J. M., and Kostarelos, K. (2009). Tumor targeting of functionalized quantum dot–liposome hybrids by intravenous administration, *Mol. Pharm.*, **6**, pp. 520–530.

44. Park, J. H., Gu, L., von Maltzahn, G., Ruoslahti, E., Bhatia, S. N., and Sailor, M. J. (2009). Biodegradable luminescent porous silicon nanoparticles for in vivo applications, *Nat. Mater.*, **8**, pp. 331–336.

45. Poon, Z., Lee, J. B., Morton, S. W., and Hammond, P. T. (2011). Controlling in vivo stability and biodistribution in electrostatically assembled nanoparticles for systemic delivery, *Nano Lett.*, **11**, pp. 2096–2103.

46. Bessière, A., Sharma, S. K., Basavaraju, N., Priolkar, K. R., Binet, L., Viana, B., Bos, A. J. J., Maldiney, T., Richard, C., Scherman, D., and Gourier, D. (2014). Storage of visible light for long-lasting phosphorescence in chromium-doped zinc gallate, *Chem. Mater.*, **26**, pp. 1365–1373.

47. Sharma, S. K., Gourier, D., Viana, B., Maldiney, T., Teston, E., Scherman, D., and Richard, C. (2014). Persistent luminescence of AB_2O_4:Cr^{3+} (A=Zn, Mg, B=Ga, Al) spinels: new biomarkers for in vivo imaging, *Opt. Mater.*, **36**, pp. 1901–1906.

48. Pan, Z., Lu, Y.-Y., and Liu, F. (2012). Sunlight-activated long-persistent luminescence in the near-infrared from Cr^{3+}-doped zinc gallogermanates, *Nat. Mater.*, **11**, pp. 58–63.

49. Allix, M., Chenu, S., Veron, E., Poumeyrol, T., Kouadri-Boudjelthia, E. A., Alahrache, S., Porcher, F., Massiot, D., and Fayon, F. (2013). Considerable improvement of long-persistent luminescence in germanium and tin substituted $ZnGa_2O_4$, *Chem. Mater.*, **25**, pp. 1600–1606.

50. Basavaraju, N., Sharma, S., Bessière, A., Viana, B., Gourier, D., and Priolkar, K. R. (2013). Red persistent luminescence in $MgGa_2O_4$: Cr^{3+}; a new phosphor for in vivo imaging, *J. Phys. D: Appl. Phys.*, **46**, pp. 375401-5.

51. Maldiney, T., Doan, B.-T., Alloyeau, D., Bessodes, M., Scherman, D., and Richard, C. (2015). Gadolinium-doped persistent nanophosphors as versatile tool for multimodal in vivo imaging, *Adv. Funct. Mater.*, **25**, pp. 331–338.

52. Li, Y., Li, Y.-Y., Sharafudeen, K., Dong, G.-P., Zhou, S.-F., Ma, Z.-J., Peng, M.-Y., and Qiu, J.-R. (2014). A strategy for developing near infrared long-persistent phosphors: taking $MAlO_3$:Mn^{4+},Ge^{4+} (M = La, Gd) as an example, *J. Mater. Chem. C*, **2**, pp. 2019–2027.

53. Liu, F., Yan, W., Chuang, Y.-J., Zhen, Z., Xie, J., and Pan, Z. (2013). Photostimulated near-infrared persistent luminescence as a new optical read-out from Cr^{3+}-doped $LiGa_5O_8$, *Sci. Rep.*, **3**, pp. 1554–1562

54. Maldiney, T., Richard, C., Seguin, J., Wattier, N., Bessodes, M., and Scherman, D. (2011). Effect of core diameter, surface coating, and PEG chain length on the biodistribution of persistent luminescence nanoparticles in mice, *ACS Nano*, **5**, pp. 854–862.

55. Shi, J., Sun, X., Li, J., Man, H., Shen, J., Yu, Y., and Zhang, H. (2015). Multifunctional near infrared-emitting long-persistence luminescent nanoprobes for drug delivery and targeted tumor imaging, *Biomaterials*, **37**, pp. 260–270.

56. Abdukayum, A., Chen, J. T., Zhao, Q., and Yan, X. P. (2013). Functional near infrared-emitting Cr^{3+}/Pr^{3+} co-doped zinc gallogermanate persistent luminescent nanoparticles with superlong afterglow for in vivo targeted bioimaging, *J. Am. Chem. Soc.*, **135**, pp. 14125–14133.

57. Katayama, Y., Viana, B., Gourier, D., Xu, J., and Tanabe, S. (2016). Photostimulation induced persistent luminescence in $Y_3Al_2Ga_3O_{12}$:Cr^{3+},*Opt. Mater. Express*, **6**, pp. 1405–1413.

58. Sharma, S. K., Bessière, A., Basavaraju, N., Priolkar, K. R., Binet, L., Viana, B., and Gourier, D. (2014). Interplay between chromium content and lattice disorder on persistent luminescence of $ZnGa_2O_4$:Cr^{3+} for in vivo imaging, *J. Lumin.*, **155**, pp. 251–256.

59. Gourier, D., Bessière, A., Sharma, S. K., Binet, L., Viana, B., Basavaraju, N., and Priolkar, K. R. (2014). Origin of the visible light induced persistent luminescence of Cr^{3+}-doped zinc gallate, *J. Phys. Chem. Solids*, **75**, pp. 826–837.

60. Binet, L., Sharma, S. K., and Gourier, D. (2016). Interaction of Cr^{3+} with valence and conduction bands in the long persistent phosphor $ZnGa_2O_4$:Cr^{3+}, studied by ENDOR spectroscopy, *J. Phys.: Condens. Matter*, **28**, pp. 385501-6.

61. Nie, W., Michel-Calendini, F. M., Linares, C., Boulon, G., and Daul, C. (1990). New results on optical properties and term-energy calculations in Cr^{3+}-doped $ZnAl_2O_4$, *J. Lumin.*, **46**, pp. 177–190.

62. Mikenda, W., and Preisinger, A. (1981). N-lines in the luminescence spectra of Cr^{3+} -doped spinels (II) origins of N-lines, *J. Lumin.*, **26**, pp. 67–83.

63. Zhang, W., Zhang, J., Chen, Z., Wang, T., and Zheng, S. (2010). Spectrum designation and effect of Al substitution on the luminescence of Cr^{3+} doped $ZnGa_2O_4$ nano-sized phosphors, *J. Lumin.*, **130**, pp. 1738–1743.

64. Hund, F. (1925). Atomtheoretische Deutung des Magnetismus der seltenen. Erden, *Z. Phys.*, **33**, pp. 855–859.

65. Lang, R. J. (1935). The spectrum of trebly ionized cerium, *Can. J. Res.*, **13A**, pp. I.

66. Lang, R. J. (1936). The spectrum of trebly ionized cerium, *Can. J. Res.*, **14A**, pp. 127–130.

67. Lang, R. J. (1936). The F Terms of Ce IV, *Phys. Rev.*, **49**, pp. 552–552.

68. George, N. C., Denault, K. A., and Seshadri, R. (2013). Phosphors for solid-state white lighting, *Annu. Rev. Mater. Res.*, **43**, 481–501.

69. Dorenbos, P. (2002). Relating the energy of the [Xe] $5d^1$ configuration of Ce^{3+} in inorganic compounds with anion polarizability and cation electronegativity, *Phys. Rev. B*, **65**, pp. 235110-15.

70. Dorenbos, P. (2000). 5d-level energies of Ce^{3+} and the crystalline environment. I. Fluoride compounds, *Phys. Rev. B*, **62**, pp. 15640–15649.

71. Blasse, G., and Bril, A. (1967). Investigation of some Ce^{3+}-activated phosphors, *J. Chem. Phys.*, **47**, pp. 5139–5142.

72. Dorenbos, P. (2002). Crystal field splitting of lanthanide $4f^{n-1}5d$-levels in inorganic compounds, *J. Alloys Compd.*, **341**, pp. 156–159.

73. Dorenbos, P. (2013). A review on how lanthanide impurity levels change with chemistry and structure of inorganic compounds luminescence and display materials, devices, and processing, *ECS J. Solid State Sci. Technol.*, **2**(2), pp. R3001–R3011.

74. Dorenbos, P. (2007). Absolute location of lanthanide energy levels and the performance of phosphors, *J. Lumin.*, **122–123**, pp. 315–317.

75. Jia, D. (2006). Relocalization of Ce^{3+} 5d electrons from host conduction band, *J. Lumin.*, **117**, pp. 170–178.

76. Xu, X., Wang, Y., Yu, X., Li, Y., and Gong, Y. (2011). Investigation of Ce–Mn energy transfer in $SrAl_2O_4$:Ce^{3+},Mn^{2+}, *J. Am. Ceram. Soc.*, **94**, pp. 160–163.

77. Gutiérrez-Martín, F., Fernández-Martinez, F., Díaz, P., Colón, C., and Alonso-Medina, A. (2010). Persistent UV phosphors for application in photo catalysis, *J. Alloys Compd.*, **501**, pp. 193–197.

78. Jia, D., Meltzer, R. S., Yen, W. M., Jia, W., and Wang, X. (2002). Green phosphorescence of $CaAl_2O_4$:Tb^{3+},Ce^{3+} through persistence energy transfer, *Appl. Phys. Lett.*, **80**, pp. 1535–1540.

79. Jia, D., Wang, X. J., Jia, W., and Yen, W. M. (2003). Persistent energy transfer in $CaAl_2O_4$:Tb^{3+},$CaAl_2O_4$:Tb^{3+}, Ce^{3+}, *J. Appl. Phys.*, **93**, pp. 148–152.

80. Jia, D., and Yen, W. M. (2003). Trapping mechanism associated with electron delocalization and tunneling of $CaAl_2O_4:Ce^{3+}$, a persistent phosphor, *J. Electrochem. Soc.*, **150**, pp. H61–H65.

81. Kodama, N., Takahashi, T., Yamaga, M., Tanii, Y., Qiu, J., and Hirao, K. (1999). Long-lasting phosphorescence in Ce^{3+}-doped $Ca_2Al_2SiO_7$ and $CaYAl_3O_7$ crystals, *Appl. Phys. Lett.*, **75**, pp. 1715–1717.

82. Kodama, N., Tanii, Y., and Yamaga, M. (2000). Optical properties of long-lasting phosphorescent crystals Ce^{3+}-doped $Ca_2Al_2SiO_7$ and CaYAl3O7, *J. Lumin.*, **87–89**, pp. 1076–1078.

83. Yamaga, M., Tanii, Y., Kodama, N., Takahashi, T., and Honda, M. (2002). Mechanism of long-lasting phosphorescence process of Ce^{3+}-doped $Ca_2Al_2SiO_7$ melilite crystals, *Phys. Rev. B*, **65**, pp. 235108-12.

84. Wu, H., Hu, Y., Ju, G., Chen, L., Wang, X., and Yang, Z. (2011). Photoluminescence and thermoluminescence of Ce^{3+} and Eu^{2+} in $Ca_2Al_2SiO_7$ matrix, *J. Lumin.*, **131**, pp. 2441–2445.

85. Dorenbos, P., Vaneijk, C. W. E., Bos, A. J. J., and Melcher, C. L. (1994). Afterglow and thermoluminescence properties of Lu_2SiO_5:Ce scintillation crystals, *J. Phys.: Condens. Matter*, **6**, pp. 4167–4180.

86. Yamaga, M., Ohsumi, Y., Nakayama, T., and Han, T. P. J. (2012). Persistent phosphorescence in Ce-doped Lu_2SiO_5, *Opt. Mater. Express*, **2**, pp. 413–419.

87. Jia, D., Wang, X.-j., van der Kolk, E., and Yen, W. M. (2002). Site dependent thermoluminescence of long persistent phosphorescence of $BaAl_2O_4:Ce^{3+}$, *Opt. Commun.*, **204**, pp. 247–251.

88. Qiu, J., Miura, K., Inoue, H., Kondo, Y., Mitsuyu, T., and Hirao, K. (1998). Femtosecond laser-induced three-dimensional bright and long-lasting phosphorescence inside calcium aluminosilicate glasses doped with rare earth ions, *Appl. Phys. Lett.*, **73**, pp. 1763–1765.

89. Kodama, N., Takahashi, T., Yamaga, M., Tanii, Y., Qiu, J., and Hirao, K. (1999). Long-lasting phosphorescence in Ce^{3+}-doped $Ca_2Al_2SiO_7$ and $CaYAl_3O_7$ crystals, *Appl. Phys. Lett.*, **75**, pp. 1715–1717.

90. Ning, L. X., Zhou, C. C., Chen, W. P., Huang, Y. C., Duan, C. K., Dorenbos, P., Tao, Y., and Liang, H. B. (2015). Electronic properties of Ce^{3+}-doped $Sr_3Al_2O_5Cl_2$: a combined spectroscopic and theoretical study, *J. Phys. Chem. C*, **119**, pp. 6785–6792.

91. Sharma, S. K., Pitale, S. S., Malik, M. M., Dubey, R. N., and Qureshi, M. S. (2009). Luminescence studies on the blue–green emitting $Sr_4Al_{14}O_{25}:Ce^{3+}$ phosphor synthesized through solution combustion route, *J. Lumin.*, **129**, pp. 140–147.

92. Ueda, J., Aishima, K., Nishiura, S., and Tanabe, S. (2011). Afterglow luminescence in Ce^{3+}-doped $Y_3Sc_2Ga_3O_{12}$ ceramics, *Appl. Phys. Express*, **4**, pp. 042602-4.

93. Jia, D., Meltzer, R. S., and Yen, W. M. (2002). Ce^{3+} energy levels relative to the band structure in CaS: evidence from photoionization and electron trapping, *J. Lumin.*, **99**, pp. 1–6.

94. Mu, Z. F., Wang, Y. H., Hu, Y. H., Wu, H. Y., Deng, L. Y., Xie, W., Fu, C. J., and Liao, C. X. (2011). The afterglow and thermoluminescence properties of $Y_3Al_5O_{12}$:Ce^{3+}, *Acta Physica Sinica*, **60**, pp. 013201-5.

95. Zhang, S., Li, C., Pang, R., Jiang, L., Shi, L., and Su, Q. (2011). Long-lasting phosphorescence study on $Y_3Al_5O_{12}$ doped with different concentrations of Ce^{3+},*J. Rare Earths*, **29**, pp. 426–430.

96. Ju, G., Hu, Y., Chen, L., and Wang, X. (2012). Persistent luminescence and its mechanism of $Ba_5(PO_4)_3Cl$:Ce^{3+},Eu^{2+}, *J. Appl. Phys.*, **111**, pp. 113508-6.

97. Ito, Y., Komeno, A., Uematsu, K., Toda, K., and Sato, M. (2006). Luminescence properties of long-persistence silicate phosphors, *J. Alloys Compd.*, **408–412**, pp. 907–910.

98. Gong, Y., Wang, Y., Li, Y., and Xu, X. (2010). Ce^{3+}, Dy^{3+} co-doped white-light long-lasting phosphor: $Sr_2Al_2SiO_7$ through energy transfer, *J. Electrochem. Soc.*, **157**, pp. J208–J211.

99. Pan, W., Ning, G., Lin, Y., and Yang, X. (2008). Sol-gel processed Ce^{3+}, Tb^{3+} codoped white emitting phosphors in $Sr_2Al_2SiO_7$, *J. Rare Earths*, **26**, pp. 207–210.

100. Gong, Y., Xu, H., Zeng, W., Wu, C. J., and Wang, Y. H. (2012). Ce^{3+}, Mn^{2+} co-doped red–light long–lasting phosphor: $BaMg_2Si_2O_7$ through energy transfer, *Physics Procedia*, **29**, pp. 86–90.

101. Ju, G. F., Hu, Y. H., Chen, L., and Wang, X. J. (2012). Persistent luminescence and its mechanism of $Ba_5(PO_4)_3Cl$:Ce^{3+},Eu^{2+}, *J. Appl. Phys.*, **111**, pp. 113508.

102. Xu, J., Ueda, J., Kuroishi, K., and Tanabe, S. (2015). Fabrication of Ce^{3+}–Cr^{3+} co-doped yttrium aluminium gallium garnet transparent ceramic phosphors with super long persistent luminescence, *Scr. Mater.*, **102**, pp. 47–50.

103. Wang, X.-J., Jia, D., and Yen, W. M. (2003). Mn^{2+} activated green, yellow, and red long persistent phosphors, *J. Lumin.*, **102–103**, pp. 34–37.

104. Xu, J., Ueda, J., Zhuang, Y. X., Viana, B., and Tanabe, S. (2015). $Y_3Al_{5-x}Ga_xO_{12}$:Cr^{3+}: A novel red persistent phosphor with high brightness, *Appl. Phys. Express*, **8**, pp. 042602-6.

105. Ueda, J., Kuroishi, K., and Tanabe, S. (2014). Yellow persistent luminescence in Ce^{3+}–Cr^{3+}-codoped gadolinium aluminum gallium garnet transparent ceramics after blue-light excitation, *Appl. Phys. Express*, **7**, pp. 062201-5.

106. Jia, D., Meltzer, R. S., Yen, W. M., Jia, W., and Wang, X. (2002). Green phosphorescence of $CaAl_2O_4$:Tb^{3+},Ce^{3+} through persistence energy transfer, *Appl. Phys. Lett.*, **80**, pp. 1535–1537.

107. Jia, D., Wang, X. J., Jia, W., and Yen, W. M. (2003). Persistent energy transfer in $CaAl_2O_4$:Tb^{3+}, Ce^{3+}, *J. Appl. Phys.*, **93**, pp. 148–152.

108. Xu, J., Tanabe, S., Sontakke, A. D., and Ueda, J. (2015). Near-infrared multi-wavelengths long persistent luminescence of Nd^{3+} ion through persistent energy transfer in Ce^{3+}, Cr^{3+} co-doped $Y_3Al_2Ga_3O_{12}$ for the first and second bio-imaging windows, *Appl. Phys. Lett.*, **107**, pp. 081903-7.

109. Dexter, D. L. (1953). A theory of sensitized luminescence in solids, *J. Chem. Phys.*, **21**, pp. 836–838.

Chapter 3

Structural and Luminescence Characteristics of $LiNa_{3-x}P_2O_7{:}xRE^{3+}$ Tricolor-Emitting Phosphors for White-Light Emission

B. Deva Prasad Raju,[a] N. John Sushma,[b] and Vibha Chopra[c]
[a]*Department of Future Studies, Sri Venkateswara University, Tirupati 517502, India*
[b]*Department of Biotechnology, Sri Padmavathi Mahila Viswavidyalayam Tirupati 517502, India*
[c]*PG Department of Physics & Electronics, DAV College, Amritsar 143001, Punjab, India*
drdevaprasadraju@gmail.com, johnsushma@gmail.com, vibhachopra04@gmail.com

This chapter deals with the synthesis and structural and luminescence characteristics of a novel class of inorganic phosphate-based single-host $LiNa_{3-x}P_2O_7{:}xRE^{3+}$ tricolor-emitting phosphors, with independent RE = Tb, Eu, and Dy dopants. Also it focuses on applications of developed phosphors in the solid-state lighting industry and display panel systems.

3.1 Introduction

In the 21st century, lighting sources are phosphor-converted white-light-emitting diodes (pc-WLEDs), which have superior

Phosphors: Synthesis and Applications
Edited by Sanjay J. Dhoble, B. Deva Prasad Raju, and Vijay Singh

ISBN 978-981-4774-49-9 (Hardcover), 978-042-9460-99-9 (eBook)
www.panstanford.com

qualities of high efficiency, ecofriendliness, and long lifetime compared to fluorescence and conventional incandescent lamps [1–3]. Commercial pc-WLEDs are generally fabricated by coating $Y_3Al_5O_{12}:Ce^{3+}$ (YAG:Ce) yellow phosphor on a blue GaN light-emitting diode (LED) chip of wavelength 450–470 nm. However, this type of WLEDs face problems such as a low color-rendering index (CRI < 70) and a high correlated color temperature (CCT > 47,000 K) due to the absence of the red component [4–7]. To get rid of these problems and enhance the optical properties of pc-WLEDs, a process of combining tricolor (blue, green, and red) phosphors upon near-ultraviolet (NUV) chips was adopted to obtain warm WLEDs [8]. These phosphors have been synthesized by doping rare earth (RE) ions in host materials. A suitable host matrix and a dopant ion are important for obtaining a single-phase white-lighting phosphor. Among various hosts, pyrophosphates have gained more attention due to their thermal and chemical stability and structural diversity. Phosphate-based phosphors are an important family of luminescent materials because of their excellent properties, such as a large bandgap, high absorption of PO_4^{3-} in the ultraviolet (UV) region (300–400 nm), cheaper raw materials, and simple synthesis conditions, and so these properties have great potential in optoelectronic applications. Phosphate-based phosphors play good host materials for RE ion transitions. Electronic transition of RE ions and interaction are independent of surroundings in phosphate matrices because the RE ions are separated from each other by the phosphate groups. Therefore, good optical properties are expected from these materials [9–15]. Among other RE ions, particularly terbium (Tb)-doped phosphors are quite efficient to emit narrow-band green light due to effective 4f–5d and 4f–4f transitions under NUV excitation. Thus, the Tb^{3+} ion serves as an activator in green-emitting phosphors for fabrication of WLEDs. The emission spectrum of Tb^{3+} consists of emissions due to $^5D_3 \rightarrow {}^7F_J$ transitions, mainly in the blue region, and $^5D_4 \rightarrow {}^7F_J$ transitions (J = 0–6), mainly in the green region. The blue/green ratio strongly depends on the Tb^{3+} concentration due to the cross-relaxation effect between two adjacent Tb^{3+} pairs: $^5D_3 \rightarrow {}^7F_6$, $^5D_4 \rightarrow {}^7F_0$. This phenomenon has been observed in many host crystals. [16, 17]. Trivalent Eu^{3+} ions are expected to be one of the promising species in different host materials that undergo optical transitions and give out radiation in the red region. Eu^{3+}

ions usually occupy the sites that have no inversion symmetry in the host crystals. The strong red emission line at around 610 to 630 nm is due to the electric dipole (ED) transition of $^5D_0 \rightarrow {}^7F_2$. In some cases, Eu^{3+} ions can also occupy the sites of inversion symmetry; when it occurs the emission line at ~600 nm due to the magnetic dipole (MD) transition of $^5D_0 \rightarrow {}^7F_1$ becomes relatively stronger and dominates [18]. Trivalent dysprosium ions (Dy^{3+}) also have attracted significant attention since dysprosium-doped phosphors can be used in fluorescent lamps as a tricolor component and single-phase full-color phosphors in phosphor-converted light-emitting diodes (pc-LEDs). Under NUV excitation, there are generally three consistent bands in the visible region—blue (470–500 nm), yellow (570–600 nm), and feeble red (670 nm) emissions—which are achieved when Dy^{3+} ions are incorporated in wide-bandgap host materials. These bands are associated with the hypersensitive ED $^4F_{9/2} \rightarrow {}^6H_{15/2,\,13/2}$ transition and the optical $^4F_{9/2} \rightarrow {}^6H_{11/2}$ transition in Dy^{3+} ions [19–21]. This combination of three phosphors can produce white light with an excellent CRI. Therefore, it is urgent to develop novel single-composition (or single-phase) phosphors that can produce highly efficient white-light emission. This chapter reviews the synthesis, photoluminescence (PL), and applications of phosphate-based novel $LiNa_{3-x}P_2O_7{:}xRE^{3+}$ (RE = Tb, Eu, Dy) phosphors that have been reported recently.

3.2 Synthesis of Phosphate-Based Phosphors

Polycrystalline $LiNa_{3-x}P_2O_7{:}xRE^{3+}$ phosphors with different concentrations of Tb^{3+} (0.01, 0.03, 0.05, 0.07, 0.09, and 0.11)-, Eu^{3+} (0.01, 0.03, 0.05, 0.07 and 0.09)-, and Dy^{3+} (0.03, 0.05, 0.07, 0.09, and 0.11)-doped $LiNa_3P_2O_7$ are prepared by a solid-state reaction [18, 22–23]. The materials are synthesized at the same experimental conditions. In brief, analytical reagent-grade high-purity (99.99%) Li_2CO_3, Na_2CO_3, $NH_4H_2PO_4$, Tb_4O_7, Eu_2O_3, and Dy_2O_3 are taken as starting materials. To obtain homogeneity, stoichiometric amounts of reactive mixtures have to be finely ground in an agate mortar for 1 hr. Each of the mixtures was placed in a separate silica crucible and gradually heated in a muffle furnace from room temperature (RT) to 400°C for 4 hr. This process eliminates the residual water, CO_2, and NH_3 from the mixtures. The samples were reground after cooling down to RT and then sintered at 540°C for 12 hr with several intermediate grindings. Finally, the samples were furnace-cooled, ground again, and used for characterization.

3.3 Characterization and Analysis of Phosphors

The crystal structure of $LiNa_3P_2O_7$:xRE^{3+} (RE) phosphors were analyzed by X-ray diffraction (XRD) analysis using a PANalytical diffractometer (Siemens, AXS D5005) with Cu-Kα radiation (λ = 0.1540 nm) at a scanning step of 0.02° in the 2θ range from 10° to 80° at 40 kV and 20 mA. The infrared spectrum was recorded by the diffuse reflection technique using a Bruker IFS Equinox 55 Fourier transform infrared (FTIR) spectrometer in the range of 4000 and 400 cm^{-1}. The morphology of the sample was analyzed using a field emission scanning electron microscope (FE-SEM) (S-4200, Hitachi, Japan). Ultraviolet-visible (UV-Vis) diffuse reflectance spectra (DRS) were recorded on a UV-Vis-NIR (near-infrared) spectrophotometer (Jobin Varian Cary 5000, USA). The reflectance was measured in the range of 800 to 200 nm using polytetrafluoroethylene (PTFE) as a standard. PL emission and photoluminescence excitation (PLE) spectra were recorded using a fluorescence spectrophotometer (Jobin Vyon Fluorolog-3, USA) with a xenon lamp as an excitation source.

3.3.1 XRD Analysis

The XRD patterns of $LiNa_{3-x}P_2O_7$:xRE^{3+} Tb^{3+} (0.01, 0.03, 0.05, 0.07, 0.09, and 0.11), Eu^{3+} (0.01, 0.03, 0.05, 0.07, and 0.09) and Dy^{3+} (0.03, 0.05, 0.07, 0.09, and 0.11) phosphors are shown in Fig. 3.1. From the figure, it is noticed that the single-phase samples were successfully obtained by the solid-state reaction. All the diffraction peaks obtained in the phosphors were well matched with the reported in the Inorganic Crystal Structure Database (ICSD): 424375 and also reported in literature [24, 25]. According to the reports, $LiNa_3P_2O_7$ was crystallized in the orthorhombic space group C2221 with lattice parameters a = 0.54966 nm, b = 0.91365 nm, and c = 1.22764 nm. From the XRD pattern, no extra peaks were observed due to the increase of RE^{3+} ion concentration in the host matrix. The possible reason might be that the doped Tb^{3+}, Eu^{3+}, and Dy^{3+} ions could occupy Na^+ ions sites, since the ionic radii of Tb^{3+} and Eu^{3+} (0.092 nm and 0.0947 nm) are smaller and the ionic radius of Dy^{3+} (0.103 nm) is nearly equal to that of Na^+ (0.102 nm, coordination number [CN] = 6) and greater than Li^+ (0.076 nm, CN = 6) [26–28]. From the above results, it is suggested that the dopant, Tb^{3+}, Eu^{3+}, and Dy^{3+} ions, will preferentially substitute at the sodium sites in the host. Thus, the charge loss is compensated by Na^+ vacancies (V_{Na}) by the following process:

$$3Na^{3+} \rightarrow RE^{3+} + V_{Na} \tag{3.1}$$

The substitution of Tb^{3+}, Eu^{3+} and Dy^{3+} ions in the host cation sites induces lattice expansion, resulting in the shift of the XRD peaks to the lower angle side. To prove the lattice expansion of the phosphors, typically Tb^{3+} ion-doped host has been taken and orthorhombic crystal structure parameters such as unit cell constants (a, b, and c) and volume (V) were calculated from the crystal geometry equations using the following relations:

$$\frac{1}{d^2} = \frac{h^2}{a^2} + \frac{k^2}{b^2} + \frac{l^2}{c^2} \text{ and } V = abc, \tag{3.2}$$

where d is the distance between adjacent planes (nm) and (hkl) the Miller indices. To determine the lattice parameters, we adopted the dominant intensity lattice planes of (0 2 0), (0 2 1), and (1 3 0) for interplanar distances d_1, d_2, and d_3, respectively. The lattice parameters of the phosphors are listed in Table 3.1. As shown in the table, the unit cell volume was increased as a function of Tb^{3+} ion concentration, indicating lattice expansion in the host material.

Table 3.1 Structural parameters of $LiNa_{3-x}P_2O7{:}xTb^{3+}$ (x = 0–9 mol%) phosphors

Tb^{3+} concentration in	**Lattice parameters**			
$LiNa_3P_2O_7$ (mol%)	***a* (nm)**	***b* (nm)**	***c* (nm)**	***V* (nm^3)**
Undoped	0.5494	0.9143	1.2257	0.61573
1	0.5485	0.9133	1.2306	0.61665
3	0.5522	0.9111	1.2308	0.61933
5	0.5532	0.9099	1.2356	0.62204
7	0.5568	0.9089	1.2361	0.62567
9	0.5528	0.9087	1.2484	0.62727

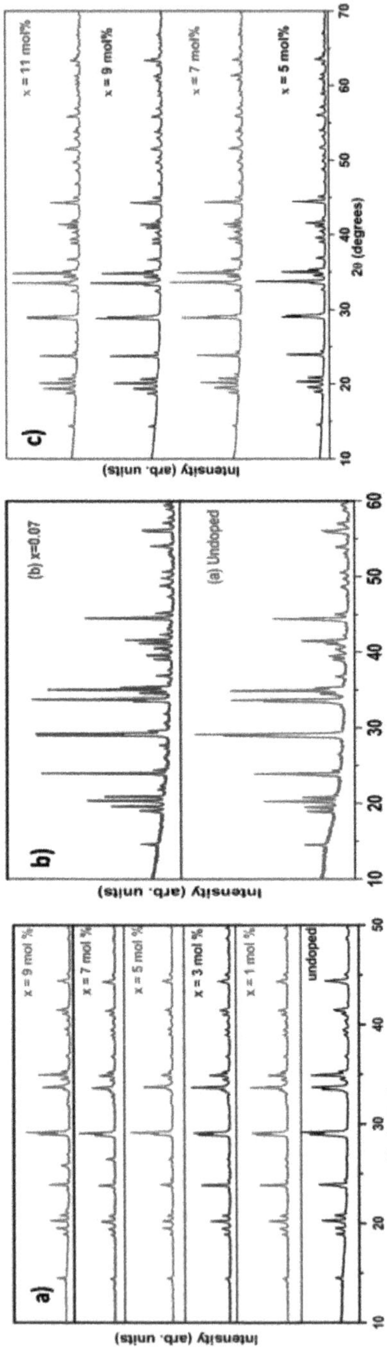

Figure 3.1 XRD patterns of (a) $LiNa_{3-x}P_2O_7$:xTb^{3+} (x = 0–9 mol%) phosphors, (b) $LiNa_{3-x}P_2O_7$:xEu^{3+} (x = 0, 0.07), and (c) $LiNa_3P_2O_7$:Dy^{3+} (x = 5–11 mol%).

3.3.2 FTIR Analysis

To investigate the nature of chemical bonding in phosphate materials, the FTIR spectrum of the $LiNa_{3-x}P_2O_7$:xRE^{3+} (RE = Tb^{3+}, 9 mol%; Eu^{3+}, x = 0.09; Dy^{3+}, x = 0.05, 0.07, 0.09, and 0.11) phosphor were recorded and are shown in Fig. 3.2a–c. According to Corbridge et al. [29], the infrared (IR) spectrum of pyrophosphate is divided mainly into four frequency regions: at ν_1 = 3654–3535 cm^{-1} (the O–H stretching of absorbed water), ν_2 = 2667.66–2337.9 cm^{-1} (C–O vibration of CO2 in the air), ν_3 = 1811.79–1432.65 cm^{-1} (the vibration of PO_4^{3-}), and ν_4 = 1249.65–412.58 cm^{-1} (vibration of the P_2O_7 group (O_3P–O–PO_3)) corresponding to the vibrational bands of the –OH, PO_3, PO_4, and P–O–P, respectively. From the spectrum, it is identified that there are two distinct bands near 1115 and 1017 cm^{-1}, attributed to the symmetric and asymmetric vibrational stretching modes of the PO_3 species. The band appeared at around 745 cm^{-1} due to the symmetric stretching vibrational mode of (P–O–P) bridges in the pyrophosphate (P_2O_7) group. It reveals that the material belongs to the diphosphate family [24]. Further, in the range from 600 to 400 cm^{-1}, two bands are noticed at 570 cm^{-1} and 462 cm^{-1}, owing to the vibrational modes of PO_3 groups. In addition, the spectrum consists of a weak absorption band at 3422 cm^{-1}, indicating the presence of –OH group in the sample. This band is the characteristic vibrations of water, physically absorbed on the sample surface during the preparation time under nonvacuum condition. It is known that a minimal of the –OH band decreases the optical losses, thereby increasing the quantum efficiency of the phosphors [30]. However, for the present phosphors, the intensity of the IR band linked to the –OH group is extremely low, maybe suitable for practical applications. A typical FTIR spectrum of the $LiNa_{3-x}P_2O_7$:xEu^{3+} (x = 0.09) phosphor is shown in Fig. 3.2b. We note from the spectrum that it consists of two distinct bands around 745 and 924 cm^{-1}, which are assigned to the symmetric and asymmetric vibrational stretching modes of the P–O–P bridges. These bands are the characteristic of pyrophosphate groups $(P_2O_7)^{4-}$ (termed as the ν_4 region) [31]. The strong band centered at 1658 cm^{-1} corresponds to the stretching vibration of PO_4^{3-} groups (termed as the ν_3 region). FTIR spectra of $LiNa_3P_2O_7$:Dy^{3+} phosphors (Dy^{3+} = 5–11 mol%) are shown in Fig. 3.2c. For all samples, the spectra have similar

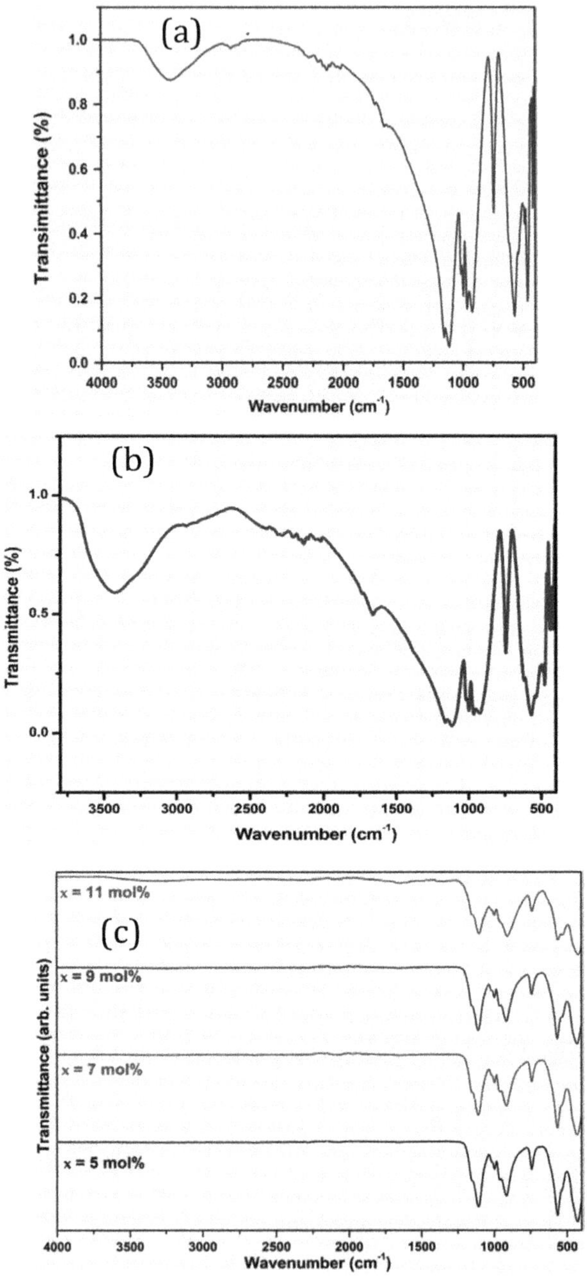

Figure 3.2 FTIR spectrum of (a) $LiNa_{3-x}P_2O_7$:$x$$Tb^{3+}$, (b) $LiNa_{3-x}P_2O_7$:$x$$Eu^{3+}$($x$ = 0.09), and (c) $LiNa_{3-x}P_2O_7$:$x$$Dy^{3+}$ (x = 0.05, 0.07, 0.09, and 0.11) phosphors.

profiles without any other phases, suggesting that the variation of Dy concentration did not have much effect on the structure of the host. The two peaks at 747 and 911 cm^{-1} are due to the symmetric and asymmetric vibrational stretching modes of P–O–P bridges. The small peak at 1633 cm^{-1} corresponds to the stretching vibration of PO_4^{3-} groups, which revealed that the phosphors have the molecular structure of pyrophosphates. The absorption peaks showed good agreement with those reported for the $LiNa_3P_2O_7$ host compound [32]. Peaks are negligible or nonexistent in the range of 3655–3535 cm^{-1}, which indicates the absence of water molecules (–OH) in the samples [33]. The identified band assignments confirmed the presence of diphosphate groups in the $LiNa_3P_2O_7:Dy^{3+}$ phosphors. Therefore, the obtained results suggest that the prepared material belongs to the diphosphate family.

3.3.3 Morphological Studies

To investigate the surface morphology of phosphors, scanning electron microscopy (SEM) analysis was carried out. The SEM images of the samples, $LiNa_{3-x}P_2O_7{:}xRE^{3+}$ (RE = Tb^{3+}, 9 mol%; Eu^{3+}, $x = 0.09$; Dy^{3+}, $x = 0.09$) are shown in Fig. 3.3a–c. Figure 3.3a–c shows the morphology of Tb^{3+}-doped samples, and it is found that they have an agglomerated surface, probably due to the liberation of a huge amount of NH_3 and CO_2 during synthesis. The high-magnified SEM image of the prepared Eu^{3+}-doped phosphor is shown in Fig. 3.3b. We note from the figure that the particles possess irregular morphology and are aggregated during the high calcination. A closer look at the high-magnified SEM microgram certainly indicates the orthorhombic crystal structure of the synthesized phosphor (indicated in Fig. 3.3b by white circles). From the micrograph, it is clear that various particles have different sizes of the orthorhombic structure and some of them are agglomerated to form the phosphor. However, the average sizes of the particles are in the range of a few micrometers. Thus, the defined structure with the micrometer dimension of the phosphor warrants considerable attention in the field of display and lighting. A typical SEM image of the Dy^{3+} (9 mol%) phosphor is represented in Fig. 3.3c. It was found that the phosphor has irregular morphology due to the calcination of the samples at high temperature and intermediate grinding between the calcinations. The particles are formed by several rocklike structures with non-uniform sizes and shapes. A closer look indicates that the grains vary in size from a few microns to several tens of microns, which could be suitable for WLED applications.

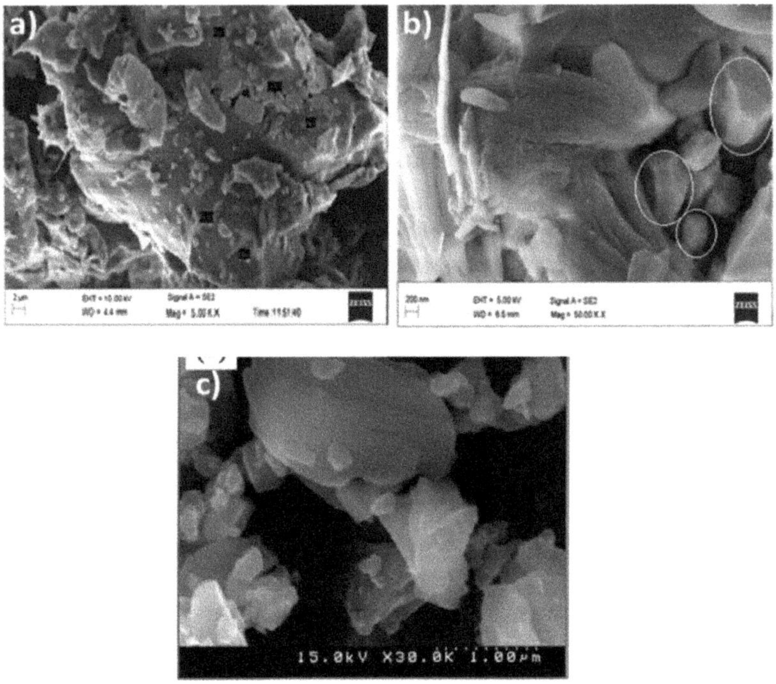

Figure 3.3 SEM images of $LiNa_{3-x}P_2O_7$:xRE^{3+} phosphor: (a) Tb^{3+} (x = 0.09), (b) Eu^{3+} (x = 0.09), and (c) Dy^{3+} (x = 0.09).

3.3.4 Photoluminescence Studies on Tb^{3+}-Doped $LiNa_{3-x}P_2O_7$ Phosphors

The typical PLE spectrum of the $LiNa_{3-x}P_2O_7$:xTb^{3+} (x = 9 mol%) phosphor, recorded at 545 nm emission, is shown in Fig. 3.4. The excitation spectrum is composed of several sharp bands at 308 nm ($^7F_6 \rightarrow {}^3H_6$), 319 nm ($^7F_6 \rightarrow {}^5D_0$), and 378 nm ($^7F_6 \rightarrow {}^5G_6$), respectively. These sharp bands are due to the 4f–4f transitions of Tb^{3+} [34]. Among the sharp lines, the broad band at about 378 nm ($^7F_6 \rightarrow {}^5G_6$) has the strongest intensity and is used further to measure the emission spectra of phosphors. From the figure, it is evidenced that the prepared sample can effectively excited at NUV wavelength.

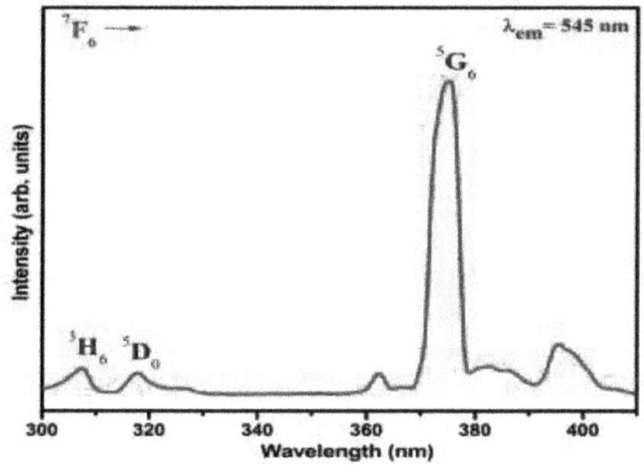

Figure 3.4 Photoluminescence excitation spectrum of the $LiNa_{3-x}P_2O_7$:xTb^{3+} (x = 9 mol%) phosphor.

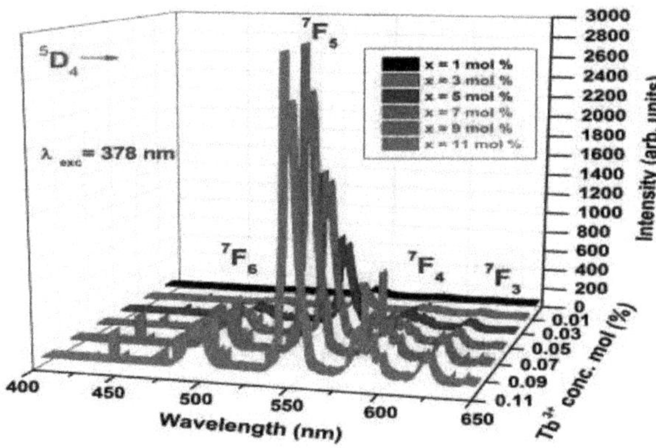

Figure 3.5 Emission spectra of $LiNa_{3-x}P_2O_7$:xTb^{3+} (x = 0.01–0.09) phosphors.

The emission spectra of $LiNa_{3-x}P_2O_7$:xTb^{3+} (x = 1 to 11 mol%) phosphors monitored at excitation wavelength of 378 nm are shown in Fig. 3.5. It can be seen that the emission spectra consist of several sharp peaks at around 482 nm, 545 nm, 588 nm, and 620 nm, which are due to the transitions of $^5D_4 \rightarrow {}^7F_{J=6,5,4,3}$, respectively. Among several sharp peaks, the band at 545 nm corresponds to the $^5D_4 \rightarrow {}^7F_5$ transition of Tb^{3+} ions and is the strongest in the

$LiNa_3P_2O_7$:Tb^{3+} phosphors. Further, to find the effect of dopant concentration on the emission intensities of the host material, the Tb^{3+} ions in the host were varied. We note from the figure that the PL intensity of phosphor was enhanced by increasing the dopant (Tb^{3+}) concentration from 1 to 9 mol% and decreased on further increase of dopant concentration due to the concentration quenching effect.

The concentration quenching phenomena will occur due to the nonradiative energy transfer between energy levels of RE ions. The critical distance (R_c) is an important parameter that determines the concentration quenching phenomena. It can be estimated according to the formula given by Blasse and Grabmaier [35]

$$R_c = 2\left(\frac{3V}{4\pi x_c Z}\right)^{1/3}, \tag{3.3}$$

where V is the volume of the unit cell, x_c is the critical concentration of Tb^{3+} ions, and Z is the number of cations in the unit cell. In the current case, the values of V, x_c, and Z are 0.6272 nm, 0.09, and 4, respectively. The R_c is found to be 1.1849 nm. Thus, the above results indicated that the concentration quenching mechanism of Tb^{3+} emission could be a common effect of nearest-neighbor interaction in the host material. The emission intensity is enhanced in the $LiNa_{3-x}P_2O_7$:xTb^{3+} ($x = 0.09$) phosphor and evidences that the phosphor may be a promising candidate for emitting green light under NUV excitation.

A typical energy-level diagram of Tb^{3+} ions in a $LiNa_3P_2O_7$ phosphor is shown in Fig. 3.6.

Commission International de l'Eclairage (CIE) chromatic color coordinates refer to some specifications of light sources, which represents kinds of color. In general, the color of any light source can be represented by the (x, y) coordinate in color space [36]. From the emission spectra, the CIE coordinates were calculated using the CIE software under 378 nm excitation wavelength. The CIE diagram is shown in Fig. 3.7.

We note from the figure that the CIE values were located in the green light region. For instance, the CIE chromaticity coordinates of the $LiNa_{3-x}P_2O_7$:xTb^{3+} ($x = 0.09$) phosphor is found to be (0.341, 0.506), which is in the green region. Thus, the results demonstrate that the prepared phosphors may have potential application for green light emitting in the designing of WLEDs under NUV excitation.

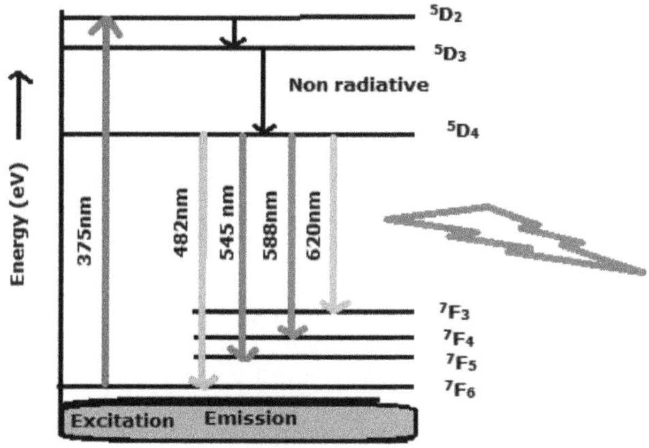

Figure 3.6 Schematic energy transfer mechanism of Tb^{3+} in $LiNa_3P_2O_7$.

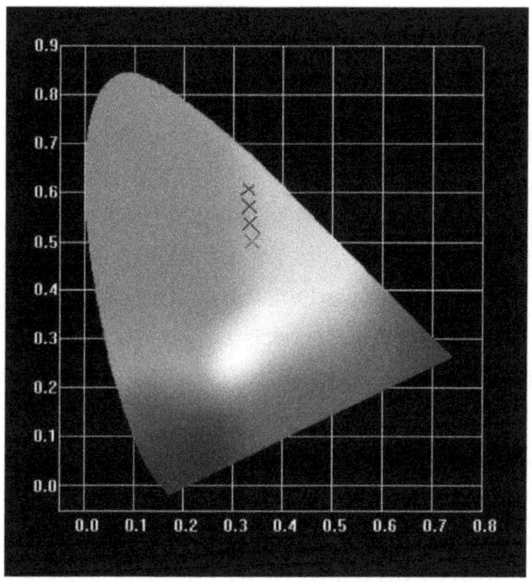

Figure 3.7 CIE chromaticity diagram of $LiNa_{3-x}P_2O_7$:xTb^{3+} (x = 1–11 mol%) phosphors.

3.3.5 Photoluminescence Studies on Eu^{3+}-Doped $LiNa_{3-x}P_2O_7$ Phosphors

The excitation spectrum of the $LiNa_{3-x}P_2O_7$:xEu^{3+} (x = 9 at.%) phosphor is shown in Fig. 3.8. The spectrum showed a weak broad band in the range from 250 to 350 nm, which is assigned to the Eu–O charge transfer band (CTB) transition, and several sharp absorption peaks at around 366, 386, 395, and 420 nm correspond to $^7F_0 \rightarrow {}^5D_4$, $^7F_0 \rightarrow {}^5L_7$, $^7F_0 \rightarrow {}^5L_6$, and $^7F_0 \rightarrow {}^5D_3$ transitions of Eu^{3+} ions, respectively [37].

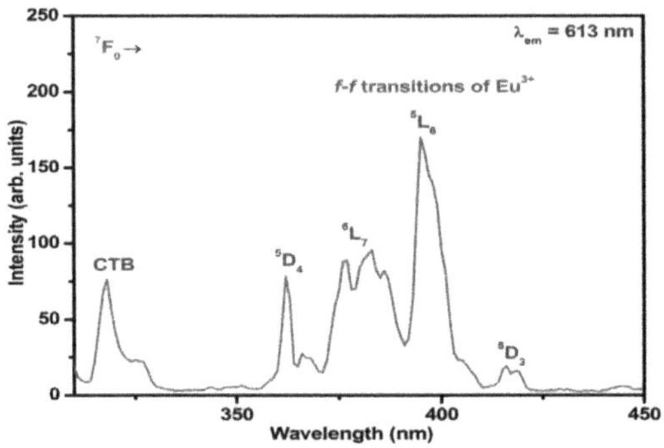

Figure 3.8 Excitation spectrum of the $LiNa_{3-x}P_2O_7$:xEu^{3+} (x = 0.09) phosphor.

Among the sharp lines, the absorption band at 395 nm is the strongest, which is assigned to the $^7F_0 \rightarrow {}^5L_6$ transition of Eu^{3+} ions and is used to measure the emission spectra of the phosphors. Therefore, it clearly indicates that the prepared phosphor could strongly absorb the UV light, which is matched well with the characteristic emission of the NUV LED chips [38]. To further investigate the effect of Eu^{3+} concentration on $LiNa_{3-x}P_2O_7$:xEu^{3+} (x = 1–9 at.%) phosphors, the PL emission spectra were recorded.

Figure 3.9 shows the emission spectra of phosphors under 395 nm excitation. The emission band covers a region from 580 to 630 nm and includes several distinct emission bands. These bands were ascribed to the transition from $^5D_0 \rightarrow {}^7F_J$ (J = 0, 1, 2) of Eu^{3+} ions [39]. From the emission spectra, it was observed that the orange emission peaks at 585 and 595 nm due to the MD transition ($^5D_0 \rightarrow {}^7F_1$) is dominant than a weak shoulder around 581 nm corresponds to the $^5D_0 \rightarrow {}^7F_0$ transition. The emission bands at 615 and 623 nm (in the red region) were attributed to the $^5D_0 \rightarrow {}^7F_2$ ED transitions [40]. The emission due to $^5D_0 \rightarrow {}^7F_2$ ED transition is hypersensitive to the symmetry of the crystal field environment, while the $^5D_0 \rightarrow {}^7F_1$ MD transition is insensitive to the site symmetry and hardly varies with the environment, because it is parity-allowed [41]. It is noteworthy that if the Eu^{3+} ions occupy the inversion symmetry center, the MD transition ($^5D_0 \rightarrow {}^7F_1$) dominates the ED transition $^5D_0 \rightarrow {}^7F_2$. In the present case, the $^5D_0 \rightarrow {}^7F_1$ transition is dominant, indicating that Eu^{3+} ions are located at high-symmetry positions. The luminescence intensity ratio (R), $I_e(^5D_0 \rightarrow {}^7F_2)/I_m(^5D_0 \rightarrow {}^7F_1)$ (Eq. 3.3) values, widely known as the *asymmetric ratio*, allows for the estimation of the covalent nature, polarization of the surrounding of the Eu^{3+} ions by short-range effects, and centrosymmetry distortion of the Eu^{3+} ion site. The lower the value of R, the higher the symmetry around the Eu^{3+} ions, and the higher the Eu–O covalence, and vice versa [42]. The R values of $LiNa_{3-x}P_2O_7{:}xEu^{3+}$ phosphors were calculated and found to be 0.42, 0.31, 0.25, 0.24, and 0.23 for x = 0.01, 0.03, 0.05, 0.07, and 0.09, respectively. A decrease in the asymmetric ratio revealed a relatively high local symmetry around the Eu^{3+} ions in $LiNa_3P_2O_7$. Further, it could be concluded that the prepared phosphors will convert NUV radiation to more orange emission due to MD transitions than red emission due to ED transitions. The transitions of Eu^{3+} ions in the host $LiNa_3P_2O_7$ phosphor is represented by the energy-level diagram and is shown in Fig. 3.10. Further, to find the effect of Eu^{3+} dopant concentration on the $LiNa_{3-x}P_2O_7{:}xEu^{3+}$ phosphor, the percentage of Eu was varied in the host from x = 1 to 9 at%. Unfortunately, we could not find any quenching up to 9 at%, which may be due to greater bandgap energy of the host matrix.

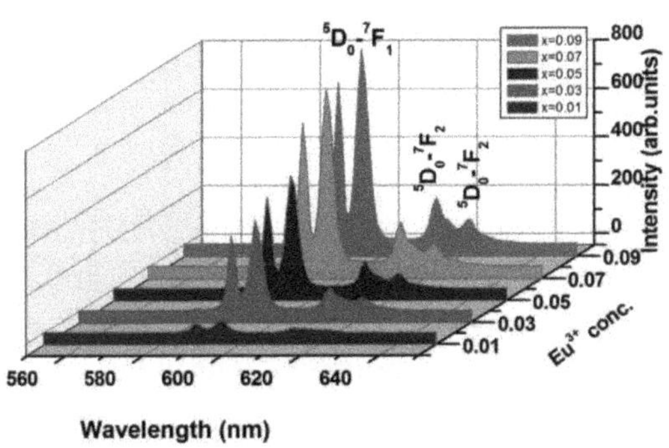

Figure 3.9 PL emission spectra of the $LiNa_{3-x}P_2O_7{:}xEu^{3+}$ (x = 0.01–0.09) phosphor at 395 nm excitation.

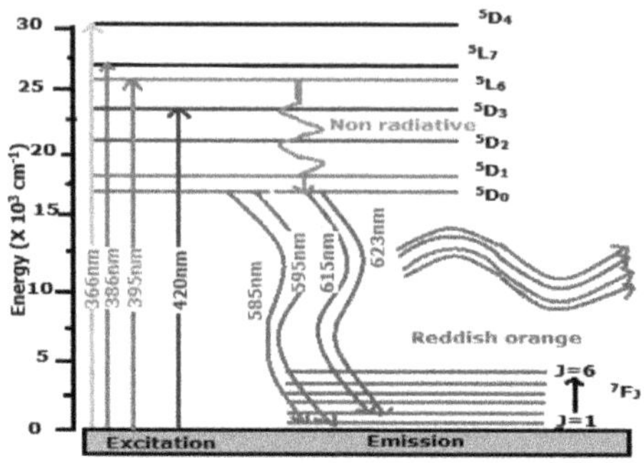

Figure 3.10 Schematic energy-level diagram of Eu^{3+} ions in the $LiNa_3P_2O_7$ host.

The CIE chromaticity coordinates (x, y) of $LiNa_3P_2O_7$ phosphors and CCT values were calculated to characterize the color emission from their corresponding emission spectra, excited by 395 nm. The obtained color coordinates of these samples lie in the orange-red regions shown in Fig. 3.11 and compared to the commercially available red phosphor of $Y_2O_2S{:}Eu^{3+}$ (0.622, 0.351) [43].

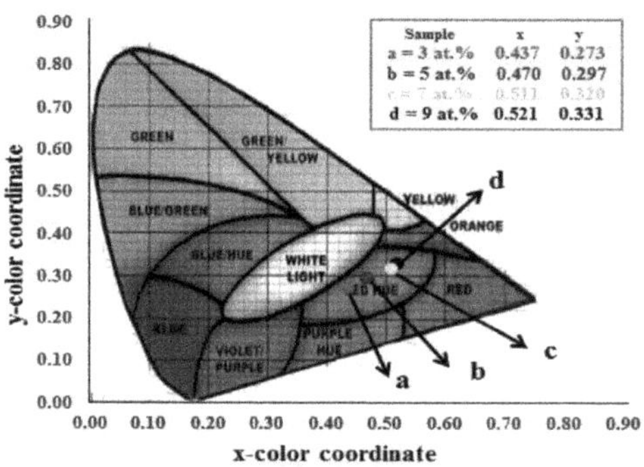

Figure 3.11 CIE chromaticity diagram of $LiNa_{3-x}P_2O_7{:}xEu^{3+}$ (x = 0.03–0.09).

We note from the figure that the color coordinates traversed a wide range from the orange to a deep-red region, and the changes in the color coordinates may be due to the variation of the asymmetric ratios of various concentrations of Eu^{3+} ions in $LiNa_3P_2O_7$ phosphors. The CCT values are calculated by using the McCamy empirical formula given by

$$CCT = -437n^3 + 3601n^2 - 6861n + 5514.31, \quad (3.4)$$

where $n = (x - x_e)/(y - y_e)$ and the chromaticity epicenter is at x_e = 0.3320 and y_e = 0.1858. The estimated color coordinates and CCT values are presented in the Table 3.2. The CCT values of the studied phosphors were found to be 1620 K.

Table 3.2 CIE coordinates and CCT values of the Eu^{3+}-doped $LiNa_3P_2O_7$ phosphor

Dopant concentration (at%)	Color coordinates		CCT (K)
	x	*y*	
3	0.437	0.273	1630
5	0.470	0.297	1624
7	0.511	0.320	1622
9	0.521	0.331	1619

3.3.6 Photoluminescence Studies on Dy^{3+}-Doped $LiNa_{3-x}P_2O_7$ Phosphors

The excitation spectra of $LiNa_3P_2O_7:Dy^{3+}$ phosphors are shown in Fig. 3.12. In the range of 300 to 500 nm, the excitation spectra consist of a series of characteristic bands that are attributed to the transitions of $^6H_{15/2} \rightarrow {}^4L_{19/2}$ (328 nm), $^6H_{15/2} \rightarrow {}^6P_{7/2}$ (351 nm), $^6H_{15/2} \rightarrow {}^6P_{5/2}$ (367 nm), $^6H_{15/2} \rightarrow {}^4I_{13/2}$ (388 nm), $^6H_{15/2} \rightarrow {}^4F_{7/2}$ (395 nm), $^6H_{15/2} \rightarrow {}^4G_{11/2}$ (427 nm), $^6H_{15/2} \rightarrow {}^4I_{15/2}$ (453 nm), and $^6H_{15/2} \rightarrow {}^4F_{9/2}$ (471 nm), respectively [44]. The band at 351 nm is the predominant peak and was used for further emission spectra measurements. The excitation spectra reveal that Dy^{3+} ions could be excited effectively by NUV sources with a wavelength of 351 nm, offering new potential for white-light phosphors. The PL emission spectra were measured under excitation at 351 nm, as shown in Fig. 3.13.

There were two strong characteristic peaks centered at 485 nm (blue) and 575 nm (yellow). These bands correspond to the MD $^4F_{9/2} \rightarrow {}^6H_{15/2}$ transition and the ED $^4F_{9/2} \rightarrow {}^6H_{13/2}$ transition, respectively [45, 46]. The transitions of ions in the dopant (Dy^{3+}) in excitation and emission states are clearly presented in the energy-level diagram shown in Fig. 3.14. To optimize the critical dopant concentration, the Dy^{3+} ion dopant concentration in the host was varied (as described in the experimental section). With increased dopant concentration in the host, the emission spectra appear to have similar profiles with the same peak positions of peaks but with variations in theirs, from 5 to 9 mol%, beyond which the intensity decreased due to the concentration quenching effect. In many inorganic materials, the luminescence intensity remarkably decreases because of nonradiative transition between the similar dopant Dy^{3+} ions before being emitted to the ground level and reduction in the average distance between Dy^{3+}, which leads to the quenching effect [47].

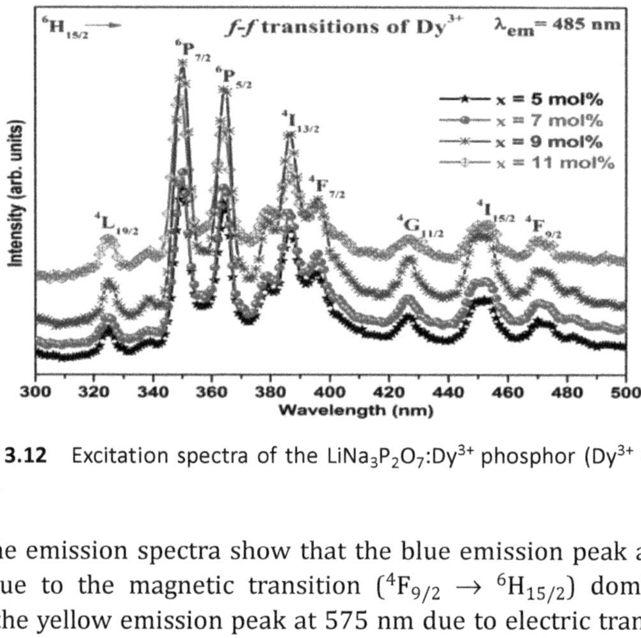

Figure 3.12 Excitation spectra of the $LiNa_3P_2O_7:Dy^{3+}$ phosphor (Dy^{3+} = 5–11 mol%).

The emission spectra show that the blue emission peak at 485 nm due to the magnetic transition ($^4F_{9/2} \rightarrow {}^6H_{15/2}$) dominates over the yellow emission peak at 575 nm due to electric transition ($^4F_{9/2} \rightarrow {}^6H_{13/2}$). This illustrates that Dy^{3+} ions occupied the high-symmetry sites with inversion centers in the host matrix [48]. Further information can be revealed by calculating the asymmetric ratio, which is the ratio of the luminescence intensities due to ED transition and MD transition. The relative intensity ratio of the yellow (575 nm) to blue (481 nm) emission can be used to analyze the structural distortion around Dy^{3+} ions: $(R) = I_e(^4F_{9/2} \rightarrow {}^6H_{13/2})/I_m(^4F_{9/2} \rightarrow {}^6H_{15/2})$ [49]. The R values were found to be 0.87, 0.88, 0.93, and 0.83 for $LiNa_3P_2O_7:Dy^{3+}$ (Dy^{3+} = 0.05, 0.07, 0.09, and 0.11), respectively. These values indicate that the ratio increased with increasing Dy^{3+} ion concentration from 0.87 to 0.93 and then decreased thereafter. Lower *R* values lead to higher symmetry of the dopant ions at the host cation sites, and since the obtained *R* values were small, it can be concluded that Dy^{3+} ions occupy high-symmetry sites [50].

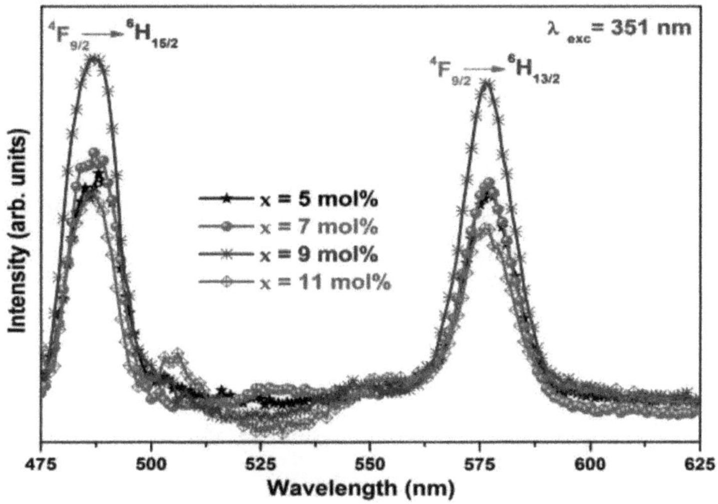

Figure 3.13 Emission spectra of the $LiNa_3P_2O_7:Dy^{3+}$ phosphor (Dy^{3+} = 5–11 mol%).

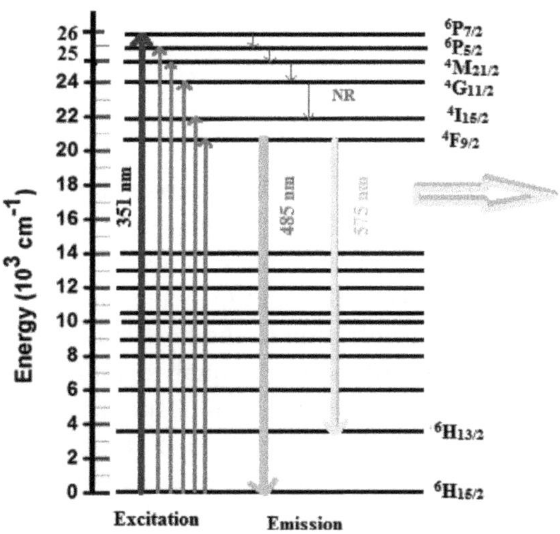

Figure 3.14 Energy-level diagram of f–f transitions of Dy^{3+} ions in the $LiNa_3P_2O_7$ host.

The *x* and *y* values of the 1931 CIE chromaticity coordinates of the $LiNa_3P_2O_7:Dy^{3+}$ phosphor (Dy^{3+} = 9 mol%) were calculated using color coordinate calculation software and are presented in Fig. 3.15. CIE values of the phosphor were calculated in three regions of the emission spectra. The color coordinates are *a* (0.103, 0.127) (461 to 500 nm), *b* (0.469, 0.501) (560 to 600 nm), and *c* (0.281, 0.338) (461 to 650 nm). These CIE values are located in the blue, yellow, and near-white regions, respectively (see Fig. 3.15).

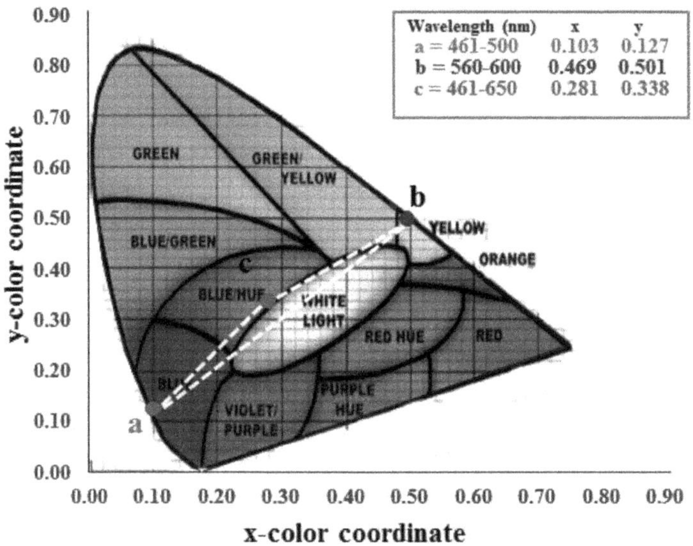

Figure 3.15 CIE diagram of the $LiNa_3P_2O_7:Dy^{3+}$ phosphor (Dy^{3+} = 9 mol%).

The white dashed lines in the figure show that a suitable combination of the blue and yellow emission will produce white-light emission. It is known that suitable ratios of yellow to blue (Y/B) emissions could produce white light in single-phase host materials. By varying the Dy concentration in the host matrix, the Y/B ratios could also vary. The Y/B value is equal to unity at a particular concentration, which is one of the significant criteria for obtaining white light in Dy-doped phosphors. Similar results were published previously [51]. However, in the present case, the Y/B ratio is 0.93 for 9 mol%, which produces near-white light.

3.4 Conclusion

RE-doped phosphate phosphors have emerged as a new class of luminescent materials, showing promising photoluminescent properties, such as significantly NUV excitation and emission spectra, abundant emission colors, small thermal quenching, and high quantum efficiency. These interesting properties enable phosphate-based phosphors to be suitable for use as promising down-conversion luminescent materials in WLEDs with a high CRI and a low CCT.

References

1. Schubert, E. F., and Kim, J. K. (2005). Solid-state light sources getting smart, *Science*, **308**, pp. 1274–1278.
2. Pimputkar, S., Speck, J. S., Den Baars, S. P., and Nakamura, S. (2009). Prospects for lighting, *Nat. Photonics*, **3**, pp. 179–181.
3. Tonzani, S. (2009). Lighting technology: time to change the bulb, *Nature*, **459**, pp. 312–314.
4. Jang, H. S., Won Y. H., and Jeon, D. Y. (2009). Improvement of electroluminescent property of blue LED coated with highly luminescent yellow-emitting phosphors, *Appl. Phys. B: Lasers Opt.*, **95**, pp. 715–720.
5. Hecht, F. S. C., Schmidt, P. J., Auf der Gunne, J. S., Baumann, V., and Schnick, W. (2009). $SrAlSi_4N_7:Eu^{2+}$: a nitridoalumosilicate phosphor for warm white light (pc)LEDs with edge-sharing tetrahedra, *Chem. Mater.*, **21**, pp. 1595–1601.
6. Jung, K. Y., Lee, H. W., and Jung, H.-K. (2006). Luminescent properties of $(Sr, Zn)Al_2O_4:Eu^{2+},B^{3+}$ particles as a potential green phosphor for UV LEDs, *Chem. Mater.*, **18**, pp. 2249–2255.
7. Zhu, H., Lin, C. C., Luo, W., Shu, S., Liu, Z., Liu, Y., Kong, J., Ma, E., Cao, Y., Liu, R. S., and Chen, X. (2014). Highly efficient non-rare-earth red emitting phosphor for warm white light-emitting diodes, *Nat. Commun.*, **5**, pp. 4312–4316.
8. Jiao, M., Jia, Y., Lu, W., Lv, W., Zhao, Q., Shao, B., and You, H. (2014). $Sr_3GdNa(PO_4)_3F:Eu^{2+},Mn^{2+}$: a potential color tunable phosphor for white LEDs, *J. Mater. Chem. C*, **2**, pp. 90–97.
9. Hamady, A., Zid, M. F., and Jouini, T. (1994). Structure cristalline of KYP_2O_7 *J. Solid State Chem.*, **113**, pp. 120–124.

10. Novais, S. M. V., and Macedo, Z. S. (2013). Optical properties of Eu-doped $NaYP_2O_7$ under UV and X-ray excitation, *Phys. Status Solidi C*, **10**, pp. 185–188.

11. Yuan, J. L., Wang, J., Zhang, Z. J., Zhao, J. T., and Zhang, G. B. (2008). Synthesis and luminescent properties of $AGd_{1-x}Ln_xP_2O_7$ (A = Rb, Cs; Ln = Ce, Eu, Tb) under VUV and X-ray excitations, *Opt. Mater.*, **30**, pp. 1327–1334.

12. Yuan, J. L., Zhang, H., Chen, H. H., Yang, X. X., Zhao, J. T. and Gu, M. (2007). Synthesis, structure and X-ray excited luminescence of Ce^{3+}-doped $AREP_2O_7$-type alkali rare earth diphosphates (*A*=Na, K, Rb, Cs; RE=Y, Lu), *J. Solid State Chem.*, **180**, pp. 3381–3387.

13. Khay, N., and Ennaciri, A. (2001). Vibrational spectra of double diphosphates $CsLnP_2O_7$ (Ln = Gd, Tb, Dy, Ho, Y, Er, Tm, Yb), *J. Alloys Compd.*, **323**, pp. 800–805.

14. Khay, N., Ennaciri, A., and Harcharras, M. (2001). Vibrational spectra of double diphosphates $RbLnP_2O_7$ (Ln = Dy, Ho, Y, Er, Tm, Yb), *Vib. Spectrosc.*, **27**, pp. 119–126.

15. Yuan, J. L., Wang, J., Xiong, D. B., Zhao, J. T., Fu, Y. B., Bin Zhang, G., Shi, C. S., Wang X. J., and Duan, C. J. (2007). VUV spectroscopic properties of Ce^{3+} and Pr^{3+}-doped $AREP_2O_7$-type alkali rare earth diphosphates (A=Na, K, Rb, Cs; RE=Y, Lu), *J. Lumin.*, **126**, pp. 130–134.

16. Kumar, V., Som, S., Kumar, V., Kumar, V., Ntwaeaborwa, O. M., Coetsee, E., and Swart, H.C. (2014). Tunable and white emission from $ZnO:Tb^{3+}$ nanophosphors for solid state lighting applications, *Chem. Eng. J.*, **255**, pp. 541–552.

17. Berdowski, P. A. M., Lammers, M. J. J., and Blasse, G. (1985). $^5D_3$-$^5D_4$ cross relaxation in Tb^{3+} pairs in $CsCdBr_3$ crystals, *J. Chem. Phys.*, **83**(2), pp. 476–479.

18. Munirathnam, K., Dillip, G. R., Deva Prasad Raju, B., Joo, S. W., Dhoble, S. J., Nagabhushana, B. M., Hari Krishna, R., Ramesh, K. P., Varadharaj Perumal, S., and Prakashbabu, D. (2015). Synthesis, photoluminescence and Judd–Ofelt parameters of $LiNa_3P_2O_7:Eu^{3+}$ orthorhombic microstructures, *Appl. Phys. A*, **120**, pp. 1615–1623.

19. Liu, B., Shi, C., and Qi, Z. (2005). Potential white-light long-lasting phosphor: Dy^{3+}-doped aluminate, potential white-light long-lasting phosphor: Dy^{3+}-doped aluminate, *Appl. Phys. Lett.*, **86**, pp. 191111–191113.

20. Gupta, S. K., Mohapatra, M., Natarajan, V., and Godbole, S. V. (2013). Photoluminescence investigations of the near white light emitting

perovskite ceramic $SrZrO_3:Dy^{3+}$ prepared via gel-combustion route, *Int. J. Appl. Ceram. Technol.*, **10**, pp. 593–602.

21. Dillip, G. R., Dhoble, S. J., and Raju, B. D. P. (2013). Luminescence properties of $Na_3SrB_5O_{10}:Dy^{3+}$ plate-like microstructures for solid state lighting applications, *Opt. Mater.*, **35**, pp. 2261–2266.

22. Munirathnam, K. , Dillip, G. R., Chaurasia, S., Joo, S. W., Raju, Prasad Deva B., and Sushma, John N. (2016). Investigations on surface chemical analysis using X-ray photoelectron spectroscopy and optical properties of Dy^{3+}-doped $LiNa_3P_2O_7$ phosphor, *J. Mol. Struct.*, **1118**, pp. 117–123.

23. Munirathnam, K., Dillip, G. R., Ramesh, B., Joo, S. W., and Raju, Prasad Deva B. (2015). Synthesis, photoluminescence and thermoluminescence properties of $LiNa_3P_2O_7:Tb^{3+}$ green emitting phosphor, *J. Phys. Chem. Solids*, **86**, pp. 170–176.

24. Shi, Y., Wang, Y., Pan, S., Yang, Z., Dong, X., Wu, H., Zhang, M., Cao, J., and Zhou, Z. (2013). Synthesis, crystal structures and optical properties of two congruent-melting isotypic diphosphates: $LiM_3P_2O_7$ (M = Na, K), *J. Solid State Chem.*, **197**, pp. 128–133.

25. Zaafouri, A., Megdiche, M., and Gargouri, M. (2014). AC conductivity and dielectric behavior in lithium and sodium diphosphate $LiNa_3P_2O_7$, *J. Alloys Compd.*, **584**, pp. 152–158.

26. Shannon, R. D. (1976). Revised effective ionic radii and systematic studies of interatomic distances in halides and chalcogenides, *Acta Crystallogr., Sect. A*, **32**, pp. 751–756.

27. Jia, Y. Q. (1991). Crystal radii and effective ionic radii of the rare earth ions, *J. Solid State Chem.*, **95**, pp. 184–187.

28. Guan, L., Wang, Y., Chen, W., Jin, L., Li, X., Guo, Q., Yang, Z., and Fu, G. (2010). Fabrication and luminescent properties of red phosphor $M_3BO_6:Eu^{3+}$(M = La, Y), *J. Rare Earths*, **28**(10), pp. 295–298.

29. Corbridge, D. E. (1966). *Topics in Phosphorous Chemistry*; Griffith, E., and Grayson, M., eds. (John Wiley Interscience, New York), p. 275.

30. Dillip, G. R. AND Raju, B. D. P. (2012). A study of the luminescence in near UV-pumped red-emitting novel Eu^{3+}-doped $Ba_3Ca_3(PO_4)_4$ phosphors for white light emitting diodes, *J. Alloys Compd.*, **540**, pp. 67–74.

31. Kumar, V., Som, S., Kumar, V., Kumar, V., Ntwaeaborwa, O. M., Coetsee, E., and Swart, H. C. (2014). Tunable and white emission from ZnO: Tb^{3+} nanophosphors for solid state lighting applications, *Chem. Eng. J.*, **255**, pp. 541–552.

32. Hen, S., Wang, Y., Zhang, J., Zhao, L., Wang, Q., and Han, L. (2014). Luminescent properties of novel $K_3R(PO_4)_2$:Tb^{3+} (R=Y and Gd) phosphors for displays and lightings, *J. Lumin.*, **150**, pp. 46–49.

33. Dillip, G. R., Dhoble, S. J., and Raju, B. D. P. (2013). Luminescence properties of $Na_3SrB_5O_{10}$: Dy^{3+} plate-like microstructures for solid state lighting applications, *Opt. Mater.*, **35**, pp. 2261–2266.

34. Wang, Z., Yang, Z., Li, P., Guo, Q., and Yang, Y. (2010). *J. Rare Earths*, **28**, pp. 30–34.

35. Blasse, G., and Grabmaier, B. C. (1994). *Luminescent Materials* (Springer, Berlin).

36. Liu, Q., Liu, Y., Yang, Z., Li, X., and Han, Y. (2012). UV-excited red-emitting phosphor Eu^{3+}-activated $Ca_9Y(PO_4)_7$, *Spectrochim. Acta Part A*, **87**, pp. 190–193.

37. Yu, P., Su, L., and Xu, J. (2014). Synthesis and luminescence properties of Eu^{3+}, Bi^{3+}-doped $BaWO_4$ phosphors, *Opt. Rev.*, **21**, pp. 455–460.

38. Han, B., Zhang, J., Wang, Z., Liu, Y., and Shi, H. (2014). Investigation on the concentration quenching and energy transfer of red-light-emitting phosphor Y_2 MoO_6: Eu^{3+}, *J. Lumin.*, **149**, pp. 150–154.

39. Wang, R., Zhou, L., and Wang, Y. (2011). Luminescent properties of $MMgP_2O_7$: Eu^{3+}(M=Ca, Sr, Ba) phosphor, *J. Rare Earths*, **29**, pp. 1045–1048.

40. Dillip, G. R., Dhoble, S. J., Manoj, L., Reddy, C. M., and Raju, B. D. P. (2012). A potential red emitting $K_4Ca(PO_4)_2$: Eu^{3+} phosphor for white light emitting diodes, *J. Lumin.*, **132**, pp. 3072–3076.

41. Song, E., Zhao, W., Zhou, G., Dou, X., Yi, C., and Zhou, M. (2011). Luminescence properties of red phosphors $Ca_{10}Li(PO_4)_7$: Eu^{3+}, *J. Rare Earths*, **29**, pp. 440–443.

42. Parchur, A. K., Prasad, A. I., Rai, S. B., Tewari, R., Sahu, R. K., Okram, G. S., Singh, R. A., and Ningthoujam, R. S. (2012). Observation of intermediate bands in Eu^{3+} doped YPO_4 host: Li^+ ion effect and blue to pink light emitter, *AIP Adv.*, **2**, p. 032119.

43. Dillip, G. R., and Raju, B. D. P. (2012). A study of the luminescence in near UV-pumped red-emitting novel Eu^{3+}-doped $Ba_3Ca_3(PO_4)_4$ phosphors for white light emitting diodes, *J. Alloys Compd.*, **540**, pp. 67–74.

44. Watras, A., Dereń, P. J., and Pązik, R. (2014). Luminescence properties and determination of optimal RE^{3+} (Sm^{3+}, Tb^{3+} and Dy^{3+}) doping levels in the KYP_2O_7 host lattice obtained by combustion synthesis, *New J. Chem.*, **38**, pp. 5058–5068.

45. Liu, B., Shi, C., and Qi, Z. (2005). Potential white-light long-lasting phosphor: Dy^{3+}-doped aluminate, *Appl. Phys. Lett.*, **86**, p. 191111.

46. Gupta, S. K., Mohapatra, M., Natarajan, V., and Godbole, S. V. (2013). Photoluminescence investigations of the near white light emitting perovskite ceramic $SrZrO_3$:Dy^{3+} prepared via gel-combustion route, *Int. J. Appl. Ceram. Technol.*, **10**, pp. 593–602.

47. Blasse, G. (1986). Energy transfer between inequivalent Eu^{2+} ions, *J. Solid State Chem.*, **211**, pp. 207–211.

48. Xu, Q., Sun, J., Cui, D., Di, Q., and Zeng, J. (2015). Synthesis and luminescence properties of novel $Sr_3Gd(PO_4)_3$:Dy^{3+} phosphor, *J. Lumin.*, **158**, pp. 301–305.

49. Dillip, G. R., Ramesh, B., Reddy, C. M., Mallikarjuna, K., Ravi, O., Dhoble, S. J., Joo, S. W., and Deva Prasad Raju, B. (2014). X-ray analysis and optical studies of Dy^{3+} doped $NaSrB_5O_9$ microstructures for white light generation, *J. Alloys Compd.*, **615**, pp. 719–727.

50. Zhang, Z. W., Sun, X. Y., Liu, L., Peng, Y., Shen, X., Zhang, W. G., and Wang, D. J. (2013). Synthesis and luminescence properties of novel $LiSr_4(BO_3)_3$:Dy^{3+} phosphors, *Ceram. Int.*, **39**, pp. 1723–1728.

51. Dillip, G. R., Dhoble, S. J., and Raju, B. D. P. (2013). Luminescence properties of $Na_3SrB_5O_{10}$:Dy^{3+} plate-like microstructures for solid state lighting applications, *Opt. Mater.*, **35**, pp. 2261–2266.

Chapter 4

Recent Advances in Sulfate- and Sulfide-Based Phosphors Used in Versatile Applications

Govind B. Nair and S. J. Dhoble
Department of Physics, R. T. M. Nagpur University, Nagpur 440033, Maharashtra, India
govind1291@yahoo.com

The recent developments achieved in the phosphor industry have been highlighted along with a brief introduction to dosimetry and lighting applications. The discussion is strongly focused on the synthesis and properties of sulfate and sulfide phosphors. The discussion has been extended to some mixed sulfates, owing to their possession of well-desired characteristics that are required by a phosphor. Some of the phosphors listed in this article are crucially known for their thermoluminescence (TL), lyoluminescence (LL), mechanoluminescence (ML), optically stimulated luminescence (OSL), and photoluminescence (PL) properties.

Phosphors: Synthesis and Applications
Edited by Sanjay J. Dhoble, B. Deva Prasad Raju, and Vijay Singh

ISBN 978-981-4774-49-9 (Hardcover), 978-042-9460-99-9 (eBook)
www.panstanford.com

4.1 Introduction

Remarkable progress in sulfate- and sulfide-based phosphors has been witnessed in the last few decades. The choice of a host material is a crucial part in the study of luminescence [1]. Several factors have encouraged researchers to rely on sulfates, sulfides, halosulfates, etc., as the host matrix for luminescent materials. They have a wide range of applications, which are highly beneficial to the growing phosphor market. The fields that highly depend on phosphors are radiation dosimetry, display devices, lighting, etc., and each of these fields has much to talk about sulfate- and sulfide-based phosphors.

Radiation dosimetry can be performed using phenomena like thermoluminescence (TL), optically stimulated luminescence (OSL), lyoluminescence (LL), and mechanoluminescence (ML). Each of these phenomena shows an entirely different mechanism to measure the dose absorbed by the luminescent material. However, TL is the most widely accepted phenomena for dosimetric applications. Nowadays, OSL has also been adopted for radiation dosimetry due to certain advantages, which will be discussed later in this review. The availability of limited TL phosphors for dosimetry has posed a major problem for high-dose measurements. This has led to the development of more dependable radiation dosimetric procedures such as LL and ML, which are applicable over a wide range of dose levels.

A high spectral yield is a must for phosphors if they are to be considered for lighting and display devices. Generally, phenomena like photoluminescence (PL), cathodoluminescence (CL), and electroluminescence (EL) are responsible for emission of light from a luminescent material. The relationship between crystalline structures and light emission plays a crucial role in controlling the spectral yield of phosphors [2]. By exploiting this feature, more advanced phosphors for lighting and display panels could be brought to existence, and their efficiency will directly result in huge energy savings.

The present chapter decides upon the efficient sulfate- and sulfide-based phosphors for their applications in diverse fields. The discussion has been parted into two sections: one for radiation dosimetry and the other for lighting and display devices.

4.2 Radiation Dosimetry Phosphors

Radiations are omnipresent in the atmosphere in the form of energy that traverses as electromagnetic waves or high-speed particles. The ionizing radiations are often used for the treatment of cancerous tumors and other heterogeneous classes of diseases. However, incorrect exposure to ionizing radiations may prove fatal, and hence, knowledge of the absorbed dose of radiation is much important for increasing the efficiency of the treatment. Heavy-ion therapy is the use of particles more massive than protons or neutrons, such as carbon ions [3]. The dosimetry of charged particle beams in cancer diagnosis and therapy has taken a decisive lead to meet the demands of an accurate calibration of radiotherapy sources. TL is the radiation-analyzing process that is often implemented to determine the amount of dose absorbed for a specific duration of the radiation exposure. This phenomenon includes heating of the irradiated material, due to which the energy stored in the crystal/material is released with the emission of light, and the intensity of the emitted light as a function of temperature is related to the dose absorbed by material before readout [4]. TL is the most common technique used for the dosimetry of different ionizing radiations, while considering the objectives like radiation protection monitoring. Lattice defects in the crystals can be studied by TL analysis [5]. Thermoluminescent dosimeters (TLDs) have been in service, owing to their small size and easy handling. The suitability and applicability of a TL material for dosimetric purposes depends on dosimetric characteristics such as the nature of the glow curve, TL response as a function of dose, energy dependence, and stability of the TL signal as a function of time [6]. TLD materials are expected to have a low photon energy dependence of response. Tissue-equivalent phosphors (effective atomic number of tissue $Z_{eff} = 7.4$), or approximated tissue equivalent, should be used for personal and medical applications to avoid energy corrections [7]. Although fluorides are the most commonly used TLDs, their susceptibility to heat treatments and complicated glow curve structures posed some difficulties in dosimetry. The next in category are the sulfate-based TL materials, which are widely studied because of their well-desired characteristics like a high-temperature glow peak and high sensitivity toward the absorbed dose. The major problem with sulfate-based dosimeters is that they are not tissue equivalent (low *Z*) and so efforts are being made to improve their tissue equivalence [8].

A number of technical reasons have made OSL dosimeters more attractive as compared to TLDs, while considering the facts that the optical readout method is fast and relatively simple and the OSL technique lends itself to repeat readout of samples when an isolated optical simulation does not result in complete clearing of stored signal [9]. The OSL technique has its advantage over TL as it rules out the loss of reusability of the phosphor material caused due to heating. In OSL, only a fraction of traps generated on irradiation to a particular dose are used on each readout and the doses could be re-estimated unlike in the case of TL.

LL is the phenomenon of light emission when a certain substance irradiated with ionizing radiation is dissolved in a suitable solvent, and it owes its origin to the transfer of energy from the radiation-induced free radicals to the solvent molecules, a part of which gets transposed to visible light. The quantum of light yield produced is generally proportional to the absorbed dose in the solid. The parameters that influence the LL intensity are grain size, mass of the irradiated phosphor, pH of the solvent, temperature of the solvent, and irradiation dose [10, 11]. Although alkaline halides, saccharides, and amino acids are the most widely studied LL materials, the spectral LL of irradiated alkali sulfates such as Li, Na, K, and Rb has also been investigated [12].

ML is another kind of luminescence that can be excited by the mechanical actions on solids. ML technology is a versatile tool for the study of the fracture mechanism. It has potential for use in low-cost safety diagnosis of structural components or visualization of microelectromechanical system operation [13].

Here, radiation dosimetry phosphors have been conveniently classified into the following three classes: sulfates and oxysulfates, halosulfates, and sulfides and oxysulfides.

4.2.1 Sulfates and Oxysulfates

Calcium sulfate ($CaSO_4$) with different dopants is one of the most investigated and widely acclaimed materials to be used as a TLD. It is capable of accommodating a great variety of defect types located at a number of different crystallographic locations [14]. $CaSO_4$ has become so popular among researchers that almost every feature related to it has been investigated and reported. The TL intensity of $CaSO_4$:Dy is found to increase by about 50% after incorporating P [15]. Further attempts were made to increase the TL intensity by codoping Cu and other rare earth ions together in the $CaSO_4$:Dy,P host [16]. However, these attempts failed to yield remarkable results, although codoping was able to produce a slight increase in their TL intensities. Self-agglomerating Eu-doped $CaSO_4$ phosphors, synthesized by a cheap, easy, and environment-friendly chemical route, exhibited a main peak intensity about 2.5 times greater than that of TLD-100 [17]. These phosphors showed a maximum peak intensity at a temperature close to 200°C when a 5°C/sec heating rate was used. This phosphor has good reusability and thermal stability, and it is suitable to avoid interference of TL with black-body radiation. $K_2Ca_2(SO_4)_3$:Cu nanophosphors also gave exceptional properties with a potential to be useful as a γ-ray dosimeter. The TL response curve of the $K_2Ca_2(SO_4)_3$:Cu nanophosphor, annealed at 700°C for 2 hr, exhibited linearity over a wide range of γ-ray doses [18]. The TL response for this nanophosphor is linear from a 0.01 Gy to a 300 Gy dose of γ-rays and is sublinear from a 300 Gy to a 1 kGy dose, beyond which it gets saturated. A similar linear TL response was observed for 0.1 Gy to 300 Gy exposure of a 150 MeV proton beam [19]. On the other hand, its OSL response is found to be linear from a 0.01 Gy to 1 kGy γ-ray dose. Another sensitive nanophosphor was tried in the form of $K_3Na(SO_4)_2$:Eu by Sahare et al. [20]. $K_3Na(SO_4)_2$:Eu nanophosphors revealed high sensitivity, low fading, and saturation at a dose as high as 70 kGy. Although the sensitivity of the $K_3Na(SO_4)_2$:Eu nanophosphor for 100 Gy γ-ray exposure is 10 times less than TLD-700H, it is found to be 4 times more than TLD-100. The TL glow curve for this nanophosphor comprises of a prominent peak at 423 K along with three smaller peaks positioned at 382, 460, and 509 K on the temperature scale. Exceptional reusability and low fading were also observed for this nanophosphor. This nanophosphor has broad prospects to be used for the estimation of very high-energy ionizing radiations.

Recently, Bahl et al. introduced a multi-utility barite ($BaSO_4$) nanophosphor that was prepared by the co-precipitation method [21]. Samples in the form of pellets were subjected to two different exposures, that is, 1.25 MeV γ-rays as well as to 150 MeV proton beams for varying doses. These two irradiations are different in energies and type, the former being electromagnetic and the latter being particulate. However, the TL glow curve structure was similar for both types of irradiation. The nanophosphor exhibits a linear TL response up to 1000 Gy for gamma and 305 Gy for proton doses.

4.2.2 Halosulfates

The TL glow curve of the $Na_3Ca_2(SO_4)_3F:Ce^{3+}$ phosphor showed a single broad peak at 147°C [22]. The calculation of kinetic parameters for the maximum-intensity glow peak by Chen's peak shape method showed that it followed second-order kinetics with an energy (E) value of 0.64 ± 0.002 eV and a frequency factor (s) value (1.43 ± 0.014) × 10^7 sec^{-1}, respectively. Effect of gamma irradiation on the TL properties in rare earth–doped $NaMgSO_4Cl$ phosphors, synthesized by the wet-chemical method, has been reported recently [23]. The $NaMgSO_4Cl$:Tb phosphors (between the range of 257°C and 284°C), $NaMgSO_4Cl$:Dy (173°C), and $NaMgSO_4Cl$:Dy,Eu (156°C) have a single prominent peak, whereas $NaMgSO_4Cl$:Ce,Tb has two peaks located at 154°C and 233°C, indicating single and double trapping sites, respectively. $NaMg(SO_4)Cl$:Ce phosphors, prepared by the wet-chemical method, exhibited good sensitivity, even for low concentrations of rare earth ions and a low γ-ray dose [24]. TL glow curves for this phosphor peaked at about 241°C, for 1 Gy γ-ray exposure at a rate of 0.995 kGy/h. For the same γ-ray exposure, the TL glow curve for $NaMg(SO_4)Cl$:Ce,Dy peaked between 247°C and 312°C. However, these phosphors are highly soluble in water and have a poor shelf-life, thereby making them unsuitable for routine dosimetry purposes. The TL glow curve for $NaMg(SO_4)F:Dy^{3+}$ shows a single peak at 170°C, and in $NaMg(SO_4)F$:Ce it is located at 172°C when they are exposed to γ-rays for 5 Gy at the rate of 0.995 $kGyh^{-1}$ [25]. $NaMg(SO_4)F$:Dy is found to be more sensitive than standard $CaSO_4$:Dy, whereas $NaMg(SO_4)F$:Ce is found to be less sensitive in its comparison. The TL glow curves of gamma-irradiated $KMgSO_4F$:Ce and $KMgSO_4F$:Mn phosphors, synthesized by the wet-chemical method, showed simple structures with a prominent single TL glow peak at 167°C and 175°C, respectively, when deconvoluted [26]. The $NaZnSO_4Cl$:Ce and $NaZnSO_4Cl$:Mn phosphors exhibited a simple TL glow curve structure with a single prominent peak at around 170°C and 190°C, whereas $NaZnSO_4Cl$:Dy has two peaks located at 175°C and 298°C, indicating single and double trapping sites, respectively. The sensitivity of the $NaZnSO_4Cl$:Ce,Dy or $NaZnSO_4Cl$:Ce,Mn phosphors was found to be higher than that of $CaSO_4$:Dy, and their glow curves showed broadness for various concentrations of Ce, Dy, and Mn and different γ-ray doses [27]. Another interesting phosphor with high sensitivity toward γ-rays was found in the form of $Na_{21}Mg(SO_4)_{10}Cl_3:Dy^{3+}$ [28]. Its low fading and linear dose response characteristics are highly crucial for TL dosimetry purpose. It has 1.23 times more sensitivity for γ-rays compared to the commercial $CaSO_4$:Dy phosphor. However, this phosphor developed shallow traps when irradiated with beta rays. The glow curve obtained for the beta-irradiated $Na_{21}Mg(SO_4)_{10}Cl_3:Dy^{3+}$ phosphor is entirely different than the one irradiated by γ-rays.

4.3 Phosphors for Lighting and Display Devices

Phosphors must exhibit high quantum efficiency, maintain their luminescence efficiency while working at differing temperatures, have a high thermal quenching temperature, and must be physically and chemically stable [29–31]. The efficiency and stability of phosphors is affected by many factors such as chemical reactions, crystallization of solid, migration and incorporation of ionic species, size of the particle, fine and agglomeration-free morphology, atmosphere, humidity, temperature, and heating effect [5].

There is a huge demand for an advanced visual display with a flat screen, low power consumption, and high definition (HD) [32].

4.3.1 Sulfates and Oxysulfates

The PL intensity of $CaSO_4$:Dy is found to increase by about 150% after incorporating P [15]. A further increase in PL emission was observed after codoping Cu and other rare earth ions together in the $CaSO_4$:Dy,P host [16]. Eu^{2+}-activated barite ($BaSO_4$) nanophosphors, prepared by the co-precipitation method, exhibited a consistent broad-band emission at 374 nm, under 250 nm excitation wavelength [21]. A typical character of this phosphor was detected when its PL emission intensity increased on irradiating with γ-rays. The PL intensity increased up to a dose of 800 Gy, and beyond this dose, a decrease was observed in the intensity, as shown in Fig. 4.1. $KNaSO_4$:Ce [33], prepared by the wet-chemical method, showed PL emission at around 327 nm due to a 5d–4f transition under 279 nm excitation. Efficient energy transfer was also reported in this host from $Ce^{3+} \rightarrow Dy^{3+}$ ions that led to the peaking of the Dy^{3+} emissions at 483 nm and 556 nm under excitation at 280 nm. Interestingly, no such Dy^{3+} emission was observed in this phosphor in the absence of Ce^{3+}. However, this phosphor suffers from a serious disadvantage of lower shelf-life and instability in moist conditions due to its high solubility in water, and hence, special care must be taken before considering this phosphor for lamp-based applications. A lot of research has been done to study the energy transfer process from Ce^{3+} to different activator ions in different host lattices [34]. The characteristic emission of Ce^{3+} in the near-ultraviolet (NUV) spectral region originates from parity-allowed electric–dipole transitions between the excited 5d and ground 4f states [35] $Ce^{3+} \rightarrow Eu^{3+}$ and $Ce^{3+} \rightarrow Dy^{3+}$ energy transfer mechanisms were reported in a $BaCa(SO_4)_2$ host lattice [36]. The phosphors, prepared by the co-precipitation method, exhibited efficient individual emissions at 327 nm corresponding to Ce^{3+}; 339 nm, 396 nm, and 419 nm corresponding to Eu^{3+}; and 486 nm and 574 nm corresponding to Dy^{3+}. Attractive characteristics such as large Stokes shift values, narrow emission bandwidths, and long emission lifetimes make europium-activated hosts suitable as candidates for diodes and displays [37, 38].

The rare earth oxysulfates (general formula $Ln_2O_2SO_4$) first came into the limelight due to their interesting and unusual magnetic properties at low temperature [39]. These materials have proven themselves to be outstanding phosphors by acting as excellent hosts for other rare earths. Several methods have been employed for the synthesis of rare earth oxysulfates. These phosphors can be synthesized by a solid-state reaction; but there is a serious difficulty to obtain a pure-phase and fine powder of phosphors due to the formation of some impurity phases of rare earth oxides [40]. A novel scheme was described to obtain a pure phase of rare earth oxysulfates by reducing the rare earth oxide with an excess amount of CS_2 [41]. A combined method of electrospinning and heat treatment

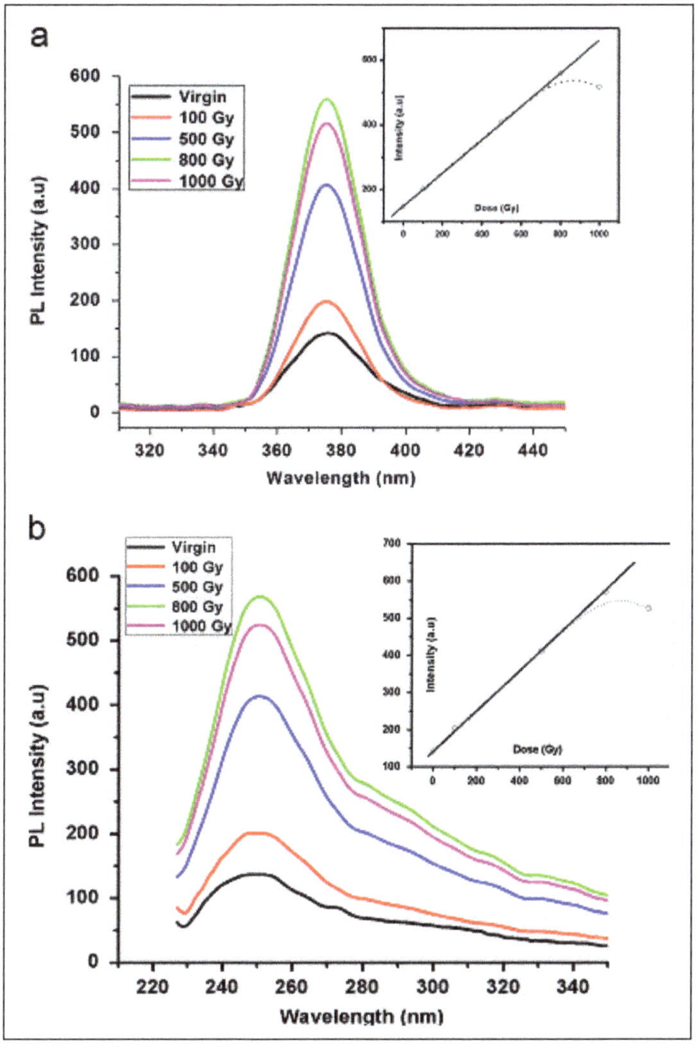

Figure 4.1 PL emission spectra for gamma-irradiated $BaSO_4$:Eu nanocrystalline phosphor irradiated to different doses at an excitation wavelength of (a) 250 nm and (b) 374 nm. The insets show the PL excitation response as a function of dose. Reprinted from Ref. [21], Copyright (2014), with permission from Elsevier.

in a mixed gas of sulfur dioxide and air was proposed to synthesize nanopieces of $Y_2O_2SO_4$:Eu^{3+} [42]. The thicknesses and in-plane sizes of the nanopieces were 40–90 nm and 200–900 nm, respectively. Using the same method, it was possible to synthesize nanopieces of

$Gd_2O_2SO_4$:Tb^{3+}, which exhibited excellent luminescent properties under UV excitation [43]. For preparing the electrospinning solution, the precursors (in the form of nitrates) were dissolved in deionized water along with the stalk solution of poly(vinyl pyrrolidone) (PVP) and a small quantity of acetic acid. The nanofibers were calcined at 1000°C in a mixed-gas atmosphere with excess of sulfur dioxide gas to remove PVP. Initially, the samples were nanoribbons with widths of 200–1000 nm, which later got reduced to 130–450 nm due to the removal of PVP and the decomposition of the nitrate during calcination. This trend was observed in both samples prepared by this method. Figure 4.2 shows the scanning electron microscopy (SEM) images of $Gd_2O_2SO_4$:Tb^{3+} nanopieces before and after calcination. The SEM images for $Y_2O_2SO_4$:Eu^{3+} nanoribbons also depict an appearance similar to $Gd_2O_2SO_4$:Tb^{3+} nanopieces. The transmission electron microscopy (TEM) images of $Y_2O_2SO_4$:Eu^{3+} nanoribbons, as shown in Fig. 4.3, make an obvious statement about the existence of multilayers in a single nanopiece. $Gd_2O_2SO_4$:Eu^{3+} nanophosphors were synthesized by the co-precipitation method [44], although a higher luminescence intensity and a longer lifetime were obtained for the $Gd_2O_2SO_4$:Eu^{3+} submicrophosphors synthesized by the homogeneous precipitation method [45]. $Gd_2O_2SO_4$:Eu^{3+} phosphors have good thermal stability and an ideal red-light emission performance with a peak maximum at around 618 nm, making them suitable candidates in the field of high-resolution red-light emission. $Gd_2O_2SO_4$:Eu^{3+} nanoparticles, synthesized in the presence of Gd^{3+} ions and sodium dodecyl sulfate (SDS) by the simple complexation-thermal decomposition (CTD) method, showed an intense-red PL associated with the $^5D_0 \rightarrow {}^7F_2$ transition in the presence of Eu^{3+} [46]. Emission intensities of the $La_2O_2SO_4$:Eu^{3+} phosphors, prepared by the homogeneous precipitation method, depend on the calcination temperature, and the strongest emission peak was located at 620 nm under 280 nm UV excitation [47]. Similar emission was observed for $La_2O_2SO_4$:Eu^{3+} phosphors prepared by the hydrothermal method [48]. An abnormally enhanced Eu^{3+} emission was observed for a nanostructure-mediated synthesis approach of $Y_2O_2SO_4$:Eu^{3+}. The abnormally enhanced emission was likely to be based on specific deformation of sulfate groups induced through the conversion of concentric dodecylsulfate layers to straight sulfate layers in the oxysulfate framework upon calcination [49].

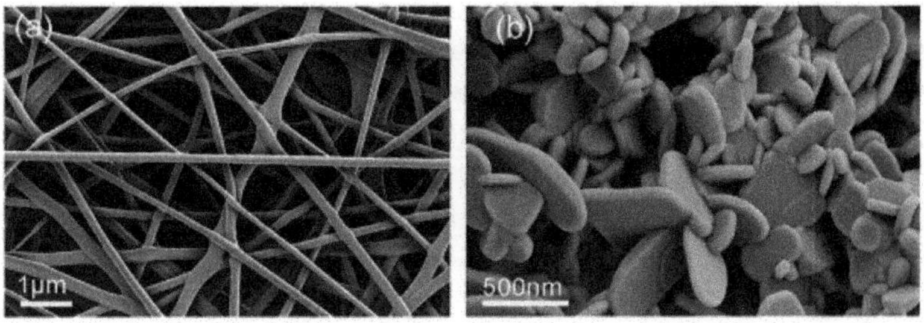

Figure 4.2 Field emission scanning electron microscopy images of $Gd_2O_2SO_4:Tb^{3+}$ samples: (a) as-spun nanofibers and (b) nanofibers after calcination. Reprinted from Ref. [43], Copyright (2014), with permission from Elsevier.

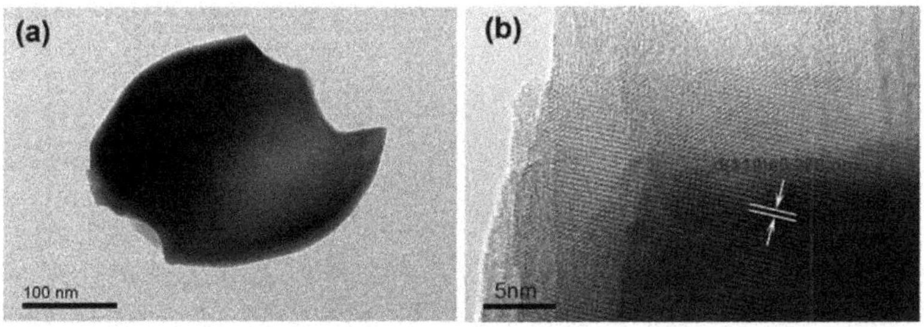

Figure 4.3 TEM images of $Y_2O_2SO_4:Eu^{3+}$ nanopieces. Reprinted from Ref. [42], Copyright (2013), with permission from Elsevier.

Under 275 nm excitation, the strongest peak of $Y_2O_2SO_4:Eu^{3+}$ nanopieces was recorded at 619 nm due to the $^5D_0 \rightarrow {^7F_2}$ transition of Eu^{3+}, and other minor peaks were recorded at 581, 589, 597, 653, and 697 nm, due to the $^5D_0 \rightarrow {^7F_J}$ (J = 0, 1, 3, 4) transitions. The peaks at 589 and 597 nm are due to the $^5D_0 \rightarrow {^7F_1}$ transition of Eu^{3+}. The critical concentration reported for $Y_2O_2SO_4:Eu^{3+}$ nanopieces was 11%, and these nanopieces exhibited stronger excitation and emission peaks in comparison to the $Y_2O_3:Eu^{3+}$ nanoribbons and commercial $Y_2O_3:Eu^{3+}$ phosphors. It was observed that the concentric layered $Y_2O_2SO_4:RE^{3+}$ (RE = Eu, Tb) exhibited the strongest emission intensity than other layered $Y_2O_2SO_4:RE^{3+}$ nanostructures [50, 51]. Monitored at 545 nm, the PL excitation spectra of the bulk $Gd_2O_2SO_4:Tb^{3+}$ and $Gd_2O_2SO_4:Tb^{3+}$ nanopieces exhibited a sharp strong band of 230 nm due to the parity- and spin-allowed transition

$4f^8 \rightarrow 4f^7\,5d^1$ of Tb^{3+} and some broad bands around 303 nm, owing to the $^{8}S_{7/2} \rightarrow {}^6I_J$ transitions of Gd^{3+} and the f–f transitions of Tb^{3+}. The emission spectrum, recorded at 230 nm excitation, consists of peaks at 417 nm ($^5D_3 \rightarrow {}^7F_5$), 438 nm ($^5D_3 \rightarrow {}^7F_4$), 487 nm ($^5D_4 \rightarrow {}^7F_6$), 544 nm ($^5D_4 \rightarrow {}^7F_5$), 587 nm ($^5D_4 \rightarrow {}^7F_4$), and 621 nm ($^5D_4 \rightarrow {}^7F_3$), as shown in Fig. 4.4. The emission peaks had greater intensity for the nanopieces as compared to the bulk samples of $Gd_2O_2SO_4:Tb^{3+}$. The intensity of the peak arising from the $^5D_4 \rightarrow {}^7F_5$ transition in $Gd_2O_2SO_4:Tb^{3+}$ nanopieces increases by about 34%, 27%, and 18% with the alkali metal ions (Li^+, Na^+, and K^+) codoping, respectively, as shown in Fig. 4.5. These nanopieces are expected to have potential applications in field emission display (FEDs) devices, light-emitting diodes (LEDs), and other optically functional devices.

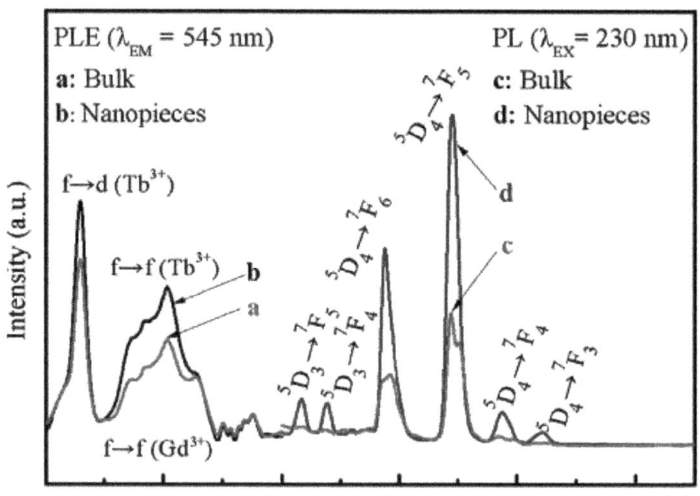

Figure 4.4 PL excitation (a, b) and emission spectra (c, d) of bulk $Gd_2O_2SO_4:Tb^{3+}$ and $Gd_2O_2SO_4:Tb^{3+}$ nanopieces, respectively. Reprinted from Ref. [43], Copyright (2014), with permission from Elsevier.

4.3.2 Halosulfates

Besides sulfates and mixed sulfates, investigations on halosulfate-based phosphors are also in great progression.

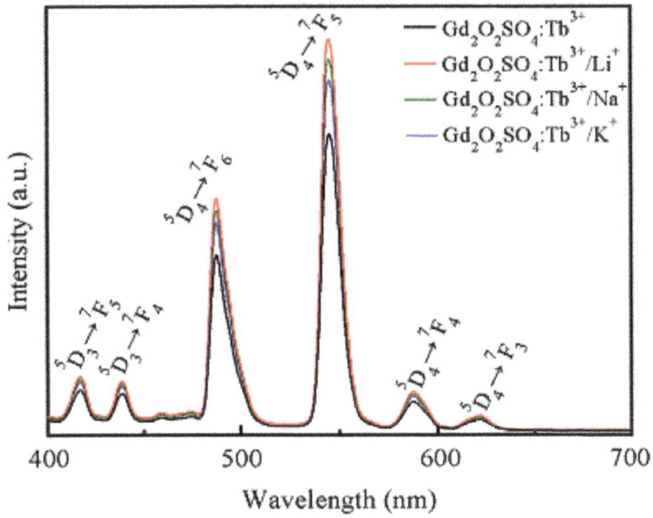

Figure 4.5 PL emission spectra of $Gd_2O_2SO_4:Tb^{3+}$ nanopieces and $Gd_2O_2SO_4:Tb^{3+}/M^+$ (M = Li, Na, and K) nanopieces with a molar ratio of M/Gd of 0.06. Reprinted from Ref. [43], Copyright (2014), with permission from Elsevier.

An efficient $Ce^{3+} \rightarrow Dy^{3+}$, $Eu^{2+} \rightarrow Dy^{3+}$, and $Eu^{2+} \rightarrow Eu^{3+}$ energy transfer process was reported in Na_3SO_4Cl-X (X = Ce^{3+}, Eu^{3+}, or Dy^{3+}) prepared by the wet-chemical method [52]. the $Na_3Ca_2(SO_4)_3F:Ce^{3+}$ phosphor, synthesized by a solid-state diffusion method, showed a single high-intensity emission peak at a 307 nm wavelength when excited by UV light of wavelength 278 nm [22]. An unresolved emission peak was also observed in the PL spectra at about 357 nm wavelength, which clearly indicated the splitting of the ground energy level for the Ce^{3+} ion as $^2F_{7/2}$ and $^2F_{5/2}$. The $NaMgSO_4Cl$:Ce phosphor showed emissions at 346 nm having a small shoulder at 326 nm due to the 4f $\rightarrow$ 5d transition when excited at 270 nm. Energy transfer from Ce^{3+} to Dy^{3+} was indicated by the appearance of Dy^{3+} emissions when excited in the Ce^{3+} absorption band. Under 270 nm excitation, a strong PL emission of Dy^{3+} ions was observed at 485 nm and 576 nm in $NaMgSO_4Cl$:Ce,Dy phosphors due to $^4F_{9/2} \rightarrow {}^6H_{15/2}$ and $^4F_{9/2} \rightarrow {}^6H_{13/2}$ transitions of the Dy^{3+} ion, respectively, although individual Dy^{3+} emission in the $NaMgSO_4Cl$ host was not observed [24]. $K_3Ca_2(SO_4)_3Cl:Eu^{3+}$ phosphors, prepared by the wet-chemical method, showed efficient red emission at 594 nm and 615 nm on keeping the excitation wavelength constant at 396 nm [53].

The energy transfer process from Dy^{3+} to Eu^{3+} ions was observed in this host [54]. $K_3Ca_2(SO_4)_3Cl$ codoped with Eu/Dy is a suitable candidate for NUV-based white-light-emitting phosphor technology. $K_3Ca_2(SO_4)_3Cl:Dy^{3+}$ exhibits its usual emission at blue (484 nm) and yellow (575 nm) when excited by a wavelength of 351 nm. However, codoping these samples with Eu^{3+} increases the intensity of the emission peaks at 351 nm excitation. At an excitation wavelength of 396 nm, this phosphor showed four different peaks at 484, 575, 594, and 617 nm. Also the Commission International de l'Eclairage (CIE) chromaticity coordinates of this phosphor were (0.35, 0.31), and these were found to be close to the standard white-light chromaticity coordinates (0.33, 0.33).

The PL properties of $CaAl(SO_4)_2Br:RE$ (RE = Dy^{3+}, Ce^{3+}, Eu^{3+}) phosphors, synthesized by the wet-chemical method, were investigated recently [55]. Further, Eu^{3+}-activated $MAl(SO_4)_2Br$ phosphors (M = Mg or Sr), prepared using a wet-chemical reaction technique, were also studied and were found to have suitable applications as efficient red-emitting phosphors [56].

4.3.3 Sulfides and Oxysulfides

The luminescence characteristics of sulfide-based phosphors often degrade due to their thermal instability and sensitivity to moisture [57]. Sulfides have a tendency to decompose into carbonates or sulfates when they are exposed to moisture. Also, high-energy UV or electron beams pose a great threat of degrading sulfides, which ultimately may result in the loss of genuine luminescence properties of sulfide phosphors. However, such problems do not hamper the chances of using sulfides for LED applications, since these problems can be avoided by proper seal adhesion and encapsulation within the LED module. On the other hand, rare earth oxysulfides exhibit high chemical and thermal stability, in addition to high absorption of light and efficient energy transfer, thereby making them feasible optical functional materials [58].

Alkali earth sulfide phosphors such as CaS and SrS doped with Ce^{3+} or Eu^{2+} ions have strong absorption in the blue region, making them excellent candidates for LED applications that employ blue-LED pumping. $SrS:Eu^{2+}$ and $CaS:Eu^{2+}$ show strong broad-band emission in the red region, with their peaks centered at 606 nm and

648 nm, respectively. On introducing Ce^{3+} ions as sensitizers, these phosphors were able to exhibit emission enhanced by 28% and 18%, respectively. When SrS and CaS were individually doped with Ce^{3+} ions, PL emissions were obtained at 482 nm (blue) and 515 nm (green) [59].

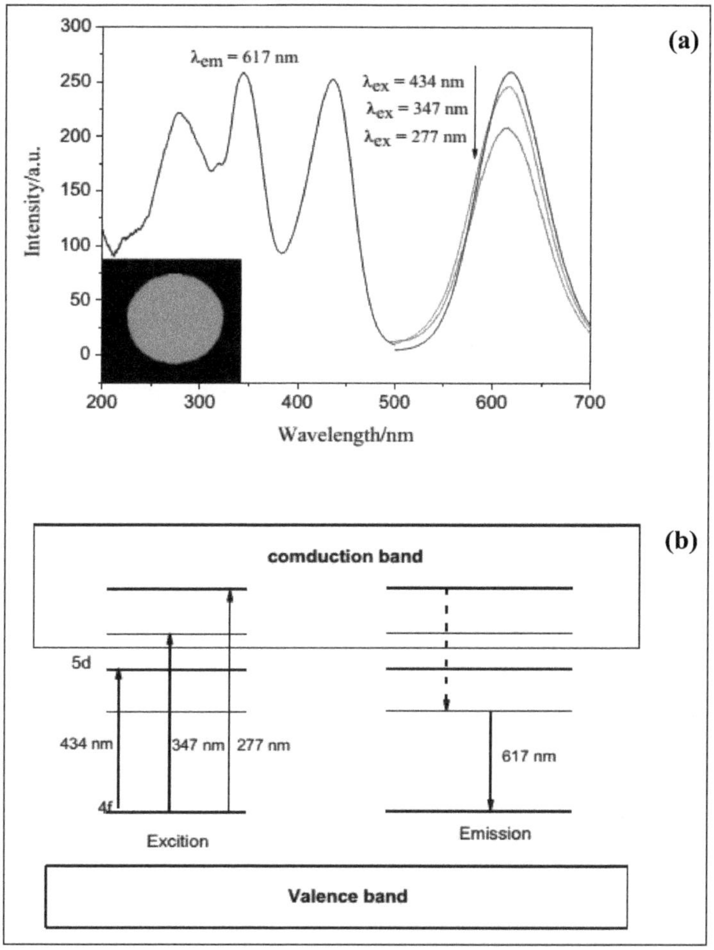

Figure 4.6 (a) PL emission and excitation spectra of BaS:0.1 mol% Yb^{2+}, and (b) luminescent mechanism for BaS:Yb^{2+}. Reprinted from Ref. [60], Copyright (2014), with permission from Elsevier.

Lots of investigation has been carried out on the luminescence properties of Eu^{2+}-doped sulfides. On the other hand, sparse reports can be found on the luminescence spectra of Yb^{2+} ions. However, Yb_2O_3 is much cheaper than Eu_2O_3, and hence, sulfides doped with Yb^{2+} ions must have a greater potential application than Eu^{2+}-doped sulfides in the lighting industry. BaS:Yb^{2+} phosphors are efficient orange-emitting sulfide phosphors with an energy gap of 3.57 eV [60]. When the concentration of Yb^{2+} is 0.1%, the emission spectrum shows a broad band extending from 500 nm to 700 nm, with the maximum intensity at 617 nm. The excitation spectrum monitored at 617 nm consists of three bands peaking at 277, 347, and 434 nm. The PL excitation and emission spectra of this phosphor along with a model presenting its mechanism is depicted in Fig. 4.6.

4.4 Conclusions

Although sulfate- and sulfide-based phosphors are highly efficient, they suffer from a disadvantage due to their possibility to degrade because of their sensitivity to moisture and thermally instability. Definitely, the best phosphors for radiation dosimetry are sulfates. Sulfate-based phosphors have undoubtedly shown excellent characteristics required for radiation dosimetry. Oxysulfates such as $Y_2O_2SO_4$ and $Gd_2O_2SO_4$ doped with rare earth ions were found to be better suited for lamp phosphors. Halosulfate phosphors have also exhibited some good luminescence properties. However, they have a tendency to absorb moisture, thereby diminishing their luminescence properties, and this makes it difficult to store them properly.

References

1. Shinde, K., Pawade, V., Dhoble, S. and Hakeem, A. (2014). Impact of temperature on $Na_2Sr(PO_4)F:Eu^{3+}$ phosphor, *Luminescence*, **29**(4), pp. 352–356.
2. Bhagat, S., Borghate, S. V., Kalyani, N. T. and Dhoble, S. J. (2014). Novel Na^+ doped Alq_3 hybrid materials for organic light-emitting diode (OLED) devices and flat panel displays, *Luminescence*, **30**, pp. 251–256.
3. Oza, A. H., Dhoble, N. S., Lochab, S. P. and Dhoble, S. J. (2015). Luminescence study of Dy or Ce activated $LiCaBO_3$ phosphor for γ-ray and C^{5+} ion beam irradiation, *Luminescence*, **30**, pp. 967–977.

4. Kore, B. P., Dhoble, N. S. S., Lochab, S. P. P. and Dhoble, S. J. (2014). Photoluminescence and thermoluminescence study of $KCaSO_4Cl$ doped with Dy and Ce synthesized by acid distillation method, *J. Lumin.*, **145**, pp. 299–306.

5. Pimpalshende, D. M. and Dhoble, S. J. (2015). Stability of luminescence in $LaPO_4$, $LaPO_4$:RE^{3+} (RE = Dy, Eu) nanophosphors, *Luminescence*, **30**(2), pp. 144–154.

6. Bajpai, N., Tiwari, A., Khan, S. A., Kher, R. S., Dhoble, S. J. and Bramhe, N. (2014). Effects of rare earth ions (Tb, Ce, Eu, Dy) on the thermoluminescence characteristics of sol-gel derived and γ-irradiated SiO_2 nanoparticles, *Luminescence*, **29**, pp. 669–673.

7. Oza, A. H., Dhoble, N. S., Park, K. and Dhoble, S. J. (2014). Synthesis and thermoluminescence characterizations of $Sr_2B_5O_9Cl:Dy^{3+}$ phosphor for TL dosimetry, *Luminescence*, **30**, pp. 768–774.

8. Sahare, P. D., Bakare, J. S., Dhole, S. D. and Kumar, P. (2012). Effect of phase transition and particle size on thermoluminescence characteristics of nanocrystalline $K2Ca_2(SO_4)_3:Cu^+$ phosphor, *Radiat. Meas.*, **47**(11–12), pp. 1083–1091.

9. Kearfott, K. J., Geoffrey, W. and Ra, M. (2015). The optically stimulated luminescence (OSL) properties of LiF:Mg,Ti, $Li_2B_4O_7$:Cu, $CaSO_4$:Tm, and CaF2:Mn thermoluminescent (TL) materials, *Appl. Radiat. Isot.*, **99**, pp. 155–161.

10. Nayar, V., Chowdhary, P. S., Bhujbal, P. M. and Dhoble, S. J. (2011). Studies on the mass and temperature dependence of lyoluminescence intensity of microcrystalline powder of KCl, *Luminescence*, **26**(5), pp. 324–330.

11. Sahu, A. K., Kadukar, M. R., Chowdhary, P. S., Nayar, V. and Dhoble, S. J. (2014). Experimental and theoretical study of mechanoluminescence and lyoluminescenc=e of Li_3PO_4:RE (RE=Dy and Tb) phosphors, *Luminescence*, **29**, pp. 1082–1094.

12. Sahu, A. K., Kore, B. P., Chowdhary, P. S., Nayar, V. and Dhoble, S. J. (2014). Systematic study of photoluminescence, lyoluminescence and mechanoluminescence in Ce^{3+} - and Eu3+ -activated Li_3PO_4 phosphors, *Luminescence*, **29**, pp. 58–64.

13. Kim, J. S., Kwon, Y. N., Shin, N. and Sohn, K. S. (2005). Visualization of fractures in alumina ceramics by mechanoluminescence, *Acta Mater.*, **53**(16), pp. 4337–4343.

14. Lakshmanan, A. R. (1999). Photoluminescence and thermostimulated luminescence processes in rare-earth-doped $CaSO_4$ phosphors, *Prog. Mater. Sci.*, **44**, pp. 1–187.

15. Atone, M. S., Dhoble, S. J., Moharil, S. V, Dhopte, S. M., Muthal, P. L. and Kondawar, V. K. (1993). Sensitization of Luminescence of $CaSO_4$: Dy, *Phys. Status Solidi A*, **299**, pp. 299–305.
16. Wani, J. A., Dhoble, N. S. and Dhoble, S. J. (2014). Mixed influence of copper and some f-block elements on thermoluminescence intensity of $CaSO_4$: Dy, P phosphors, *Adv. Mater. Lett.*, **5**(1), pp. 52–58.
17. Bernal, R., García-Haro, A. R., Machi, L., Brown, F., Pérez-Salas, R., Castaño, V. M. and Cruz-Vázquez, C. (2008). Advances in the synthesis of new Europium doped $CaSO_4$ phosphors and their thermoluminescence characterization, *Radiat. Meas.*, **43**(2–6), pp. 371–374.
18. Mandlik, N., Sahare, P. D., Kulkarni, M. S., Bhatt, B. C., Bhoraskar, V. N. and Dhole, S. D. (2014). Study of TL and optically stimulated luminescence of $K_2Ca_2(SO_4)_3$:Cu nanophosphor for radiation dosimetry, *J. Lumin.*, **146**, pp. 128–132.
19. Bahl, S., Lochab, S. P., Pandey, A., Aleynikov, V. E., Molokanov, A. G. and Kumar, P. (2012). Thermoluminescence and photoluminescence study on 150MeV proton beam irradiated $K_2Ca_2(SO_4)_3$:Eu phosphor, *Radiat. Phys. Chem.*, **81**(11), pp. 1683–1687.
20. Sahare, P. D., Ranjan, R., Salah, N. and Lochab, S. P. (2007). $K_3Na(SO_4)_2$: Eu nanoparticles for high dose of ionizing radiation, *J. Phys. D: Appl. Phys.*, **40**(3), pp. 759–764.
21. Bahl, S., Lochab, S. P. P., Pandey, A., Kumar, V., Aleynikov, V. E. E., Molokanov, A. G. G. and Kumar, P. (2014). Characterization and luminescence studies of Eu doped Barite nanophosphor, *J. Lumin.*, **149**, pp. 176–184.
22. Nikhare, G. N., Gedam, S. C. and Dhoble, S. J. (2015). Luminescence characteristics of $Na_3Ca_2(SO_4)_3F:Ce^{3+}$ fluoride material, *Luminescence*, **30**(5), pp. 703–706.
23. Choubey, S. R., Gedam, S. C. and Dhoble, S. J. (2015). Effect of γ-radiation on thermoluminescence in rare earths doped $NaMgSO_4Cl$ material, *Radiat. Eff. Defects Solids*, **170**(1), pp. 7–17.
24. Choubey, S. R., Gedam, S. C. and Dhoble, S. J. (2014). Study of luminescence and effect of Dy^{3+} on $NaMgSO_4Cl$:Ce chlorosulphate phosphor, *Luminescence*, **29**(8), pp. 1199–1201.
25. Choubey, S. R., Gedam, S. C. and Dhoble, S. J. (2013). PL and TL study of $NaMgSO_4F$:X (X=Dy or Ce) fluoride phosphor by SSD method, *J. Lumin.*, **142**, pp. 48–51.
26. Poddar, A., Gedam, S. C. and Dhoble, S. J. (2015). Photo and thermoluminescence of $KMgSO_4F$:Ce and :Mn phosphors, *Luminescence*, **30**(4), pp. 425–431.

27. Gedam, S. C. and Dhoble, S. J. (2013). Effect of γ-radiation on $NaZnSO_4Cl$:Ce, Dy or Mn ions and retrapping of charges in TL, *J. Lumin.*, **142**, pp. 139–143.

28. Kore, B. P., Dhoble, N. S. and Dhoble, S. J. (2014). Glow curve analysis of beta-particles irradiated $Na_{21}Mg(SO_4)_{10}Cl_3$:Dy phosphor, *J. Lumin.*, **145**, pp. 888–894.

29. Nair, G. B. and Dhoble, S. J. (2015). A perspective perception on the applications of light-emitting diodes, *Luminescence*, **30**, pp. 1167–1175.

30. Tamboli, S., Kadam, R. M. and Dhoble, S. J. (2016). Photoluminescence and electron paramagnetic resonance properties of a potential phototherapic agent: $MMgF_4:Gd^{3+}$ (M = Ba, Sr) sub-microphosphors, *Luminescence*, **31**(7), pp. 1321–1328.

31. Tamboli, S., Rajeswari, B. and Dhoble, S. J. (2015). Investigation of UV-emitting Gd^{3+} -doped $LiCaBO_3$ phosphor, *Luminescence*, **31**, pp. 551–556.

32. Pimpalshende, D. M. and Dhoble, S. J. (2014). Synthesis and characterization of diphenyl quinoline and bromine-activated diphenyl quinoline organic phosphors, *Luminescence*, **29**, pp. 451–455.

33. Manik, U., Gedam, S. C. and Dhoble, S. J. (2015). Photoluminescence of Dy^{3+} activator sensitized by Ce^{3+} in $KNaSO_4$ microphosphor, *Luminescence*, **30**(6), pp. 910–913.

34. Shinde, K. N. and Dhoble, S. J. (2014). Energy transfer in $M_5(PO_4)_3$ $F:Eu^{2+},Ce^{3+}$ (M = Ca and Ba) phosphors, *Luminescence*, **29**(5), pp. 549–552.

35. Nikhare, G. N., Gedam, S. C. and Dhoble, S. J. (2015). Luminescence in $Sr_4Al_{14}O_{25}$: Ce^{3+} aluminate phosphor, *Luminescence*, **30**, pp. 163–167.

36. Kongre, V. C., Gedam, S. C. and Dhoble, S. J. (2013). Photoluminescence and thermoluminescence characteristics of $BaCa(SO_4)_2$:Ce mixed alkaline earth sulfate, *J. Lumin.*, **135**, pp. 55–59.

37. Kalyani, N. T., Atram, R. G. and Dhoble, S. J. (2014). Synthesis and characterization of $Y_{(1-x)}Eu_{(x)}(TTA)_3(Phen)$ organic luminescent thin films, *Luminescence*, **29**(6), pp. 674–678.

38. Raju, S. H., Sudhakar, B. M., Reddy, B. S., Dhoble, S. J., Thyagarajan, K. and Raju, C. N. (2014). Luminescence properties of barium - gadolinium-titanate ceramics doped with rare-earth ions (Eu3+ and Tb3+), *Luminescence*, **29**, pp. 861–867.

39. Lian, J., Sun, X., Li, J.-G., Xiao, B. and Duan, K. (2010). Characterization and optical properties of $(Gd_{1-x}, Pr_x)_2O_2S$ nano-phosphors synthesized

using a novel co-precipitation method, *Mater. Chem. Phys.*, **122**(2–3), pp. 354–361.

40. Srivastava, A. M., Setlur, A. A., Comanzo, H. A., Gao, Y., Hannah, M. E., Hughes, J. A. and Happek, U. (2008). Optical spectroscopy and thermal quenching of the Ce^{3+} luminescence in yttrium oxysulfate, *Opt. Mater.*, **30**, pp. 1499–1503.
41. Shoji, M. and Sakurai, K. (2006). A versatile scheme for preparing single-phase yttrium oxysulfate phosphor, *J. Alloys Compd.*, **426**(1–2), pp. 244–246.
42. Song, L., Shao, X., Du, P., Cao, H., Hui, Q., Xing, T. and Xiong, J. (2013). A facile preparation and the luminescent properties of Eu3+-doped $Y_2O_2SO_4$ nanopieces, *Mater. Res. Bull.*, **48**(11), pp. 4896–4900.
43. Song, L., Du, P., Jiang, Q., Cao, H. and Xiong, J. (2014). Synthesis and luminescence of high-brightness $Gd_2O_2SO_4:Tb^{3+}$ nanopieces and the enhanced luminescence by alkali metal ions co-doping, *J. Lumin.*, **150**, pp. 50–54.
44. Lian, J., Sun, X., Liu, Z., Yu, J. and Li, X. (2009). Synthesis and optical properties of $(Gd_{1-x},Eu_x)_2O_2SO_4$ nano-phosphors by a novel co-precipitation method, *Mater. Res. Bull.*, **44**(9), pp. 1822–1827.
45. Lian, J., Sun, X. and Li, X. (2011). Synthesis, characterization and photoluminescence properties of $(Gd_{1-x},Eu_x)_2O_2SO_4$ sub-microphosphors by homogeneous precipitation method, *Mater. Chem. Phys.*, **125**(3), pp. 479–484.
46. Manigandan, R., Giribabu, K., Suresh, R., Munusamy, S., Praveen kumar, S., Muthamizh, S., Dhanasekaran, T., Padmanaban, A. and Narayanan, V. (2014). Synthesis, growth and photoluminescence behaviour of $Gd_2O_2SO_4:Eu^{3+}$ nanophosphors: the effect of temperature on the structural, morphological and optical properties, *RSC Adv.*, **5**(10), pp. 7515–7521.
47. Lian, J., Liang, P., Wang, B. and Liu, F. (2014). Homogeneous precipitation synthesis and photoluminescence properties of $La_2O_2SO_4:Eu^{3+}$ quasi-spherical phosphors, *J. Ceram. Proc. Res.*, **15**(6), pp. 382–388.
48. Wang, X., Li, J.-G., Zhu, Q., Li, X., Sun, X. and Sakka, Y. (2014). Synthesis, characterization, and photoluminescent properties of $(La_{0.95}Eu_{0.05})_2O_2SO_4$ red phosphors with layered hydroxyl sulfate as precursor, *J. Alloys Compd.*, **603**, pp. 28–34.
49. Kijima, T., Shinbori, T., Sekita, M., Uota, M. and Sakai, G. (2008). Abnormally enhanced Eu^{3+} emission in $Y_2O_2SO_4:Eu^{3+}$ inherited from their precursory dodecylsulfate-templated concentric-layered nanostructure, *J. Lumin.*, **128**(3), pp. 311–316.

50. Liu, G., Hong, G., Wang, J. and Dong, X. (2007). Hydrothermal synthesis of spherical and hollow $Gd_2O_3:Eu^{3+}$ phosphors, *J. Alloys Compd.*, **432**(1–2), pp. 200–204.

51. Uhlich, D., Plewa, J. and Jüstel, T. (2008). Phase formation and characterization of $Sr_3Y_2Ge_3O_{12}$, $Sr_3In_2Ge_3O_{12}$, and $Ca_3Ga_2Ge_3O_{12}$ doped by trivalent europium, *J. Lumin.*, **128**(10), pp. 1649–1654.

52. Shinde, N., Dhoble, N. S., Gedam, S. C. and Dhoble, S. J. (2015). Photoluminescence enhancement in $Na_3SO_4Cl:X$ (X = Ce^{3+}, Eu^{3+} or Dy^{3+}) material, *Luminescence*, **30**, pp. 898–903.

53. Baig, N., Dhoble, N. S., Pawade, V. B., Kokode, N. S. and Dhoble, S. J. (2014). Photoluminescence in $K_3Ca_2(SO_4)_3Cl$-doped Eu^{3+} phosphor, *Luminescence*, **29**, pp. 695–697.

54. Baig, N., Dhoble, N. S., Park, K., Kokode, N. S. and Dhoble, S. J. (2014). Enhanced luminescence and white light emission from Eu^{3+} -co-doped $K_3Ca_2(SO_4)_3Cl:Dy^{3+}$ phosphor with near visible ultraviolet excitation for white LEDs, *Luminescence*, **30**, pp. 479–484.

55. Deshmukh, P. B., Puppalwar, S. P., Dhoble, N. S. and Dhoble, S. J. (2014). Optical properties of $CaAl(SO_4)_2$ Br:RE (RE = Dy, Eu, Ce) novel phosphors, *Luminescence*, **29**, pp. 609–613.

56. Deshmukh, P. B., Puppalwar, S. P., Dhoble, N. S. and Dhoble, S. J. (2015). Luminescent properties of $MAl(SO_4)_2Br:Eu^{3+}$ (M = Sr or Mg) red phosphors for near-UV light-emitting diodes, *Luminescence*, **30**(1), pp. 118–121.

57. Nair, G. B. and Dhoble, S. J. (2015). Highly enterprising calcium zirconium phosphate [$CaZr_4(PO_4)_6:Dy^{3+},Ce^{3+}$] phosphor for white light emission, *RSC Adv.*, **5**, pp. 49235–49247.

58. Mikami, M. and Oshiyama, A. (1998). First-principles band-structure calculation of yttrium oxysulfide, *Phys. Rev. B*, **57**(15), pp. 8939–8944.

59. Jia, D. and Wang, X. J. (2007). Alkali earth sulfide phosphors doped with Eu^{2+} and Ce^{3+} for LEDs, *Opt. Mater.*, **30**(3), pp. 375–379.

60. Yang, Y., Su, X., Li, X., Yu, F., Mi, C. and Li, G. (2014). A novel orange emitting $BaS:xYb^{2+}$ phosphor for white light LEDs, *Mater. Res. Bull.*, **51**, pp. 202–204.

Chapter 5

Luminescent Down-Conversion Materials as Spectral Convertors for Photovoltaic Applications

Swati Bishnoi, Vinay Gupta, and D. Haranath
Advanced Materials and Devices Division, CSIR-National Physical Laboratory, Dr. K. S. Krishnan Road, New Delhi 110012, India
haranath@nplindia.org

One of the most proficient methods for harnessing abundant solar energy and converting it to electrical energy is through photovoltaic (PV) conversion. Though the concept of capturing solar energy is not new, this technology is still not cost-effective due to the poor power conversion efficiencies (PCEs) of existing solar cells. The most critical factor hindering the improvement in PCE is the so-called spectral mismatch between the incident solar photon energy distribution and the bandgap of the PV material. This has motivated global researchers to focus on the development of efficient luminescent materials that are capable of converting the broad spectrum of incident sunlight into photons of a suitable wavelength corresponding to PV cells. This can minimize significant losses such as thermalization (<550 nm) and transmission (>1100 nm) losses

Phosphors: Synthesis and Applications
Edited by Sanjay J. Dhoble, B. Deva Prasad Raju, and Vijay Singh

ISBN 978-981-4774-49-9 (Hardcover), 978-042-9460-99-9 (eBook)
www.panstanford.com

pertaining to solar-cell-based energy conversion technologies and hence can contribute to the enhancement of the overall efficiency of PV cells. In this chapter, we will be highlighting recent progress in the development of lanthanide-based down-conversion phosphor materials for spectral modification and efficiency enhancement in PV cells.

5.1 Introduction

Harnessing the abundantly available, renewable, and free solar energy has been recognized as one of the efficient ways to meet the rapidly growing energy demands and also concerns about increasing greenhouse gas emissions in the environment by increasing fossil fuel consumption. Therefore, solar energy harvesting is a leading research area for environmental safety and supply of clean renewable energy. A wide range of methods are proposed and studied extensively for renewable solar energy harvesting, including solar–thermal technology, photocatalytic and photoelectrochemical water splitting, and photovoltaic (PV) cells [1–6]. It is worth noting that the solar energy reaching the earth's surface is 10^3 times more than what we are able to harness with the existing PV technologies [7]. Thus, solar energy is considered a potential source to meet a large fraction of future energy requirements if we are able to harness it to its fullest extent. Despite significant advancements in the area of PVs during the past two decades, efficient solar energy conversion through PV cells remains a grand challenge. For maximizing solar energy harvesting, efforts are needed to be concentrated on the development of cost-effective and efficient PV devices. Currently, among all PV cells available, silicon, inorganic semiconductor, and thin-film solar cells are dominating the market, due to their high stability and power conversion efficiency (PCE). In recent years, polymer solar cells (PSCs) have also emerged as a potential alternative to the commercial silicon-based solar cells, owing to their advantages of being flexible, low cost, and environment friendly [8, 9]. Although significant improvement in the PCE of existing PV devices has been made, many issues are yet to be resolved for improving device performance. Experts have suggested that for achieving further enhancement in the PCE of existing solar cells, spectral conversion of the incident solar light is the best alternative

[10, 11]. This is because the major factor responsible for limiting the conversion efficiency of PV cells is their insensitivity to the complete solar spectrum. The spectral distribution of sunlight reaching the earth's surface (AM 1.5G) comprises photons of wavelengths ranging from UV to IR (280–2500 nm, 0.5–4.4 eV), as shown in Fig. 5.1a.

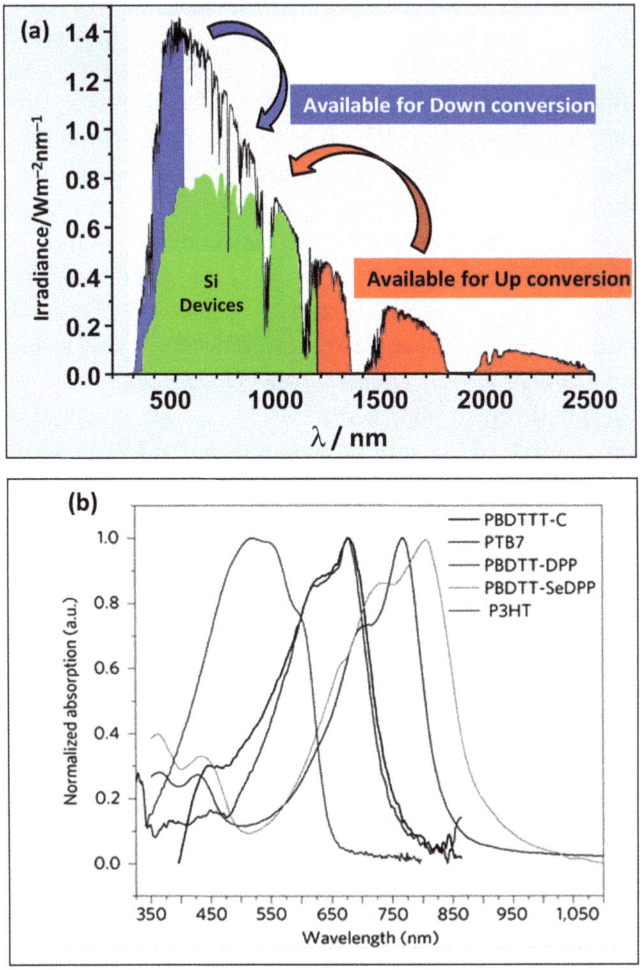

Figure 5.1 (a) AM 1.5G spectrum showing the fraction (highlighted in green) absorbed by a Si-based PV cell. The spectral regions marked as blue and red are available for utilization by down- and up-conversion processes. (b) Spectra showing the absorption range of commonly used polymers in organic solar cells. Reproduced with permission from Ref. [12], Copyright © 2015 Nature Publishing Group.

But, only solar photons with energy higher than the bandgap are absorbed, and excess energy is released as heat. This thermalization along with nonabsorption (transmission) losses, that is, photons with energy less than the bandgap, accounts for approximately 50% loss of the incident solar energy in silicon-based solar cell conversion to electricity. However, the maximum theoretical efficiency calculated for crystalline silicon (c-Si) with bandgap energy (E_g) of 1.1 eV is approximately 31% or 41%, depending on the concentration ratio [13]. This is also designated as the Shockley–Queisser limit [13]. Similarly, in organic solar cells (OSCs), the various polymers used for fabrication of an active layer (which plays a major role in light absorption and exciton generation) have the maximum absorption in the visible and near-infrared (NIR) regions, as shown in Fig. 5.1b. Hence, in this case, excess photon energy of UV light is also released as heat. These inherent losses of current PV devices like thermalization and nonabsorption losses can be resolved by employing luminescent materials (phosphors) as spectral convertors. This can be achieved by means of three luminescence processes: up-conversion (UC), quantum cutting (QC), and down-shifting (DS). The photon UC process involves absorption of two or more lower-wavelength photons and emission of a photon of shorter wavelength than the incident photon. DC is the process by which a high-energy photon, which is inefficiently absorbed by the PV cell due to thermalization losses, is converted to a lower-energy photon. In QC a single high-energy photon is converted into multiple low-energy photons. DS is similar to QC, but its conversion efficiency does not exceed 100%.

5.2 Luminescent Materials (Phosphors) as Spectral Convertors

Luminescence is a process that corresponds to the emission of electromagnetic radiation beyond thermal equilibrium. Luminescence can be generated from different types of energy sources, including high-energy photons, electromagnetic radiations, electric fields, X-rays, and charged particles from radioactive decay. Depending upon the nature of the excitation source, luminescence can be subdivided into several categories, like in the case of photoexcitation, the luminescence is called *photoluminescence*. A phosphor is generally a substance that exhibits the phenomenon of luminescence. A typical inorganic phosphor consists of a host lattice doped with a few mol% or less of an activator ion, such as a lanthanide ion, which is responsible for the luminescence. Lanthanide ions are the most widely explored candidates for achieving spectral conversion because of their rich and unique energy level structure arising from the $4f^n$ inner shell configuration. With a partially filled 4f shell, where the number of electrons (n) ranges from 0 to 14, they possess a large number of different energy levels that are labeled by the so-called term symbols; transitions corresponding to these energy levels lead to sharp emission lines. They also exhibit the energy transfer (ET) mechanism and hence help in incident photon energy conversion. ET from one rare earth ion (or the host lattice) to another rare earth ion is a sequential phenomenon involving (i) charge carrier migration (electrons and holes), (ii) exciton migration, (iii) resonance between atoms with sufficient overlap integrals, and (iv) reabsorption of photons emitted by another activator ion or sensitizer, as shown in Fig. 5.2a–c.

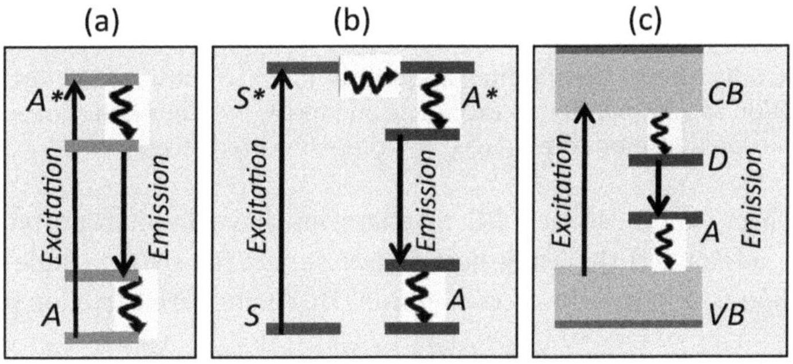

Figure 5.2 Luminescence mechanisms: (a) emission from a luminescence activator upon excitation, (b) sensitized emission from an activator through ET from a sensitizer to the activator upon excitation of the sensitizer, and (c) emission from a semiconductor after band-to-band excitation. A and A* represent the ground and excited states of the activator, respectively. S and S* represent the ground and excited states of the sensitizer, respectively. VB and CB represent the valence and conduction bands of the semiconductor, while D and A represent the donor and acceptor energy levels, respectively.

ET is often required to overcome the small absorption coefficients of the rare earth ions, and energy absorbed in the host material can be transferred to the desired activator. However, this may not be a one-way process, and energy already absorbed in a rare earth ion can also migrate to host lattice defects, where it can recombine nonradiatively (Fig. 5.3). Therefore, luminescent host materials should be highly crystalline and have as few lattice defects and impurities as possible, and the surface area of the crystals should be minimized.

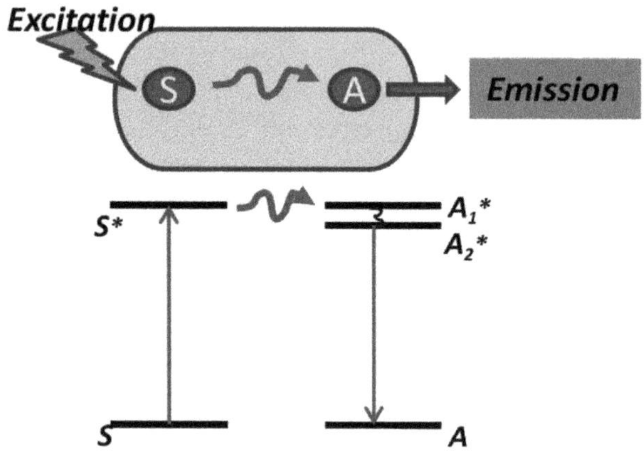

Figure 5.3 Energy transfer from a sensitizer (S) to an activator (A). The S–S* transition corresponds to the excitation, and the A_2*–A transition corresponds to emission transition. The A_1*–A_2* transition is nonradiative.

Three luminescence DC mechanisms have been reported till date, namely (i) QC using host lattice states, (ii) QC on single rare earth ions, and (iii) down-conversion (DC) using rare earth ion pairs.

5.3 Down-Conversion Mechanisms

5.3.1 Quantum Cutting Using Host Lattice States

QC using host lattice states is the process in which an electron–hole pair with an energy at least twice the energy of the bandgap of the host material results in two electron–hole pairs, both with the bandgap energy (Fig. 5.4). This process is known as impact ionization (II). This process has great potential for improving the efficiency of solar cells [14].

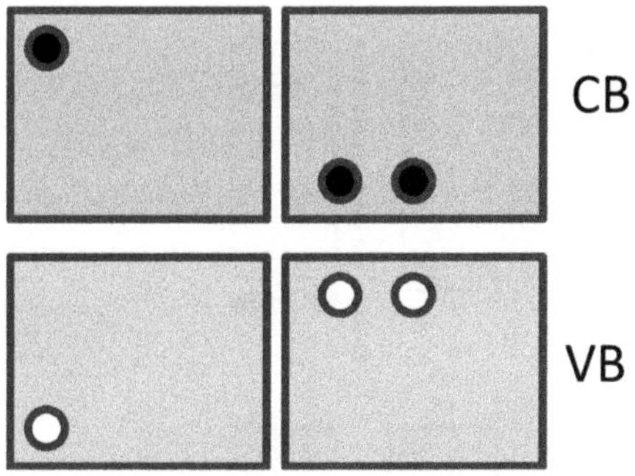

Figure 5.4 Host lattice–based quantum-cutting phenomenon. One electron–hole pair with an energy at least twice the energy of the bandgap of the host material results in two electron–hole pairs, both with the bandgap energy.

Theoretically, for II to occur, the threshold energy required should be either equal to 1 or 1.5 times the bandgap of the semiconductor for a conduction band (CB) electron to be able to excite another electron from the valence band (VB), depending on the effective masses of electrons and holes. In the case of silicon ($E_g \approx 1.12$ eV), the generation of a secondary electron–hole pair is not observed for an initial electron energy less than an E_{II} of 2.6 eV [15]. Correspondingly, the gains observed in the internal quantum efficiency (IQE) of silicon solar cells by the II process are comparatively low; at $\lambda_{exc} = 350$ nm the value of the IQE was 105%, and at $\lambda_{exc} = 250$ nm the IQE was 150% [16, 17]. Various alternatives have been reported to improvise the IQE, like bandgap engineering, minimizing nonradiative processes, and fabrication of semiconducting nanocrystal–based host materials [18, 19]. An experimental evidence of this has been reported experimentally using lead selenide (PbSe) nanocrystals [20, 21]. An extension of the II phenomenon is the Auger recombination mechanism. Traditionally, Auger recombination is viewed as a three-particle interaction where a CB electron and a VB hole recombine, with the excess energy being transferred to a third free electron or hole [21]. This mechanism is host lattice mediated, and it occurs when the excitation energy of the incident photon, E_{exc}, is more than $2E_g$, such that two electron–hole pairs each with energy approximately equal to E_g are generated. In the interband Auger effect (Fig. 5.5), an electron is excited from the VB to the far CB such that excess energy can excite another electron over the bandgap.

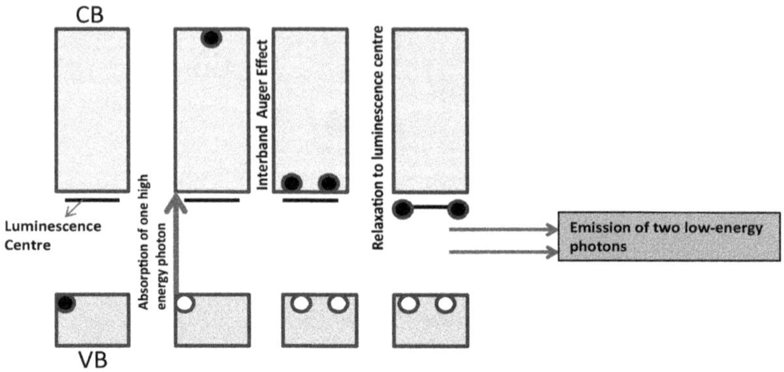

Figure 5.5 Interband Auger mechanism for the generation of two low-energy photons from one high-energy incident photon within a host doped with luminescent centers.

As a result, the resultant electron–hole pair recombines at the luminescent center, thus yielding two emitted photons for every photon absorbed. For lighting applications such as lamps, luminescent materials should fulfill other criteria:

- The host material should necessarily have a bandgap larger than 3.0 eV so that it is transparent to visible light.
- Moreover, the exciting photon energy should be at least 6.0 eV, for example, the Xe discharge radiation source with emission in the spectral range of 6–8 eV.

Certain rare earth–based luminescent materials, like europium-doped yttrium oxide ($Y_2O_3:Eu^{3+}$), exhibit high visible external quantum efficiencies (EQEs) of up to 240%, as they require extremely high excitation energy ~23 eV [22]. Surface recombination effects are responsible for the limited performance of phosphors in the near-bandgap region [23]. Moreover, at a lower penetration depth of the incident radiation, nonradiative recombination rates enhance and at about $2.5E_g$ a minimum EQE is observed, whereas for $E_{exc} > 2.5E_g$ the EQE increases with increasing photon energy due to Auger interband transitions [22].

5.3.2 Quantum Cutting on Single Rare Earth Ions

The second QC mechanism relies on single rare earth ions with well-separated energy levels that can be used to generate more than one visible photon out of one absorbed UV photon (Fig. 5.6a). Preliminary work on this was reported by two groups in the 1970s [24, 25] using praseodymium (Pr^{3+}) as the activator ion in a YF_3 host. Representative examples have been reported for single ions like Ho^{3+}, Tm^{3+}, and Er^{3+} [26–30]. However, the IQE of the phosphor cannot exceed 100%, due to the unwanted emissions in the IR and UV ranges and nonradiative recombination that competes with the desired emission of two visible photons.

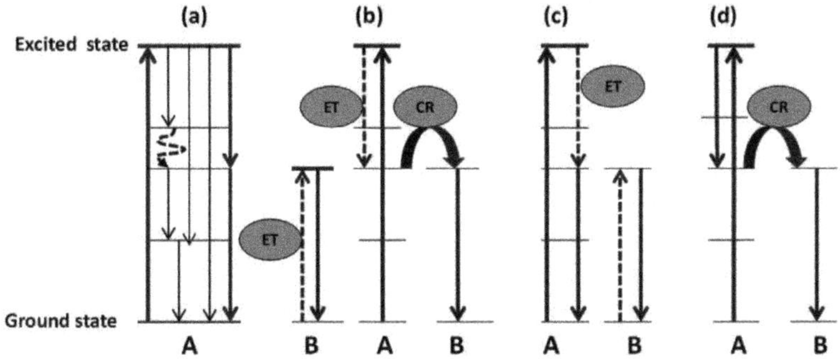

Figure 5.6 Quantum-cutting mechanisms for rare earth ions: (a) two-photon emission from a single ion; down-conversion with ion pairs by (b) cross-relaxation from ion A to ion B (A) and energy transfer from ion A to ion B (B) with emission from ion B; (c, d) cross-relaxation followed by emission from ions A and B. Reprinted from Ref. [31], Copyright (2000), with permission from Elsevier.

5.3.3 Down-Conversion Using Rare Earth Ion Pairs

The third QC mechanism is based on physically interacting rare earth ions: an ion pair showing the cascade emission. In such materials, one of the ions in the pair shows QC and transfers part of its energy to the other ion, thus minimizing the undesirable UV and IR emissions. By interaction with the other ion in the pair, part of the energy is transferred to this latter ion. In this manner, the previous IR and UV losses encountered with a single ion can be avoided by selecting the appropriate ions. The two-photon emission process has been abbreviated as DC in the rest of the chapter. Figure 5.6b–d shows the energy level diagrams for three DC mechanisms involving ET between rare earth ion pairs (A and B). Ion A exhibits emission from a high-energy level, while ion B supports ET. Figure 5.6b indicates the two-photon emission process by a cross-relaxation mechanism from ions A to B (denoted by CR) and ET from ions A to B (denoted by ET) with emission from ion B.

5.4 Down-Shifting

DS or photoluminescence is similar to DC, but the quantum efficiency is lower than unity, owing to the nonradiative relaxations. DS can be employed to overcome the poor blue response of solar cells due to, for example, noneffective front surface passivation for silicon solar cells. In 2005 a research group proposed that shifting the incident spectrum to wavelengths where the IQE of the solar cell is higher can effectively enhance the overall conversion efficiency by ~10% [11].

The DS materials absorb short-wavelength light, typically in the 300–500 nm range, and re-emit it at a longer wavelength where the EQE of the PV device is high. While luminescent DS could potentially enhance the solar cell efficiency, it is important to note that the design will not be able to overcome the Shockley–Queisser efficiency limit, as the absorption of a high energy photon by the DS materials can only result in the generation of one electron–hole pair in the solar cell. To date, two main classes of DS-based PV devices have been investigated: luminescent solar concentrators and planar DS layers. A luminescent solar concentrator is used to concentrate sunlight onto PV cells as a simple means to reduce an amount of expensive PV materials required. The concentrator, typically composed of polymer sheets doped with luminescent species, converts a fraction of the incident sunlight into longer-wavelength light that is subsequently guided toward a PV cell. On the other hand, a planar DS layer is generally placed directly onto the front surface of a solar cell to improve the device performance by overcoming the poor spectral response of the solar cell to short-wavelength light.

5.5 Down-Conversion Mechanisms for PV Applications

All the three aforementioned DC mechanisms based on host lattice states, single ions, and ion pairs have potential scope in enhancing

PV cell performance. Luminescent DC can be utilized in many PV devices that exhibit a poor spectral response to short-wavelength light. Such DC materials can absorb the high-energy photon (300–500 nm) and re-emit a longer-wavelength photon to which the PV device is more sensitive. Various type of DC materials has been investigated for this purpose, like quantum dots (QDs), organic dyes, rare earth ions/complexes, and luminescent glasses.

5.5.1 Down-Conversion in Si-Based PV Cells

This concept of spectral DC has been investigated in wafer-based silicon (c-Si, mc-Si, and Si ribbon) devices as they are dominating the PV market. The c-Si solar cells show significant differences in spectral response for short-wavelength (UV–blue) and long-wavelength (red–NIR) sunlight (Fig. 5.1a), making them ideal systems for exploring the DC process. Figure 5.7 shows a schematic design of a DC layer placed above a solar cell. The layer absorbs a high-energy photon and converts it into a longer-wavelength photon where the spectral response of solar cell is high. Among various Ln ions, NIR-emitting Nd^{3+} and Yb^{3+} have attained remarkable attention as DC materials for Si-based solar cells [32–35].

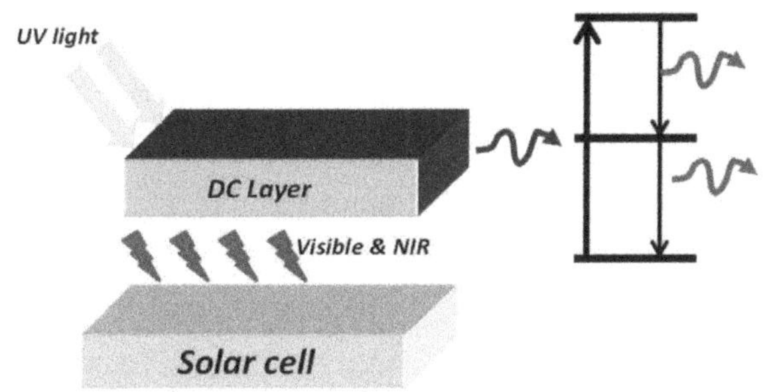

Figure 5.7 Schematic design of a solar cell device with a luminescent down-shifting layer directly placed onto the surface of the solar cell. The down-shifting layer absorbs short-wavelength photons and then re-emits them at a longer wavelength where the device exhibits a better spectral response.

Their typical NIR emission around 1000 nm is just above the band edge of c-Si where the solar cell exhibits the greatest spectral response. However, the luminescence of Nd^{3+} and Yb^{3+} is relatively weak due to parity-forbidden 4f–4f transitions [36]. To solve this problem, several research groups have used other luminescent species with higher absorption coefficients to sensitize the Nd ions [37–39]. In addition, host sensitization via ET from an excited host (such as ZnO and TiO_2) to Ln^{3+} (Ln = Nd and Yb) ions also offers an effective way to boost the luminescence signal [40–45]. As alternatives, ions such as Ce^{3+}, Eu^{2+}, Bi^{3+}, and Mn^{2+} and/or combinations with intrinsic absorption of the host itself (e.g., YVO_4 and ZnO) have also been studied for Yb^{3+} activation [46–50]. In recent years, various DC phosphors have been explored for spectral modification in silicon-based PV devices. Table 5.1 summarizes the recent phosphors and glasses that have been explored for efficiency enhancement in commercial Si solar cells.

The phosphors investigated principally convert the UV–blue radiation into red–NIR emission by utilizing Stokes-shifted photoluminescence, where the spectral response of the Si solar cell is maximum.

Table 5.1 Various phosphors and glasses explored as down-shifting layers for Si-based solar cells

Phosphor	Excitation (nm)	Emission (nm)	Reference
Y_2O_3:Eu^{3+}	250	570–700	[51]
La_2O_3:Eu^{3+}	285	570–700	[52]
YVO_4:Bi^{3+},Eu^{3+}	250–400	570–700	[53]
YAG:Ce^{3+},Cr^{3+}	400–500	688	[54]
$MgSrAl_{10}O_{17}$:Eu^{2+},Cr^{3+}	250–450	693	[55]
SrB_4O_7:Sm^{2+},Eu^{2+}	300–600	685	[56]
YAG:Ce^{3+},Nd^{3+}	400–500	888, 1064	[57]
YAG:Ce^{3+},Nd^{3+}	400–500	850–950, 1062	[58]
Cr^{3+}-Yb^{3+}-codoped fluorosilicate glass	300–700	950–1100	[37]
ZnO-$LiYbO_2$	200–400	950–1100	[41]
ZnO:Yb^{3+}-doped glass rods	200–400	950–1100	[43]
$LiYbMO_2O_8$: Pb^{2+}	300–450	950–1100	[40, 42]
$CaLaGa_3S_6O$:Yb^{3+}	250–500	950–1100	[59]

Phosphor	Excitation (nm)	Emission (nm)	Reference
$LaSi_3N_5:Yb^{3+}$	300–400	950–1100	[39]
Ag,Yb^{3+}-codoped oxyfluoride glass	300–500	950–1100	[60]
$Ba_2TiSi_2O_8$,Yb^{3+}-codoped glass	250–350	950–1100	[61]
Yb^{3+}-doped SrO-TiO_2-SiO_2	250–350	950–1100	[62]
Dy^{3+}-Tb^{3+}-codoped glass ceramics with YF_3 nanocrystals	300–500	450–650	[63]
Cu^+-doped glass	250–320	495	[64]
Dy^{3+}-doped $GdVO_4$	200–500	480–840	[26, 27, 29]
$KCaGd(PO_4)_2:Eu^{3+}$ phosphor	395	580	[65]
$YVO_4:Eu^{3+}$	250–350	580–700	[66]
$(Ba_{1-x}Eu_x)_2SiO_4$+Ag NPs	350	505	[67]
Eu^{3+}-doped $Y(OH)_3$	325	550–700	[68]

5.5.2 Down-Conversion in Dye-Sensitized Solar Cells

Dye-sensitized solar cells (DSSCs) have also attracted extensive attention as one of the promising alternatives to silicon solar cells. However, DSSCs have a maximum absorption in the visible range of the solar spectrum, which confines their PCE. Besides, for the practical use of DSSCs, chemical stability is as important as PCE. Thermal degradation of the dye or electrolyte components on UV exposure affects the chemical stability of DSSCs. An attractive alternative to resolve this stability issue is to introduce DC phosphors as a UV filter in DSSC, which absorbs the incident UV light and converts it to visible light, which is further absorbed by the dye. Schematically, three categories of experiments have been conducted on DSSCs in this direction for enhancement of solar light harvesting:

- By inserting a scattering layer on top of the TiO_2 electrode [69]
- By doping the TiO_2 photoelectrode with a rare earth ion, which modifies its VB potential and improvises the efficiency [70]

- By adding wavelength DS materials, either by depositing a layer of rare earth–doped TiO_2 nanorods (NRs) on top of the cell or by doping the cathode with a rare earth complex with an organic ligand [71]

Though DSSCs made up of wide-bandgap semiconductors have good sensitivity to short-wavelength photons, no significant improvement was expected in their case on conjugation with DC phosphors. But this was not exactly the case. Liu et al. [72] noted that DC phosphors can be efficient UV filters in DSSCs to enhance their stability. They reported an enhanced lifetime and improved efficiency 23.3% in dye-sensitized TiO_2 solar cells with a DC $LaVO_4:Dy^{3+}$ film compared to DSSCs without the DC film. The DC phosphor down-converts the UV (250–320 nm) light into visible (450–700 nm) emission, which is subsequently absorbed by the N3 dyes of the device, thus imparting stability by preventing dye degradation. Moreover, the DC film's emission lies in the visible range emission (450–700 nm), which is additionally absorbed by the N3 dye. The stability of the device with the introduction of $LaVO_4:Dy^{3+}$ luminescent film was improved, as presented in Fig. 5.8. A major limitation of using Dy^{3+}-doped $LaVO_4$ nanocrystals is that the UV absorption range is quite low; hence all UV radiations are not captured. Therefore, an ideal DC material for DSSCs should exhibit broad UV absorption spectra and their emission should overlap with the absorption range of the dye.

In a similar way, rare earth–doped materials have also been used for bandgap engineering of energy levels in DSSCs. The drawback of the DC phosphor used in the above case was its sensitivity to a narrow UV (250–320 nm) range. Ideally, a DC phosphor for DSSCs should absorb the whole UV range up to 400 nm and should re-emit photons in the range where the sensitivity of the dye is at its maximum. In 2010, Wu et al. [73] introduced europium-doped yttria ($Y_2O_3:Eu^{3+}$) into a TiO_2 film electrode in a DSSC, and the PCE was improved by a factor of 1.14 compared to that of a cell without $Y_2O_3:Eu^{3+}$ doping.

In 2011, another group reported the use of modified Sm^{3+}- and Eu^{3+}-doped TiO_2 electrodes for DSSCs, and efficiency improved by 1.58% in the case of Sm^{3+} and 0.93% in the case of Eu^{3+} compared to pure TiO_2 photoactive electrode–based DSSCs [74].

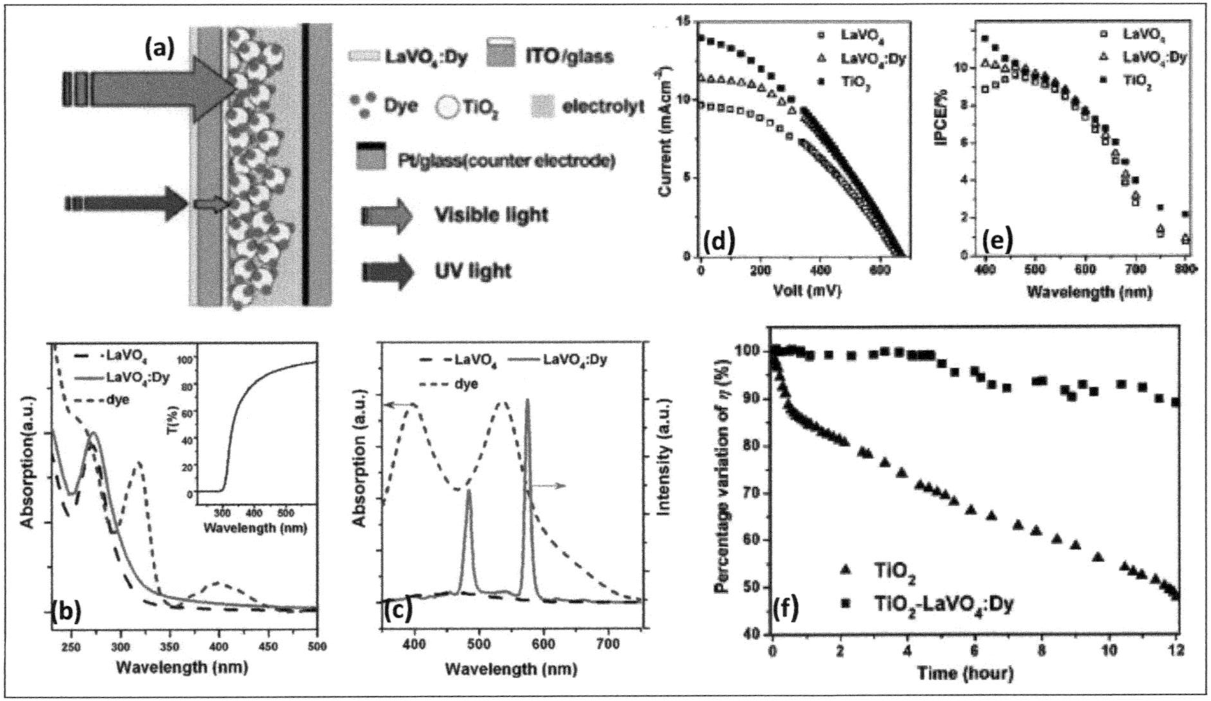

Figure 5.8 (a) Structure of DSSCs with a luminescent UV filtration film. UV light could be absorbed and changed to visible light. (b) The absorption and (c) emission spectra of nanocrystals compared with the absorption spectra of N3 dye. The inset shows the transmittance spectra of the $LaVO_4$:Dy film. (d) Current–voltage curves and (e) action spectra of dye-sensitized TiO_2 solar cells with or without a $LaVO_4$:Dy film. (f) Stability of as-prepared DSSCs. Reprinted with permission from Ref. [72]. Copyright © 2006, AIP Publishing LLC.

Similarly, DC material $ZnO:Eu^{3+},Dy^{3+}$ synthesized by the precipitation method was used to prepare the photoanode of DSSCs to extend the spectral response range to the UV region [75]. As a result, an enhanced short-circuit current and PCE were obtained. In 2012, Hafez at al. [76] reported the synthesis of $TiO_2:Eu^{3+}$ NRs by the hydrothermal route and used them to fabricate a bilayer-film electrode in a DSSC. The PCE of the DSSC with the DC phosphor layer comprising the bilayer electrode was improved to about 1 % due to the DS of Eu^{3+} ions from UV light to visible and the enhanced absorption of dye in the visible region.

In 2014, Lai et al. [77] embedded submicron-size $YVO_4:Eu^{3+},Bi^{3+}@SiO_2$ core-shell particles in a nanostructured TiO_2 layer of DSSCs. The broad excitation of $YVO_4:Eu^{3+},Bi^{3+}@SiO_2$ core-shell particles ranging from 220 nm to 380 nm and a strong emission peak at 618 nm result in the enhanced harvesting of short-wavelength photons. Along with that, the DC core-shell particles also acted as light-scattering particles, and an overall enhancement of 64% in PCE with these dual-functional particles was achieved.

A different approach was adapted by Diau and coworkers [78] in order to enhance the light-harvesting efficiency by placing a reflective luminescent down-shifting (R-LDS) layer outside the DSSCs. This R-LDS layer serves a dual functionality as a luminescent DS layer as well as a back-reflective layer. This approach also resulted in enhanced efficiency of the DSSC device. Further research in this direction was to improvise the stability of DSSCs under UV radiation; for this methodology a layer of these UV-absorbing DC phosphors can be employed as an optical filter [72, 79].

In 2013 another group reported the introduction of a $Gd_2O_3:Sm^{3+}$ rare earth oxide into a TiO_2 film of DSSCs for improving light harvesting via photon energy DS, which enhances the photocurrent. Along with that, $Gd_2O_3:Sm^{3+}$ acts as a p-type dopant and hence elevates the energy level of the oxide film and increases the photovoltage of the DSSCs. Due to this, a PCE enhancement of 12.6% compared to a DSSC lacking $Gd_2O_3:Sm^{3+}$ was achieved (Fig. 5.9) [80].

A hierarchical structure composed of porous $TiO_2:Al_2O_3:Eu^{3+}$ nanoparticles (NPs) and vertically grown one-dimensional $TiO_2:Er^{3+},Yb^{3+}$ NRs on fluorine-doped tin oxide (FTO) substrates coated with a TiO_2:graphene (G) seed layer was investigated for use in photoanodes for DSSCs. The DSSCs comprising this hierarchical structure exhibit an outstanding PCE of 4.58%, which is superior to that of devices based on pure TiO_2 [81].

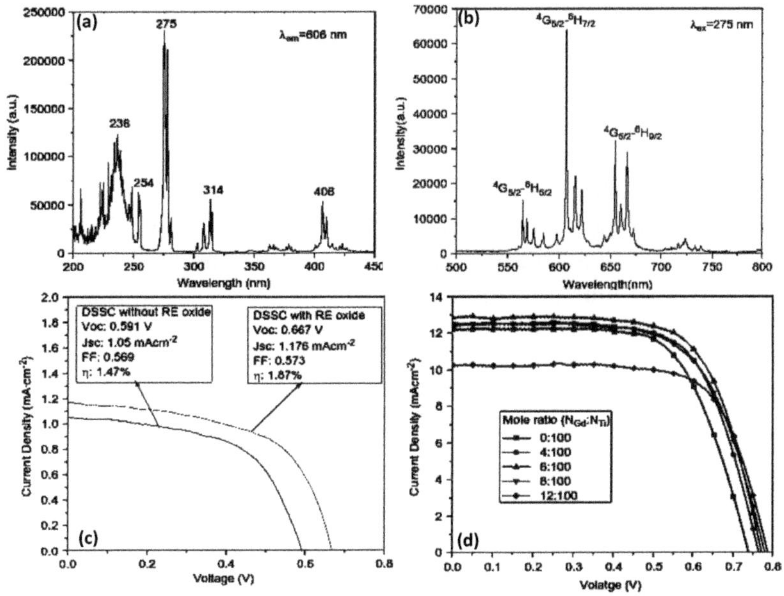

Figure 5.9 $Gd_2O_3:Sm^{3+}$ (a) excitation spectrum, (b) emission spectrum, and (c) current–voltage curves of DSSCs under UV irradiation (24 mW cm^2). (d) Current–voltage curves of DSSCs with different amounts of $Gd_2O_3:Sm^{3+}$ in a TiO_2 electrode under simulated solar light irradiation (100 mW cm^2). Reprinted from Ref. [80], Copyright (2013), with permission from Elsevier.

5.5.3 Down-Conversion in Organic Solar Cells

Bulk heterojunction (BHJ) solar cells based on an organic donor and fullerene-based acceptor phase-separated blends are promising low-cost alternatives to conventional solid-state solar cells, but issues like lack of photostability under prolonged UV irradiation and limited absorption range of active layer polymers have restricted them from widespread commercial use. Both issues can be resolved by employing luminescent DC phosphors, which can successfully down-convert the harmful UV photons into visible light, thus enhancing both the lifetime and the PCE [82, 83].

UV-absorbing DC materials with suitable emission matching that of active-layer polymers can successfully enhance light absorption in PSCs; moreover, these nanophosphors absorb UV light, which the polymers fail to absorb. In 2011, a group used a $YVO_4:Eu^{3+}/Bi^{3+}$ luminescent nanofilm as a UV-blocking layer to enhance the photostability of the polymer poly(3-hexylthiophene) (P3HT), which is the most widely used polymer in OSCs.

The luminescent UV-blocking layer has broad UV absorption ranging from 220–400 nm, and its emission lies around 621 nm. Due to UV absorption by the film, the photostability of P3HT increased by three times with the $YVO_4:Eu^{3+}/Bi^{3+}$ nanofilm as opposed to that measured without the nanofilm [83]. There are also a few reports on UC phosphors into OSCs [84].

5.5.3.1 Organic-inorganic hybrid solar cells

Another group reported the incorporation of hydrothermally synthesized dysprosium ion–decorated yttrium oxide ($Dy^{3+}:Y_2O_3$) nanocrystals (DC phosphors into TiO_2 acceptor thin films in order to enhance light harvesting, charge separation, and transfer in hybrid solar cells [HSCs]). This led to narrowing of the energy level offset between the donor (P3HT) and the acceptor ($Dy^{3+}:Y_2O_3$–TiO_2) and thus enhanced electron and hole transport properties (Fig. 5.10).

The photocurrent–voltage (*J–V*) characteristics and the PV performances of HSCs with different loading amounts of $Dy^{3+}:Y_2O_3$ were recorded (Fig. 5.10f). It was observed that with an increase of the $Dy^{3+}:Y_2O_3$ amount in HSCs from 0 wt% to 6 wt%, both V_{oc} and J_{sc} gradually increase. Thus a maximum PCE of 2.97% was achieved under the optimized loading amount 6.0 wt% of the $Dy^{3+}:Y_2O_3$ phosphor in the doping layer. Under identical experimental conditions, a $Dy^{3+}:Y_2O_3$-doped TiO_2/P3HT HSC exhibits a better PCE than a pure TiO_2/P3HT HSC by a factor of 2.02% [85].

In 2014, another group incorporated $SmPO_4$ NPs into a TiO_2/P3HT BHJ, which significantly enhanced the PV performances of the inorganic/organic HSCs due to the improved hot electrons, less energetic electrons, and hole transports at the interface of the BHJ apart from DC photoluminescence of $SmPO_4$ NPs. The $SmPO_4$ NPs synthesized by the hydrothermal route were single crystalline. Figure 5.11b shows the photoluminescence emission spectra of $SmPO_4$ NPs under UV (393 nm) excitation, consisting of sharp peaks located at 550, 600, 636, and 659 nm, corresponding to the transitions of the Sm^{3+} ion.

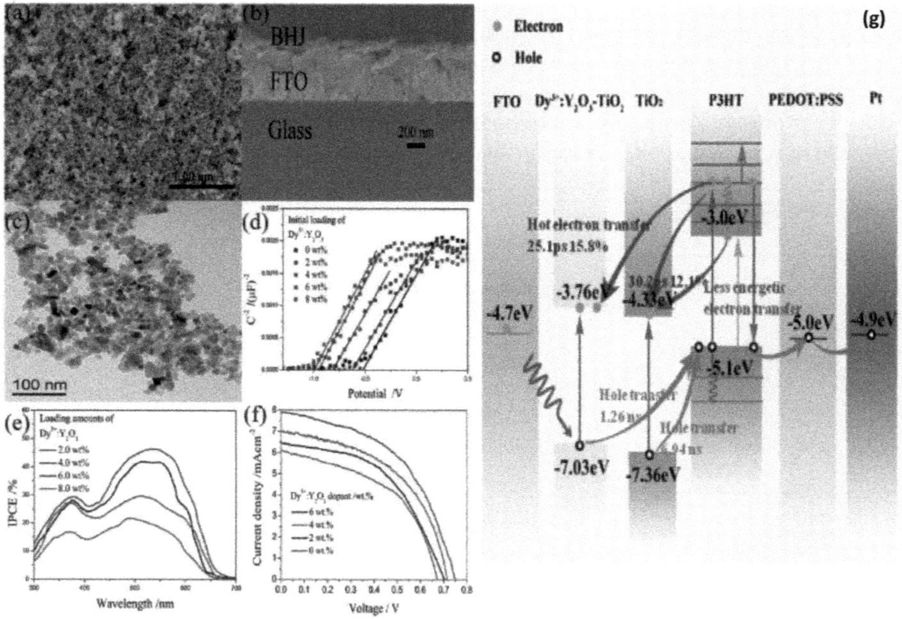

Figure 5.10 Planar (a) cross-sectional (b) SEM images and (c) TEM image of the Dy^{3+}:Y_2O_3–TiO_2 acceptor layer, (d) Mott–Schottky plots of different films, (e) IPCE, and (f) photocurrent–voltage (*J*–*V*) characteristics of HSCs made from different BHJs with different doping amounts of Dy^{3+}:Y_2O_3 nanophosphor. (g) Energy level diagram of acceptor/donor showing optical excitation (up arrow), nonradiative relaxation (curved line), radiative relaxation (down arrow), and electron and hole transfer times at either Dy^{3+}:Y_2O_3-doped TiO_2/P3HT or TiO_2/P3HT interfaces. The VB and CB energy levels of pure TiO_2 are also shown for comparison. The lowest unoccupied molecular orbital (LUMO) and the highest occupied molecular orbital (HOMO) of P3HT are 3.0 and 5.1 eV, respectively. Reproduced from Ref. [85] with permission of the PCCP Owner Societies.

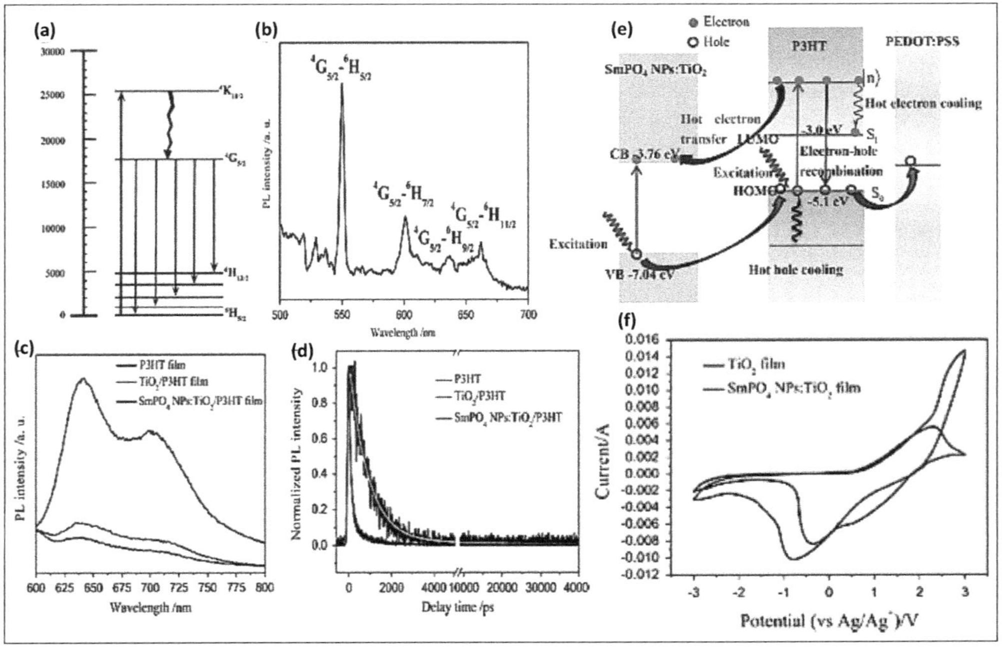

Figure 5.11 (a) Energy level diagram of trivalent Sm^{3+} ions with dominant visible emission transitions designated by arrows representing the approximate color of the fluorescence. (b) Fluorescence emission spectrum of $SmPO_4$ NPs excited at 393 nm. (c) Steady-state PL and (d) transient PL spectra for P3HT, TiO_2/P3HT, and $SmPO_4$ NP:TiO_2/P3HT films. The solid lines are the double-exponential fits of the PL decay transients. (e) Optical excitation and subsequent relaxation process at the interface of $SmPO_4$ NPs:TiO_2 and P3HT (f) Current–voltage characteristics of TiO_2 and $SmPO_4$NPs:TiO_2 film in acetonitrile containing 0.1 M $TBAPF_6$ at a sweep rate of 50 mV s^{-1}. Reprinted from Ref. [86], Copyright (2014), with permission from Elsevier.

The energy band model shown in Fig. 5.11e interprets the optical excitation and subsequent relaxation process at the interface of $SmPO_4$ NPs:TiO_2 and P3HT. The current–voltage characteristics (Fig. 5.11f) show that $SmPO_4$ NPs lead to variations of the positions of CB and VB energy bands of TiO_2, leading to PCE enhancement of the prepared solar cell with $SmPO_4$ NPs (1.98%) than its counterpart without $SmPO_4$ NPs under identical experimental conditions [86].

5.6 Conclusions

In this chapter, luminescent DC materials for spectral modification were presented in the context of efficiency enhancement of existing PV cells. The major drawback with existing PV cells is their inability to absorb the whole incident solar spectra, which results in various losses. DC phosphors allow for efficient utilization of high-energy (above-bandgap) photons, which otherwise leads to thermalization losses. Thus, such materials can effectively reduce spectral mismatch losses and, as a consequence, boost the efficiency of solar cells. Lanthanide-doped phosphors and glasses, semiconductor QDs, etc., are the most widely explored candidates for the realization of luminescent solar concentrators and planar DS layers for enhancing the PCE of existing solar cells. In conclusion, DC materials as spectral converters are potential candidates for enhancing the PCE of existing PV cells by more efficient utilization of high-energy (UV) photons. However, most of the improvements based on DC phosphors are related to Si-based solar cells. There are few reports on the applications of DC for efficiency and stability enhancement in OSCs; in particular their UV-absorbing capability can resolve the stability issue of existing OSCs and can enhance the lifetime of existing organic PV devices without any significant changes in their underlying architecture. Hence DC materials have a bright future in the area of OSCs.

References

1. Hisatomi, T., Kubota, J. and Domen, K. (2014). Recent advances in semiconductors for photocatalytic and photoelectrochemical water splitting, *Chem. Soc. Rev.*, **43**, pp. 7520–7535.

2. Chen, X., Li, C., Gratzel, M., Kostecki, R. and Mao, S. S. (2012). Nanomaterials for renewable energy production and storage, *Chem. Soc. Rev.*, **41**, pp. 7909–7937.
3. Chen, H. M., Chen, C. K., Liu, R. S., Zhang, L., Zhang, J. and Wilkinson, D. P. (2012). Nano-architecture and material designs for water splitting photoelectrodes, *Chem. Soc. Rev.*, **41**, pp. 5654–5671.
4. Braga, A. F. B., Moreira, S. P., Zampieri, P. R., Bacchin, J. M. G. and, Mei, P. R. (2008). New processes for the production of solar-grade polycrystalline silicon: a review, *Sol. Energy Mater. Sol. Cells,* **92**, pp. 418–424.
5. Huang, J., Yin, Z. and Zheng, Q. (2011). Applications of ZnO in organic and hybrid solar cells, *Energy Environ. Sci.*, **4**, pp. 3861–3877.
6. He, F. and Yu, L. (2011). How far can polymer solar cells go? In need of a synergistic approach, *J. Phys. Chem. Lett.*, **2**, pp. 3102–3113.
7. Aguas, H., Ram, S. K., Araujo, A., Gaspar, D., Vicente, A., Filonovich, S. A., Fortunato, E., Martins, R. and Ferreira, I. (2011). Silicon thin film solar cells on commercial tiles, *Energy Environ. Sci.*, **4**, pp. 4620–4632.
8. Darling, S. B. and You, F. (2013). The case for organic photovoltaics, *RSC Adv.*, **3**, pp. 17633–17648.
9. Lizin, S., Van Passel, S., De Schepper, E., Maes, W., Lutsen, L., Manca, J. and Vanderzande D. (2013). Life cycle analyses of organic photovoltaics: a review, *Energy Environ. Sci.*, **6**, pp. 3136–3149.
10. Strumpel, C., Mccann, M., Beaucarne, G., Arkhipov, V., Slaoui, A., Svrcek, V., Canizo, C. D. and Tobias, I. (2007). Modifying the solar spectrum to enhance silicon solar cell efficiency—an overview of available materials, *Sol. Energy Mater. Sol. Cells*, **91**, p. 238.
11. Van Sark, W. G. J. H. M. (2005). Enhancement of solar cell performance by employing planar spectral converters, *App. Phys. Lett.*, **87**, p. 151117
12. Yang, Y., Chen, W., Dou, L., Chang, W.-H., Duan, H.-S., Bob, B. and Li, G. (2015). High-performance multiple-donor bulk heterojunction solar cells, *Nat. Photonics*, **9**, pp. 190–198.
13. Shockley, W. and Queisser H. J. (1961). Detailed balance limit of efficiency of p-n junction solar cells, *J. Appl. Phys.*, **32**, pp. 510–519.
14. Green, M. A. (2004). *Third Generation Photovoltaics* (Springer, Berlin), Chapter 7.
15. Boer, K. W. (1990). *Survey of Semiconductor Physics: Electrons and Other Particles in Bulk Semiconductors* (Van Nostrand Reinhold, New York, USA).

16. Wilkinson, F. J., Farmer, A. J. D. and Geist, J. (1983). The near ultraviolet quantum yield of silicon, *J. Appl. Phys.*, **54**, pp. 1172–1174.

17. Wolf, M., Brendel, R. and Werner, J. H. (1996). *Tech. Digest of the International PVSEC-9* (Miyazaki, Japan), pp. 519–520.

18. Kolodinski, S., Werner, J. H. and Quessier, H. J. (1993). Potential of $Si_{1-x}Ge_x$ alloys for Auger generation in highly efficient solar cells, *Appl. Phys. A*, **61**, pp. 535–539.

19. Nozik, A. J. (2002). Quantum dot solar cells, *Physica E*, **14**, pp. 115–120.

20. Schaller, R. D. and Klimov, V. I. (2004). High-efficiency carrier multiplication in PbSe nanocrystals: implications for solar energy conversion, *Phys. Rev. Lett.*, **92**(18), p. 186601.

21. Hangleiter, A. and Hacker, R. (1990). Enhancement of band-to-band Auger recombination by electron-hole correlations, *Phys. Rev. Lett.*, **65**, p. 215.

22. Berkowitz, J. K. and Olsen, J. A. (1991). Investigation of luminescent materials under ultraviolet excitation energies from 5 to 25 eV, *J. Lumin.*, **50**, pp. 111–121.

23. Benitez, E. L., Husk, D. E., Schnatterly, S. E. and Tarrio, C. (1991). A surface recombination model applied to large features in inorganic phosphor efficiency measurements in the soft x-ray region, *J. Appl. Phys.*, **70**, pp. 3256–3260.

24. Piper, W. W., De Luca, J. A. and Ham, F. S. (1974). Cascade fluorescent decay in Pr^{3+}-doped fluorides: achievement of a quantum yield greater than unity for emission of visible light, *J. Lumin.*, **8**(4), pp. 344–348.

25. Sommerdijk, J. L., Bril, A. and De Jager, A. W. (1974). Two-photon luminescence with ultraviolet excitation of trivalent praseodymium, *J. Lumin.*, **8**(4), pp. 341–343.

26. Yu, D. C., Huang, X., Ye, S., Peng, M. Y., Zhang, Q. Y. and Wondraczek, L. (2011). Three-photon near-infrared quantum splitting in β-$NaYF_4$:Ho^{3+}, *Appl. Phys. Lett.*, **99**, p. 161904.

27. Yu, D. C., Huang, X., Ye, S., Zhang, Q. Y. and Wang, J. (2011). A sequential two-step near-infrared quantum splitting in Ho^{3+} singly doped $NaYF_4$, *AIP Adv.*, **1**, p. 042161.

28. Yu, D. C., Ye, S., Peng, M. Y., Zhang, Q. Y. and Wondraczek, L. (2012). Sequential three-step three-photon near-infrared quantum splitting in β-$NaYF_4$:Tm^{3+}, *Appl. Phys. Lett.*, **100**, p. 191911.

29. Yu, D. C., Ye, S., Peng, M. Y., Zhang, Q. Y., Qiu, J. R., Wang, J. and Wondraczek, L. (2011). Efficient near-infrared down-conversion in $GdVO_4$:Dy^{3+}

phosphors for enhancing the photo-response of solar cells, *Sol. Energy Mater. Sol. Cells*, **95**, pp. 1590–1593.

30. Zhang, W. J., Yu, D. C., Zhang, J. P., Qian, Q., Xu, S. H., Yang, Z. M. and Zhang, Q. Y. (2012). Near-infrared quantum splitting in Ho^{3+}:LaF_3 nanocrystals embedded germanate glass ceramic, *Opt. Mater. Express*, **2**, pp. 636–643.

31. Wegh, R. T., Donker, H., van Loef, E. V. D., Oskamand, K. D. and Meijerink, A. (2000). Quantum cutting through down-conversion in rare-earth compounds, *J. Lumin.*, **87**, pp. 1017–1019.

32. Hebbink, G. A., Stouwdam, J. W., Reinhoudt, D. N. and van Veggel, F. C. J. M. (2002). Lanthanide(III)-doped nanoparticles that emit in the near-infrared, *Adv. Mater.*, **14**, pp. 1147–1150.

33. Stouwdam, J. W. and van Veggel, F. C. J. M. (2002). Near-infrared emission of redispersible Er^{3+}, Nd^{3+}, and Ho^{3+} doped LaF_3 nanoparticles, *Nano Lett.*, **2**, pp. 733–737.

34. Sun, L. N., Zhang, H. J., Meng, Q. G., Liu, F. Y., Fu, L. S., Peng, C. Y., Yu, J. B., Zheng, G. L. and Wang, SB. (2005). Near-infrared luminescent hybrid materials doped with lanthanide (Ln) complexes (Ln=Nd, Yb) and their possible laser application, *J. Phys. Chem. B*, **109**, pp. 6174–6182.

35. Zhang, M. F., Shi, S. G., Meng, J. X., Wang, X. Q., Fan, H., Zhu, Y. C., Wang, X. Y. and Qian, Y. T. (2008). Preparation and characterization of near-infrared luminescent bifunctional core/shell nanocomposites, *J. Phys. Chem. C*, **112**, pp. 2825–2830.

36. Sun, L. N., Mai, W. P., Dang, S., Qiu, Y. N., Deng, W., Shi, L. Y., Yan, W. and Zhang, H. J. (2012). Near-infrared luminescence of periodic mesoporous organosilica grafted with lanthanide complexes based on visible-light sensitization, *J. Mater. Chem.*, **22**, pp. 5121–5127.

37. Fu, H., Cui, S., Luo, Q., Qiao, X., Fan, X. and Zhang, X. (2012). Broadband downshifting luminescence of Cr^{3+}/Yb^{3+}-codoped fluorosilicate glass, *J. Non-Cryst. Solids*, **358**, pp. 1217–1220.

38. Zhang, G. G., Liu, C. M., Wang, J., Kuang, X. J. and Su, Q. (2011). An intense charge transfer broadband sensitized near-infrared emitting $CaLaGa_3S_6O$:Yb^{3+} phosphor suitable for the solar spectral convertor, *Opt. Express*, **19**, pp. 24314–24319.

39. Kate, O. M. ten, Hintzen, H. T., Dorenbos, P. and van der Kolk, E. (2011). Yb^{3+} doped $LaSi_3N_5$ and YSi_3N_5 with low energy charge transfer for the near-infrared light-emitting diode and solar cell application, *J. Mater. Chem.*, **21**, pp. 18289–18294.

40. Ye, S., Jiang, N., He, F., Liu, X., Zhu, B., Teng, Y. and Qiu, J. R. (2010). Intense near-infrared emission from ZnO-$LiYbO_2$ hybrid phosphors

through efficient energy transfer from ZnO to Yb^{3+}, *Opt. Express*, **18**, pp. 639–644.

41. Ye, S., Jiang, N., Zhou, J., Wang, D. and Qiu, J. (2012). Optical property and energy transfer in the ZnO-$LiYbO_2$ hybrid phosphors under the indirect near-UV excitation, *J. Electrochem. Soc.*, **159**, pp. H11–H15.
42. Ye, S., Yu, D. C., Huang, X. Y., Yang, Z. M. and Zhang, Q. Y. (2010). Ultrabroadband sensitization of near infrared emission through energy transfer from Pb to Yb ions in $LiYbMo_2O_8$:Pb, *J. Appl. Phys.*, **108**, p. 083528.
43. Luo, Q., Qiao, X. S., Fan, X. P. and Zhang, X. H. (2011). Near-infrared emission of Yb^{3+} through energy transfer from ZnO to Yb^{3+} in glass ceramic containing ZnO nanocrystals, *Opt. Lett.*, **36**, pp. 2767–2769.
44. Liu, Y. S., Luo, W. Q., Li, R. F. and Chen, X. Y. (2010). Optical properties of Nd^{3+} ion-doped ZnO nanocrystals, *J. Nanosci. Nanotechnol.*, **10**, pp. 1871–1876.
45. Luo, W. Q., Li, R. F. and Chen, X. Y. (2009). Host-sensitized luminescence of Nd^{3+} and Sm^{3+} ions incorporated in anatase titania nanocrystals, *J. Phys. Chem. C*, **113**, pp. 8772–8777.
46. Gao, G. and Wondraczek, L. (2013). Near-infrared down-conversion in Mn^{2+}–Yb^{3+} co-doped Zn_2GeO_4, *J. Mater. Chem. C*, **1**, pp. 1952–1958.
47. Li, Y., Wang, J., Zhou, W., Zhang, G., Chen, Y. and Su, Q. (2013). Ultraviolet–visible–near-infrared luminescence properties and energy transfer mechanism of a novel 5d broadband sensitized Sr_3SiO_5:Ce^{3+}, Yb^{3+} suitable for solar spectral converter, *Appl. Phys. Express*, **6**, p. 082301.
48. Huang, X. Y. and Zhang, Q. Y. (2010). Near-infrared quantum cutting via cooperative energy transfer in Gd_2O_3:Bi^{3+},Yb^{3+} phosphors, *J. Appl. Phys.*, **107**, p. 063505.
49. Wei, X., Huang, S., Chen, Y., Guo, C., Yin, M. and Xu, W. (2010). Energy transfer mechanisms in Yb^{3+} doped YVO_4 near-infrared down-conversion phosphor, *J. Appl. Phys.*, **107**, p. 103107.
50. Balestrieri, M., Ferblantier, G., Colis, S., Schmerber, G., Ulhaq-Bouillet, C., Muller, D., Slaoui, A. and Dinia, A. (2013). On the growth process of $Cu_2ZnSn(S,Se)_4$ absorber layer formed by selenizing Cu–ZnS–SnS precursors and its photovoltaic performance, *Sol. Energy Mater. Sol. Cells*, **117**, pp. 363–371.
51. Chung, P., Chung, H. H. and Holloway, P. H. (2007). Phosphor coatings to enhance Si photovoltaic cell performance, *J. Vac. Sci. Technol. A*, **25**, pp. 61–66.

52. Mendez, M., Cesteros, Y., Marsal, L. F., Martinez-Ferrero, E., Salagre, P., Formentin, P., Pallares, J., Aguilo, M., Diaz, F. and Carvajal, J. J. (2011). Polymer composite P3HT:Eu^{3+} doped La_2O_3 nanoparticles as a down-converter material to improve the solar spectrum energy, *Opt. Mater.*, **33**, pp. 1120–1123.

53. Takeshita, S., Nakayama, K., Isobe, T., Sawayama, T. and Niikura, S. (2009). Optical properties of transparent wavelength-conversion film prepared from YVO_4:Bi^{3+},Eu^{3+} nanophosphors, *J. Electrochem. Soc.*, **156**, pp. J273–J277.

54. Shao, L. M. and Jing, X. P. (2011). Energy transfer and luminescent properties of Ce^{3+}, Cr^{3+} co-doped $Y_3Al_5O_{12}$, *J. Lumin.*, **131**, pp. 1216–1221.

55. Shao, L. M., Zhao, J. Q., Xia, Z. G. and Jing, X. P. (2011). Photoluminescence and energy transfer of Eu^{2+}, Cr^{3+} co-doped $MgSrAl_{10}O_{17}$, *J. Electrochem. Soc.*, **158**, pp. J300–J304.

56. Boer, D. K. G. de, Broer, D. J., Debije, M. G., Keur, W., Meijerink, A., Ronda, C. R. and Verbunt, P. P. C. (2012). Progress in phosphors and filters for luminescent solar concentrators, *Opt. Express*, **20**, pp. A395–A405.

57. Meng, J. X., Li, J. Q., Shi, Z. P. and Cheah, K. W. (2008). Efficient energy transfer for Ce to Nd in Nd/Ce codoped yttrium aluminum garnet, *Appl. Phys. Lett.*, **93**, p. 221908.

58. Zhou, J., Teng, Y., Liu, X., Ma, Z. and Qiu, J. (2011). Broadband spectral conversion of visible light to near-infrared emission via energy transfer from Ce^{3+} to Nd^{3+}/Yb^{3+} in YAG, *J. Mater. Res.*, **26**, pp. 689–692.

59. Zhang, G. G., Liu, C. M., Wang, J., Kuang, X. J. and Su, Q. (2011). An intense charge transfer broadband sensitized near-infrared emitting $CaLaGa_3S_6O$:Yb^{3+} phosphor suitable for the solar spectral convertor, *Opt. Express*, **19**, pp. 24314–24319.

60. Tikhomirov, V. K., Vosch, T., Fron, E., Rodriguez, V. D., Velazquez, J. J., Kirilenko, D., Tendeloo, G. van, Hofkens, J., van der Auweraer, M. and Moshchalkov, V. V. (2012). Luminescence of oxyfluoride glasses co-doped with Ag nanoclusters and Yb^{3+} ions, *RSC Adv.*, **2**, pp. 1496–1501.

61. Ye, S., Zhu, B., Luo, J., Teng, Y., Chen, J., Lakshminarayana, G., Qian, G. and Qiu, J. (2008). Spectral modification and quantum cutting in RE^{3+}-Yb^{3+} (R=Pr, Tm, and Tb) co-doped transparent glass-ceramics containing CaF_2 nanocrystals, *Appl. Phys. Lett.*, **93**, p. 181110.

62. Ye, S., Zhu, B., Liu, Y., Teng, Y., Lin, G., Lakshminarayana, G., Fan, X. P. and Qiu, J. R. (2009). Conversion of near-ultraviolet radiation into visible and infrared emissions through energy transfer in $Yb_2O_3Yb_2O_3$ doped SrO–TiO_2–SiO_2SrO–TiO_2–SiO_2 glasses, *J. Appl. Phys.*, **105**, p. 063508.

63. Santana-Alonso, A., Yanes, A. C., Mendez-Ramos, J., del-Castillo, J. and Rodriguez, V. D. (2011). Down-shifting by energy transfer in Dy^{3+}–Tb^{3+} co-doped YF_3-based sol–gel nano-glass-ceramics for photovoltaic applications, *Opt. Mater.*, **33**, pp. 587–591.

64. Gomez, S., Urra, I., Valiente, R. and Rodriguez, F. (2011). Spectroscopic study of Cu^{2+}/Cu^{+} doubly doped and highly transmitting glasses for solar spectral transformation, *Sol. Energy Mater. Sol. Cells*, **95**, pp. 2018–2022.

65. Chen, Y.-C., Huang, W.-Y. and Chen, T.-M. (2011). Enhancing the performance of photovoltaic cells by using down-converting $KCaGd(PO_4)_2:Eu^{3+}$ phosphors, *J. Rare Earths*, **29**(9), p. 907.

66. Chander, N., Sardana, S. K., Parashar, P. K., Khan, A. F., Chawla, S. and Komarala, V. K. (2015). Improving the short-wavelength spectral response of silicon solar cells by spray deposition of $YVO_4:Eu^{3+}$ downshifting phosphor nanoparticles, *IEEE J. Photovol.*, **5**, pp. 1373–1379.

67. Chen, J.-Y., Huang, C. K., Hung, W. B., Sun, K. W. and Chen, T. M. (2014). Efficiency improvement of Si solar cells using metal-enhanced nanophosphor fluorescence, *Sol. Energy Mater. Sol. Cells*, **120**, pp. 168–174.

68. Cheng, C.-L. and Yang, J.-Y. (2012). Hydrothermal synthesis of Eu^{3+}-doped $Y(OH)_3$ nanotubes as down-conversion materials for efficiency enhancement of screen-printed monocrystalline silicon solar cells, *IEEE Electron Device Lett.*, **33**, p. 697.

69. Zhu, G., Wang, X., Li, H., Pan, L., Sun, H., Liu, X., Lv, T. and Sun, Z. (2012). $Y_3Al_5O_{12}:Ce$ phosphors as a scattering layer for high-efficiency dye-sensitized solar cells, *Chem. Commun.*, **48**, pp. 958–960.

70. Yahav, S., Rühle, S., Greenwald, S., Barad, H. N., Shalom, M. and Zaban, A. (2012). Strong efficiency enhancement of dye-sensitized solar cells using a La-modified $TiCl_4$ treatment of mesoporous TiO_2 electrodes, *J. Phys. Chem. C*, **115**, pp. 21481–21486.

71. Oh, J. H., Song, H. M., Eom, Y. K., Ryu, J. H., Ju, M. J. and Kim, H. K. (2011). Wavelength conversion lanthanide (III)-cored complex for highly efficient dye-sensitized solar cells, *Bull. Kor. Chem. Soc.*, **32**, pp. 2743–2750.

72. Liu, J. F., Yao, Q. H. and Li, Y. D. (2006). Effects of downconversion luminescent film in dye-sensitized solar cells, *Appl. Phys. Lett.*, **88**, p. 173119.

73. Wu, J. H., Xie, G. X., Lin, J. M., Lan, Z., Huang, M. L. and Huang, Y. F. (2010). Enhancing the photoelectrical performance of dye-sensitized solar cell by doping with europium-doped yttria rare-earth oxide, *J. Power Sources*, **195**, pp. 6937–6940.

74. Hafez, H., Saif, M. and Abdel-Mottaleb, M. S. A. (2011). Down-converting lanthanide doped TiO_2 photoelectrodes for efficiency enhancement of dye-sensitized solar cells, *J. Power Sources*, **196**, pp. 5792–5796.

75. Yao, N., Huang, J., Fu, K., Liu, S., Dong, E., Wang, Y., Xu, X., Zhu, M. and Cao, B. (2014). Efficiency enhancement in dye-sensitized solar cells with down conversion material ZnO:Eu^{3+}, Dy^{3+}, *J. Power Source*, **5**, p. 135.

76. Hafez, H., Wu, J., Lan, Z., Li, Q., Xie, G., Lin, J., Huang, M., Huang, Y., Abdel-Mottaleb, M. S. (2010). Enhancing the photoelectrical performance of dye-sensitized solar cells using TiO_2:Eu^{3+} nanorods, *Nanotechnology*, **21**, p. 415201.

77. Lai, H., Wang, Y., Du, G., Li, W. and Han, W. (2014). Dual functional VO_4:Eu^{3+},Bi^{3+}@SiO_2 submicron-sized core–shell particles for dye-sensitized solar cells: light scattering and down-conversion, *Ceram. Int.*, **40**, pp. 6103–6108.

78. Hosseini, Z., Huang, W. K., Tsai, C. M., Chen, T. M., Taghavinia, N. and Diau, E. W. G. (2013). Enhanced light harvesting with a reflective luminescent down-shifting layer for dye-sensitized solar cells, *ACS Appl. Mater. Interfaces*, **5**, pp. 5397–5402.

79. Chander, N., Khan, A. F. and Komarala, V. K. (2015). Improved stability and enhanced efficiency of dye-sensitized solar cells by using europium doped yttrium vanadate down-shifting nanophosphor, *RSC Adv.*, **5**, pp. 66057–66066.

80. Li, Q., Lin, J., Wu, J., Lan, Z., Wang, Y., Peng, F. and Huang, M. (2013). Improving photovoltaic performance of dye-sensitized solar cell by down-shift luminescence and p-doping effect of Gd_2O_3:Sm^{3+}, *J. Lumin.*, **134**, pp. 59–62.

81. Yao, N., Huang, J., Fu, K., Deng, X., Ding, M., Zhang, S. and Xu, X. (2016). Improving the photovoltaic performance of dye-sensitized solar cells based on a hierarchical structure with up/down converters, *RSC Adv.*, **6**, pp. 11880–11887.

82. Gao, D. L., Zheng, H. R., Zhang, X. Y., Fu, Z. X., Zhang, Z. L., Tian, Y. and Cui, M. (2011). Efficient fluorescence emission and photon conversion of LaOF:Eu^{3+}LaOF:Eu^{3+} nanocrystals, *Appl. Phys. Lett.*, **98**, p. 011907.

83. Xu, W., Song, H. W., Yan, D. T., Zhu, H. C., Wang, Y., Xu, S., Bai, X., Dong, B. A. and Liu, Y. X. (2011). YVO_4:Eu^{3+},Bi^{3+} UV to visible conversion nano-

films used for organic photovoltaic solar cells, *J. Mater. Chem.*, **21**, pp. 12331–12336.

84. Kong, M., Hu, W., Cheng, F., Huang, Z., Zhang, J., Han, Z., Shi, N., Fan, Q., Xhen, S. C. and Huang, W. (2013). Efficiency enhancement in P3HT-based polymer solar cells with a $NaYF_4$:2%Er^{3+},18%Yb^{3+} up-converter, *J. Mater. Chem. C*, **1**, p. 5872.

85. Chen, Z., Li, Y., Li, Q., Yuan, Y., Chen, C., Qin, Y., Du, J., Wei, T., Tong, J., Sun, W. and Jin, X. (2014). Enhanced charge transport and photovoltaic performance induced by incorporating rare-earth phosphor into organic–inorganic hybrid solar cells, *Phys. Chem. Chem. Phys.*, **16**, pp. 24499–24508.

86. Li, Q., Yuan, Y., Wei, T., Li, Y., Chen, Z., Jin, X., Qin, Y. and Sun, W. (2014). The origin of efficiency enhancement of inorganic/organic Hybrid Solar Cells by robust samarium phosphate nanophosphors, *Sol. Energy Mater. Sol. Cells*, **130**, pp. 426–434.

Chapter 6

Development of Red Light–Emitting Electroluminescent Cell with a $Eu(TTA)_3$bipy Hybrid Organic Complex as an Emissive Layer

N. Thejo Kalyani[a] and S. J. Dhoble[b]

[a]*Department of Applied Physics, Laxminarayan Institute of Technology, Nagpur 440033, Maharashtra, India*

[b]*Department of Physics, R. T. M. Nagpur University, Nagpur 440033, Maharashtra, India*

thejokalyaninallapareddy@gmail.com

With the vision toward developing a red light–emitting electroluminescent cell, volatile $Eu(TTA)_3$bipy hybrid organic complexes were synthesized by the solution technique at different pH values (6 to 8 at an interval of 0.5) by maintaining the stoichiometric ratio. Among all the synthesized complexes, the emission intensity was found to be maximum for the complex with pH = 7.0, and the intensity wilted when pH was lowered to 6.0 or raised to 8.0 (at an interval of 0.5), proving that the complex is pH sensitive. Photoluminescence spectra of $Eu(TTA)_3$bipy registered an intense, sharp, and narrow

Phosphors: Synthesis and Applications
Edited by Sanjay J. Dhoble, B. Deva Prasad Raju, and Vijay Singh

ISBN 978-981-4774-49-9 (Hardcover), 978-042-9460-99-9 (eBook)
www.panstanford.com

emission peak at 614 nm, attributed to the electronic dipole $^5D_0 \rightarrow {}^7F_2$ transition. A single-layer electroluminescence cell was designed by sandwiching Eu(TTA)$_3$bipy between indium tin oxide (ITO) and aluminum (Al). Voltage–current characteristics and voltage–brightness characteristics of the developed electroluminescent cell were carried out. The electroluminescence spectra of the developed cell also emit intense red emission at 614 nm, proving its potential as an emissive layer for a multilayer OLED.

6.1 Introduction

Electricity is a basic requirement for economic development. Every sector of the global economy (agriculture, industry, transport, commercial, and domestic) needs inputs of electricity. The economic development plans implemented since the old days necessarily required increasing amounts of electricity to remain operational. As a result, consumption of electricity in all forms has been steadily rising all over the globe. In this background, there is an urgent need to develop a sustainable path of energy development. It's the need of the hour to adopt a cautious approach for the judicious use of our limited electricity resources. Promotion of electricity conservation and increased use of energy efficient sources are highly essential because energy saved is energy produced. Among all, the lighting sector consumes most of the generated energy sources because of a lack of energy-efficient sources. Hence, the current goal in the field of lighting is to replace conventional lighting sources such as incandescent and fluorescent lighting with more power-efficient semiconducting light sources such as light-emitting diodes (LEDs) and organic light-emitting diodes (OLEDs), which reduces the consumption of energy, saves resources, provides a longer lifetime, and has a good impact on the climate [1]. However, organic semiconductors have numerous advantages over inorganic ones: low-cost synthesis, good chemical compatibility, relative ease of handling, simpler manufacturing process, and even properties of plastic. Hence, lighting devices based on the organic electroluminescence (EL) phenomenon have provided a novel approach in the development of new displays, backlighting sources for liquid crystal displays, and next-generation solid-state lighting (SSL) sources with unique advantages [2–4], such as thin-film formation, high contrast, light weight, fast response, wide-view angle, low-power attributes, and the ability to cover the entire spectrum of visible light without sharp peaks, attaining high color-rendering index (CRI) values. Among all these applications, SSL is of prime importance owing to the fact that traditional lighting technologies, such as incandescent light bulbs, generally achieve poor electricity-to-light conversion efficiency [5–7]. White organic light-emitting diodes or devices (WOLEDs) represent the next-generation energy-saving light source to relieve the energy crisis. For the comprehension of white light from these OLEDs, red, green, and blue (RGB)-emitting materials with excellent performance are necessary. These newly developed organic EL devices are self-emissive with high luminescence and visibility, they are ultrathin, and they have low power consumption. OLEDs have the potential of creating lighting that enhances the visual environment and satisfies our psychological appetite for brightness. Thus OLED technology is the only lighting technology capable of creating an embedded light-emitting surface without diffusers or other functional wrappings [8]. This chapter summarizes an attempt made in our laboratory to (i) develop a red light–emitting $Eu(TTA)_3bipy$ hybrid organic complex by the solution technique, (ii) fabricate an EL cell with $Eu(TTA)_3bipy$ as the emissive layer, and (iii) study voltage–current (V–I) characteristics, brightness–voltage (B–V) characteristics, and EL of the fabricated cell.

6.2 Organic Light-Emitting Diodes

OLEDs are a class of LEDs that have emerged as one of the major display technologies, combining great colors and contrast with low power. An OLED is a solid-state device containing thin films of organic (carbon-based) molecules that create light with the application of electricity. Their performance has driven their adoption in the consumer electronics and cell phone markets. They offer the potential for very low-cost displays using continuous-film-printing techniques and transparent flexible displays. These displays use a material with a self-luminous property that eliminates the need for backlighting in displays, making the process flow easier. An OLED display is created by arranging several OLEDs in a pattern with alternating compounds to provide full color. OLEDs use one of two modes of operation to control all of the OLED pixels, either passive matrix (PMOLED) addressing or an active matrix (AMOLED) addressing. Active matrix addressing requires a backplane with transistors to control each pixel as well as to control a large number of pixels in a high-resolution or large display. OLED fabrication is typically a low-temperature process, which is suitable for flexible optoelectronics applications. An OLED device is generally fabricated on a huge glass substrate as a flat panel display, and hence it is considered a planar light source [9]. OLEDs produce light in much similar to ordinary LEDs, except that the positive and negative charges originate in organic compounds rather than in crystalline semiconductors. They emit light across the visible, ultraviolet (UV), and infrared (IR) wavelengths, with high brightness, and have the potential for energy-efficient solutions [10].

6.3 OLED Configuration

The anatomy of OLEDs can be a single layer or a double layer or a triple layer or a multilayer, as shown in Fig. 6.1. A single-layer OLED is made up of a single organic layer sandwiched between an anode and a cathode. In these devices, the organic materials serve three major functions: emission, electron, and hole transporting. Hence it should have relatively high photoluminescence (PL) quantum efficiency and good bipolar (hole and electron) charge transport properties in order to transport holes and electrons through the organic layer [11].

To forward-bias an OLED, an anode with a high-work-function metal is connected to a positive potential and a cathode with low-work-function metals is connected to a negative potential. This enhances the injection of holes and electrons, respectively. In such devices both charge carriers, namely electrons and holes, must be injected at an equal rate so as to achieve high efficiency, or else the surplus electrons will not recombine, resulting in low operation efficiency. The mobility of electrons and holes in the organic layer determines the efficiency of electron and hole transport. With the application of an electric field, electrons and holes move toward the emissive layer and recombine to emit light.

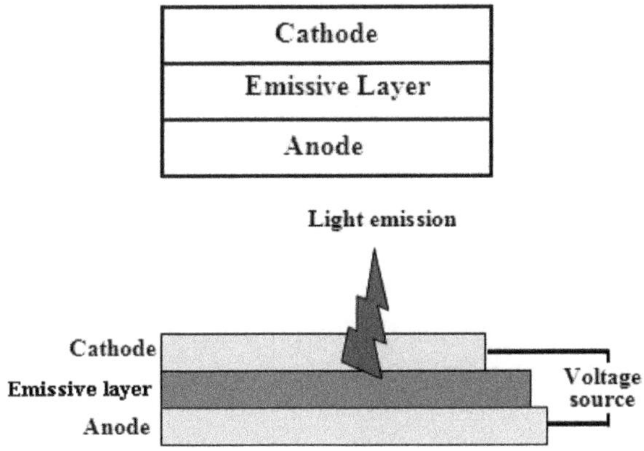

Figure 6.1 Anatomy of a single-layer OLED [12].

A schematic representation of electron–hole recombination in a single-layer OLED is shown in Fig. 6.2. It clearly reflects that the electron–hole pairs, generally called excitons, recombine randomly within the emissive layer so as to emit light. However, if these recombinations can be managed to be in a particular region of the emissive region, light from the device can be efficiently harnessed. This drawback of random site recombination can be overcome by employing a multilayer structure, which consists of an anode, a hole injection layer (HIL), a hole transport layer (HTL), an emission layer (EML), an electron transport layer (ETL), and a cathode and sometimes a hole-blocking layer (HBL). This structure also improves the charge transportation and injection of charge carriers.

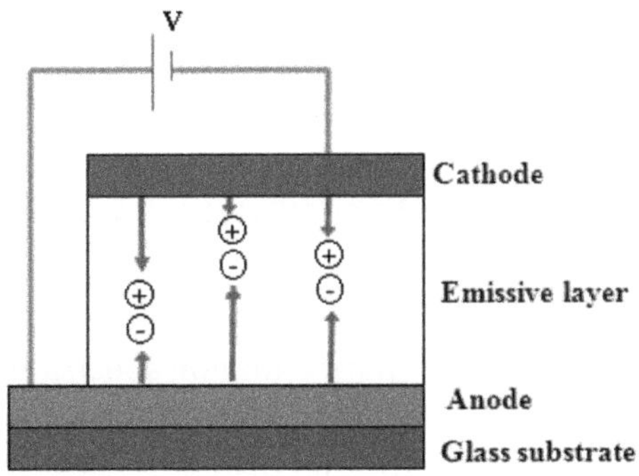

Figure 6.2 Schematic representation of electron–hole recombination in a single-layer OLED [12].

6.4 Light-Emitting Mechanism

An OLED works on the principle of EL: the phenomena in which electrical energy is converted into light energy. When current is applied to the electrodes, they emit positively and negatively charged holes and electrons. These combine in the emissive layer and create a brief, high-energy state called excitation. In an OLED, electrons are transported through the lowest unoccupied molecular orbit (LUMO), while holes are transported through the highest occupied molecular orbit (HOMO), as shown in Fig. 6.3.

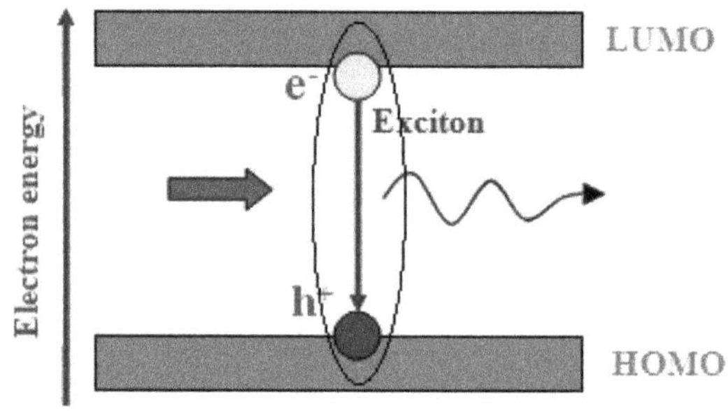

Figure 6.3 Light generation mechanism in an OLED [13].

As this layer returns to its original, stable, nonexcited state, the energy flows evenly through the organic film, causing it to give out bright light. The color of the light emission depends on the type of organic molecule in the emissive layer. Emission color is basically determined by the energy difference of the HOMO and LUMO of the emitting organic material. The intensity or brightness of the light depends on the amount of electrical current applied.

6.5 Rare Earth β-Diketonates as an Emissive Layer

Rare earth (RE) ions have sharp emission spectra (full width at half maximum [FWHM], 5 nm) [14] and hence are ideal as red light–emitting materials for OLEDs and display devices. A possible route to fabricate OLEDs based on these materials is to use the concept of

Förster energy transfer in which the films contain an RE compound and a semiconducting organic molecule or a polymer. The molecule or polymer is excited electrically, and energy is then transferred to the ion, resulting in a sharp emission. The requirement for this process to take place is that the emission spectrum of the polymer or organic overlap the absorption spectrum of the ligand. When β-diketonates are used as ligands with these REs, they form stable chelate complexes with most metal ions. β-diketonates represent one of the oldest classes of chelating ligands and have become one of the ligands of choice because of the recent industrial applications of several of their metal complexes. Hence the combination of RE and β-diketonates results in RE β-diketonates, which play a vital role in light-emitting materials and devices. They are complexes of β-diketones (1,3-diketones) with RE ions. These complexes are most popular and most intensively investigated RE coordination compounds because of the fact that they are commercially available and easy to synthesize.

Moreover, since the bandwidth of the RE compounds' emission is narrow, this would provide a route to fabricate highly monochromatic LEDs and displays. Incorporation of RE complexes in the emitting layer of OLEDs offers two main advantages: (i) improved color saturation and (ii) higher efficiency of the OLED. Because of the sharp emission bands of the trivalent lanthanide ions, luminescence from lanthanides is highly monochromatic. This results in a much better color saturation than when organic molecules are used as the emissive material. In this case the bandwidths of the emission bands are typically around 80 to 100 nm. A saturated monochromatic emission is necessary for the development of full-color displays based on OLEDs. Efficiency of OLEDs is limited to 25% by spin statistics. However, when lanthanide complexes are used, the efficiency is not limited because the excitation energy can be transferred both from an excited singlet or triplet to the lanthanide ion. Considering these facts we have synthesized a red light–emitting europium β-diketonate hybrid organic $Eu(TTA)_3bipy$ complex by the solution technique and fabricated an EL cell with $Eu(TTA)_3bipy$ as the emissive layer.

6.6 Experiment

Reagents of analytical reagent (AR) grade were purchased from Sigma-Aldrich and Acros Organics. All reactions were carried out with freshly distilled solvents under anhydrous conditions or in an inert atmosphere.

6.6.1 Reagents and Solvents

Starting materials used for the synthesis of the $Eu(TTA)_3bipy$, complex are as follows:

- Europium chloride hexahydrate ($EuCl_3{\cdot}6H_2O$): Acros Organics. Purity = 99.99%; molecular weight = 366.40 g/mol
- 2-Thenoyl trifluoroacetone ($C_8H_5F_3O_2S$): Sigma-Aldrich. Purity = 99.9%; molecular weight = 222.19 g/mol; boiling point (8 mmHg) = 96°C–98°C; melting point = 40°C–44°C
- 2,2' Bipyridyl ($C_{10}H_8N_2$): Sigma-Aldrich. Purity = 99.9%; molecular weight = 156.18 g/mol; melting point = 70°C–73°C
- Ethanol absolute (C_2H_5OH): Purity = 99.98%; molecular weight = 46.07 g/mol
- Chloroform ($CHCl_3$): Qualigens Fine Chemicals. Purity = 99.7%; molecular weight = 119.38
- Acetic acid (CH_3COOH): SD Fine Chemicals. Purity = 99.5%; molecular weight = 60.05
- Potassium hydroxide (KOH): Purity = 99.9%
- Double-distilled water

6.6.2 Synthesis Procedure

Volatile $Eu(TTA)_3bipy$ was synthesized by the solution technique according to the following steps:

1. TTA (6.63 mmol = 1.473 g) and bipy (2.21 mmol = 0.345 g) were dissolved in 20 mL ethanol in a beaker (Sol I).
2. $EuCl_3{\cdot}6H_2O$ (2.21 mmol = 0.809 g) was separately dissolved in 10 mL double-distilled water in another beaker (Sol II).
3. Both solutions were mixed, and the pH of the resultant solution was found to be between 0.25 and 0.35. To obtain a

neutral solution with pH = 7, KOH solution was added drop by drop to the resultant solution.

4. The solution thus obtained was heated at 60°C on a hot plate with continuous stirring with a magnetic stirrer for 1 h.
5. The precipitate so obtained was filtered and washed with double-distilled water twice or thrice.
6. The complex was dried at room temperature and then eventually in a hot-air oven at 80°C for 2 h [15].

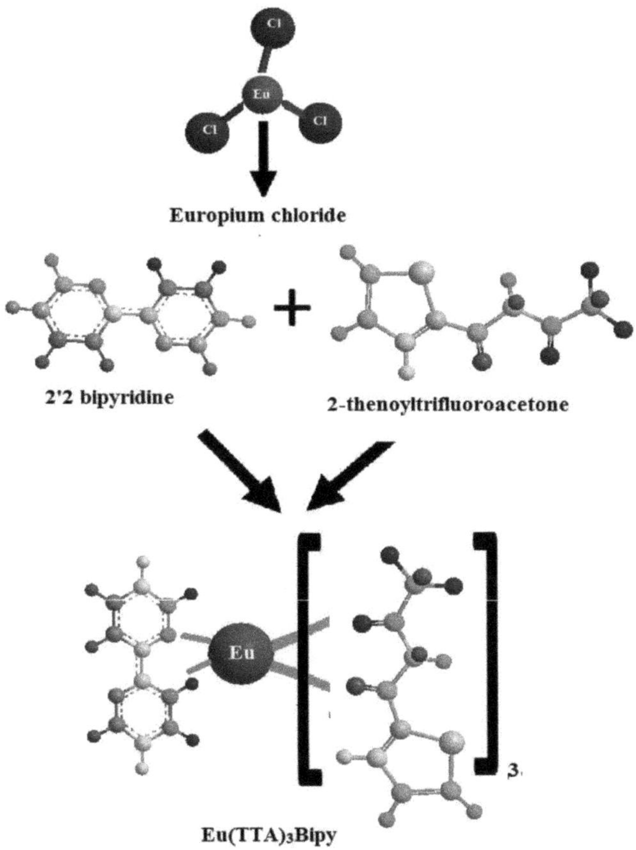

Figure 6.4 Schematic representation of synthesis procedure of $Eu(TTA)_3$bipy.

The synthesized complex was purified by the train sublimation method. In this method, the complexes were sublimed inside a quartz tube at 350°C, under high vacuum (10^{-6} Torr), and the lighter impurities diffused away, leaving the residues in the quartz boat itself. The sublimed, highly purified complexes were then collected from the walls of the quartz boat. Similarly, series of complexes were prepared by increasing pH to 8.0 from 7.0 and also by decreasing the pH to 6.0 from 7.0 at an interval of 0.5. A schematic representation of th esynthesis procedure of $Eu(TTA)_3bipy$ is given in Fig. 6.4. From the figure, it is clear that TTA and bipy are bidendate. Eu^{3+} is associated with three molecules of TTA and one molecule of bipy. Eu^{3+} has eight coordinates (six with TTA, which are shown to the right of Eu^{3+}, while the other two with bipy, which are shown to the left of Eu^{3+}). Hence, for the formation of $Eu(TTA)_3bipy$, the stoichiometry of chemical compounds must be in the 1:3:1 ratio. The raw precipitate at different pH values (6, 6.5, 7, 8) under daylight and the precipitate at pH = 7.5 under UV light are shown in Fig. 6.5.

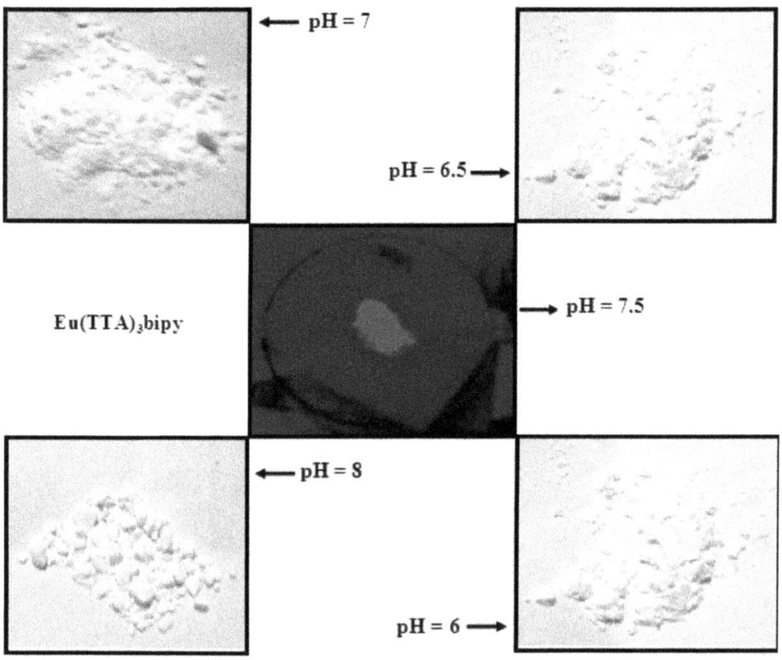

Figure 6.5 Raw precipitate at different pH values (6, 6.5, 7, 8) under daylight and precipitate at pH = 7.5 under ultraviolet light.

6.6.3 Results and Discussion

Under an excitation wavelength of 381 nm, the PL spectra of a $Eu(TTA)_3bipy$ hybrid organic complex at different pH values displays a prominent sharp red emission peak centered at 614 nm, which may be due to the $^5D_1 \rightarrow {}^7F_2$ transition, with few weak shoulder peaks at 581, 592, 598, 618, and 628 nm, as shown in Fig. 6.6. The emission from $Eu(TTA)_3bipy$ originated by the transition of the excitation energy from ligand TTA to the Eu^{3+} ion in the excited state. The synthesized Eu complex exhibits a synergistic effect in which the ligand absorbs UV light and transfers the energy to the Eu^{3+} ion, leading to enhancement of intensity. The intensity of the electric dipole $^5D_0 \rightarrow {}^7F_2$ transition is much stronger than the magnetic dipole-allowed $^5D_0 \rightarrow {}^7F_1$ transition, indicating that Eu^{3+} ions in these systems occupy low symmetric sites [16]. This hypersensitive $^5D_0 \rightarrow {}^7F_2$ transition is sensitive to the intermediate environments around Eu^{3+} ions.

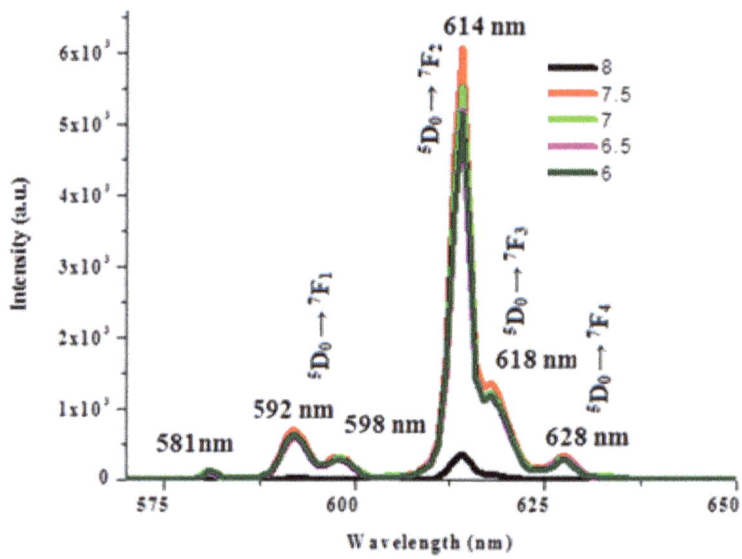

Figure 6.6 PL spectra of $Eu(TTA)_3bipy$ at different pH values.

Among all the synthesized complexes of $Eu(TTA)_3bipy$ at different pH values, the emission intensity was found to be maximum for the complex with pH = 7.0, and the intensity dropped when the pH was lowered to 6.0 or raised to 8.0 (at an interval of 0.5), proving that the complex is pH sensitive.

6.7 Fabrication of Single-Layer OLEDs

A single-layer OLED comprises an anode and a cathode with an emissive layer sandwiched in between. The anode material must be transparent with a high work function. The material of choice for anodes has traditionally been the transparent conductor ITO because of (i) good conductivity, (ii) high chemical stability, (iii) high work function, and (iv) good transparency in the visible range [17]. The emissive layer must have a characteristic emission wavelength as well as good electron and hole transport properties. Cathode materials should possess a low work function and a low threshold voltage, and they must be efficient electron injectors. Considering these facts, a single-layer OLED device was fabricated by introducing the synthesized hybrid organic $Eu(TTA)_3bipy$ complex (100 nm) between a transparent and a conducting ITO-coated glass substrate with a sheet resistance less than 10 Ω/square as the anode and an aluminum foil (150 nm) as the cathode, as shown in Fig. 6.7. During the process of fabrication, ITO is placed on a substrate holder and then transferred to a plasma chamber from the glove box for plasma treatment and then to an organic chamber to deposit an organic layer at a pressure of 6×10^{-8} Torr in a vacuum system, utilizing the deposition rate of 1 A°/s. As a low-work-function cathode and organic material is sensitive to oxygen and moisture, the device is encapsulated using a glass lid sealed to the substrate with a bead of UV-cured epoxy in a nitrogen-filled glove box (dew point ≈ 75°C) for 3 min after fabrication [18].

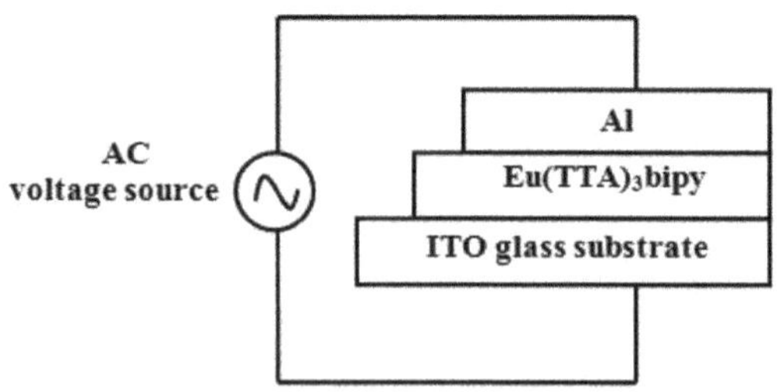

Figure 6.7 Anatomy of a single-layer OLED.

6.7.1 Characterization of an OLED Device

The output voltage of the function generator was set at 0.1 V. This output was fed to the input of the EL power supply, which is basically a wide-band amplifier (WBA). The device under study was connected to the AC power supply (WBA) through a microammeter and placed at the slit of a photomultiplier tube (PMT). Voltage was supplied to the PMT by a high-voltage power supply, and the output of the PMT was given to a Pico ammeter to measure the brightness or intensity of emitted light in terms of photocurrent. To obtain voltage–brightness characteristics of the device, the frequency of the function generator was fixed at 1 kHz, and then the AC voltage of a fixed frequency across the cell was increased and the corresponding brightness recorded with the help of a PMT by means of a Pico ammeter. The same procedure was followed by adjusting the frequency generator to 1.1 kHz and 1.2 kHz. To obtain *V–I* characteristics of the device, again the AC voltage of a fixed frequency across the cell was increased and the corresponding current through the device recorded.

6.7.1.1 Voltage–current (*V–I*) characteristics

The voltage–current characteristics for a $Eu(TTA)_3bipy$ β- diketonate complex are shown in Fig. 6.8. The figure clearly illustrates a linear relation between current and voltage, indicating an ohmic nature. Such ohmic behavior can be attributed to the hopping conductivity of electrons through a fine emissive layer.

6.7.1.2 Brightness–voltage (*B–V*) characteristics

B–V curves of a $Eu(TTA)_3bipy$ device at different frequencies of the function generator are shown in Fig. 6.9. An enhancement in the injection of electrons and holes from cathode and anode is increased by the application of an electric field.

The enhancement in the EL intensity is due to the immediate recombination of electron–hole pairs, resulting in emission of more photons. The threshold voltage was obtained at 420, 380, and 360 V for the device at 1, 1.1, and 1.2 kHz of frequency of the function generator, respectively. It was observed that the luminance increased with the increase in the frequency of the function generator. On the contrary, the threshold frequency was found to decrease with increasing frequency.

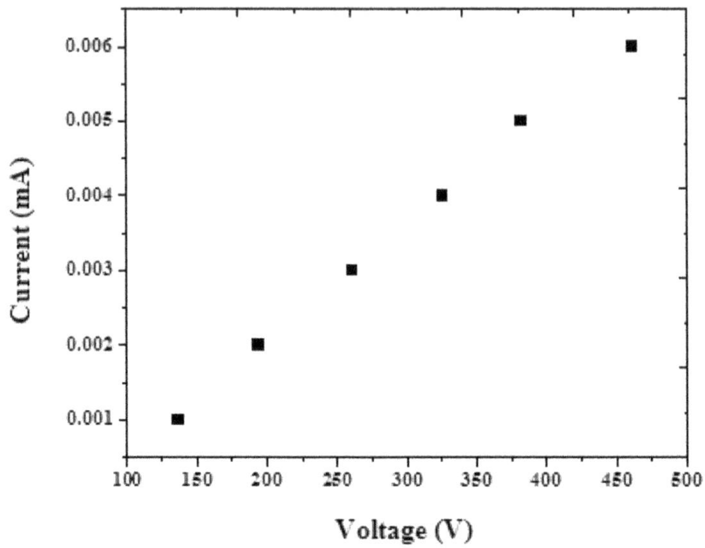

Figure 6.8 *V–I* characteristics of a $Eu(TTA)_3$bipy device.

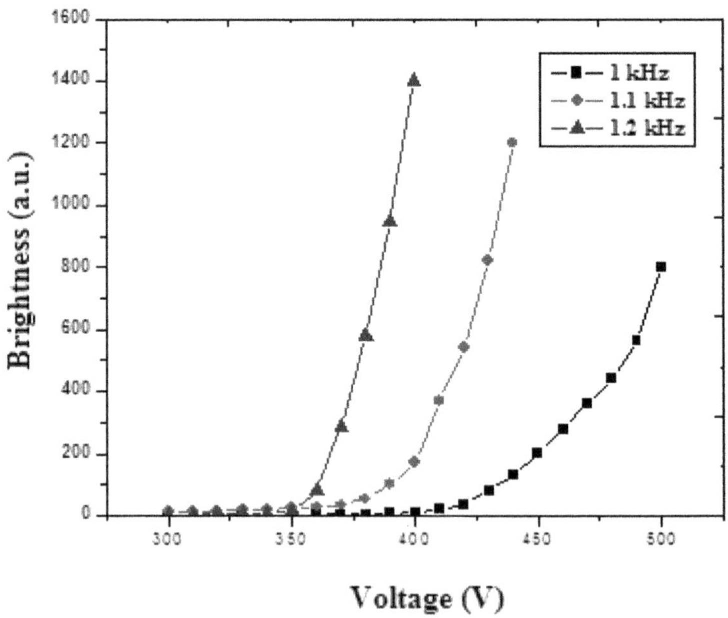

Figure 6.9 *B–V* characteristics of a $Eu(TTA)_3$bipy device.

6.7.1.3 Electroluminescence

An EL spectrum of the fabricated single-layer device at a bias voltage of 13 V is shown in Fig. 6.10. The spectrum contains four peaks at 592, 598, 614, and 630 nm, corresponding to $^5D_0 \rightarrow {}^7F_j$ transitions (j = 0, 1, 2, 3). The sharp emission peak at 614 nm is due to the electric dipole-induced $^5D_0 \rightarrow {}^7F_2$ transition, which is much stronger than the magnetic dipole-allowed $^5D_0 \rightarrow {}^7F_1$ transition, indicating that Eu^{3+} ions in these systems occupy low symmetric sites.

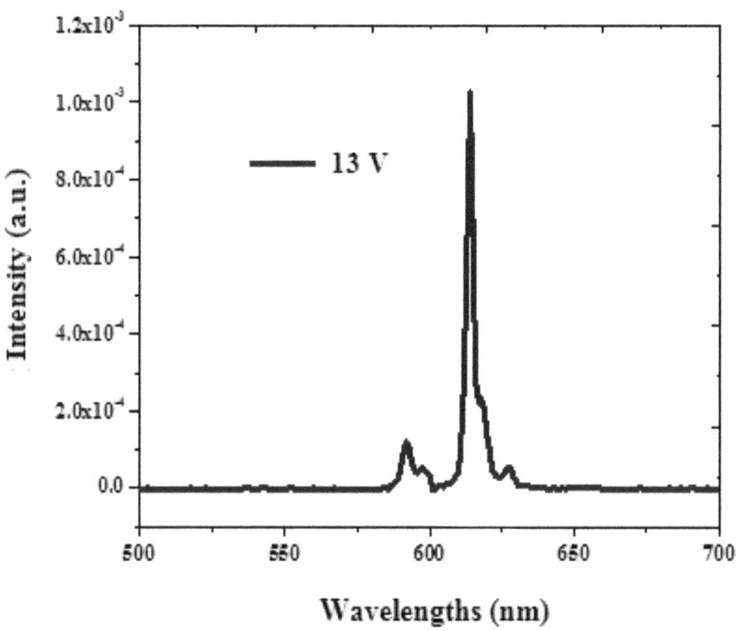

Figure 6.10 EL spectra of a single-layer OLED device.

This hypersensitive $^5D_0 \rightarrow {}^7F_2$ transition is sensitive to the intermediate environments around Eu^{3+} ions. The observed Eu^{3+} emission is due to the long lifetime of the $^5D_0$ state. Due to this long lifetime, this state can be easily filled quickly and the emission accordingly saturates. Ideally, this value of voltage at which the device emits light should be as low as possible, but in many lanthanide-based OLEDs, the values are between 5 and 10 V [19].

6.8 Techniques to Improve the Efficiency of an OLED

The research effort in the direction of a lower driving voltage and improvement in efficiency includes great progress, making WOLEDs as competitive as other lighting sources currently in use. Various techniques to improve the performance of an OLED include:

- Upgrading hole injection and transportation
- Developing light-emissive layers
- Making advances in electron injection and transportation
- Employing a multilayer structure
- Selecting suitable anode and cathode
- Choosing a material with good quantum efficiency
- Using an appropriate manufacturing process
- Encapsulating the device after fabrication

6.9 Traits of OLEDs

SSL with LEDs has already proved to be one of the vital lighting sources for existing and future challenges, such as sustainable development, energy, and community health, as well as improving the quality and standard of life. Versatile OLEDs have the following traits [18–20]:

- **Power saving**: With regard to current consumption, CO_2 emissions, and maintenance costs, OLEDs offer huge potential for savings, without making any reductions in light quality. They consume less power (2–10 V DC). When the longer service life is taken into account, the overall balance is clearly positive, even if the procurement costs are comparatively higher than conventional incandescent lamps. Depending on the system solution, clever technical solutions can reduce power costs by up to 80%. There is still the potential to increase efficacy in order to make OLEDs even more efficient light sources than they are today.
- **Better conversion efficiency**: As compared to compact fluorescent lamps, OLED-based lamps consume over 98% of

the energy supplied during operation, that is, in the generation of light. Less than 2% is used in production. As the efficiency of OLEDs is continually increasing, we can expect even better ecological balance results of OLED lamps in the future.

- **Color selectivity**: There are carbon-based light-emitting components available in a wide variety of organic materials, which create intense light in the whole spectrum of visible light.
- **Environment-friendliness**: OLEDs are environmentally friendly as they have neither UV nor IR components in their light, offering more flexibility in usage, particularly in heat-sensitive areas, such as lighting for food or cosmetics, or in applications with limited space. They are suitable for any climate.
- **Long service life**: A WOLED with more than 30,000 h will last between three and seven times longer than conventional fluorescent and energy-saving lamps and 30 times longer than an incandescent lamp. Their EL decay time is less than 1 μs.
- **Self-luminous**: OLEDs are self-luminous and thus do not require backlights, diffusers, or polarizers.
- **Ultrathin**: They are lightweight, compact, and thin devices, generally 400 times thinner than a human hair [13].
- **High brightness and high resolution**: OLEDs are bright at a low operating voltage, for example, WOLEDs can be as bright as 150,000 cd/m^2 with a viewing angle as high as 160° due to its characteristic Lambertian emission with a high contrast ratio and color-rendering index (CRI).
- **Design freedom**: OLEDs can be designed as pendant sources or spread panel light sources that emit diffuse, nonglaring illumination [20]. They are extremely small and allow great design freedom for lighting applications. They can be easily fabricated on plastic substrates paving the way for flexible electronics.

6.10 Impact of OLEDs on the Environment

- OLEDs proffer a window of opportunity for environmentally compassionate technology to thrive with SSL, which is proposed to be compact, cost-effective, energy efficient, and environmentally friendly with a wide variety of designable features.
- SSL using OLEDs is poised to reduce the electricity consumption by at least 50%, so lighting will then use less than one-tenth of all electricity generated.
- As they are energy efficient, OLEDs reduce the need for electricity (less air pollution, less depletion of nonrenewable sources of energy, and less greenhouse effect).
- They do not emit UV radiation, which causes problems ranging from insect attraction to wallpaper and artwork fading.

This green technology brings quality to light, claims revitalization, and expands excellence in the field of lighting solutions.

6.11 Limitations

With all of their advantages, OLEDs seem to be the perfect technology for all types of displays and lighting sources; however, they still face limitations:

- **Current costs**: OLED manufacturing currently requires process steps that make it extremely expensive, especially for thermal vacuum-evaporated OLEDs. Hence commercialization of large OLED screens is a big task. Nonexistence of infrastructure is also a key cause for their high cost. However, low cost and easy fabrication are possible for polymer OLEDs by roll-to-roll manufacturing, such as inkjet printing and screen printing [21].
- **Lifespan**: The biggest technical problem for OLEDs was the limited lifetime of the devices. Red and green OLED films have longer lifetimes (46,000 to 230,000 h), while blue organics currently have much shorter lifetimes (up to around 14,000 h).

- **Color balance issues**: As blue emitters degrade significantly more rapidly than green or red emitters, blue light output decreases relative to other colors. This variation in the differential color output changes the color balance of the display and is much more noticeable than a decrease in overall luminance.
- **Water damage**: Organic materials are highly susceptible to degradation by oxygen and water molecules, which can degrade the device quickly. So the main disadvantage of an OLED is the short lifetime. Therefore, improved encapsulation techniques are critical for practical manufacturing.
- **Outdoor performance**: As an emissive display technology, OLEDs rely completely upon converting electricity to light. The metallic cathode in an OLED acts as a mirror, with reflectance approaching 80%, leading to poor readability in bright ambient light such as outdoors.
- **Efficiency**: Among the three primary RGB colors, the synthetic procedures and manufacturing methods of green and red phosphors meet the essential requirements as one of the primary colors, the color blue, is crucial for full-color flat panel display applications and lighting sources through either a primary or a complementary color strategy, which has made blue emitters important in the field of OLEDs. In this aspect, the progress of blue emitters can determine the development of organic EL to a great extent, making an allowance for the practical applications of OLEDs.
- **Practical limitations**: Small molecular OLED devices possess a low glass transition temperature, and hence during fabrication, the operating temperature should not exceed the glass transition temperature. Due to the amorphous nature of the organic molecules, electrons and holes possess low mobility.

Perceptibly, these challenges lead to great difficulties, but once the problems are solved, there are great opportunities for commercialization.

6.12 Applications of OLEDs

Fascinating applications of organic OLEDs include:

- **Monochrome applications**: Small monochrome displays for hand-held electronic devices (cell phones, personal desktop assistants [PDAs], digital cameras, Global Positioning System [GPS] devices, etc.)
- **Two- or multicolour applications**: Car electronics (radios, GPS displays, maps, warning lights, etc.), instrument electronics, heads-up instrumentation for aircraft and automobiles, and rugged PDAs
- **Full-colour applications**: Liquid-crystal display (LCD) backlights and small full-color displays such as high-resolution personal communicators
- **Large displays**: Wall-mounted TV monitors and large-screen computer monitors
- **Versatile lighting sources**: White-lighting sources in the form of pendant or spread light sources for domestic use
- **Lighting panels**: For illumination of residential and commercial buildings, advertising boards, large signs, office windows, walls and partitions, color-changing lighting panels, and light walls for home and office
- **Ultrathin displays**: Ultra-lightweight, wall-size TV and computer monitors
- **Flexible displays**: Rollable and bendable displays

OLEDs can open up these new applications in the emerging future.

6.13 Conclusions

Volatile Eu(TTA)$_3$bipy hybrid organic complexes were synthesized at different pH values (6 to 8 at an interval of 0.5) by maintaining the stoichiometric ratio. Among all the synthesized complexes of Eu(TTA)$_3$bipy at different pH values, the emission intensity was found to be maximum for the complex with pH = 7.0, and the intensity dropped when the pH was lowered to 6.0 or raised to 8.0 (at an interval of 0.5), proving that the complex is pH sensitive. PL spectra exhibit an emission peak at 614 nm due to the $^5D_0 \rightarrow {}^7F_2$ transition. A single-layer EL cell was designed by sandwiching $Eu(TTA)_3bipy$ between ITO and aluminum. Voltage–current characteristics and voltage–brightness characteristics of the developed EL cell were carried out. The device emits intense red emission at 614 nm, proving its potential applications as a red light–emitting phosphor for the fabrication of OLEDs—the future versatile illumination sources.

References

1. Chitnis, D., Thejo Kalyani, N., Swart, H. C. and Dhoble, S. J. (2016). Escalating opportunities in the field of lighting, *Renewable Sustainable Energy Rev.*, **64**, pp. 727–748.
2. Van Elsbergen, V., Boerner, H., Lobl, H. P., Goldmann, C., Grabowski, S. P., Young, E., Gaertner, G. and Greiner, H. (2008). OLEDs for lighting applications, *Proc. SPIE*, **7051**, p. 70511A-1.
3. da Cunha Andrade, L. H., Lima, S. M., da Silva, R. V., Baesso, M. L., Guyot, Y. and de Oliveira Nunes, L. A. (2014). White-light-emitting $KCl:Eu^{2+}$/KCN crystal for solid-state lighting devices, *J. Mater. Chem. C*, **2**, pp. 10149–10156.
4. Thejo Kalyani, N., Dhoble, S. J. and Pode, R. B. (2013). Characterization of novel rare earth β-diketonate organic complexes for solid state lighting, *Luminescence*, **28**(2), pp. 183–187.
5. Kim, S. O., Lee, K. H., Kim, G. Y., Seo, J. H., Kim, Y. K. and Yoon, S. S. (2010). A highly efficient deep blue fluorescent OLED based on diphenylaminofluorenylstyrene-containing emitting materials, *Synth. Met.*, **160** (11–12), pp. 1259–1265.
6. Chitnis, D., Thejo Kalyani, N. and Dhoble, S. J. (2012). Lighting up lives with eco-friendly and energy efficient OLEDs for solid state lighting, *Bio Nano Front.*, **2**, pp. 346–348.
7. Thejo Kalyani, N. and Dhoble, S. J. (2014). Novel approaches for energy efficient solid state lighting by RGB organic light emitting diodes: a review, *Renewable Sustainable Energy Rev.*, **32**, pp. 448–467.
8. Robert Harris, T. (2008). Review of organic light emitting diode fabrication and recent progress, *ECE-592-S Soft Electronics Final Paper*, pp. 1–6.

9. Thejo Kalyani, N. and Dhoble, S. J. (2014). Importance of eco-friendly OLED lighting, in *Defect and Diffusion Forum* (Trans Tech Publications, Switzerland), **357**, pp. 1–27.
10. Castellano, J. A. (1992). *Handbook of Display Technology* (Gulf Professional Publishing).
11. Qi, Q. J., Wu, X. M., Hua, Y. L., Hou, Q. C., Dong, M. S., Mao, Z. Y., Yin, B. and Yin, S. G. (2010). Enhancement of performance for blue organic light emitting devices based on double emission layers, *Org. Electron.*, **11**(3), pp. 503–507.
12. Thejo Kalyani, N. and Dhoble, S. J. (2015). Novel materials for fabrication and encapsulation of OLEDs, *Renewable Sustainable Energy Rev.*, **44**, pp. 319–347.
13. www.slideshare.net; www.oledinfo.com; www.osddisplays.com
14. Nash, K. L. and Jensen, M. P. (2000). Analytical separations of the lanthanides: basic chemistry and methods, in *Handbook on the Physics and Chemistry of Rare Earths* (Elsevier Science B V, USA), **28**, pp. 311–371.
15. Thejo Kalyani, N., Dhoble, S. J. and Pode, R. B. (2012). Optical energy gap of EuxLn(1-x)$(TTA)_3$Phen organic complexes in polymer matrix, *Int. J. Knowl. Eng.*, **3**(1), pp. 47–49.
16. Diaz-Garcia, M. A. and De Avila, S. F. (2002). Energy transfer from organics to rare-earth complexes, *Appl. Phys. Lett.*, **81**(21), pp. 3924–3926.
17. Liang, C. J., Hong, Z. R., Liu, X. Y., Zhao, D. X., Zhao, D., Li, W. L., Xia, S., Mei, Y. and Wang, L. (2000). Organic electroluminescent devices using europium complex as an electron-transport emitting layer, *Thin Solid Films*, **359**, pp. 14–16.
18. Thejo Kalyani, N., Dhoble, S. J. and Gajendra, S. (2013). Development of electroluminescence cell using a $Eu_{0.5}Y_{0.5}$ $(TTA)_3$Phen organic luminescent complex, *Optik*, **124**, pp. 3614–3617.
19. Gschneiduer Jr., K. A., Bünzli, J.-C. G. and Pencharsky, V. K. (2005). *Handbook on Physics and Chemistry of Rare Earth*, Vol. 35 (Elsevier B V).
20. www.lgdisplay.com/eng/product/oled_light.jsp; https://www.quora.com
21. Stanford Resources Inc. (2000). *Organic Light Emitting Diodes (OLED's): Technology Trends and Display Market Assessment*, San Jose, CA.

Chapter 7

Optical Analysis of RE^{3+} (RE = Eu^{3+}, Tb^{3+}, Sm^{3+}, and Dy^{3+}):$Ca_2Gd_2W_3O_{14}$ Phosphors

S. Sailaja,[a] Vibha Chopra,[b] S. J. Dhoble,[c] and B. Sudhakar Reddy[c,d]

[a]*Department of Humanities and Basic Sciences, G. Pulla Reddy Engineering College (Autonomous), Kurnool 518007, India*

[b]*PG Department of Physics & Electronics, DAV College, Amritsar 143001, Punjab, India*

[c]*Department of Physics, RTM Nagpur University, Nagpur 440033, India*

[c]*Department of Physics, SV Degree College, Kadapa 516003, India*

[d]*Department of Physics, SKR & SKR Government Degree College for Women (Autonomous), Kadapa 516001, Andhra Pradesh, India*

drbsreddyphd@gmail.com; yallasailaja29@gmail.com; vibhachopra04@gmail.com; sjdhoble@gmail.com

This book chapter reports on the red, green, orange-red, and yellow colors exhibiting calcium gadolinium tungstate ($Ca_2Gd_2W_3O_{14}$) powder phosphors doped with four versatile rare earth (Eu^{3+}, Tb^{3+}, Sm^{3+}, and Dy^{3+}) ions, powders that are synthesized by the high-temperature solid-state reaction method. In this, powder samples have been characterized by different techniques, like XRD, SEM, EDAX, FTIR, PL, and ML, for their analysis. By using X-ray diffraction patterns of the powder samples, the tetragonal structure of all the

Phosphors: Synthesis and Applications
Edited by Sanjay J. Dhoble, B. Deva Prasad Raju, and Vijay Singh

ISBN 978-981-4774-49-9 (Hardcover), 978-042-9460-99-9 (eBook)
www.panstanford.com

samples is confirmed. By the PL spectra of all powder samples, their characteristic emissions are confirmed. With the help of SEM and EDAX, the size and orientation of the powder particles in the crystal structure have been analyzed. Photoluminescence intensity is plotted against the concentration of rare earth (Eu^{3+}, Tb^{3+}, Sm^{3+}, and Dy^{3+}) ions to optimize the best concentration for WLEDs, fluorescent lamps, and novel materials for luminescent applications.

7.1 Introduction

In recent times, the usage and importance of tungstates have been popularized among several inorganic phosphors because of their stable crystal structure, rigid nature, and also a charge transfer band (CTB) to absorb efficient amounts of energy [1–4]. The rare earth ion–doped tungstates are more useful in varied competent applications, especially in lighting systems such as white-light-emitting diodes (WLEDs) and fluorescent lamps, and as efficient luminescent materials [5–22], when compared with several inorganic luminescent compounds. WLEDs have generated much interest because of their advantages over the conventional incandescent and fluorescent lamps, such as high efficiency, low power consumption, lack of toxic mercury, long lifetime, low cost, fast response, high luminescence efficiency, energy saving, and environment friendliness [23–26]. Currently, WLEDs because of their advantages are thought of as the next-generation illumination sources that will take over the conventional fluorescent lamps and have been widely used in interior lighting, vehicle displays, and flashlights and can be fabricated in a straightforward manner by using red-, green-, and blue-light-emitting phosphors coated on a near-ultraviolet (UV) LED chip. Therefore, novel and superior red-, green-, and blue-light-emitting phosphors need to be prepared for WLEDs [27–40]. The selection of a suitable host material is also an important factor in the preparation of luminescent materials for diverse and useful applications. Generally, phosphor host materials are based on borates, phosphates, aluminates, silicates, tungstates, molybdates, vanadates, etc. In this context after the review of the literature after knowing the importance of tungstates, work related to calcium gadolinium tungstate ($Ca_2Gd_2W_3O_{14}$) powder phosphors has been undertaken.

Especially rare earth–doped crystal structures are booming nowadays with their admirable properties and advantageous applications. Moving on this path, firstly Eu^{3+} ions are doped in calcium gadolinium tungstate ($Ca_2Gd_2W_3O_{14}$) powder phosphors. Many researchers have reported on Eu^{3+} ions earlier as these ions have been extensively used as active ions of red phosphors for different reasons, such as the characteristic emission band at ~613 nm in the red region, and these ions can also be used to detect crystal symmetry. The red luminescence of Eu^{3+} ions has been extensively used in lighting and displays because of their 4f–4f transitions. The trivalent Eu^{3+} ion is one of the most promising species that exhibit a strong visible emission in the red regions and can be used in fluorescent lamps, field emission displays, plasma display panels, and color lighting [41–44].

To continue the sequence in the next step tungstates are doped with trivalent terbium ions. From the literature survey it is known that a Tb^{3+} ion shows an intense green emission in different hosts, and hence these ions are used for the development of efficient green-light-emitting phosphors and scintillator materials. Terbium ions with a $4f^8$ electronic configuration have complicated energy levels with a number of possible transitions between the f levels. The Tb^{3+} ion is considered as one of the excellent optical activators with a strong green emission, having good characteristics such as high luminescence output, good color purity, and great radiation stability, and these properties make terbium-doped phosphors suitable for many potential applications involving cathode ray tubes, X-ray imaging devices, fluorescence lamps, LEDs, scintillators, and field emission displays [45, 46]. The intense emission performance of Tb^{3+} ions is mainly due to the $^5D_4 \rightarrow {}^7F_J$ transitions, which occur in the green region. Sometimes, emission from higher-energy levels corresponding to $^5D_3 \rightarrow {}^7F_J$ transitions also takes place, which is mainly in the blue region [47].

After observing the results of Eu^{3+} and Tb^{3+} ion–doped tungstates, to enhance the applications of trivalent rare earth–doped calcium gadolinium tungstate powder phosphors, another important activator ion taken is Sm^{3+}, which has the characteristic emission in the orange-red region. Sm^{3+} ions belong to the $4f^5$ configuration. Sm^{3+} ions in various hosts show bright emission in the orange or the red region because of the transitions from the excited state $^4G_{5/2}$

to the ground state $^6H_{5/2}$ and also to the higher levels ^6H_j (j = 7/2, 9/2, and 11/2). The excitation spectra of Sm^{3+} ion–doped materials cover the UV, blue, and bluish-green spectral regions. Hence they have found applications in high-density optical storage, temperature sensors, undersea communication, various fluorescent devices, color displays, and visible solid-state lasers [48].

Optical properties of dysprosium-doped tungstate materials are widely studied in different host materials, for their applications in tunable laser, solid-state lighting, and electronic display devices. In general, Dy^{3+} ions have to exhibit two important emission bands. One is in the yellow region (575 nm), corresponding to the hypersensitive transition $^4F_{9/2} \rightarrow {}^6H_{13/2}$ (ΔL = 2; ΔJ = 2), and it is electric dipole allowed. The other is in the blue region (480 nm) and corresponds to the $^4F_{9/2} \rightarrow {}^6H_{15/2}$ transition, it is magnetic dipole allowed, and its intensity is not considerably altered by the local environment of Dy^{3+} in the crystal lattice. Moreover, the intensity of the hypersensitive $^4F_{9/2} \rightarrow {}^6H_{13/2}$ transition depends on the local symmetry around Dy^{3+} ions. The relative intensities of these two emission bands depend on the host composition, doping concentration, and excitation wavelength. [49–64].

From the literature survey, it is noticed that there are no reports on rare earth ion–doped $Ca_2Gd_2W_3O_{14}$ phosphors; therefore, we have undertaken the present work and reported the synthesis, structural, photoluminescence (PL), and mechanoluminescence (ML) properties of Eu^{3+}, Tb^{3+}, Sm^{3+}, and Dy^{3+} ion–doped $Ca_2Gd_2W_3O_{14}$ novel phosphors.

7.2 Experimental

7.2.1 Synthesis

$Ca_2Gd_2W_3O_{14}$ phosphors doped with four different trivalent rare earth (Eu^{3+}, Tb^{3+}, Sm^{3+}, and Dy^{3+}) ions were prepared by the conventional solid-state reaction method. From Sigma-Aldrich starting materials such as $CaCO_3$, Gd_2O_3, WO_3, and RE_2O_3 with 99.99% purity were purchased (these were used as received, without any further purification) and were taken in the appropriate stoichiometric ratio for the preparation of the powder phosphors.

These chemicals were mixed and ground thoroughly in an agate mortar manually, and then the mixtures were collected in an alumina crucible. They were heated in an electric furnace at a temperature of 1000°C for 2 h. The final samples were white powders and were used for their characterization.

7.2.2 Characterization

Eu^{3+}, Tb^{3+}, Sm^{3+}, and Dy^{3+} ion-doped $Ca_2Gd_2W_3O_{14}$ phosphors were initially characterized to know about their structures. The characterization was carried out using X-ray powder diffraction measurements on an X-ray diffraction (XRD) 3003TT Seifert diffractometer with CuKα radiation (λ = 1.5406 Å) at 40 kV and 20 mA, and the 2θ range was varied between 20° and 60°. The morphology of the powder phosphors was examined on a ZEISS-EVO-MA15 ESEM. A scanning electron microscopy (SEM) image was obtained for samples by using a 35mm camera attached to a high-resolution recording system. An elemental analysis was carried out by energy-dispersive X-ray analysis (EDAX) using an X-ray detector attached to the SEM instrument. The Fourier transform infrared (FTIR) spectroscopy spectrum (4000–450 cm^{-1}) was recorded on a Perkin Elmer Spectrum1 spectrometer with KBr pellets.

Eu^{3+} and Tb^{3+}:$Ca_2Gd_2W_3O_{14}$ phosphors were then characterized for PL, and their PL characteristic spectra were studied for 2 mg of the sample each time using an RF-5301PC SHIMADZU spectrofluorophotometer, at room temperature, using a range of 220–750 nm. To study the effect of gamma irradiation on the photoluminescent properties of the prepared phosphors, the samples were exposed to γ-rays from a ^{60}Co source at room temperature at the rate of 0.995 kGy/h.

After PL, Eu^{3+} and Tb^{3+}:$Ca_2Gd_2W_3O_{14}$ phosphors were excited impulsively by dropping a load onto a 1 mg gamma-irradiated phosphor placed on a Lucite plate (surface area 2.826×10^{-3} m^2) from various heights using a guiding cylinder (area of cross section 1.256×10^{-3} m^2). The ML was monitored using an RCA 931 photomultiplier tube positioned below the Lucite plate and connected to a storage oscilloscope. The ML spectra were recorded using a series of optical band pass filters. Seven different filters were used, and all the filters had bandwidths in the range of 15–30 nm and transmission from

40% to 60%. Sm^{3+} and Dy^{3+}:$Ca_2Gd_2W_3O_{14}$ phosphors were also characterized for their PL; both excitation and emission spectra of Sm^{3+} and Dy^{3+}:$Ca_2Gd_2W_3O_{14}$ phosphors were obtained on a SPEX Fluorolog-2 Fluorimeter (Model II) with Datamax software to acquire the data with a Xe arc lamp (150 W) as the excitation source.

7.3 Results and Discussion

7.3.1 Eu^{3+}:$Ca_2Gd_2W_3O_{14}$ Phosphors

7.3.1.1 X-ray diffraction patterns

The maximum diffraction peaks that are in good agreement with the standard JCPDS card no.41-0186 having tetragonal phase are obtained in the XRD pattern of the Eu^{3+}:$Ca_2Gd_2W_3O_{14}$ phosphor prepared by the solid-state reaction method, as shown in Fig. 7.1.

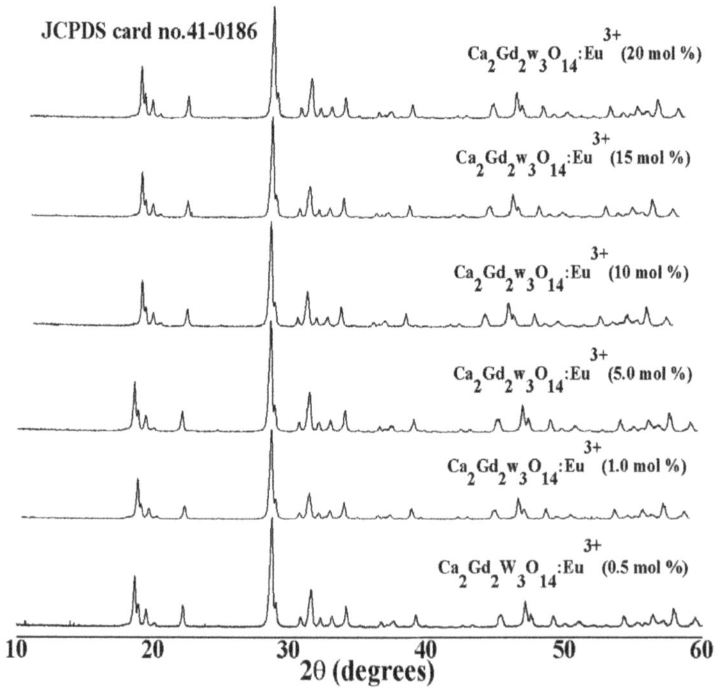

Figure 7.1 XRD profiles of Eu^{3+}:$Ca_2Gd_2W_3O_{14}$ phosphors. Reprinted by permission from Ref. [16], Springer Nature, 2011.

The pattern also indicates that the doping concentration of Eu^{3+} ions does not influence the intensity or the crystal structure of the phosphors.

7.3.1.2 SEM and EDAX analyses

The SEM image of a 10 mol% Eu^{3+}:$Ca_2Gd_2W_3O_{14}$ phosphor is shown in Fig. 7.2, and it reveals that the particles are randomly distributed, they have various shapes and sizes, and the average grain size is 300–500 nm. The EDAX spectrum of a 10 mol% Eu^{3+}:$Ca_2Gd_2W_3O_{14}$ phosphor is shown in Fig. 7.3, and it confirms the presence of Eu^{3+} ions.

Figure 7.2 SEM image of a 10 mol% Eu^{3+}:$Ca_2Gd_2W_3O_{14}$ phosphor. Reprinted by permission from Ref. [16], Springer Nature, 2011.

7.3.1.3 FTIR analysis

Different types of bondings at different wave numbers are observed from the FTIR analysis of a 10 mol% of Eu^{3+}:$Ca_2Gd_2W_3O_{14}$ phosphor, shown in Fig. 7.4. From the figure it is observed that the bending vibration of the H_2O molecule and the stretching vibrations of O–H are present at about 1480 cm^{-1} and are in the range of 3800 to 2700 cm^{-1} [65]. The band at 809 cm^{-1} can be assigned to the stretching mode of W–O bonds, and the band at 607 cm^{-1} may correspond to the W–O vibrations in the range of 800 to 600 cm^{-1}.

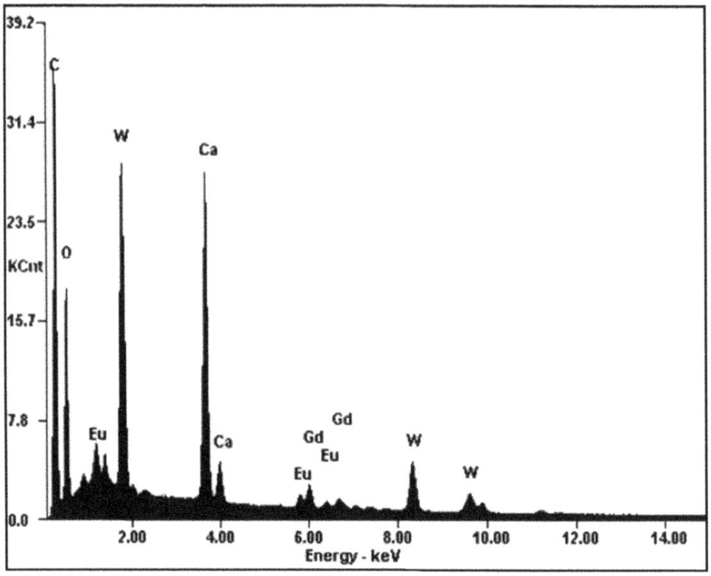

Figure 7.3 EDAX spectrum of a 10 mol% Eu^{3+}:$Ca_2Gd_2W_3O_{14}$ phosphor. Reprinted by permission from Ref. [16], Springer Nature, 2011.

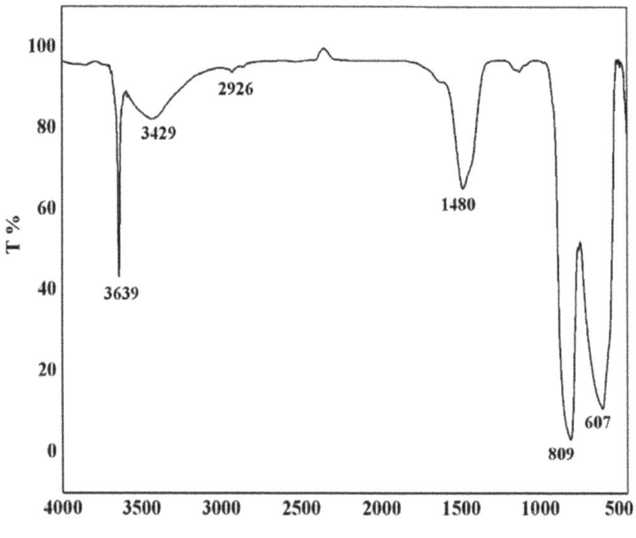

Figure 7.4 FTIR spectrum of a 10 mol% Eu^{3+}:$Ca_2Gd_2W_3O_{14}$ phosphor. Reprinted by permission from Ref. [16], Springer Nature, 2011.

7.3.1.4 Photoluminescence studies

The Eu^{3+}:$Ca_2Gd_2W_3O_{14}$ phosphor was monitored at an intense emission of 615 nm to record its excitation spectrum, as represented in Fig. 7.5. The most intense peak in this spectrum was located at 392 nm and has been chosen for the measurement of the emission spectra of Eu^{3+}:$Ca_2Gd_2W_3O_{14}$ phosphors. The spectrum also contains an intense broad band centered at 304 nm in the wavelength region of 250–350 nm, which is attributed to the CTB between $O^{2-} \rightarrow W^{6+}$ and $O^{2-} \rightarrow Eu^{3+}$ [66]. The presence of the most intense broad band of $WO_{24}{}^{2-}$ in the excitation spectrum of Eu^{3+} indicates that there is an efficient energy transfer from the tungstate groups to Eu^{3+} ions in $Ca_2Gd_2W_3O_{14}$ phosphors [16]. The remaining excitation peaks are observed in the wavelength region of 350–450 nm and are located at 359 nm ($^7F_0 \rightarrow {}^5D_4$), 379 nm ($^7F_0 \rightarrow {}^5L_7$), 392 nm ($^7F_0 \rightarrow {}^5L_6$), 414 nm ($^7F_0 \rightarrow {}^5D_3$), 463 nm ($^7F_0 \rightarrow {}^5D_2$), and 533 nm ($^7F_0 \rightarrow {}^5D_1$), which are attributed to f–f transitions of Eu^{3+}. Moreover, the CTB is sensitive to changes in the crystal field, unlike the f–f transitions of Eu^{3+} ions [67].

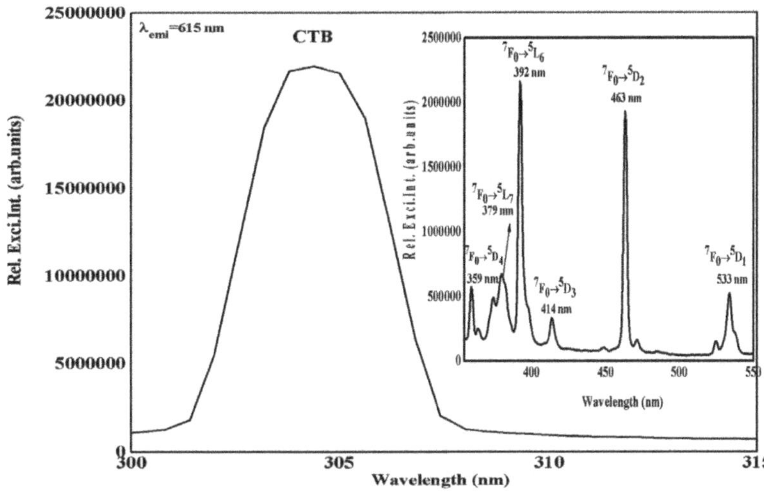

Figure 7.5 Excitation spectrum of a 10 mol% Eu^{3+}:$Ca_2Gd_2W_3O_{14}$ phosphor. Reprinted by permission from Ref. [16], Springer Nature, 2011.

The emission spectra shown in Fig. 7.6 contain an intense emission peak at 615 nm, which confirms the characteristic emission of the Eu^{3+}:$Ca_2Gd_2W_3O_{14}$ phosphor and is assigned to the electric dipole transition of $^5D_0 \rightarrow {}^7F_2$, while the weak emission peaks are located at 579 nm, 591 nm, 645 nm, and 702 nm and are assigned to the transitions $^5D_0 \rightarrow {}^7F_0$, $^5D_0 \rightarrow {}^7F_1$, $^5D_0 \rightarrow {}^7F_3$, and $^5D_0 \rightarrow {}^7F_4$, respectively [16]. Luminescence originating from transitions between 4f levels is predominant due to electric dipole or magnetic dipole interactions [68, 69], and the energy has been transferred from $W_4O_{24}{}^{-24}$ to Eu^{3+}. From the emission spectra, it is also observed that the $^5D_0 \rightarrow {}^7F_1$ and $^5D_0 \rightarrow {}^7F_2$ emission bands split into two Stark components, which depends on the local environment of Eu^{3+} ions in the crystalline structure.

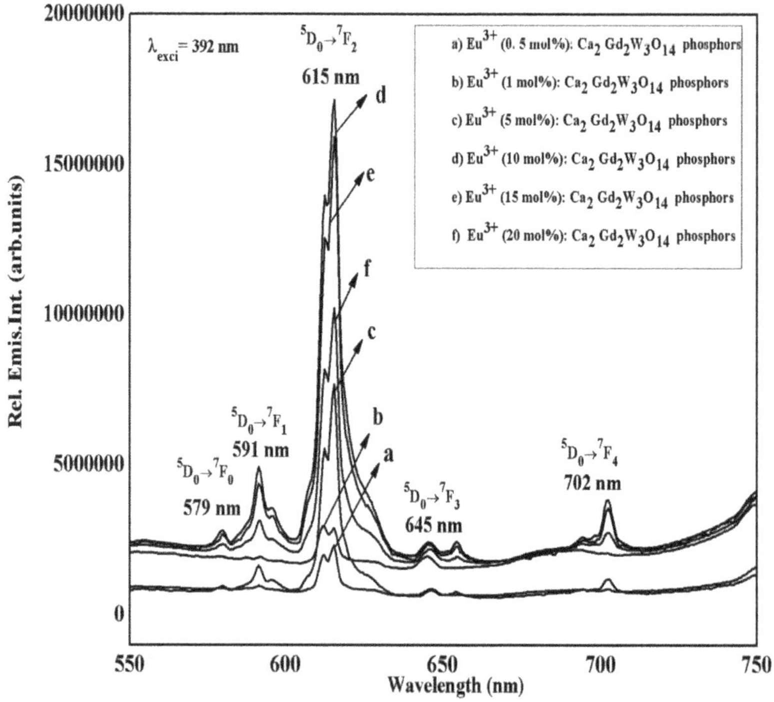

Figure 7.6 Emission spectra of Eu^{3+}:$Ca_2Gd_2W_3O_{14}$ phosphors. Reprinted by permission from Ref. [16], Springer Nature, 2011.

In other words, the local environment of rare earth ions plays an important role in the luminescence performance. Particularly, the $^5D_0 \rightarrow {}^7F_2$ transition is sensitive to variations in the local crystal field, which causes the splitting and shifting of the emission band

[70, 71]. Moreover, the luminescence performance of powder phosphors depends on the concentration of dopant, so the study of the optimum dopant concentration is necessary. Figure 7.7 shows the emission intensity of Eu^{3+}:$Ca_2Gd_2W_3O_{14}$ as a function of its doping concentration (mol%).

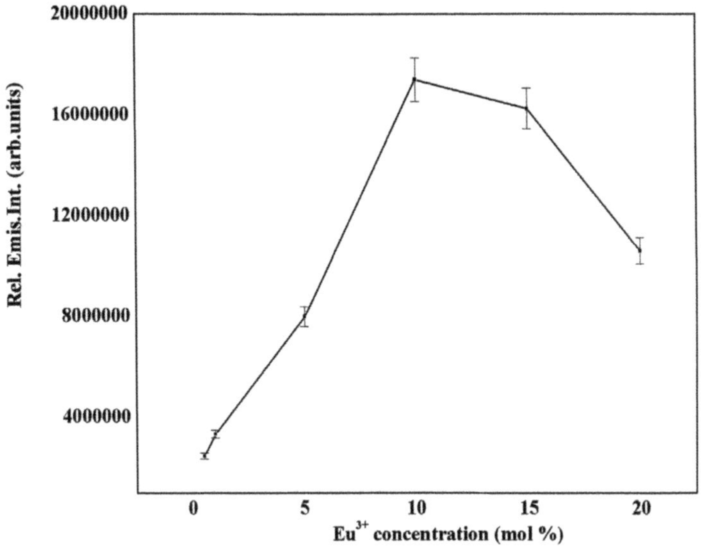

Figure 7.7 Emission intensity of Eu^{3+} as a function of its doping concentration of Eu^{3+}:$Ca_2Gd_2W_3O_{14}$ phosphors. Reprinted by permission from Ref. [16], Springer Nature, 2011.

It is observed from the emission spectra that by increasing the dopant (Eu^{3+}) concentration, the PL intensity increases up to 10 mol%; beyond that concentration, the PL intensity decreases. This phenomenon is known as concentration quenching; in general, by increasing the dopant concentration, the number of Eu^{3+} ions will increase and the distance between activator ions will decrease, which causes more interaction between Eu^{3+} ions and also energy migration among rare earth ions, which brings the excitation energy to defect sites in the crystal lattice [72].

7.3.1.5 Mechanoluminescence studies

Figure 7.8 shows the ML studies of a gamma-irradiated $Ca_2Gd_2W_3O_{14}$:Eu^{3+} (5 mol%) phosphor for 30 min. irradiation time. It is observed that initially the ML intensity increases with time, attains a maximum, then decreases with time, and finally disappears. The ML intensity is due to radiative recombination of dislocation holes with the recombination centers (defect centers containing electrons). The ML process can be explained by the following equations [16]:

During irradiation,

$$Eu^{3+} + e^- = Eu^{2+}$$

$$\text{Positive ion vacancy} + \text{Hole} = V_k$$

During deformation,

$$V_k + D = D^+ + \text{Positive ion vacancy}$$

$$Re^{2+} + D^+ = Re^{3+} + D$$

$$Re^{3+}* = Re^{3+} + h\nu$$

Here, V_k and D represent the V_k center and dislocation, respectively; D^+ is the dislocation that contains holes.

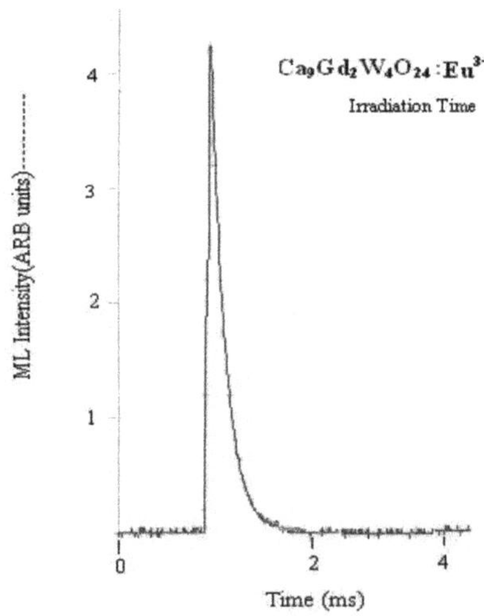

Figure 7.8 Mechanoluminescence of a Eu^3:$Ca_2Gd_2W_3O_{14}$ phosphor (5 mol%) for 30 min. irradiation time. Reprinted by permission from Ref. [16], Springer Nature, 2011.

Further Fig. 7.9 shows the ML of a gamma-irradiated $Ca_2Gd_2W_3O_{14}$:Eu^{3+} phosphor for different concentrations of Eu^{3+}. It is seen that the ML intensity initially increases with increasing concentration of the dopant, is maximum at 10 mol% concentration, and gets saturated at 1 kGy/h due to aggregation of color centers formed by irradiation. And after that the ML intensity decreases with increasing concentration of the dopant. This is due to concentration quenching at higher concentrations.

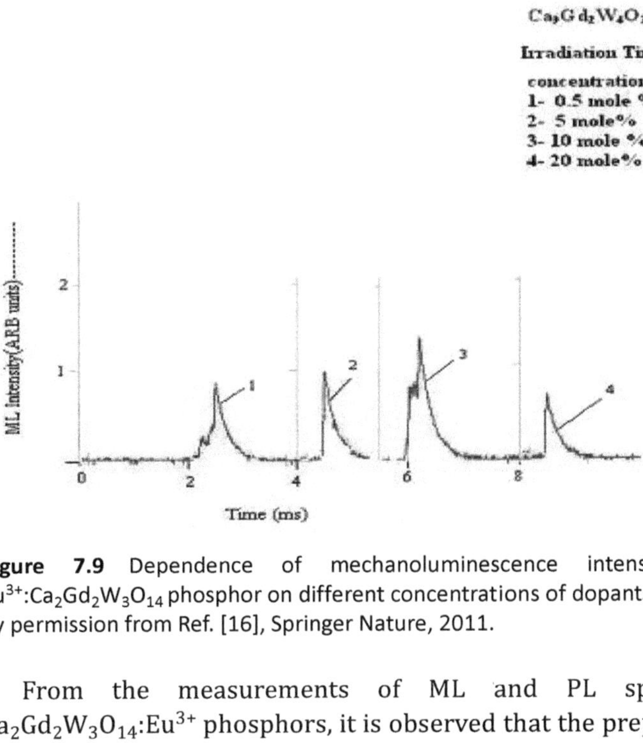

Figure 7.9 Dependence of mechanoluminescence intensity of a Eu^{3+}:$Ca_2Gd_2W_3O_{14}$ phosphor on different concentrations of dopants. Reprinted by permission from Ref. [16], Springer Nature, 2011.

From the measurements of ML and PL spectra of $Ca_2Gd_2W_3O_{14}$:Eu^{3+} phosphors, it is observed that the prepared materials can be used as a ML dosimetry phosphor for radiation measurements up to a higher exposure of 1 kGy and are also found to have applications in WLEDs.

7.3.2 Tb^{3+}:$Ca_2Gd_2W_3O_{14}$ Phosphor

7.3.2.1 Structural, morphological, elemental, and FTIR studies of Sm^{3+}: and Dy^{3+}:$Ca_2Gd_2W_3O_{14}$ phosphors

The XRD spectrum of a $Ca_2Gd_2W_3O_{14}$: Tb^{3+} (5 mol%) phosphor is shown in Fig. 7.10. From this figure it is revealed that the phosphor exists in the tetragonal phase and the diffraction peaks are in good agreement with the standard JCPDS card no.41-0186 and moreover the doping of Tb^{3+} ions does not influence the crystal structure of the prepared phosphors. Figure 7.11 shows the SEM image of a $Ca_2Gd_2W_3O_{14}$:Tb^{3+} phosphor. The image shows that the grain size is around 300 nm and the particles are distributed randomly, indicating the phenomenon of agglomeration.

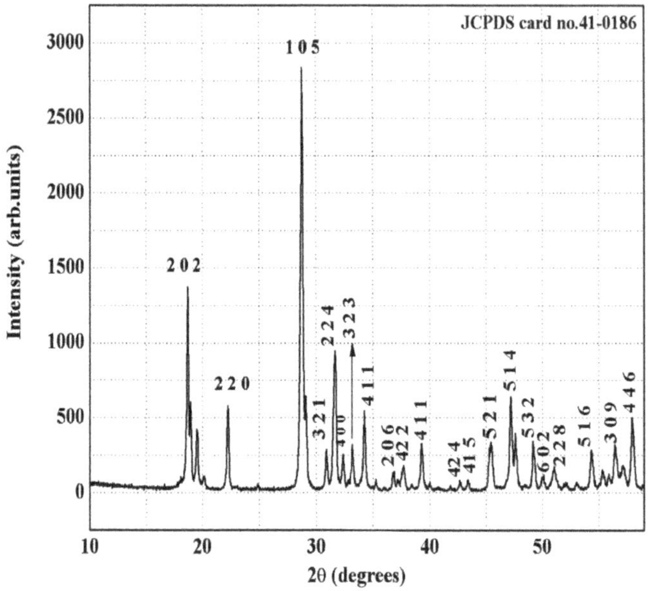

Figure 7.10 XRD profile of 5 mol% Tb^{3+}:$Ca_2Gd_2W_3O_{14}$ phosphors. Reprinted by permission from Ref. [17], Springer Nature, 2011.

Figure 7.12 represents the EDAX spectrum of 5 mol% Tb^{3+}:$Ca_2Gd_2W_3O_{14}$ phosphors, which confirms the presence of Ca, Gd, W, O, and Tb elements in the prepared phosphor.

Figure 7.13 represents the FTIR spectrum of 5 mol% Tb^{3+}:$Ca_2Gd_2W_3O_{14}$ phosphors.

From the FTIR spectrum it is observed that in the first region, 1487 cm^{-1} and 1627 cm^{-1} represent the bending vibration of the H_2O molecule and O–H stretching vibrations, respectively, in the range of 3800 to 2700 cm^{-1}.

In the second region, the band at 809 cm^{-1} can be assigned to the stretching mode of W–O bonds and the band at 607 cm^{-1} corresponds to the W–O vibrations in the range of 800 to 600 cm^{-1}.

Figure 7.11 SEM image of 5 mol% Tb^{3+}:$Ca_2Gd_2W_3O_{14}$ phosphors. Reprinted by permission from Ref. [17], Springer Nature, 2011.

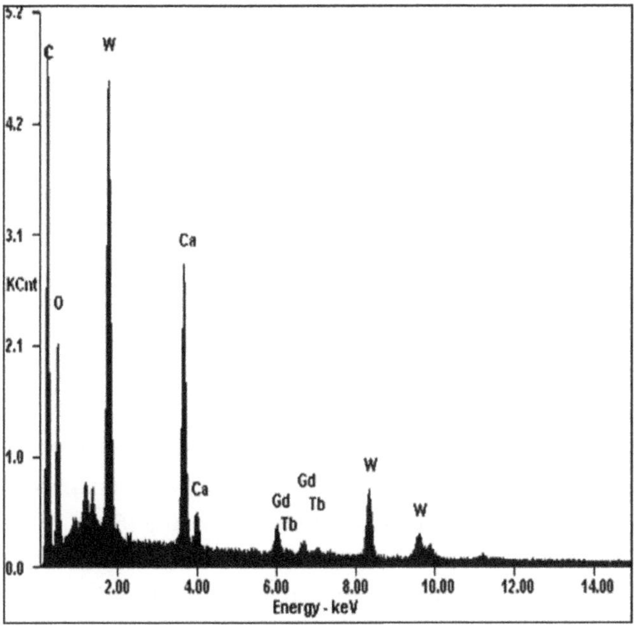

Figure 7.12 EDAX spectrum of 5 mol% Tb^{3+}:$Ca_2Gd_2W_3O_{14}$ phosphors. Reprinted by permission from Ref. [17], Springer Nature, 2011.

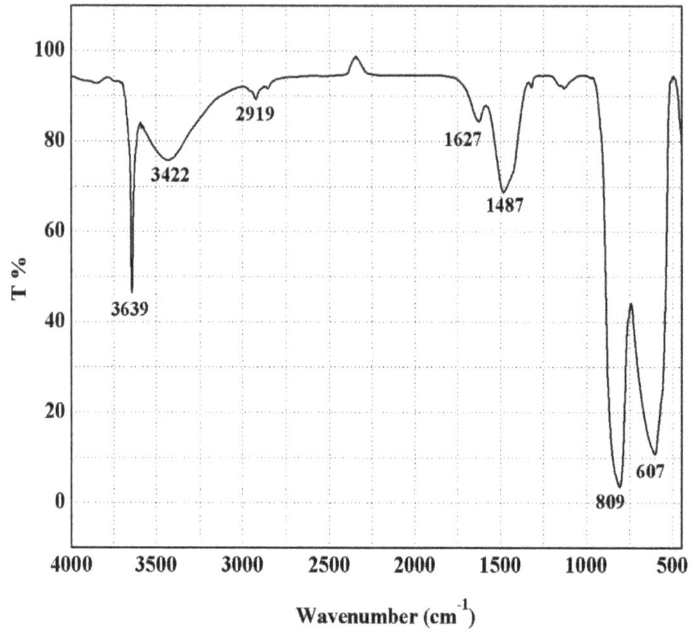

Figure 7.13 FTIR spectrum of 5 mol% Tb^{3+}:$Ca_2Gd_2W_3O_{14}$ phosphors. Reprinted by permission from Ref. [17], Springer Nature, 2011.

7.3.2.2 Photoluminescence studies

Figure 7.14 represents the excitation spectrum of $Ca_2Gd_2W_3O_{14}$:Tb^{3+} (5 mol%) phosphors by monitoring the emission at 545 nm ($^5D_4 \rightarrow {}^7F_5$). In the wavelength region 300–400 nm, the f–f transitions of Tb^{3+} ions are observed at 316 nm, 337 nm, 350 nm, and 374 nm and are assigned to the transitions ($^7F_6 \rightarrow {}^5H_7$), ($^7F_6 \rightarrow {}^5D_1$) ($^7F_6 \rightarrow {}^5D_2$), and ($^7F_6 \rightarrow {}^5G_6$), respectively, in the near-UV region of the excitation spectrum of fluorescent lamp materials. The most prominent excitation band, at 374 nm, has been chosen to measure the emission spectra of Tb^{3+}:$Ca_2Gd_2W_3O_{14}$ nanophosphors and is shown in Figure 7.15. The most intense peak, at 545 nm, arises from the $^5D_4 \rightarrow {}^7F_5$ transition, while the other peaks, at 488, 584, and 621 nm, correspond to the transitions $^5D_4 \rightarrow {}^7F_6$, $^5D_4 \rightarrow {}^7F_4$, and $^5D_4 \rightarrow {}^7F_3$, respectively. The obtained emission transitions are due to the f–f transitions of Tb^{3+} within the $4f^8$ electronic configuration [73].

Further the emission transitions of Tb^{3+} ions are mainly from the two excited energy levels $^5D_3$ and $^5D_4 \rightarrow {}^7F_J$ ground states because the large energy gap between $^5D_3$, $^5D_4$ states and ^7F_J ground states causes the relaxation process from these two ($^5D_3$ and $^5D_4$) excited energy states to the ^7F_J ground states, and this occurs radiatively, so the spectra in the visible region are obtained. Hence, the bright-green emission at 545 nm due to the excitation wavelength of 374 nm of $Ca_2Gd_2W_3O_{14}$: Tb^{3+}: phosphors makes them useful candidates for fluorescent lamps.

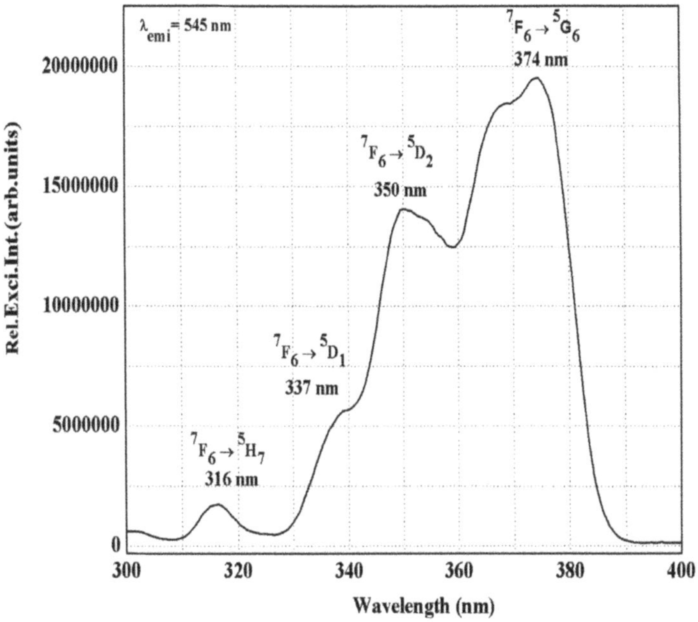

Figure 7.14 Excitation spectrum of 5 mol% Tb^{3+}:$Ca_2Gd_2W_3O_{14}$ phosphors. Reprinted by permission from Ref. [17], Springer Nature, 2011.

Figure 7.15 also shows that the spectral distribution mainly depends on the concentration of Tb^{3+} ions. If the concentration of Tb^{3+} ions is low, transitions from both excited energy states ($^5D_3$ and $^5D_4$) are possible, and by increasing the Tb^{3+} ion concentration, most of the terbium ions in the $^5D_3$ state relax nonradiatively to the $^5D_4$ level because of multipolar interactions with Tb^{3+} ions in the ground state [17]. This phenomenon is known as cross relaxation and is represented as Tb^{3+} ($^5D_3$) + Tb^{3+} ($^7F_6$) → Tb^{3+} ($^5D_4$) + Tb^{3+} ($^7F_0$) [74–76]. Therefore, the emission transitions from the $^5D_4$ state are predominant at higher concentrations of Tb^{3+} ions. Since the luminescent behavior of synthesized phosphors depends mainly on the concentration of the dopant, it is necessary to study the optimum dopant concentration. The emission spectra for different doping concentrations are shown in Fig. 7.16, and it is observed that the emission intensity increases by increasing the concentration of Tb^{3+} ions from 0.5 mol% to 5 mol% and beyond this concentration the PL intensity decreases because of concentration quenching and so the optimum concentration is found to be 5 mol%. As the concentration of Tb^{3+} ions increases, the distance between them decreases and the ions come closer to each other, which leads to the transfer of energy between terbium ions, and then this energy is released through some nonradiative processes [77].

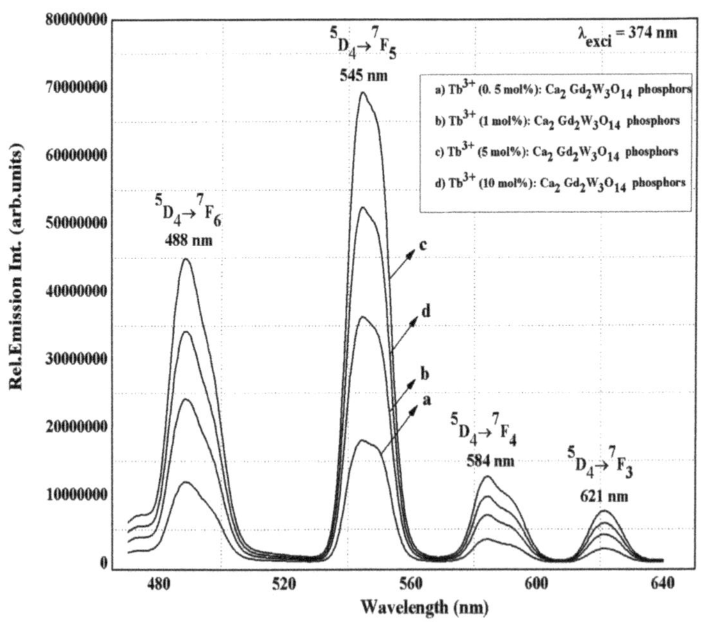

Figure 7.15 Emission spectra of Tb^{3+}:$Ca_2Gd_2W_3O_{14}$ phosphors. Reprinted by permission from Ref. [17], Springer Nature, 2011.

By comparing Eu^{3+} and Tb^{3+}:$Ca_2Gd_2W_3O_{14}$ phosphors, it was observed that Tb-activated phosphors attain their maximum luminescence intensity for smaller dopant concentrations, that is, 5 mol%, as compared to Eu^{3+}:$Ca_2Gd_2W_3O_{14}$ phosphors.

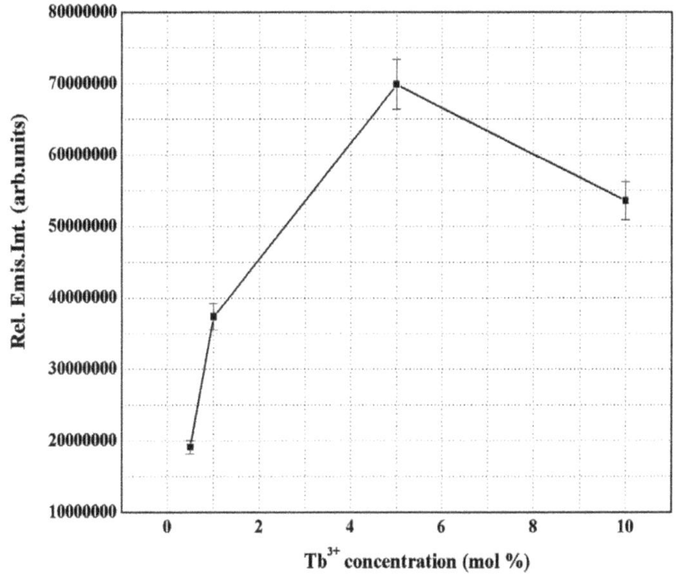

Figure 7.16 Emission intensity of Tb^{3+} as a function of its doping concentration of Tb^{3+}:$Ca_2Gd_2W_3O_{14}$ phosphors. Reprinted by permission from Ref. [17], Springer Nature, 2011.

ML properties of a Tb^{3+}:$Ca_2Gd_2W_3O_{14}$ phosphor have also been studied. Figure 7.17 shows the ML of a gamma-irradiated (5 mol%) Tb^{3+}:$Ca_2Gd_2W_3O_{14}$ phosphor for different γ-ray exposures.

It has been observed that the ML intensity increases initially with increase in γ-ray exposure, attains a maximum value for 1 h (exposure rate = 1 kGy/h) irradiation time, and for further higher irradiation time, it seems to be saturated. These results show that the prepared nanophosphors are sensitive to high radiation doses. As the irradiation dose increases, the color centers increase, resulting in the increase of intensity [17]. At higher doses the color centers get saturated and the recombination possibility of the color centers during the fractures is less, so the ML intensity is saturated.

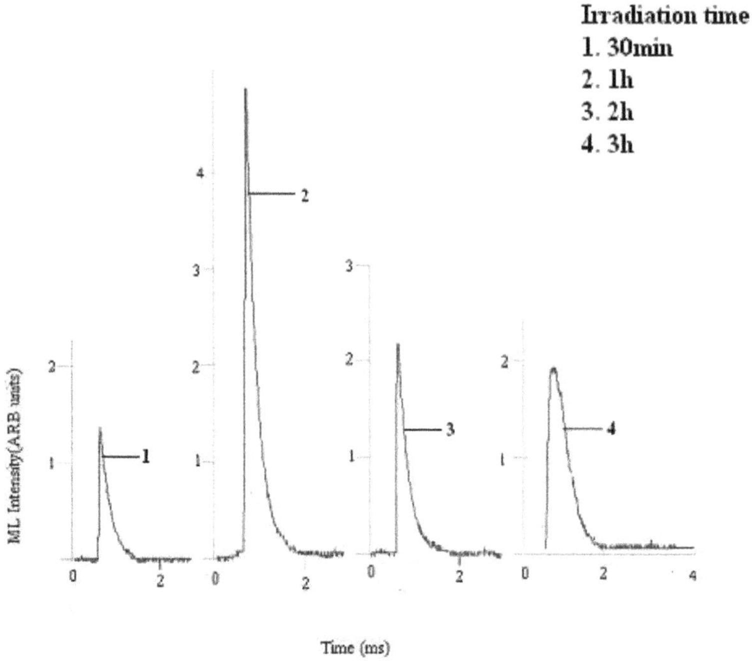

Figure 7.17 Dependence of the mechanoluminescence intensity of a 5 mol% Tb^{3+}:$Ca_2Gd_2W_3O_{14}$ phosphor at different irradiation doses. Reprinted by permission from Ref. [17], Springer Nature, 2011.

7.3.3 Sm^{3+}: and Dy^{3+}:$Ca_2Gd_2W_3O_{14}$ Phosphors

7.3.3.1 Structural, morphological, elemental, and FTIR studies of Sm^{3+}: and Dy^{3+}:$Ca_2Gd_2W_3O_{14}$ phosphors

To study the crystal-phase structure of 5 mol% Sm^{3+}: and Dy^{3+}:$Ca_2Gd_2W_3O_{14}$ phosphors prepared by the solid-state reaction method at different temperatures, from 800°C to 1400°C, the XRD profiles were measured and are shown in Figs. 7.18 and 7.19, respectively. From both figures, it has been observed that the intensity is influenced by the synthesis temperature and it shows best results at 1000°C. Thus 1000°C can be used as the optimized temperature for all the remaining measurements. The diffraction peaks in the measured patterns are consistent with the standard JCPDS card no.41-0186, and the prepared phosphors exist in a tetragonal-phase structure irrespective of the doping of rare earth (Sm^{3+} and Dy^{3+}) ions.

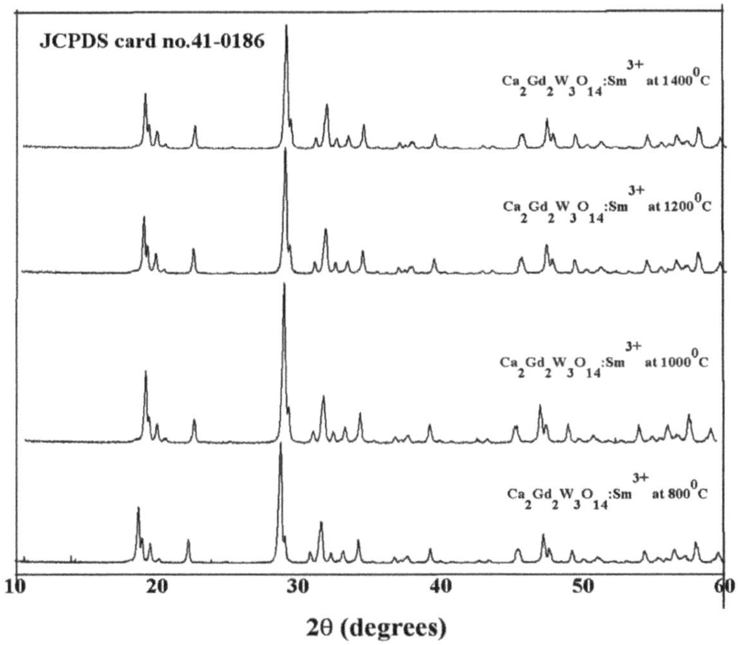

Figure 7.18 XRD profiles of 5 mol% of Sm^{3+}:$Ca_2Gd_2W_3O_{14}$ phosphors at different temperatures. Reprinted from Ref. [18], Copyright (2011), with permission from Elsevier.

The size and shape of the prepared phosphors were studied by SEM. Figures 7.20 and 7.21 show the SEM images of 5 mol% Sm^{3+}: and Dy^{3+}:$Ca_2Gd_2W_3O_{14}$ phosphors, respectively. It is revealed that for both samples, the particles are randomly distributed, having various shapes and sizes, and the average grain size is expected to be 100–250 nm. EDAX analysis was used to confirm the presence of respective dopants in the prepared phosphors.

EDAX profiles of Sm^{3+}: and Dy^{3+}:$Ca_2Gd_2W_3O_{14}$ phosphors are shown in Figs. 7.22 and 7.23, respectively, which confirm the presence of Ca, Gd, W, O, and both of the rare earth ions (Sm^{3+} and Dy^{3+}) in their respective profiles. Further FTIR spectroscopy was used to identify the functional groups present in the prepared phosphors. FTIR spectra of 5 mol% Sm^{3+}: and Dy^{3+}:$Ca_2Gd_2W_3O_{14}$ phosphors are shown in Fig. 7.24.

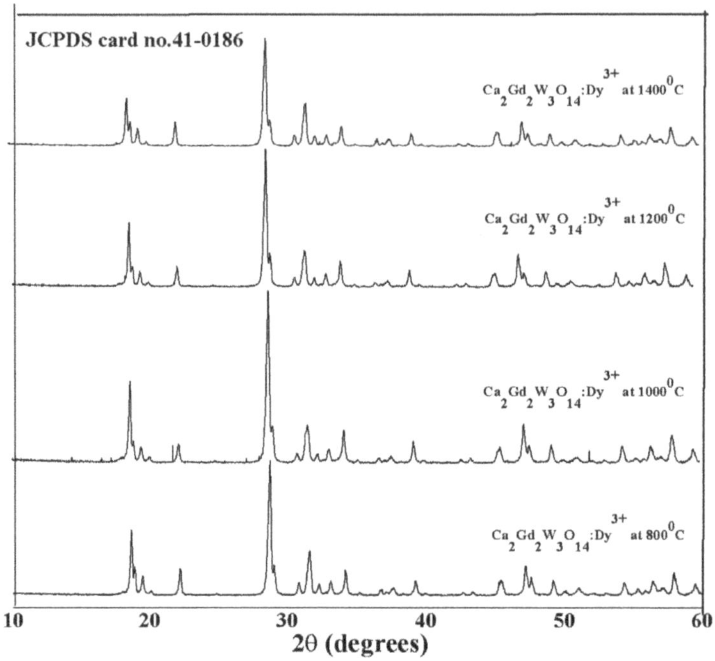

Figure 7.19 XRD profiles of 5 mol% of Dy^{3+}:$Ca_2Gd_2W_3O_{14}$ phosphors at different temperatures. Reprinted from Ref. [18], Copyright (2011), with permission from Elsevier.

Figure 7.20 SEM image of 5 mol% of Sm^{3+}:$Ca_2Gd_2W_3O_{14}$ phosphors. Reprinted from Ref. [18], Copyright (2011), with permission from Elsevier.

Figure 7.21 SEM image of 5 mol% of Dy^{3+}:$Ca_2Gd_2W_3O_{14}$ phosphors. Reprinted from Ref. [18], Copyright (2011), with permission from Elsevier.

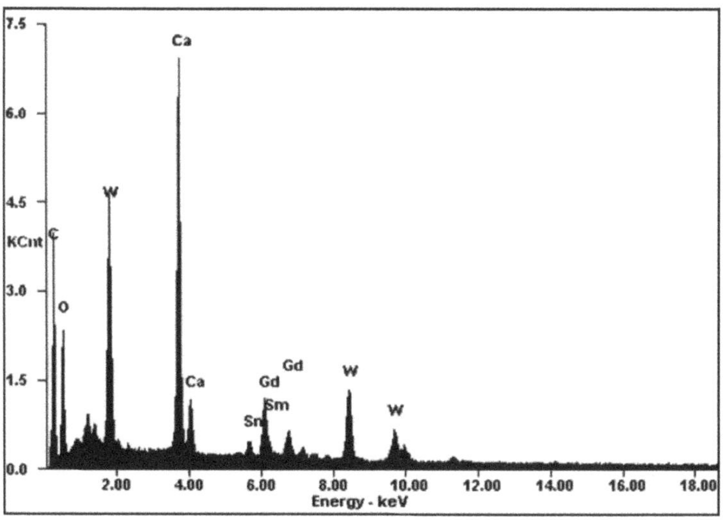

Figure 7.22 EDAX profile of 5 mol% of Sm^{3+}:$Ca_2Gd_2W_3O_{14}$ phosphors. Reprinted from Ref. [18], Copyright (2011), with permission from Elsevier.

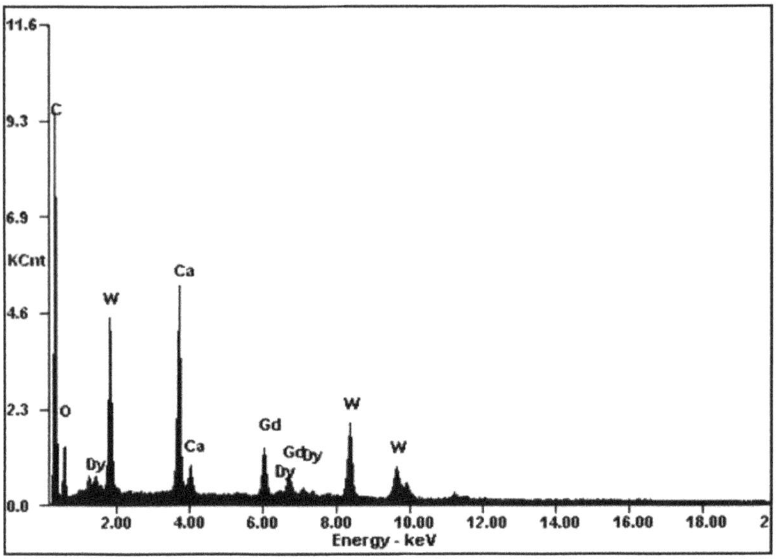

Figure 7.23 EDAX profile of 5 mol% of Dy^{3+}:$Ca_2Gd_2W_3O_{14}$ phosphors. Reprinted from Ref. [18], Copyright (2011), with permission from Elsevier.

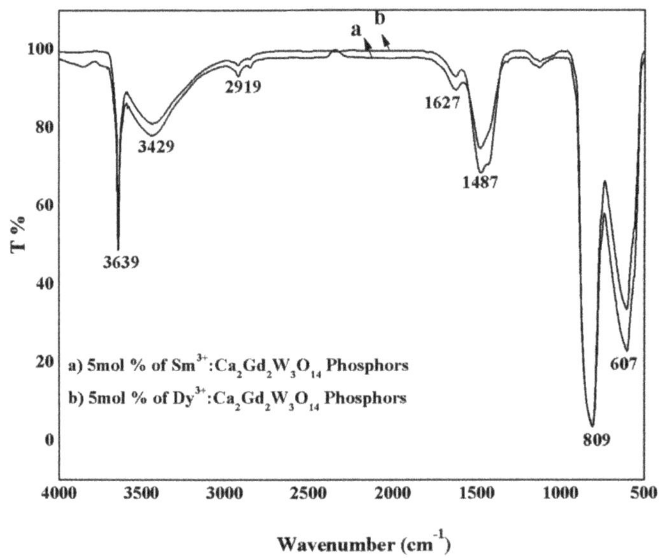

Figure 7.24 FTIR spectra of 5 mol% of Sm^{3+}: and Dy^{3+}:$Ca_2Gd_2W_3O_{14}$ phosphors. Reprinted from Ref. [18], Copyright (2011), with permission from Elsevier.

It shows that the band at 809 cm^{-1} can be assigned to the stretching mode of W–O bonds and the band at 607 cm^{-1} in the range 800 to 600 cm^{-1} corresponds to the W–O vibrations. The bending vibration of the H_2O molecule and O–H stretching vibrations are present at about 1480 cm^{-1} and in the range 3800 to 2700 cm^{-1} [18].

7.3.3.2 Photoluminescence studies

7.3.3.2.1 *Sm^{3+}:$Ca_2Gd_2W_3O_{14}$ phosphors*

The excitation spectrum of a 5 mol% Sm^{3+}:$Ca_2Gd_2W_3O_{14}$ phosphor in the wavelength region of 380–450 nm is shown in Fig. 7.25. In this spectrum, three excitation peaks, located at 402 nm ($^6H_{5/2} \rightarrow {}^4F_{7/2}$), 418 nm [$^6H_{5/2} \rightarrow (^6P, {}^4P)_{5/2}$], and 436 nm ($^6H_{5/2} \rightarrow {}^4G_{9/2}$), have been observed and are attributed to f–f transitions of Sm^{3+}. The CTB of Sm^{3+}–O^{2-} interaction or host absorption band is not observed because the interaction of Sm^{3+} ions with the host lattice is very weak; therefore, energy transfer does not occur between Sm^{3+} and the host [78].

Since the intensity of the f–f transition at 402 nm is high compared to the other transitions, it has been chosen for the measurement of emission spectra of Sm^{3+}:$Ca_2Gd_2W_3O_{14}$ phosphors. Further the emission spectra of Sm^{3+}:$Ca_2Gd_2W_3O_{14}$ phosphors are observed and shown in Fig. 7.26.

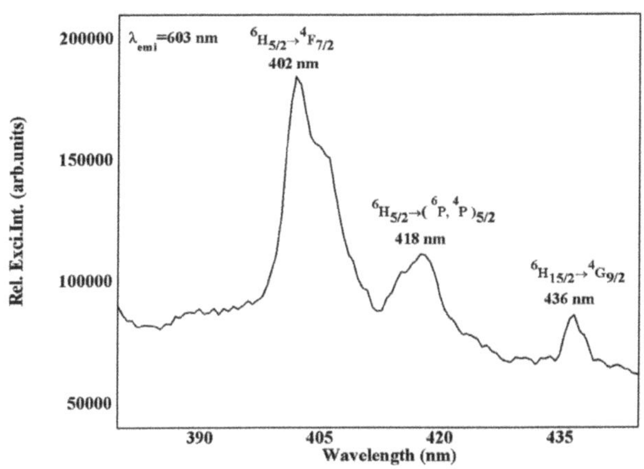

Figure 7.25 Excitation spectrum of 5 mol% of Sm^{3+}:$Ca_2Gd_2W_3O_{14}$ phosphors. Reprinted from Ref. [18], Copyright (2011), with permission from Elsevier.

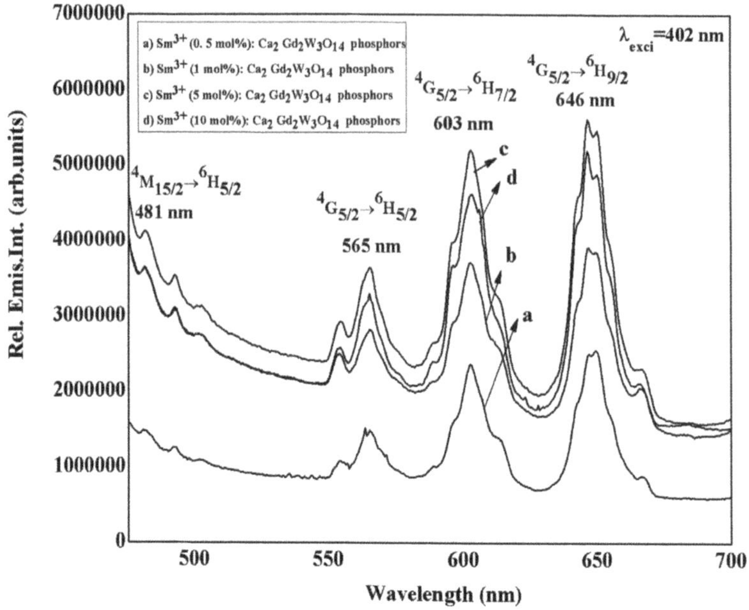

Figure 7.26 Emission spectra of Sm^{3+}:$Ca_2Gd_2W_3O_{14}$ phosphors. Reprinted from Ref. [18], Copyright (2011), with permission from Elsevier.

From the emission spectra, four emission peaks are observed, at 481 nm, 565 nm, 603 nm, and 646 nm, which are assigned to $^4M_{15/2} \rightarrow {}^6H_{5/2}$, $^4G_{5/2} \rightarrow {}^6H_{5/2}$, $^4G_{5/2} \rightarrow {}^6H_{7/2}$, and $^4G_{5/2} \rightarrow {}^6H_{9/2}$ transitions, respectively. Among these, the transition at 646 nm ($^4G_{5/2} \rightarrow {}^6H_{9/2}$) has the maximum intensity, which corresponds to the red emission of the Sm^{3+}:$Ca_2Gd_2W_3O_{14}$ phosphor. The transition at 646 nm in the red region is in agreement with the previously reported literature [79]. From the emission spectra, it is also observed that the emission band at 646 nm ($^4G_{5/2} \rightarrow {}^6H_{9/2}$) has split into two Stark components, which depends upon the local environment of Sm^{3+} ions in the crystalline structure [18].

Further the emission intensity of Sm^{3+} as a function of its doping concentration (mol%) in Sm^{3+}:$Ca_2Gd_2W_3O_{14}$ phosphors is shown in Fig. 7.27.

From the emission spectra, it is observed that the emission intensities increase gradually from 0.5 to 5 mol% and beyond this concentration the emission intensities decrease due to the concentration-quenching effect [80]. Hence 5 mol% of Sm^{3+} is the optimum dopant concentration.

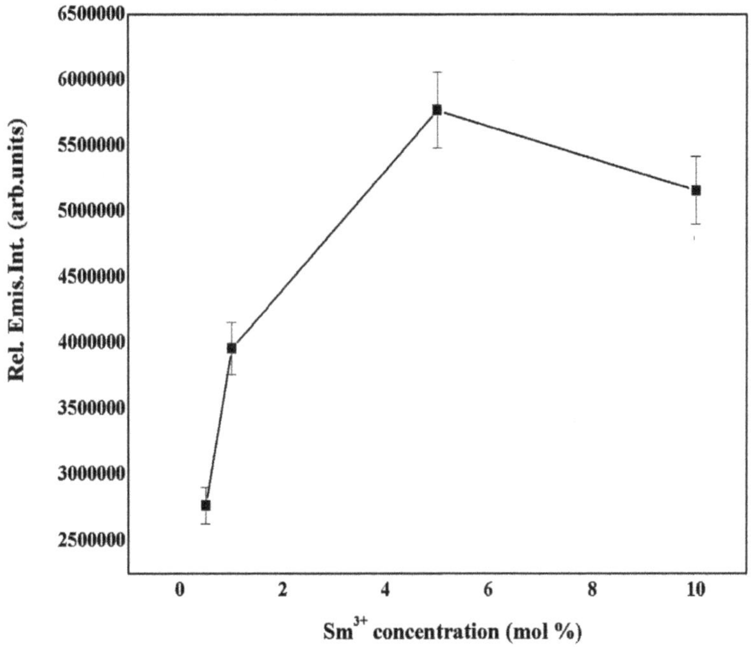

Figure 7.27 Emission intensity of the transition at 646 nm of Sm^{3+} as a function of its doping concentration in Sm^{3+}:$Ca_2Gd_2W_3O_{14}$ phosphors. Reprinted from Ref. [18], Copyright (2011), with permission from Elsevier.

7.3.3.2.2 *Dy^{3+}:$Ca_2Gd_2W_3O_{14}$ phosphors*

The excitation spectrum of a 5 mol% Dy^{3+}:$Ca_2Gd_2W_3O_{14}$ phosphor is shown in Fig. 7.28.

From the excitation spectrum, three excitation bands are observed, at 363 nm ($^6H_{15/2} \rightarrow {}^6P_{5/2}$), 385 nm ($^6H_{15/2} \rightarrow {}^4I_{13/2}$), and 419 nm ($^6H_{15/2} \rightarrow {}^4G_{11/2}$). The most intense excitation peak, at 385 nm, has been selected for the measurement of the emission spectrum of Dy^{3+}:$Ca_2Gd_2W_3O_{14}$ phosphors. Figure 7.29 shows the emission spectrum of Dy^{3+}:$Ca_2Gd_2W_3O_{14}$ phosphors.

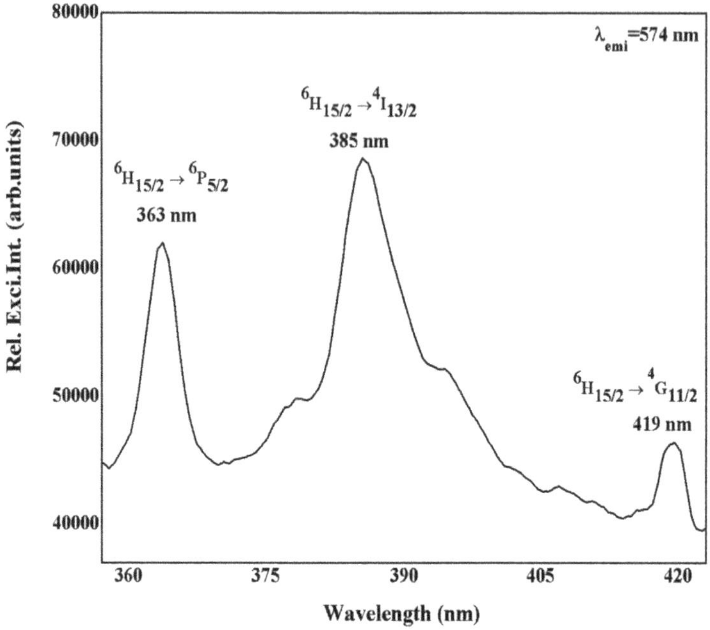

Figure 7.28 Excitation spectrum of 5 mol% of Dy^{3+}:$Ca_2Gd_2W_3O_{14}$ phosphors. Reprinted from Ref. [18], Copyright (2011), with permission from Elsevier.

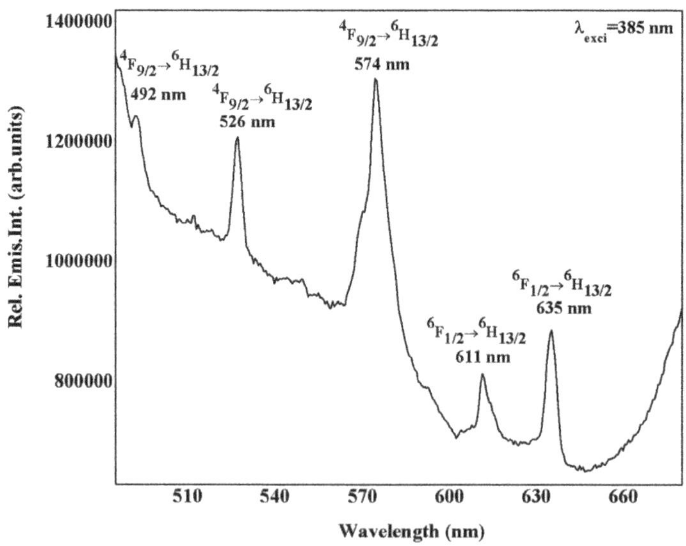

Figure 7.29 Emission spectrum of 5 mol% of Dy^{3+}:$Ca_2Gd_2W_3O_{14}$ phosphors. Reprinted from Ref. [18], Copyright (2011), with permission from Elsevier.

The emission spectrum shows five emission bands, located at 492 nm ($^4F_{9/2} \rightarrow {}^6H_{13/2}$), 526 nm ($^4F_{9/2} \rightarrow {}^6H_{13/2}$), 574 nm ($^4F_{9/2} \rightarrow {}^6H_{13/2}$), 611 nm ($^6F_{1/2} \rightarrow {}^6H_{13/2}$), and 635 nm ($^6F_{1/2} \rightarrow {}^6H_{13/2}$) [81]. From the emission spectrum, it is clear that the emission was observed in three regions, namely blue, yellow, and red. Among these, the yellow region was the most intense, with the band located at 574 nm ($^4F_{9/2} \rightarrow {}^6H_{13/2}$). It is well known that the $^4F_{9/2} \rightarrow {}^6H_{13/2}$ transition at 574 nm with a bright-yellow emission is a forced electric dipole transition following the selection rule $\Delta J = 2$, which is allowed only when Dy^{3+} is not at an inversion center, and it is also called hypersensitive transition because its intensity is sensitive to the outside crystal field environment surrounding Dy^{3+} [82]. Thus 5 mol% Dy^{3+}:$Ca_2Gd_2W_3O_{14}$ phosphors with bright-yellow emissions can be used as novel optical luminescent materials.

7.4 Conclusions

Eu^{3+}, Tb^{3+}, Sm^{3+}, and Dy^{3+} ion–doped $Ca_2Gd_2W_3O_{14}$ powder phosphors have been synthesized by using the solid-state reaction method. The excitation spectrum of the Eu^{3+}:$Ca_2Gd_2W_3O_{14}$ phosphor shows energy transfer from the WO^{2-} group of the host lattice to the Eu^{3+} ions. Emission spectra of Eu^{3+}:$Ca_2Gd_2W_3O_{14}$ phosphors have shown strong red emissions and may be promising candidates as red-emitting phosphors for WLEDs. The ML characterization shows that the prepared phosphors can be used as ML dosimetry phosphors for radiation measurement up to 1 kGy exposure. The emission spectrum of 5 mol% of Tb^{3+}:$Ca_2Gd_2W_3O_{14}$ phosphors has shown a bright-green emission at 545 nm by fluorescent lamp excitation wavelength, that is, 374 nm near-UV excitation. This makes them potential candidates for fluorescent lamps. Intense red and yellow emissions are observed from Sm^{3+} and Dy^{3+} ion–doped $Ca_2Gd_2W_3O_{14}$ powder phosphors synthesized by the solid-state reaction method. After considering the luminescent performance, it can be said that 5 mol% Sm^{3+}: and Dy^{3+}:$Ca_2Gd_2W_3O_{14}$ phosphors are potential candidates for applications as novel luminescent materials in optical systems.

References

1. Wang, B., Li, X., Zeng, Q., Yang, G., Luo, J., He, X. and Chen, Y. (2018). Efficiently enhanced photoluminescence in Eu^{3+}-doped $Lu_2(MoO_4)_3$ by Gd^{3+} substituting, *Mater. Res. Bull.*, **100**, pp. 97–101.
2. Chien, T. C., Hwang, C. S., Nien, Y. T. and Yoshimura, M. (2017). Tungstate flux-grown $Y_{10}W_2O_{21}$: Eu^{3+} phosphors with enhanced particle size and luminescence properties, *Ceram. Int.*, **43**, pp. 9437–9442.
3. Zhou, Y. and Yan, B. (2013). $RE_2(MO_4)_3$:Ln^{3+} (RE = Y, La, Gd, Lu; M = W, Mo; Ln = Eu, Sm, Dy) microcrystals: controlled synthesis, microstructure and tunable luminescence, *CrystEngComm*, **15**, pp. 5694–5702.
4. Candelaria, G. A., Cristina, G. S., Javier, G. P., Maueul, E. T., Antonio Diego, L. D., Nanci, S., Victor, S. F., Javier, C. and Juan, J. R. (2011). Structural investigation of the negative thermal expansion in yttrium and rare earth molybdates, *J. Phys. Condens. Matter*, **23**, pp. 325402–325411.
5. Du, P. and Yu, J. (2017). Self-activated multicolor emissions in $Ca_2NaZn_2(VO_4)_3$:Eu^{3+} phosphors for simultaneous warm white light-emitting diodes and safety sign, *Dyes Pigm.*, **147**, pp. 16–23.
6. Song, E., Chen, Z., Wu, M., Ding, S., Ye, S. and Zhou, S. (2016). Upconversion: room-temperature wavelength-tunable single-band upconversion luminescence from Yb^{3+}/Mn^{2+} codoped fluoride perovskites ABF_3, *Adv. Opt. Mater.*, **4**, pp. 798–806.
7. Li, X., Cao, J., Hu, F., Wei, R. and Guo, H. (2017). Transparent $Na_5Gd_9F_3$:Er^{3+} glass-ceramics: enhanced up-conversion luminescence and applications in optical temperature sensors, *RSC Adv.*, **7**, pp. 35147–35153.
8. Wang, Y., Wu, W., Fu, X., Liu, M., Cao, J. and Shao, C. (2017). Metastable scheelite $CdWO_4$: Eu^{3+} nanophosphors: solvothermal synthesis, phase transitions and their polymorph dependent luminescence properties, *Dyes Pigm.*, **147**, pp. 283–290.
9. Huang, X. (2015). Dual-model upconversion luminescence from $NaGdF_4$: Nd/Yb/Tm@ $NaGdF_4$: Eu/Tb core–shell nanoparticles, *J. Alloys Compd.*, **628**, pp. 240–244.
10. Li, J., Yan, J., Wen, D., Khan, W., Shi, J. and Wu, M. (2016). Advanced red phosphors for white light-emitting diodes, *J. Mater. Chem. C*, **4**, pp. 8611–8623.
11. Huang, X. (2015). Giant enhancement of upconversion emission in (NaYF: Nd^{3+}/Yb^{3+}/ Ho^{3+}) /(NaYF: Nd^{3+}/Yb^{3+}) core/shell nanoparticles excited at 808 nm, *Opt. Lett.*, **40**, pp. 3599–3602.

12. Huang, X. and Lin, J. (2015). Active-core/active-shell nanostructured design: an effective strategy to enhance Nd^{3+}/Yb^{3+} cascade sensitized upconversion luminescence in lanthanide-doped nanoparticles, *J. Mater. Chem. C*, **3**, pp. 7652–7657.

13. Huang, X., Ji, X. and Zhang, Q. (2011). Broadband downconversion of ultraviolet light to nearinfrared emission in Bi^{3+}-Yb^{3+}-co-doped Y_2O_3 phosphors, *J. Am. Ceram. Soc.*, **94**(3), pp. 833–837.

14. Chen, Z., Song, E., Wu, M., Ding, S., Ye, S. and Zhang, Q. (2016). Bidirectional energy transfer induced single-band red upconversion emission of Ho^{3+} in $KZnF_3$:Mn^{2+}, Yb^{3+}, Ho^{3+} nanocrystals, *J. Alloys Compd.*, **667**, pp. 134–140.

15. Lia, B., Huanga, X., Guoa, H. and Zengb, Y. (2018). Energy transfer and tunable photoluminescence of $LaBWO_6$:Tb^{3+}, Eu^{3+} phosphors for near-UV white LEDs, *Dyes Pigm.*, **150**, pp. 67–72.

16. Sailaja, S., Dhoble, S. J., Brahme, N. and Reddy, S. B. (2011). Synthesis, photoluminescence and mechanoluminescence properties of Eu^{3+} ions activated $Ca_2Gd_2W_3O_{14}$ phosphors, *J. Mater. Sci.*, **46**, pp. 7793–7798.

17. Sailaja, S., Dhoble, S. J., Brahme, N. and Reddy, B. S. (2011). Synthesis, structural, photoluminescence and mechanoluminescence properties of Tb^{3+}: $Ca_2Gd_2W_3O_{14}$ novel green nanophosphors, *J. Mater. Sci.*, **47**, pp. 2359–2364.

18. Sailaja, S., Dhoble, S. J. and Reddy, B. S. (2011). Synthesis and photoluminescence properties of Sm^{3+} and Dy^{3+} ions activated $Ca_2Gd_2W_3O_{14}$ phosphors, *J. Mol. Struct.*, **1003**, pp. 115–120.

19. Lei, F., Yan, B. and Chen, H. H. (2008). Solid-state synthesis, characterization and luminescent properties of Eu^{3+}-doped gadolinium tungstate and molybdate phosphors: $Gd_{(2-x)}MO_6$:Eu_x^{3+} (M=W, Mo), *J. Solid State Chem.*, **181**, pp. 2845–2851.

20. Guzik, M., Tomaszewicz, E., Kaczmarek, S. M., Cybińska, J. and Fuks, H. (2010). Spectroscopic investigations of $Cd_{0.25}Gd_{0.50\text{-}0.25}WO4$:$Eu^{3+}$: a new promising red phosphor, *J. Non-Cryst. Solids*, **356**, pp. 1902–907.

21. Liao, J., Qiu, B. and Lai, H. (2009). Synthesis and luminescence properties of Tb^{3+}:$NaGd(WO_4)_2$ novel green phosphors, *J. Lumin.*, **129**, pp. 668–671.

22. Yan, S. A., Chang, Y. S., Wang, J. W., Hwang, W. S. and Chang, Y. H. (2011). Synthesis and luminescence properties of color-tunable $Ba_{1\text{-}y}Sr_yLa_{4\text{-}x}Tb_x(WO_4)_7$(x = 0.02-1.2, y = 0-0.4) phosphors, *Mater. Res. Bull.*, **46**, pp. 1231–1236.

23. Shi, S., Gao, J. and Zhou, J. (2008). Effects of charge compensation on the luminescence behavior of Eu^{3+} activated $CaWO_4$ phosphor, *Opt. Mater.*, **30**, pp. 1616–1620.

24. Wei, Z. G., Sun, L. D., Liao, C. S., Yan, C. H. and Huang, S. H. (2002). Fluorescence intensity and color purity improvement in nanosized. YBO_3: Eu, *Appl. Phys. Lett.*, **80**, pp. 1447–1449.

25. Li, Z. H., Zeng, J. H. and Li, Y. D. (2007). Solvothermal route to synthesize well-dispersed YBO_3: Eu nanocrystals, *Small*, **3**, pp. 438–443.

26. Hirosaki, N., Xie, R. J., Kimoto, K., Sekiguchi, T., Yamamoto, Y., Suehiro, T. and Mitomo, M. (2005). Characterization and properties of green-emitting β-SiAlON: Eu^{2+}β-SiAlON: Eu^{2+} powder phosphors for white light-emitting diodes, *Appl. Phys. Lett.*, **86**, pp. 211905–211910.

27. Yang, C., Som, S., Das, S. and Lu, C. (2017). Synthesis of $Sr_2Si_5N_8$:Ce^{3+} phosphors for white LEDs via an efficient chemical deposition, *Sci. Rep.*, **7**, pp. 45832–45838.

28. Huang, X. (2014). Red phosphor converts white LEDs, *Nat. Photonics*, **8**, pp. 748–749.

29. Zhong, J. (2017). Synthesis and spectroscopic investigation of $Ba_3La_6(SiO_4)_6$:Eu^{2+} green phosphors for white light-emitting diodes, *Chem. Eng. J.*, **309**, pp. 795–801.

30. Shang, M., Liang, S., Qu, N., Lian, H. and Lin, J. (2017). Influence of anion/cation substitution (Sr^{2+}, Ba^{2+}, Al^{3+}, Si^{4+}, N^{3-} , O^{2-}) on phase transformation and luminescence properties of $Ba_3Si_6O_{15}$:Eu^{2+} phosphors, *Chem. Mater.*, **29**, pp. 1813–1829.

31. Zhang, N., Guo, C., Zheng, J., Su, X. and Zhao, J. (2014). Synthesis, electronic structures and luminescent properties of Eu^{3+} doped $KGdTiO_4$, *J. Mater. Chem.*, **2**, pp. 3988–3994.

32. Zhu, C., Xiao, S., Ding, J., Yang, X. and Qiang, R. (2008). Synthesis and photoluminescent properties of Eu^{3+}-doped (1 - x)CaO-xLi_2O-WO_3 phosphors, *Mater. Sci. Eng. B*, **150**, pp. 95–98.

33. Laufer, S., Strobel, S., Schleid, T., Cybinska, J., Mudring, A. V. and Hartenbach, I. (2013). Yttrium (III) oxomolybdates(VI) as potential host materials for luminescence applications: an investigation of Eu^{3+}-doped $Y_2[MoO_4]_3$ and $Y_2[MoO_4]_2[Mo_2O_7]$, *New J. Chem.*, **37**, pp. 1919–1926.

34. Li, G., Tian, Y., Zhao, Y. and Lin, J. (2015). Recent progress in luminescence tuning of Ce^{3+} and Eu^{2+}-activated phosphors for pc-WLEDs, *Chem. Soc. Rev.*, **44**, pp. 8688–8713.

35. Deng, H. J., Xue, N. and Hei, Z. F. (2015). Close-relationship between the luminescence and structural characteristics in efficient nanophosphor $Y_2Mo_4O_{15}$:Eu^{3+}, *Opt. Mater. Express*, **5**, pp. 490–496.

36. Jia, Z. and Xia, M. (2016). Blue-green tunable color of Ce^{3+}/Tb^{3+} coactivated $NaBa_3La_3Si_6O_{20}$ phosphor via energy transfer, *Sci. Rep.*, **6**, pp. 33283–33290.

37. Li, H. Y., Yang, H. K. and Moon, B. K. (2012). Structure dependence of the photoluminescence properties of Eu^{3+} ion-activated $R_2Mo_4O_{15}$ (R =Y, La, and Gd), *J. Electrochem. Soc.*, **159**, pp. J61–J66.

38. Xia, Z., Xu, Z., Chen, M. and Liu, Q. (2015). Recent developments in the new inorganic solid-state LED phosphors, *Dalton Trans.*, **45**, pp. 11214–11232.

39. Kuzmanoski, A., Pankratov, V. and Feldmann, C. (2015). Microwave-assisted ionic-liquid based synthesis of highly crystalline $CaMoO_4$:RE^{3+} (RE = Tb, Sm, Eu) and $Y_2Mo_4O_{15}$:Eu^{3+} nanoparticles, *Solid State Sci.*, **41**, pp. 56–62.

40. Deng, H. J., Zhao, Z. and Wang, J. (2015). Photoluminescence properties of a new orange-red emitting Sm^{3+}-doped $Y_2Mo_4O_{15}$ phosphor, *J. Solid State Chem.*, **228**, pp. 110–116.

41. Wei, Q. and Chen, D. (2009). Luminescence properties of Eu^{3+} and Sm^{3+} coactivated Gd(III) tungstate phosphor for light-emitting diodes, *Opt. Laser Technol.*, **41**, pp. 783–787.

42. Liu, Q., Li, X., Zhang, B., Wang, L., Zhang, Q. and Zhang, L. (2016). Structure evolution and delayed quenching of the double perovskite $NaLaMgWO_6$:Eu^{3+} phosphor for white LEDs, *Ceram. Int.*, **42**(14), pp. 15294–15300.

43. Wang, L., Liu, Q., Shen, K., Zhang, Q., Zhang, L., Song, B. and Wong, C. (2017). A high quenching content red-emitting phosphor based on double perovskite host $BaLaMgSbO_6$ for white LEDs, *J. Alloys Compd.*, **696**, pp. 443–449.

44. Liu, Q., Wang, L., Huang, W., Li, X., Yu, M. and Zhang, Q. (2018). Thermally stable double perovskite $CaLaMgSbO_6$:Eu^{3+} phosphors as a tunable LED-phosphor material, *Ceram. Int.*, **44**, pp. 1662–1667.

45. Popovici, E. J., Nazarov, M., Muresan, L., Noh, D. Y., Tudoran, L. B., Bica, E. and Indrea, E. (2010). Synthesis and characterisation of terbium activated yttrium tantalate phosphor, *J. Alloys Compd.*, **497**, pp. 201–209.

46. Yang, H., Shi, J., Gong, M. and Liang, H. (2010). The UV and VUV luminescence properties of the phosphor Mg_2GeO_4:Tb^{3+}, *Mater. Lett.*, **64**, pp. 1034–1036.

47. Zhang, X. and Seo, H. J. (2011). Luminescence properties of novel Sm^{3+}, Dy^{3+} doped $LaMoBO_6$ phosphors, *J. Alloys Compd.*, **509**, pp. 2007–2010.

48. Meng, J. X., Yang, C. T. and Chen, Q. Q. (2010). Photoluminescence characterization of Ce^{3+} and Dy^{3+} doped Li_2CaGeO_4 phosphors, *J. Lumin.*, **130**, pp. 1320–1323.

49. Pisarski, W.A., Zur, L., Goryczka, T., Soltys, M. and Pisarska, J. (2014). Structure and spectroscopy of rare earth–Doped lead phosphate glasses, *J. Alloys Compd.*, **587**, pp. 90–98.

50. Caldino, U., Lira, A., Meza-Rocha, A. N., Pasquini, E., Pelli, S., Speghini, A., Bettinelli, M. and Righini, G. C. (2015). White light generation in $Dy^{3+}$and Ce^{3+}/Dy^{3+}doped zinc -sodium-aluminosilicate glasses, *J. Lumin.*, **167**, pp. 327–332.

51. Vijaya, N., Upendra Kumar, K. and Jayasankar, C. K. (2013). Dy^{3+}-doped zinc fluorophosphate glasses for white luminescence applications, *Spectrochim. Acta A*, **113**, pp. 145–153.

52. Tian, Y., Chen, B., Tian, B., Hua, R., Sun, J. and Cheng, L., (2011). Concentration-dependent luminescence and energy transfer of flower-like $Y_2(MoO_4)_3$:Dy^{3+} phosphor, *J. Alloys Compd.*, **509**, pp. 6096–6101.

53. Zhou, Y., Zhu, C., Zhang, M. and Liu, J. (2016). Optical properties of Eu- and Dy-doped calcium aluminoborosilicate glasses for LED applications, *J. Alloys Compd.*, **688**, pp. 715–720.

54. Ramteke, D. D., Kumar, V. and Swart, H. C. (2016). Spectroscopic studies of Sm^{3+}/Dy^{3+} co-doped lithium boro-silicate glasses, *J. Non-Cryst. Solids*, **438**, pp. 49–58.

55. Srihari, T. and Jayasankar, C. K. (2017). Fluorescence properties and white light generation from Dy^{3+}-doped niobium phosphate glasses, *Opt. Mater.*, **69**, pp. 87–95.

56. Annapurna Devi, Ch. B., Mahamuda, Sk., Venkateswarlu, M., Swapna, K., Srinivasa Rao, A. and Vijaya Prakash, G. (2016). Dy^{3+} ions doped single and mixed alkali fluoro tungsten tellurite glasses for LASER and white LED applications, *Opt. Mater.*, **62**, pp. 569–577.

57. Kaewnuam, E., Wantana, N., Kim, H. J. and Kaewkhao, J. (2017). Development of lithium yttrium borate glass doped with Dy^{3+} for laser medium, W-LEDs and scintillation materials applications, *J. Non-Cryst. Solids*, **464**, pp. 96–103.

58. Kunti, A. K., Patra, N., Sharma, S. K., Swart, H. C. (2018). Radiative transition probability enhancement of white light emitting Dy^{3+} doped and K^+ co-doped $BaWO_4$ phosphors via charge compensation, *J. Alloys Compd.*, **735**, pp. 2410–2422.

59. Tian, B., Chen, B., Tian, Y., Sun, J., Li, X., Zhang, J., Zhong, H., Cheng, L. and Hua, R. (2012). Concentration and temperature quenching mechanisms of Dy^{3+} luminescence in $BaGd_2ZnO_5$ phosphors, *J. Phys. Chem. Solid*, **73**, pp. 1314–1319.

60. Neharika, Kumar, V., Singh, V. K., Sharma, J., Ntwaeaborwa, O. M. and Swart, H. C. (2016). Synthesis and photoluminescence study of a single dopant near white light emitting $Li_4CaB_2O_6$:Dy^{3+} nanophosphor, *J. Alloys Compd.*, **688**, pp. 939–945.

61. Zhijun, W., Panlai, L., Zhiping, Y., Qinglin, G., Xu, L. and Feng, T. (2015). A novel white emitting phosphor Ca_2PO_4Cl:Dy^{3+}: luminescence, concentration quenching and thermal stability, *J. Rare Earths*, **33**, pp. 1137–1141.

62. Zaifa, Y., Yumei, S., Qiguang, X. and Jiayue, S. (2015). Preparation and photoluminescence properties of Dy^{3+}-doped $Ba_3Lu(PO_4)_3$ phosphors, *J. Rare Earths*, **33**, pp. 1251–1255.

63. Zhai, Y., Han, Y., Zhang, W., Yin, Y., Zhao, X., Wang, J. and Liu, X. (2016). Influence of doping alkali metal ions on the structure and luminescent properties of microwave synthesized $CaMoO_4$:Dy^{3+} phosphors, *J. Alloys Compd.*, **688**, pp. 241–247.

64. Ogugua, S. N., Swart, H. C. and Ntwaeaborwa, O. M. (2016). White light emitting $LaGdSiO_5$:Dy^{3+} nanophosphors for solid state lighting applications, *Physica B*, **480**, pp. 131–136.

65. Chong, M. K., Pita, K. and Kam, C. H. (2005). Photoluminescence of Y_2O_3:Eu^{3+} thin film phosphors by sol–gel deposition and rapid thermal annealing, *J.Phys. Chem. Solids*, **66**, pp. 213–217.

66. Qin, C., Huang, Y., Chen, G., Shi, L., Qiao, X., Gan, J. and Seo, H. J. (2009). Luminescence properties of a red phosphor europium tungsten oxide Eu_2WO_6, *Mater. Lett.*, **63**, pp. 1162–1164.

67. Tian, Y., Chen, B., Hua, R., Zhong, H., Cheng, L. and Sun, J. (2009). Synthesis and characterization of novel red emitting nanocrystal Gd_6WO_{12}:Eu^{3+} phosphors, *Physica B*, **404**, pp. 3598–3601.

68. Zhang, Y., Jiao, H. and Du, Y. (2011). Luminescent properties of HTP $AgGd_{1-x}W_2O_8$:Eu_x^{3+} and $AgGd_{1-x}(W_{1-y}Mo_y)_2O_8$:$Eu_x^{3+}$ phosphor for white LED, *J. Lumin.*, **131**, pp. 861–865.

69. Cai, G. M., Zheng, F., Yi, D. Q., Jin, Z. P. and Chen, X. L. (2010). New promising phosphors $Ba_3InB_9O_{18}$ activated by Eu^{3+}/Tb^{3+}, *J. Lumin.*, **130**, pp. 910–916.

70. Bouajaj, A., Ferrari, M. and Montagna, M. (1997). Crystallization of silica xerogels: a study by Raman and fluorescence spectroscopy, *J. Sol-Gel Sci. Technol.*, **8**, pp. 391–395.

71. Campostrini, R., Carturan, G., Ferrari, M., Montagna, M. and Pilla, O. (1992). Luminescence of Eu^{3+} ions during thermal densification of SiO_2 gel, *J. Mater. Res.*, **7**, pp. 745–753.

72. Ju, Z., Wei, R., Gao, X., Liu, W. and Pang, C. (2011). Red phosphor $SrWO_4$:Eu^{3+} for potential application in white LED, *Opt. Mater.*, **33**, pp. 909–913.

73. Yadav, R. S. and Pandey A. C. (2010). Enhanced efficiency in quantum confined YBO_3:Tb^{3+} nanophosphor, *J. Alloys Compd.*, **494**, pp. L15–L19.

74. Liao, J., Qiu, B., Wen, H., You, W. and Xiao, Y. (2010). Synthesis and optimum luminescence of monodispersed spheres for $BaWO_4$-based green phosphors with doping of Tb^{3+}, *J. Lumin.*, **130**, pp. 762–766.

75. Singh, V., Watanabe, S., Rao, T. K. G. and Kwak, H. Y. (2010). Luminescence and defect centres in Tb^{3+} doped $LaMgAl_{11}O_{19}$ phosphors, *Solid State Sci.*, **12**, pp. 1981–1987.

76. Zhang, F., Wang, Y., Liu, B., Wen, Y. and Tao, Y. (2011). Investigation of $Na_3GdP_2O_8$:Tb^{3+} as a potential green-emitting phosphor for plasma display panels, *Mater. Res. Bull.*, **46**, pp. 722–725.

77. Mari, B., Singh, K. C., Sahal, M., Khatkar, S. P., Taxak, V. B. and Kumar, M. (2010). Preparation and luminescence properties of Tb^{3+} doped ZrO_2 and $BaZrO_3$ phosphors, *J. Lumin.*, **130**, pp. 2128–2132.

78. Tu, D., Liang, Y., Liu, R., Cheng, Z., Yang, F. and Yang, W. (2011). Photoluminescent properties of $LiSr_xBa_{1-x}PO_4$:RE^{3+} (RE = Sm^{3+}, Eu^{3+}) f-f transition phosphors, *J. Alloys Compd.*, **509**, pp. 5596–5599.

79. He, X., Zhou, J., Lian, N., Sun, J. and Guan, M. (2010). Sm^{3+}-activated gadolinium molybdate: an intense red-emitting phosphor for solid-state lighting based on InGaN LEDs, *J. Lumin.*, **130**, pp. 742–747.

80. Lei, B., Man, S. Q., Liu, Y. and Yue, S. (2010). Luminescence properties of Sm^{3+}-doped $Sr_3Sn_2O_7$ phosphor, *Mater. Chem. Phys.*, **124**, pp. 912–915.

81. Carnall, W. T., Fields, P. R. and Rajnak, K. (1968). Spectral intensities of the trivalent lanthanides and actinides in solution. II. Pm^{3+}, Sm^{3+}, Eu^{3+}, Gd^{3+}, Tb^{3+}, Dy^{3+}, and Ho^{3+}, *J. Chem. Phys.*, **49**, pp. 4412–4423.

82. Tiana, Y., Chen, B., Tian, B., Hu, R., Sun, J., Cheng, L., Zhong, H., Li, X., Zhang, J., Zheng, Y., Yu, T., Huanga, L. and Meng, Q. (2011). Concentration-dependent luminescence and energy transfer of flower-like $Y_2(MoO_4)_3$:Dy^{3+} phosphor, *J. Alloys Compd.*, **509**, pp. 6096–6101.

Chapter 8

Eu^{3+}-Based Orange-Red-Emitting Inorganic Color Convertors: An Overview

V. Sivakumar and S. Kasturi
Department of Chemistry, National Institute of Technology Rourkela, Rourkela – 769008, Odisha, India
vsiva@nitrkl.ac.in

This chapter's emphasis is on the recent progress of inorganic color convertors, providing special attention to narrow-band red-emitting phosphors for phosphor-converted white-light-emitting diodes (pc-WLEDs) based on trivalent europium (Eu^{3+})-activated oxide phosphors, with special emphasis on structure–composition–property correlations.

8.1 Introduction

Climatic and political changes demand a decreasing reliance on fossil fuels, and energy-efficient lighting systems (including white-light-emitting diodes [WLEDs]) are considered key for a sustainable energy future (large quantities of energy are consumed for lighting and during operation of displays). Phosphors are a pivotal

Phosphors: Synthesis and Applications
Edited by Sanjay J. Dhoble, B. Deva Prasad Raju, and Vijay Singh

ISBN 978-981-4774-49-9 (Hardcover), 978-042-9460-99-9 (eBook)
www.panstanford.com

component or integral part of several display devices (including white-light solid-state lighting [SSL]). There is a drastic change encountered in the illumination technology due to discoveries of new inorganic phosphor materials [1]. Discovery of the highly efficient blue LED by S. Nakamura is considered a triumph in the field of materials chemistry. The first WLED was fabricated by combining the InGaN-based blue chip with a commercially available yellow-emitting ($Y_3Al_5O_{12}:Ce^{3+}$) phosphor. In the year 1997, Nichia Chemical Co. commercialized the first WLED [2, 3]. LED lighting has several advantages over conventional lighting sources and was paid much attention during this period due to its high efficiency, enduring lifetime, energy consumption, stability, eco-friendly nature, and freedom from harmful pollutants. Unlike incandescent and fluorescent light sources, WLEDs lack of high energy losses and a large Stoke's shift. It is highly expected that by the year 2020, the luminescence efficacy of WLEDs should approach 200 lm/W and the cost of WLEDs should decrease gradually, which makes them available at the cheapest price as the best competitor in the market for conventional light sources. SSL devices have potential applications in a wide range of fields such as WLEDs, indicators, backlighting, traffic signals, etc. Especially, rare earth (RE)-activated phosphors gained extensive attention due to their distinctive photophysical properties. Trivalent RE ions have been demonstrated to be the activators of choice due to their weak interaction among the 4f electronic states with the host lattice and the enormous number of easily reached energy levels. The transition between the $4f^n$ orbital manifolds gives rise to luminescence phenomena in most of the RE metal ions. The $4f^-$ electrons of lanthanide ions are well shielded from the chemical environment and retain their atomic character; hence they exhibit a sharp line-like emission from 4f–4f emission spectra. The 4f–4f transitions are forbidden because of electric dipole (ED) selection rules but are partially allowed due to mixing of configurations having opposite parity [4]. Phosphors that are used in the WLED should possess strong and broad absorption (good excitability) in the near-ultraviolet (NUV, 370–410 nm) to blue spectral region of the visible spectrum. Moreover, the absorption spectrum should be sufficiently extended (oscillator strength should be high) to compensate for changes in the pumping LED emission spectrum, caused by changes in the driving current and/or the junction temperature. It is prudent to have an abstemiously flat excitation spectrum for the phosphor near the LED emission peak to sustain the color stability of the WLED. Hence, the development of thermally stable and highly efficient light-emitting inorganic oxide phosphors for WLEDs are still a bottleneck in accordance to their use in energy-efficient SSL devices beyond resolving the technical issues over conventional lighting sources. The research and development of narrow-band red emitters gained substantial attention in the last decade to improve color rendition and enhance the visual energy efficiency of WLEDs (to obtain warm WLEDs) by mimicking their white spectrum with the human eye sensitivity curve. In addition, it is well known that the narrow-band emitters contribute to obtaining warm white light with high luminous output [5]. It is well understood that warm WLEDs (for the purpose of general lighting) with an improved color-rendering index (CRI) and/or luminous efficacy of radiation (LER) can be obtained by combining a broad-band yellow or green phosphor with a narrow-band (narrow full width at half maximum [FWHM]) red phosphor and a blue InGaN LED/NUV LED, respectively. The visual energy performances along with the color of WLEDs are strongly influenced by the FWHM and the position of the peak of the red-emitting phosphor used in the same. To achieve the perfect human eye sensitivity curve, the reduction of the lower-energy red spectral region is, indeed, highly in accordance with greater visual energy efficiency and CRI properties. Even though red phosphors (spectral line) contribute only one-third of the whole white spectrum (occupy most of the warm white spectrum; Fig. 8.1), the change in the red spectral lines will obviously affect the warm WLED (<4000 K of the color correlation temperature [CCT]) performance rather than the cool one (>5000 K of the CCT) [6, 7]. Numerous research efforts have been devoted to developing potential new narrow-band as well as (Eu^{2+}, Ce^{3+}, Mn^{2+}, and Eu^{3+}) based on red-emitting phosphors with improved optical characteristics. Here, we mainly concentrated on giving an overall view of narrow-band red-emitting oxide phosphors with different crystal (scheelite related, double-perovskites, pyrochlore, garnet) structures activated with Eu^{3+} ions. In addition, it is also essential to discuss the structure–composition–property correlation of the phosphor materials. The compounds and the complex that we reviewed in this chapter are given in the form of the following flowchart:

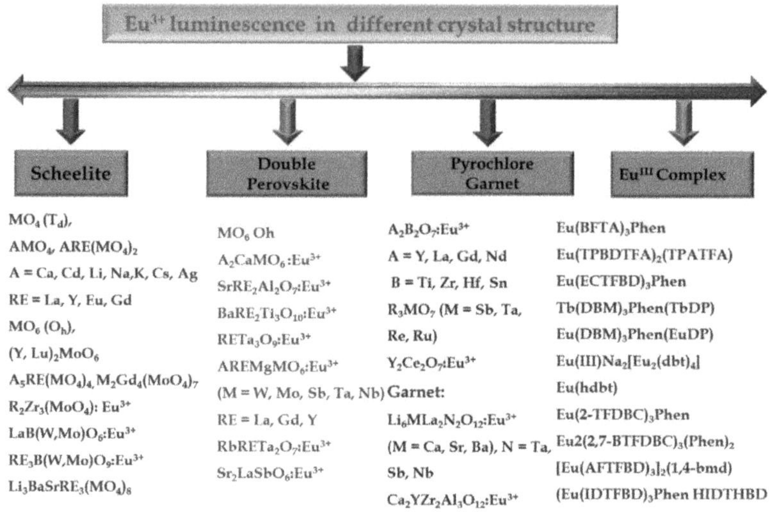

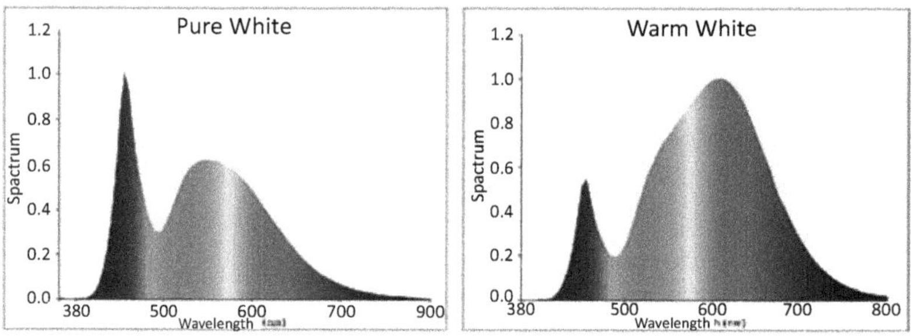

Figure 8.1 Comparison spectral distribution curves for (left) pure and (right) warm white-light spectra.

8.2 Signficance of Trivalent Europium (Eu^{3+}) Ions

Among all the RE ions, the well-documented broad-band or line-emitting activators that are capable of providing red emission are divalent europium (Eu^{2+}), trivalent europium (Eu^{3+}), trivalent praseodymium (Pr^{3+}), and trivalent samarium (Sm^{3+}) ions. The trivalent europium ion (Eu^{3+}) is one of the most interesting RE activators among all the lanthanides ions. Eu^{3+} ions have gained wide attention because of their precise spectral parameters, when used as dopants in red phosphors. Eu^{3+} ions have simple electronic energy levels and can emit either orange or red emission, depending on the host lattice (especially the local site symmetry). The line-like emission arises from the higher $^5D_0$ excited state to the lower-energy ground state of ^7F_J (J = 0–6). $^5D_0 \rightarrow {}^7F_2$ known as ED transitions centered at around 612 nm are highly hypersensitive in nature, sensitive to the local site symmetry of the Eu^{3+} ion in the crystal lattices. The $^5D_0 \rightarrow {}^7F_1$ spectral line is known as magnetic dipole (MD) transition around 590 nm, allowed according to spin and Laporte selection rules, which is insensitive to the symmetric environment of the RE ion. The ground ($^7F_0$) and first excited ($^5D_0$) levels are nondegenerated, and the spectra linked with transitions between them should contain as many lines as the number of nonequivalent sites present in the host lattice [4]. It is also worth to note that the luminescence decay time of the emitting $^5D_0$ level (ms) is longer. The selection of the host lattice plays a significant role in the luminescence properties of the Eu^{3+} ion. A small change in the structural parameters of the host lattice can directly influence the emission properties of Eu^{3+} ion–activated phosphors. The Eu^{3+} ion meets the need of WLEDs due to its major transitions of $^5D_0 \rightarrow {}^7F_1$ and $^5D_0 \rightarrow {}^7F_2$, which emit in the orange and red spectral regions, respectively. Since the $4f^6$–$4f^6$ transitions are parity forbidden, they have a poor absorption strength in the necessary excitable regions (where the LED emission occurs the NUV-to-blue region). Similar observations were also made for two other activator ions, Pr^{3+} and Sm^{3+}. The ultimate or viable solution of these problems is to use a sensitizer (host lattice or codopant). The enhancement of red emission was observed in a Bi^{3+}/Eu^{3+}, Tb^{3+}/Eu^{3+}, and $Ce^{3+}/Tb^{3+}/Eu^{3+}$ codoped system as well as a Eu^{3+}-activated tungstate or molybdate host lattice [8–10]. The light convertors' absorption ability deeply relies on host lattices as well as activator centers. Host lattices containing different anionic groups such as vanadates, niboates, tantalates, molybdates, and tungstates are found to be having good absorption (oscillator strength) due to the intense charge transfer (CT) transitions of VO_4^{3-}, NbO_4^{3-}, TaO_4^{3-}, MoO_4^{2-}, and WO_4^{2-} groups [4]. The energy that is absorbed by the MO_4 (M = V, Nb, Ta, Mo, and W) group in the lattice transfer successfully to RE ions (activators) is the well-known phenomenon *host-sensitized luminescence*. Further sensitization by the host crystal, sensitization by a sensitizer (codopant) ion, is an alternative method to augment the NUV and blue absorption of RE ions. The sharp spectral line width of the $^7F_0 \rightarrow ^5D_2$ absorption band at 465 nm, being considerably narrower than the emission band of the blue LED, limits the probability of using Eu^{3+} as an activator for the LED application. However, several absorption spectral lines are located in the region of 390–405 nm that can contribute altogether toward absorption of NUV LEDs. Traditional red-emitting phosphors such as $Y_2O_2S:Eu^{3+}$ and $Y_2O_3:Eu^{3+}$ exhibit only weak absorption in the NUV and blue regions. The forced ED transition is allowed only under the condition that the Eu^{3+} ion occupies a site without an inversion center and is sensitive to local symmetry. It leads to the formation of phosphors matchable with (National Television Standard Committee [NTSC]) standard Commission Internationale dÉclarage (CIE) chromaticity coordinates for red color. Hence, design and synthesis of a host lattice with low symmetry and a noncentrosymmetric site (Eu^{3+} activation site) will certainly lead to pure red emission. Also, efficient transfer from the host crystal to the RE activators can be exploited to realize exotic optical phenomena.

8.2.1 Importance of the Charge Transfer Band

The broad and strong absorption (or excitation) bands in the trivalent RE ions' spectra to either CT or 4f–5d transition was first identified by Jorgensen in 1962 [11]. It is also worth to note that the CT bands shift to lower energies with an increasing oxidation state, whereas Rydberg transitions (4f–5d transitions) shift to higher energies. We can use the formula proposed by Reisfeld and Jorgensen to calculate the positions of CT of the RE ions, as well as the oxygen-to-metal CT band (M = Mo, W, Nb, Ta) [12].

$$E_{ct}\,(\mathrm{cm}^{-1}) = [\chi_{opto}(X) - (\chi_{opt}(M)] \times 30000\ \mathrm{cm}^{-1}, \qquad (8.1)$$

where $\chi_{opt}(X)$ is the optical electronegativity of the anion, which is equal to the Pauling electronegativity value, 3.2; $\chi_{opt}(M)$ is the optical electronegativity of the central metal ion (the value of χ_{opt} for Nb(V) is 1.85, and the value of χ_{opt} for Ta(V) is 1.8; Mo = 2.101 and Eu = 1.74) [13–15]. The CT band calculation for the $La_2Mo_2O_9$ lattice has been done by using the above equation; the ligand-to-metal charge transfer (LMCT) of O^{2-} to Mo^{6+} is around ~303 nm and for O^{2-} to Eu^{3+}

is ~228 nm. The calculation clearly indicates that the estimations agree well with the experimental result. To further understand the process of LMCT, for the case study we have taken tungstate and molybdate host lattices over the other and analyzed. The energy absorbed by the MO_4 tetrahedra or MO_6 octahedra moieties in the lattice through LMCT transition is strongly dependent on the crystal field experienced by a number of coordinated ligands of the transition metal ion. The energy of the Eu→O CT band (E^{CT}) for many different compounds was documented by Dorenbis in 2005 [16].

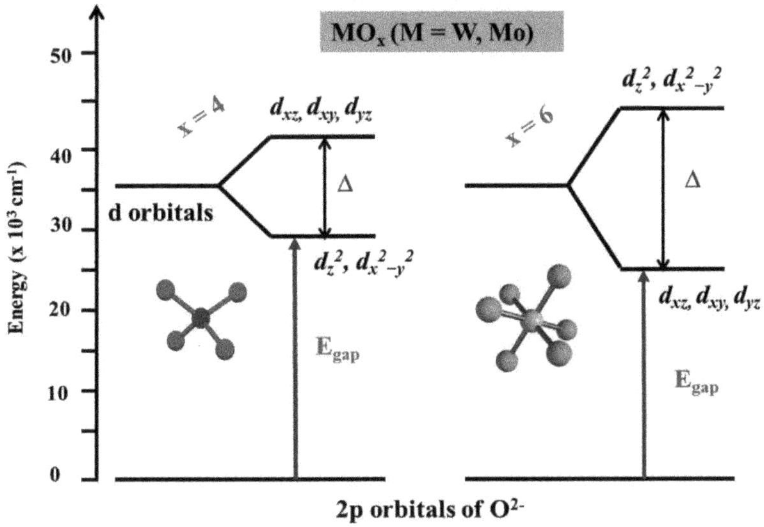

Figure 8.2 E_{LMCT} in MO_6 and MO_4 (M = Mo, W) in complexes.

The metal ions of Mo^{6+}/W^{6+} in the transition metal ion complexes (MoO_x/WO_x; $x = 5, 6$) try to attain geometry of fivefold coordination ($x = 5$; square pyramidal or trigonal bipyramidal) or octahedral coordination ($x = 6$) with O^{2-} ligands with a higher coordination number more than 4 (tetrahedral coordination). Both five- and six-coordinated moieties experience a larger crystal field splitting parameter (Δ) than a tetrahedral crystal field. Figure 8.2 shows the difference in the crystal field splitting parameter for tetrahedrally and octahedrally coordinated Mo^{6+}/W^{6+} ions. The valence band (VB) is primarily contributes by O^{2-} 2p orbitals, whereas the d orbitals of the transition metal ion contribute to the conduction band (CB). The minimum amount of energy required for excitation of the electron from the O^{2-} 2p orbital to the M^{6+} d orbital is predetermined by the bandgap of the material. Hence, the energy of the CT band (E_{LMCT}) or of the bandgap (E_g) is sincerely affected by the crystal field experienced by the transition metal ion and decreases for a stronger field, that is, for $x = 5, 6$. A smaller E_{LMCT} means longer wavelength absorption by MO_x (M = Mo, W; $x = 5, 6$) groups in comparison to MO_4 (M = Mo, W) groups. While monitoring the excitation spectrum of Eu^{3+} at $\lambda_{exc} = 615$ nm, the absorption for the host lattice (MO_x, M = Mo, W groups) was observed. The red shift in the absorption was observed for compounds with a higher coordination number ($x = 5, 6$). The broad excitation charge transfer band (CTB) ranges from 250 to 350 nm wavelength observed for the compound with tetrahedral coordinated metal ions, whereas the same is shifted further all the way from the NUV-to-blue region (300 and 450 nm) for five- or six-coordinated moieties in the case of Eu^{3+}-activated phosphors (indirect evidence of MO_x to Eu^{3+} ET; Fig. 8.3). In a double-perovskite structure, Ca_2MgMO_6 lattice, the optical E_g is dissimilar for W and Mo compounds. As mentioned earlier, Jørgensen derived the optical electronegativity χ_{opt} from electron transfer spectra of octahedral hexahalide complexes, even though the size and charge of the ions are almost similar. The only difference is the optical electronegativity of the M ion. The value of χ_{opt} for Mo(VI) is 2.16, and the value of χ_{opt} for W(VI) is 1.7. The energy difference between the Mo 4d orbital and the O 2p orbital is smaller than that between the W 5d orbital and the O 2p orbital, resulting in a smaller bandgap for Ca_2MgMoO_6 than that of Ca_2MgWO_6, in accordance with the difference in electronegativity. In addition, tilting and/or distortion of MO_6 octahedra in the lattice also influences the bandgap energy, E_g. The same is expected to eliminate the antibonding π interaction between the metal *n*d (t2g) orbital and the O 2p orbital, resulting in a decrease in the width of the CB, that is, an increase in the bandgap. Therefore, the smaller optical E_g of Ca_2MgMoO_6 than that of Ca_2MgWO_6 is principally due to the change in electronegativity between W and Mo.

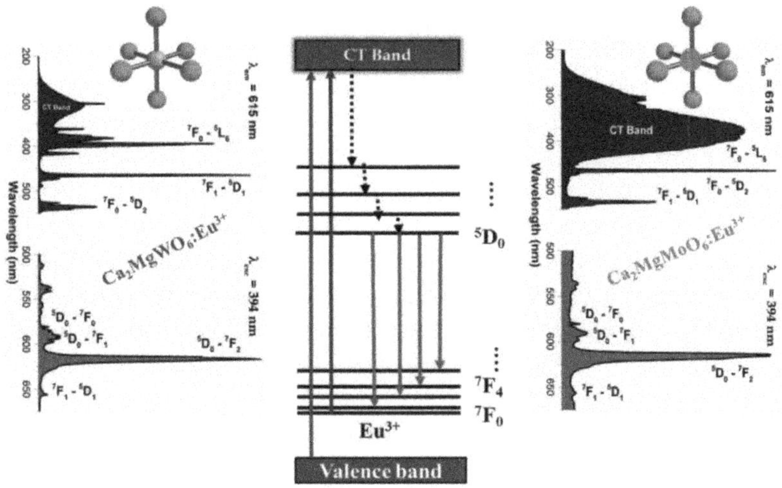

Figure 8.3 Energy transfer process from MO_6 (M = W or Mo) octahedra to Eu^{3+} ions in a double-perovskite lattice.

8.3 Importance of M–O–Eu Angle on Energy Transfer

The activator concentration (Eu^{3+}) used is about 5 at% or less. Since the concentration is very low, it is necessary that the absorbed energy by the absorbing groups transfer energy to Eu^{3+}. If this type of energy transfer through the host lattice is improbable, the Eu^{3+} luminescence cannot have a high efficiency. The necessary conditions for a Eu^{3+}-doped oxide with high efficiency are as follows [17]:

- The absorbing groups are nearest neighbors in the crystal structure, otherwise their wave function overlap is negligibly small.
- The emission of the absorbing group must overlap its absorption, otherwise the energy overlap vanishes.

The probability of energy transfer from the absorbing group to the RE ion can be determined by wave function overlap, that is, the orbitals of the central ion of the absorbing group and the RE ion using an oxygen ion as an intermediary (metal–oxygen–europium

[M–O–Eu]). The energy overlap is due to emission of the absorbing group and the excitation levels of the Eu^{3+} ion. The energy overlap requirement is satisfied, since almost all absorbing groups emissions are located in the short-wavelength visible region, where the Eu^{3+} ion has levels at 19,000, 21,500, and 23,900 cm^{-1} [18]. Even if the emission is in the long-wavelength visible region, there is some energy overlap expected with a 17,000 cm^{-1} level of the Eu^{3+} ion. This interaction is usually strong if the M–O–Eu angle is 180° (using σ bonding) and considerably weaker for 90° (using π bonding). This assumption is used to explain the exchange-regulated concentration quenching of Eu^{3+} luminescence [19]. On the basis of the above arguments, Blasse reviewed energy transfer studies on a Eu^{3+}-doped host lattice with a different crystal structure, and the results are tabulated in Table 8.1.

Table 8.1 Host ions to Eu^{3+} energy transfer using an oxygen ion as an intermediary in a different crystal structure with quantum efficiency

Compound	A–O–H (Eu^{3+})	Angle (A–O–H)	Quantum efficiency (%)	Structure type
Y_2WO_6	W–O–Y(Eu)	180°	45	Scheelite
Gd_2WO_6	W–O–Y(Eu)	180°	10	Scheelite
Y_6WO_{12}	W–O–Y(Eu)	90°	15	Fluorite
YVO_4	V–O–Y(Eu)	100°	Very efficient	Zircon
$Ca_2NaMgV_3O_{12}$	V–O–Ca(Eu)	170°	Not efficient	Garnet
$YTiTaO_6$	Ti(Ta)–O–Y(Eu)	90°/120°	12	Columbite
$GdTiTaO_6$	Ti(Ta)–O–Gd(Eu)	130°	2	ortho-rhombic $CaTa_2O_6$
$Gd_2Ti_2O_7$	Ti–O–Gd(Eu)	100°	Very low	Pyrochlore
La_2MgTiO_6	Ti–O–La(Eu)	90°	Very low	Perovskite
Gd_2GaSbO_7	Sb–O–Gd(Eu)	100°	Very low	Pyrochlore
$GdInO_2$	In–O–Gd(Eu)	90°	Very low	Ordered rock salt

8.4 Eu^{3+} Luminesence in Scheelite and Related Structures

Binary oxides with the general formula ABO_4, in which A and B cations can be multivalent and compositely substituted, have several common structure types, which include zircon, scheelite, and wolframite. Scheelite is a large family. A and B cations can be different elements with various valence states, for example, $KReO_4$ [20] and $AgIO_4$ [21], $Cd(Mo/W)O_4$ [22, 23] and $Pb(Mo/W)O_4$ [24], $Y(Nb/V)O_4$ [25, 26], and $ZrGeO_4$ [27].

ABO_4 oxides such as silicates, phosphates, arsenates, vanadates, and chromates are likely to adopt a zircon structure with the space group $I4_1/amd$. It is also evidenced that a few ABO_4 oxides such as germanates, molybdates, and tungstates crystallize in the scheelite structure with the space group $I41/a$. It is also reported that a wide range of molybdates, tungstates, and tantalates crystallize in the wolframite structure (P2/c, No. 13) and relatively few ABO_4 oxides exist as M'fergusonite structures (P21/c, No. 14). Many ABO_4 oxides transform from zircon or scheelite structures to M'fergusonite under pressure, and the fergusonite structure can be viewed as a distorted scheelite structure [28].

Figure 8.4 shows the crystal structure of $CaMO_4$ (ABO_4) with scheelite-type structure. Here the A cations are surrounded by eight O atoms with two different distances, the B cations coordinate with four oxygen atoms, and the oxygen anions are connected with two A ions and one B ion. The BO_4 tetrahedra in the scheelite structure are comparatively more distorted and are densely packed than the analog zircon structure. The distinction of the scheelite structure is the layered stacked cell, so it can be readily identified from all the related structures exhibited by ABO_4 oxides. The c axis parameter of the scheelite-type ABO_4 oxide with a layered stacked cell is larger compared to that of other nonlayered structures. The structural advantage of ABO_4 oxides makes them of potential use in theoretical as well as technological applications [29–31]. The structures are known to emit in the region from blue to green with bright luminescence [32, 33]. Scheelite-related phases have been studied as hosts for RE ions [34, 35].

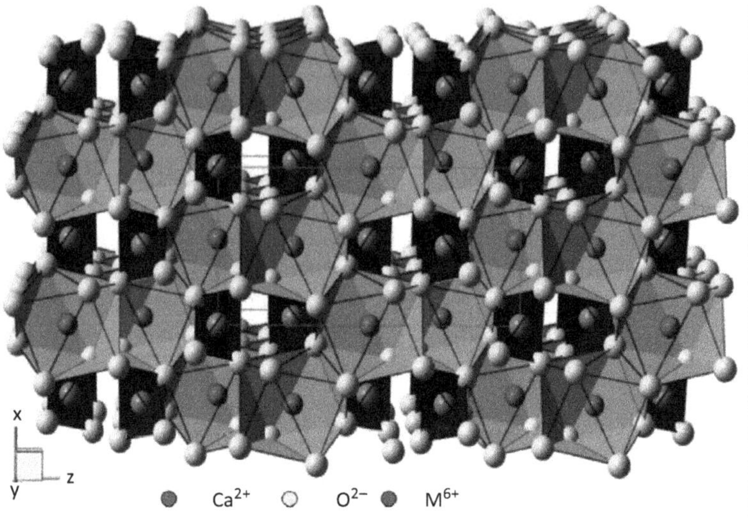

Figure 8.4 Crystal structure of $CaMO_4$ (M = Mo or W).

8.4.1 $CaMo_4$ and $CdMo_4:Eu^{3+}$(M = Mo/W)

Eu^{3+} luminescence was reported in scheelite nanostructures [$Ca_{1-2x}(Eu,Na)_{2x}WO_4$ ($0 < x \leq 0.135$)]. The phosphor shows both CT bands between oxygen and the Eu^{3+} ion as well as oxygenand the W^{6+} ion in the lattice. The ED transition is observed at around 615 nm (red emission), and significant energy transfer occurs in the lattice from the WO_4 group to the Eu^{3+} ion, which leads to enhancement in the Eu^{3+} luminescence with maximum quantum efficiencies (QEs) of 92% [36]. The effect of aliovalent substitution on structural and photoluminescence (PL) properties has been studied in $CaWO_4:Eu^{3+}$ phosphors by replacing the divalent Ca ion by the trivalent Ln ion (Ln = Y, Gd). The vibration of the O–W bond in the WO_4 tetrahedran was observed at 803 and 440 cm^{-1} in the Fourier transform infrared (FTIR) spectrum. The red emission intensity is found to be superior for $Ca_{1-x}Ln_xWO_4:Eu^{3+}$ (x = 0.4, 0.5) as compared to that of the unsubstituted lattice ($CaWO_4:Eu^{3+}$). Owing to significant absorption in the NUV and blue regions, it can be a potential red-emitting phosphor for WLEDs [37]. Enhanced PL properties of Eu^{3+} were reported by Singh et al. (2014) in a $CaMoO_4$ lattice by Gd^{3+} substitution in the Ca^{2+} site. This enhancement in the red emission is due to the substantial energy transfer from Mo–O/Gd^{3+} to Eu^{3+} (Fig. 8.5). The compound displays red emission at around 615 nm (forced ED transition), and very high asymmetric values between 12 and 16 indicate that Eu^{3+} is present in the EuO_8 polyhedran in the lattice [38]. The trivalent bismuth (Bi^{3+}) ion is used as a sensitizer to enhance the Eu^{3+} luminescence in Eu^{3+}-doped $CaMoO_4$. Influence of Bi^{3+} concentration on the phosphor was studied in detail. From PL spectral studies, red emission originating from the $^5D_0 \rightarrow ^7F_2$ transition under excitation was found either into the $^5L_6$ state with 395 nm or the $^5D_2$ state with 465 nm. Threefold enhancement in the red emission was observed by 0.2 Bi^{3+} co-doped samples due to energy transfer from Bi^{3+} to Eu^{3+} as well as increase of the $^7F_0 \rightarrow ^5L_6$ and $^7F_0 \rightarrow ^5D_2$ transition absorption strengths.

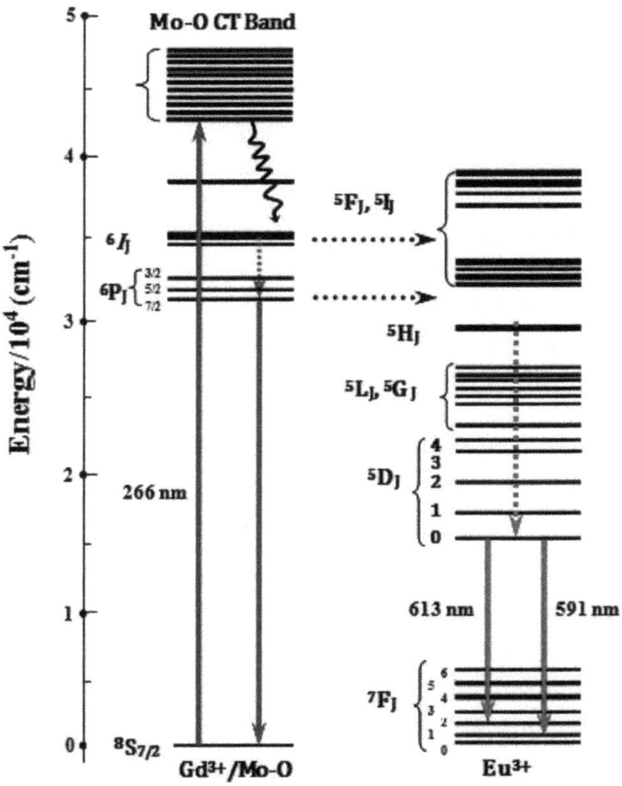

Figure 8.5 Schematic energy level diagram for the energy transfer process between Gd^{3+}/MoO_4^{2-} and Eu^{3+} ions in $CaMoO_4$.

The emission intensity is stronger than that of commercial $Y_2O_2S:Eu^{3+}$ and $Y_2O_3:Eu^{3+}$ phosphors [39]. A solid solution of the red-emitting phosphor $(Ca_{0.6},Sr_{0.16})(Mo_{0.4},W_{0.6})$:0.08$Eu^{3+}$,0.08 La^{3+} was synthesised by the sol-gel technique and its PL for WLED applications was observed. The phosphor shows a dominant emission peak at around 615 nm due to the $^5D_0 \rightarrow {}^7F_2$ transition of the Eu^{3+} ion (confirms the noncentrosymmetric site as well as local site symmetry is S_4). The red-emitting phosphor CIE color coordinates (0.67, 0.33) are close to the NTSC standard. Due to the significant absorption in the NUV-to-blue region due to the Eu^{3+} ion, this phosphor could find potential applications in WLEDs [40]. The influence of Li^+ and Eu^{3+} ion substitution in $SrMoO_4$ with a scheelite structure on their PL properties was reported by Xiangqian Shen et al. in 2012. The organic gel-thermal decomposition process was used to synthesize the $Li_xSr_{1-2x}(MoO_4)$:xEu^{3+} red-emitting phosphors. The reduction in the lattice cell and the grain growth are basically suppressed during the calcination process with the increasing dopant content from x = 0.1 to 0.5 in the $Li_xSr_{1-2x}(MoO_4)$:xEu^{3+} phosphor; in addition, the absorption and emission intensities increase along with the increase of dopant ion concentration. The CIE color coordinates well matched with NTSC standard values, and the phosphor can be effectively excited by the UV light (396 nm) and blue light (466 nm), which are well suited for current commercial NUV and blue LEDs [41]. K^+ ion substitution strengthens the absorption of the $SrMoO_4:Eu^{3+}$ phosphors around 400 nm. These were compared with Y_2O_2S:0.05Eu^{3+}, and the obtained $K_{0.25}Sr_{0.5}MoO_4$:0.25 Eu^{3+} phosphor show enhanced red emission under 393 nm excitation [42]. Recently, it was identified that the quenching concentration of Eu^{3+} in the $Ba_{0.4}Sr_{0.6}MoO_4:Eu^{3+}$ phosphor is as high as 20% by optimizing the composition in the host lattice structure in $AMoO_4:Eu^{3+}$ (A = Ca, Ba) by Sr doping or substitution ($A_{1-x}B_xMoO_4$:yEu^{3+}; A = Ca, y = 0.01, 0.08, 0.20 mol%, or A = Ba, y = 0.01, 0.08, 0.20, 0.25, 0.30 mol%; B = Sr; x = 0.4, 0.6) [43].

Aliovalent La^{3+} substitution into the Ca^{2+} site in the $CaMoO_4:Eu^{3+}$ lattice was done to enhance the red emission of the Eu^{3+} ion. Subsequently, the red emission at 615 nm ($^5D_0 \rightarrow {}^7F_2$ transition) of Eu^{3+} increased significantly. The red emission intensity of optimized the $Ca_{0.5}La(MoO_4)_2:Eu^{3+}$ phosphor was 2.67 and 5.92 times higher than that of the commercial $Y_2O_3:Eu^{3+}$ phosphor and $CaMoO_4$:$0.24Eu^{3+}$, respectively. The combined spectroscopic analysis (PL spectra, diffuse reflectance spectra [DRS], and PL life time studies) indicates that the enrichment of red emission is due to the increase of the $^7F_0 \rightarrow ^5L_6$ absorption intensity of Eu^{3+} in the distorted crystal field of the host lattice. In addition, to make use of this phosphor to UV-pumped LEDs, Bi^{3+} co-doping was executed into $Ca_{0.5}La(MoO_4)_2$ to widen the photoluminescence emission (PLE) spectra in the UV range. Nevertheless, the co-doped phosphor makes it more suitable for UV-pumped LEDs (PLE spectra in Fig. 8.6); unfortunately the substitution subsequently supports the ordered local site, where the Eu^{3+} ion resides (it decreases the distortion in the local site symmetry), and boosts the new quenching path [44].

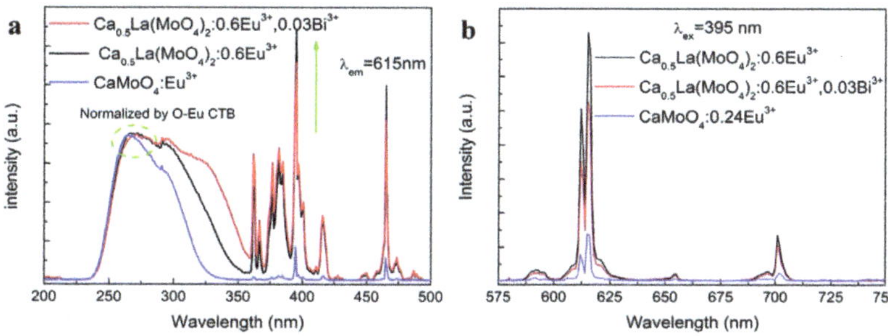

Figure 8.6 (a) PLE and (b) PL spectra recorded upon excitation at 395 nm.

Multiple ions were substituted in the $CaMO_4$ structure (both in Ca and M sites in the lattice), aiming to increase the emission intensity of the Eu^{3+} ion [45]. The solid-solution $Ca_{0.54}Sr_{0.34-1.5x}Eu_{0.08}La_x(MoO_4)_y(WO_4)_{1-y}$ red-emitting phosphors could emit red due to ED transition of the Eu^{3+} ion. It is concluded that the substitution of La^{3+} and changing Mo^{6+} or W^{6+} concentration lead to enhancement in the excitation band at 394 nm and blue light 465 nm (4f→4f transition) and emission of enhanced red light (616 nm) with a narrow-band spectrum. Cadmium-containing tungstates ($Cd_{0.25}Eu_{0.50}\Upsilon_{0.25}WO_4$ and $Cd_{0.25}$-$Gd_{0.50}\Upsilon_{0.25}WO_4$:$Eu^{3+}$; Υ = vacancy) with a scheelite structure were studied for Eu^{3+} luminescence. CT band excitation leads to blue-green emission. As usual the LMCT (O→W at 275 nm and O→Eu^{3+} at 315 nm) band along with Eu^{3+} spectral lines is observed, and all the Eu^{3+}-activated compositions show red emission (confirm the noncentrosymmetric site), and

Eu^{3+} ions resided in only one site in the compounds. Selected compositions show bright-red emission and could be of potential use in WLEDs [46].

Bi^{3+},Eu^{3+}-doped $CdWO_4$ phosphors were synthesised by the co-precipitation technique. The phosphors show a typical CT band as well as Bi^{3+} absorption ($^1S_0 \rightarrow {}^3P_1$). The blue emission was observed from the host lattice by CT band excitation. The Eu^{3+}- and Bi^{3+}-doped phosphors show broad-band emission as well as Eu^{3+} red emission. The detailed $WO_6 \rightarrow Bi^{3+}$ energy transfer as well as $Bi^{3+} \rightarrow Eu^{3+}$ energy transfer were well discussed. The calculated CIE color coordinates show that the compound is versatile to emit multiple colors. The $CdWO_4$:3%Bi^{3+},0.5%Eu^{3+} phosphor emission intensity at 150°C is about 50% of its initial value [47]. A series of solid solutions with the formula $A_{1-1.5x}Eu_x\Upsilon_{0.5x}WO_4$ and $A_{0.64-0.5y}Eu_{0.24}Li_y\Upsilon_{0.12-0.5y}WO_4$ (A = Ca, Sr) have been reported, with special attention on the influence of luminescence properties by structural changes. Successful aliovalent substitution of Eu^{3+} at the Ca^{2+} site in the lattice leads to an abnormal expansion unit cell (linear change in the cell volume confirms the doping of the Eu^{3+} ion). Eu^{3+} is located at a significantly distorted cavity, confirmed from the emission spectra of the Eu^{3+} ion. Inclusion of the Li^+ ion (charge compensation) leads to enhancement in the emission intensity compared to that of Li-free phosphors (without charge compensation, vacancy) [48].

Recently, a scheelite-type cadmium molybdate red phosphor (Eu^{3+}-doped $Cd_{1-3x}Eu_{2x}\Upsilon_xMoO_4$) was explored for WLEDs. All the compounds crystallize (Eu^{3+} ions are in the range from 0.05 mol% to 66.67 mol% with respect to Cd^{2+} ions) in the tetragonal scheelite-type structure with the space group $I4_1/a$ (local point symmetry S_4). A cation vacancy was encountered in the lattice by the substitution of larger divalent Cd^{2+} (1.10 A for CN = 8) ions by incoming smaller trivalent Eu^{3+} cations (1.066 A, CN = 8) for charge compensation ($3Cd^{2+} \rightarrow 2Eu^{3+} + \Upsilon$ vacancy).

Time-resolved laser spectroscopy (at RT and 77 K) was performed to get fine resolution in the emission spectra as well as to separate the spectral lines from the $^5D_1$ and $^5D_0$ levels, respectively. The contributions of the individual emission lines were obtained from the steady-state spectrum. Three types of substitution sites are available for Eu^{3+} in the lattice, confirmed by low-temperature spectral studies. The energy transfer process occurs either from the CT band of $Eu^{3+} \rightarrow O^{2-}$ and or from the $Mo^{6+} \rightarrow O^{2-}$ to Eu^{3+} levels, and the same is shown in Fig. 8.7. A very strong red luminescence (ED, $^5D_0 \rightarrow {}^7F_2$ transition) at 616 nm was observed by exciting with UV photons, and thus they are considered promising phosphors for WLEDs. Concentration quenching of Eu^{3+} luminescence as well as the optimum doping of optically active ions were investigated [49].

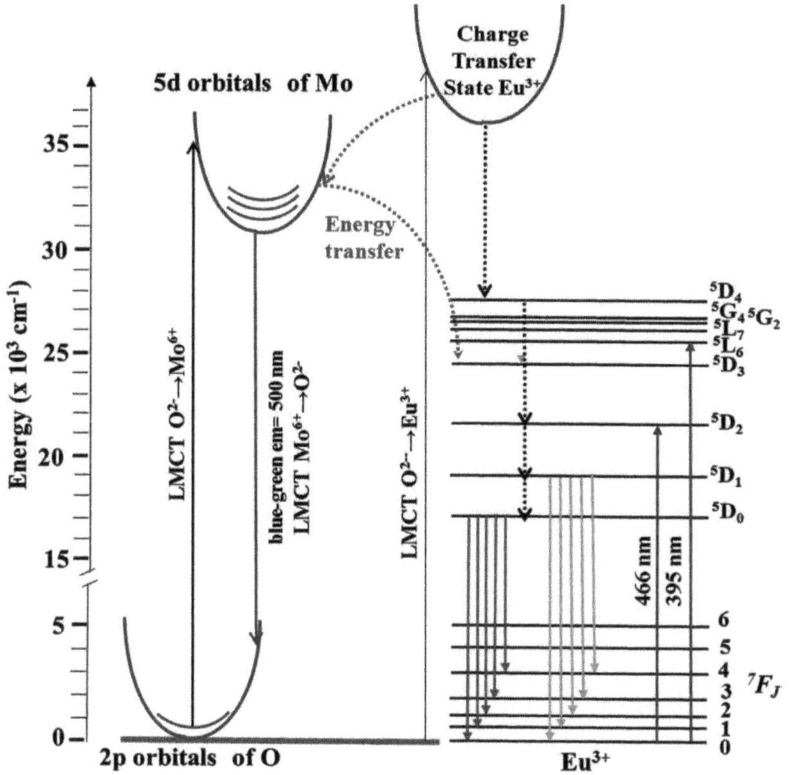

Figure 8.7 Electronic energy level diagram and optical transitions observed for $Cd_{1-3x}Eu_{2x}\Upsilon_xMoO_4$ solid solutions. Reproduced from Ref. [49] with permission of The Royal Society of Chemistry.

8.4.2 Double Tungstate and Molybdates $[AB(MO_4)_2]:Eu^{3+}$

As mentioned earlier the scheelite-related phases have been identified as a host lattice for RE ions [34, 35]. These include compositions of AMO_4 and $AB(MO_4)_2$, where A = Ca^{2+}, Sr^{2+}, Ba^{2+}, and Pb^{2+}; A = Li^+, Na^+, K^+, Rb^+, Cs^+, Ag^+, and Hg^{2+}; B = RE ions; and M = Mo^{6+} and W^{6+}, and the crystal structure and luminescence properties have been studied [50–52]. Some of the compounds crystallize in a scheelite-related structure, but others have a completely different structure. RE-substituted double tungstates/molybdates are important candidates due to expected ET from tungstate/molybdate to RE ions. Luminescence properties of AEu $(WO_4)_2$ (A = Li^+, Na^+, K^+, and Rb^+) and $AEu(MoO_4)_2$ (A = Li^+, Na^+, K^+, Rb^+, and Cs^+) have been investigated [52]. Since the crystal structures are different, the Eu^{3+} luminescence has been analyzed on the basis of the crystal structure. The observed broadening of the lines in the emission spectra of $LiEu(MoO_4)_2$ and $NaEu(MoO_4)_2$ is due to cation disorder in the scheelite structure. $KEu(MoO_4)_2$ crystallizes in the triclinic structure with an ordered scheelite structure. The site symmetry of Eu^{3+} is C_1 symmetry, and the observed emission lines in the $^5D_0 \rightarrow {}^7F_{0,1,2}$ transitions agree with this site symmetry. However, these lines are only slightly narrower than the spectral lines obtained from the disordered scheelites. This emission lines from emission spectra clearly suggest that the $KEu(MoO_4)_2$ compound also has a slightly disordered structure. The structure of $KEu(WO_4)_2$ is isomorphous with that of $-KY(WO_4)_2$. This structure contains WO_6 double chains, and RE ions are eight-coordinated with oxygen ions. The site symmetry of Eu ions is consistent with that of the emission spectrum (C_2 symmetry). $RbEu(WO_4)_2$ crystallizes in a monoclinic structure, and the Eu^{3+} polyhedra form a 1D sublattice. The observed emission spectrum is coincident with that of the C_1 symmetry. In the case of $CsEu(WO_4)_2$, the Eu^{3+} luminescence shows that the $^5D_0 \rightarrow {}^7F_0$ transition is extremely weak, the $^5D_0 \rightarrow {}^7F_1$ transition is a doublet line, and the $^5D_0 \rightarrow {}^7F_2$ contains triplet lines. S_4 symmetry is the most appropriate site symmetry for the observed emission spectrum. However, S_4 symmetry can only be an approximate site symmetry because $J = 0 \rightarrow 0$ intensity is not completely zero [51].

It has been reported that the corresponding tungstates $A^ILn(WO_4)_2$ (A^I = Li, Na, and K) do not exhibit luminesce at room temperature (RT) in contrast to Ag and Cs compounds. In the literature, it has been reported that Cs-containing double tungstates show efficient emission at RT. For example, the $CsM(WO_4)_2$ (M = Gd, Y, and Lu) compound shows efficient green emission. $CsEu(WO_4)_2$

shows efficient red emission, $CsTb(WO_4)_2$ shows green emission with medium intensity, and $CsM(WO_4)_2$ (M = La, Sm) does not show any emission at RT. In the case of $HgWO_4$, the vibrational as well as electronic spectra support tetrahedral (T_d) tungsten coordination, whereas the crystal structure determination shows it to be octahedral (O_h) coordination [53]. It has been assumed that the spectral observations were determined by the nearest oxygen neighbors only, since the octahedral coordination consists of four oxygen neighbors at shorter and two at longer distances [54, 55]. The same situation seems to exist in Cs-containing compounds. $KY(WO_4)_2$ does not show luminescence in contrast to cesium-containing compounds. This may due to the structural differences in the double chains of tungstate octahedra. In $KY(WO_4)_2$, the octahedral coordination contains four W–O, which are shorter than the remaining two W–O bonds. Delocalization of electronic charge in the tungstate double chain is the possible reason for the absence of luminescence in $KY(WO_4)_2$ [51]. It has been reported that the complexes of transition metal ions with an empty d shell often show intense broad-band emission with a large Stokes shift (10,000–20,000 cm^{-1}). This is generally called LMCT transition. Examples are VO_4^{3-}, NbO_6^{7-}, WO_4^{2-}, and WO_6^{6-} [4]. In the present study, the VB is mainly due to O 2p and the CB is made up of W (5d) orbitals. The electron is promoted from the VB to the CB by the absorption of photons, and emission occurs. The observed emission band is due to the oxygen-to-tungsten CT band. The observed large Stokes shift is due to a large change in the interatomic distance (ΔR). On the other hand, the luminescence of $AgGd(MoO_4)_2$ is quenched at RT.

8.4.3 Eu^{3+} Luminescence in $ALn(MO_4)_2$ (A = Na, Li, Ag; Ln = Y, La, Gd; M = W, Mo)

Neeraj et al. reported that the emission intensity of $NaY_{0.95}Eu_{0.05}(WO_4)(MoO_4)$ is ~7.28 times that of the $Y_2O_2S:Eu^{3+}$ commercial red phosphor under 394 nm excitation, and they suggested that $NaY_{0.95}Eu_{0.05}(WO_4)(MoO_4)$ and $NaY_{0.95}Eu_{0.05}(WO_4)_{1.25}(MoO_4)_{0.75}$ with a scheelite structure are probably promising red-emitting phosphors for GaN-based WLEDs [56]. Wang et al. reported that the $NaEu_{0.92}Sm_{0.08}(MoO_4)_2$ phosphor with a scheelite structure has a broad excitation band at ~400 nm, where the NUV LED emission occurs. The emission intensity is found to be ~2.4 times higher than that of the $NaY_{0.95}Eu_{0.05}(WO_4)(MoO_4)$ red phosphor [57]. In the $Li_{0.333}Na_{0.334}K_{0.333}Eu(MoO_4)_2$ phosphor, red emission intensity is found to be ~5.4 times higher than that of the Y_2O_2S:$Eu_{0.05}$ red phosphor [58]. The influence of fluoride ions on the luminescence of $LiEuM_2O_8$ (M = W, Mo) has been studied and the emission intensity of $LiEuMo_2O_{7.75}F_{0.5}$ is found to be ~3 times than that of the Y_2O_2S:Eu^{3+} commercial red phosphor [59].

8.4.4 Eu^{3+} Luminescence in $AgGd(MO_4)_2$ (M = W, Mo)

Eu^{3+} luminescence has been investigated systematically in $AgGd_{0.95}Eu_{0.05}(WO_4)_{2-x}(MoO_4)_x$ (x = 0–2, in steps of 0.25). Since the crystal structure of the end members (parent) is different, the tungstate crystallizes in a monoclinic structure (distorted scheelite structure), whereas the molybdate crystallizes in a tetragonal structure (scheelite structure). Tungsten is coordinated with oxygen in the tetrahedral and octahedral environments in $AgGd(WO_4)_2$, whereas molybdenum is present in the tetrahedral site in $AgGd(MoO_4)_2$. Since the Eu^{3+} luminescence depends on the next-nearest neighbors present in the lattice, changes are possible in the excitation and emission spectra of Eu^{3+} by making solid solutions of tungstate and molybdate. Eu^{3+} luminescence is expected to reveal the local site symmetry in tungstate, molybdate, and the solid solution thereof. Further, the PL properties of $AgGd_{1-y}Eu_y(WO_4/MoO_4)_2$ (x = 0.1–1 in steps of 0.1) have been studied in order to achieve high emission intensity of Eu^{3+} and find alternative red phosphors for SSL applications. The emission intensity is compared with that of the commercial red phosphor (Y_2O_2S:Eu^{3+}, Nichia) [60]. It is known that $AgGd(WO_4)_2$ crystallizes in the scheelite-related structure with monoclinic distortion (space group $C^4{}_{2h}$), and this structure is similar to that of $KLnW_2O_8$ [61, 62]. In $KLnW_2O_8$, the structure is built up of WO_6 octahedra joined by means of bridged WO_2W double bonds and a WOW single bond (Fig. 8.8a) and the Ln^{3+} site symmetry is C_2. In the case of $NaY(WO_4)_2$ the W^{6+} ions are present in the tetrahedral cavity [63], and this has been shown in Fig. 8.8b.

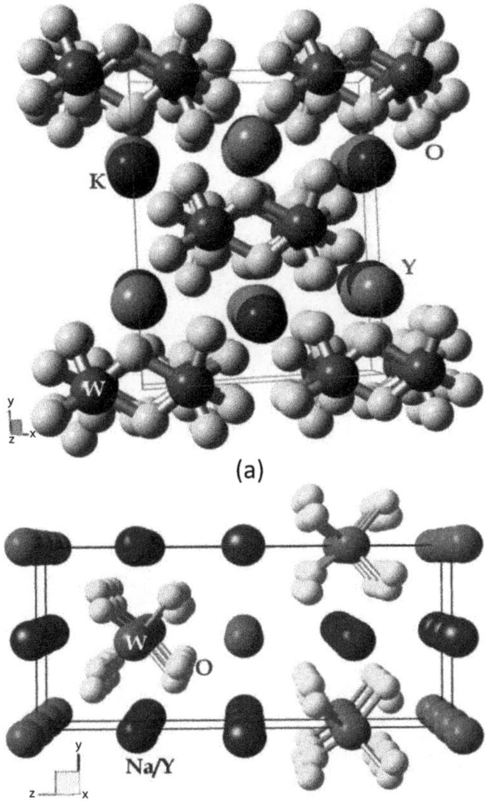

(a)

(b)

Figure 8.8 (a) Perspective view of the $KY(WO_4)_2$ crystal structure. The octahedral coordination of W atoms is shown, which includes single and double oxygen bridges. (b) Perspective view of the $NaY(WO_4)_2$ crystal structure.

The structure of $AgGd(WO_4)_2$ is shown in Fig. 8.9a. In this structure, the Gd^{3+} ion is eight-coordinated by oxygen atoms, forming a square antiprism. The polyhedra form a single chain by edge sharing. The Ag^+ ion is twelve-coordinated by oxygen atoms, forming a distorted icosahedron. By edge sharing, these polyhedra form a bidimensional layer consisting of chains. Hexavalent tungsten (W^{6+}) is located in a distorted octahedron with three shorter W–O distances ranging from 1.747 to 2.359 Å. In this structure, all the polyhedra are strongly linked by edge and corner sharing, forming a chain. GdO_8 and AgO_{12} polyhedra are shown in Fig. 8.9b.

$AgGd(MoO_4)_2$ crystallizes in the tetragonal structure (space group $I4_1/a$) [64]. The structure of $AgGd(MoO_4)_2$ is shown in Fig. 8.10a. In this structure, Ag and Gd ions are eight-coordinated with oxygen atoms, forming a square antiprism (Fig. 8.10b).

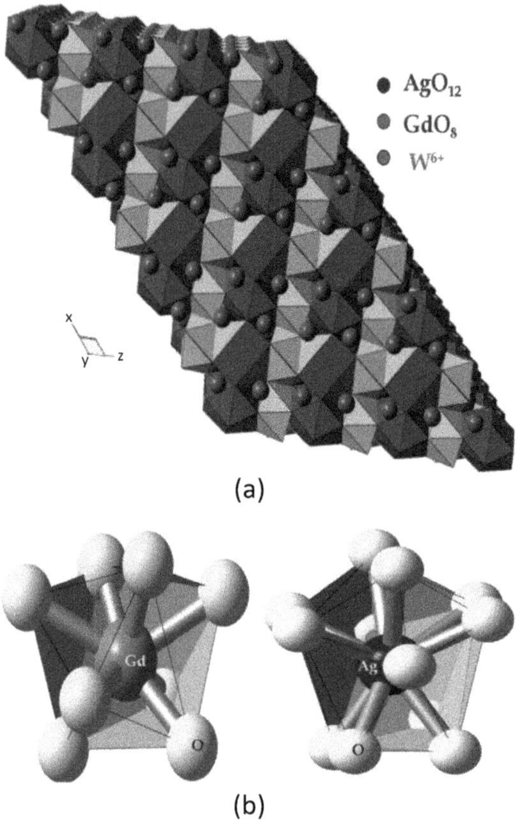

Figure 8.9 (a) Crystal structure of $AgGd(WO_4)_2$. (b) Polyhedra of GdO_8 and AgO_{12}.

Eu^{3+} luminescence in $AgGd_{0.95}Eu_{0.05}(WO_4)_{2\text{-}x}(MoO_4)_x$ (x = 0–2, in steps of 0.25) has shown that in the excitation spectra, the CT band is intense for tungstates (monoclinic) compared to molybdates because tungsten is coordinated with oxygen in an octahedral environment, whereas molybdenum is coordinated with oxygen in a tetrahedral environment. In the case of $AgLa_{0.95}Eu_{0.05}(WO_4)_{2\text{-}x}(MoO_4)_x$ (x = 0–2, in steps of 0.25), the CT band absorption is found

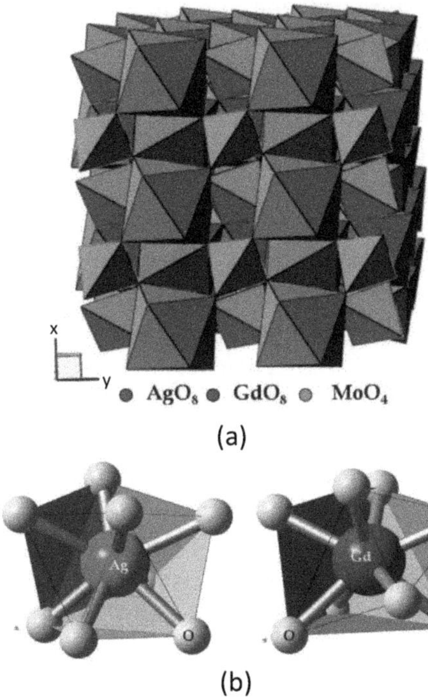

Figure 8.10 (a) Crystal structure of $AgGd(MoO_4)_2$. (b) Polyhedra of AgO_8 and GdO_8.

to be less for all compositions, since both tungsten and molybdenum are coordinated in a tetrahedral environment. All compositions show strong ED transition, since Eu^{3+} occupies C_2 and S_4 site symmetry in monoclinic and tetragonal structures, respectively. The splitting of the emission lines depends on the local site symmetry, which the Eu^{3+} ion occupies. The emission intensity of the $Eu_{0.3}$-doped molybdate phase is ~3.33 times stronger than the $Y_2O_2S:Eu^{3+}$ commercial red phosphor and that of the $Eu_{0.2}$-doped tungstate phase is comparable to the emission of the $Y_2O_2S:Eu^{3+}$ commercial red phosphor. The emission lines of all the phases studied are extremely narrow with full width at half maximum (FWHM) in the range of 3 to 4 nm, making them attractive candidates as potential red phosphors in SSL technology based on NUV/blue GaN-based LEDs. A series of solid solutions of $AgLa_{0.95}Eu_{0.05}(WO_4)_{2-x}(MoO_4)_x$ (x = 0–2, in steps of 0.25) are prepared by a conventional high-temperature solid-state reaction and the luminescence properties of Eu^{3+} are presented. The higher molybdenum content shifts the absorption maximum from 300 to 375 nm, where the LED emission occurs. A strong excitation is observed for all compositions at 465 nm, where the blue LED emission occurs. All the compositions show narrow-band red emission at around 615–620 nm, with the FWHM in the range of 2–3 nm under 465 nm excitation. $AgLa_{0.6}Eu_{0.4}(WO_4)_2$ with bright red emission can find potential application as a red phosphor for blue GaN-based LEDs [65]. The emission properties of the presently studied phases are better as compared to those of the $Y_2O_2S:Eu^{3+}$ (Nichia) phase. Particularly, the relative emission intensity of the $AgLa_{0.95}Eu_{0.05}(WO_4)_{0.75}(MoO_4)_{1.25}$ composition is ~1.9 times higher than that of $Y_2O_2S:Eu^{3+}$ (Nichia), whereas in the case of Gd-containing compounds, the $AgGd_{0.95}Eu_{0.05}(WO_4)_{0.25}(MoO_4)_{1.75}$ composition shows ~1.5 times more intensity than the commercial one. This may be because La-containing tungstate and molybdate show high absorption in the blue (465 nm) region since the tungstate emission band is observed at around 470 nm, where the Eu^{3+} excitation line is observed, and efficient energy transfer is possible from the absorbing group (WO_4/MoO_4) to Eu^{3+}.

$CaGd_{2(1-x)}Eu_{2x}(MoO_4)_{4(1-y)}(WO_4)_{4y}$ ($0 \leq x \leq 1$, $0 \leq y \leq 1$) solid solutions with a scheelite-type structure were synthesized, and detailed structural studies and optical properties were reported [66]. The detailed structural study reveals that the $CaEu_2(MoO_4)_4$, $CaGd_2(MoO_4)_4$, and $CaGd_{2(1-x)}Eu_{2x}(MoO_4)_4$ solid-solution compounds as well as their W-based analogs are not disordered scheelites but incommensurately modulated structures contrasting to the literature reports [67]. This is because of ordering of the A cations and vacancies. $CaGd_{2(1-x)}Eu_{2x}(MoO_4)_{4(1-y)}(WO_4)_{4y}$ solid solutions have either the tetragonal scheelite structure ($0 \leq y \leq 0.5$), which is (3 + 2)D modulated, or the monoclinically distorted scheelite-type structure ($0.75 \leq y \leq 1$), which is (3 + 1)D modulated. The cation and vacancy ordering was observed in the (3 + 1)D-modulated monoclinic $CaEu_2(WO_4)_4$ structure. The phosphor Eu^{3+} in $CaGd_{2(1-x)}Eu_{2x}(BO_4)_4$ (B = Mo, W) emission intensity of the $^5D_0$–$^7F_2$ transition reaches a maximum for $x = 0.5$ and is well matched with other reported scheelite-related phases [60, 65, 68, 69]. The relatively large Eu–Eu distances could arrest the concentration quenching in the lattice (varying in the range of 3.68 to 4.01 Å in the $CaEu_2(WO_4)_4$ structure). A complete structural study was established in the scheelite-related structures. Detailed crystallographic studies give information about the flexible geometry of MoO_4^{2-} and WO_4^{2-} tetrahedrals in $CaEu_2(WO_4)_4$ and $CaEu_2(MoO_4)_4$ structures. Changing the A-site by cations with a different charge/size and/or cation vacancies leads to variation in metal–oxygen (M–O) distances and bond angles [70]. A variety of crystal structures and polymorphs are encountered in the double molybdates $KR(MoO_4)_2$ and can be categorized in scheelite-type modifications (R = La–Dy) and nonscheelite modifications (R = Dy–Lu) with the $KY(MoO_4)^{2-}$type structure (space group P_{bcn}), [71]. Scheelite-type modifications can be divided into three groups: (i) tetragonal (R = La–Sm) [72, 73] or small monoclinic distortion of the scheelite subcell [70, 71], (ii) phases with the incommensurate modulated structure (R = Nd, Sm) [76, 77] (superspace group $I_{2/b}(\alpha\beta0)00$), and (iii) anorthic phases with the α-$KEu(MoO_4)_2$ structure (space group P-1) [78].

In contrast to the tetragonal $MR(MoO_4)_2$ (M = Li^+, Na^+, K^+, Ag^+; R = Ln, Y, Bi) compounds, where M^+ and R^{3+} are equally spread in the structure, the K^+ and Eu^{3+} cations in the α-$KEu(MoO_4)_2$ lattice are ordered. The A-site cations are completely ordered in $KNd(MoO_4)_2$ [76], while they are only partially ordered in $KSm(MoO_4)_2$, among the incommensurately modulated structures [77]. Recently, Yi et al. reported the synthesis and PL investigation of the $KGd_{0.75}Eu_{0.25}(MoO_4)_2$ phosphor and found it to exhibit strong and saturated red emission [79]. Morozov et al. studied the influence of the ordering of the A-cations on the luminescent properties in scheelite-related compounds (α- and β-polymorphs of $KEu(MoO_4)_2$) [80]. By combining synchrotron powder X-ray diffraction (XRD) and the electron microscopic technique, the structures of both the $KEu(MoO_4)_2$ polymorphs were studied and it was found that modulation arises due to Eu/K cation ordering at the A-site. Here a layered network of Eu^{3+} ions results in improved luminescence properties as compared to that of the 3D framework of Eu^{3+} ions.

Solid-solution $Li_{1-x}Ag_xLu_{1-y}(MoO_4)_2{:}yEu^{3+}$ phosphors with a scheelite-type structure were studied along with a change in Eu^{3+} luminescence by crystal chemical modification in the structure [81]. As mentioned earlier, these compounds, $ALn(MoO_4)_2$ (A = Li,

Ag and Ln = Lu, Eu) are crystallized in a tetragonal structure with the space group *I*41/*a*. However, some mismatch reports were made for the $LiEu(MoO_4)_2$ compound about the symmetry [78]. A-site variation in these compositions, $AgLu_{1-x}Eu_x(MoO_4)_2$ and $LiLu_{1-x}Eu_x(MoO_4)_2$, and the absorption spectra are unmatchable. Comparing both the Ag-containing compositions shows a good absorption range in the NUV region than Li-containing composition phosphors. However, $LiLu_{1-x}Eu_x(MoO_4)_2$ phosphors were proven to be a better option for NUV energy absorption and transmission properties as compared to $AgLu_{1-x}Eu_x(MoO_4)_2$ phosphors. All the compositions show red emission at around 618 nm under either CT band excitation or Eu^{3+} excitation. The strongest red emission line at 615 nm in $Li_xAg_{1-x}Lu_{0.95}(MoO_4)_2$:$0.05Eu^{3+}$ (x = 0, 0.2, 0.4, 0.6, 0.8, 1.0) clearly indicates that the Ag/Li ratio does not alter the noncentrosymmetric environment around the luminescence Eu^{3+} center in the lattice. The absolute QE of the selected $LiEu(MoO_4)_2$ is about 24% under excitation at 460 nm. The E_g values of $Li_xAg_{1-x}Lu(MoO_4)_2$ (x = 0, 1.0) were determined to be 3.08 eV, and 3.58 eV, respectively. Thus, the increasing E_g value was expected to enhance the Eu^{3+} emission due to the decrease of covalent properties of the host lattice. Detailed analysis indicates that the covalent properties of the Mo–O bond in the $AgLu_{0.95}(MoO_4)_2$:$0.05Eu^{3+}$ system is much greater than that of the $LiLu_{0.95}(MoO_4)_2$:$0.05Eu^{3+}$ system. Figure 8.11 shows the electron cloud polarizations and ionic deformations of $AgLu_{0.95}(MoO_4)_2$:$0.05Eu^{3+}$ and $LiLu_{0.95}(MoO_4)_2$:$0.05Eu^{3+}$ and energy level diagrams of $ALu_{0.95}(MoO_4)_2$:$0.05Eu^{3+}$ (A = Li or Ag). There are two main reasons for the increment in the Eu^{3+} luminescence in the solid-solution $Li_{1-x}Ag_xLu_{1-y}(MoO_4)_2$:$yEu^{3+}$ phosphors, (i) the coupling effect and the nonradiative transition between the energy levels of $Li_xAg_{1-x}Lu(MoO_4)_2$ matrices and the activator Eu^{3+} and (ii) the NUV energy absorption and transmission efficiency between the CT band of $O^{2-}\rightarrow Mo^{6+}$ and the 4f→4f emissive transitions of Eu^{3+}.

Recently, the $LiGd(WO_4)_2$:Eu^{3+} phosphor with excellent PL properties was reported [82]. The phosphor could efficiently excite with UV and yields bright red emission at 617 nm. ED transition ($^5D_0\rightarrow{}^7F_2$) of Eu^{3+} ions confirms the local site symmetry does not have a center of inversion and it is a distorted polyhedron with S_4 symmetry. The $LiGd(WO_4)_2$:80%Eu^{3+} phosphor yields excellent red emission with appropriate CIE coordinate values ($x = 0.656, y = 0.334$) and an internal QE as high as 69%. In addition, the phosphor shows good thermal stability; when heated up to 150°C, the $LiGd(WO_4)_2$:80%Eu^{3+} phosphor still has 81% of its initial emission intensity at RT.

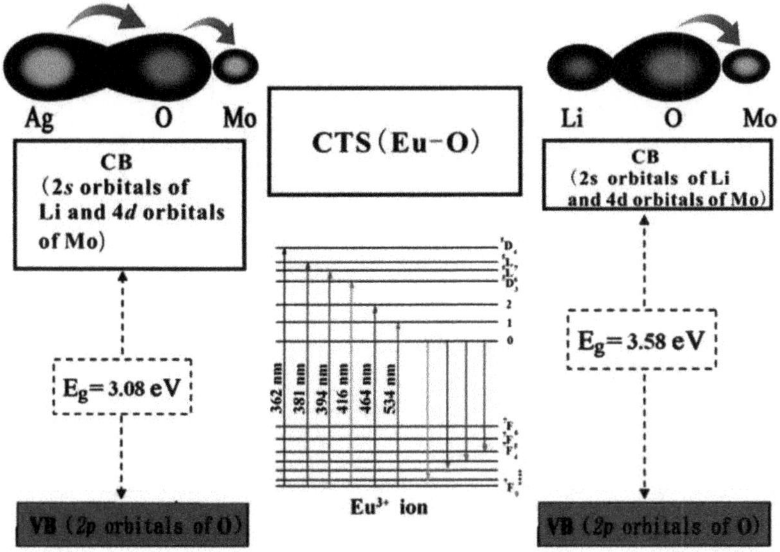

Figure 8.11 Electron cloud polarizations and ionic deformations of $AgLu_{0.95}(MoO_4)_2$:0.05Eu^{3+} and $LiLu_{0.95}(MoO_4)_2$:0.05Eu^{3+}; energy level diagrams of $ALu_{0.95}(MoO_4)_2$:0.05Eu^{3+} (A = Li or Ag). Reproduced (adapted or in part) from Ref. [81] with permission of The Royal Society of Chemistry.

Tb^{3+} excitation lines can be a mediator for the energy transfer between the MoO_4^{2-} group to the Eu^{3+} ion in $NaTb(MoO_4)_2$ [83]. Recently, the facile one-step hydrothermal synthesis route was used to synthesis $NaTb(MoO_4)_2$ microcrystals without employing any templates or surfactants. Phosphors exhibited efficient tunable emissions under UV excitation owing to the efficient energy transfer from the host lattice (MoO_4^{2-}) to Tb^{3+} and then to the Eu^{3+} ion (Fig. 8.12). Eu^{3+} doping content increment in the lattice results in the tuning of emission colors from green to yellow and then red. Theoretical simulation studies indicate that the ET mechanism ($Tb^{3+} \rightarrow Eu^{3+}$) is dominated by a dipole–quadrupole (d–q) interaction.

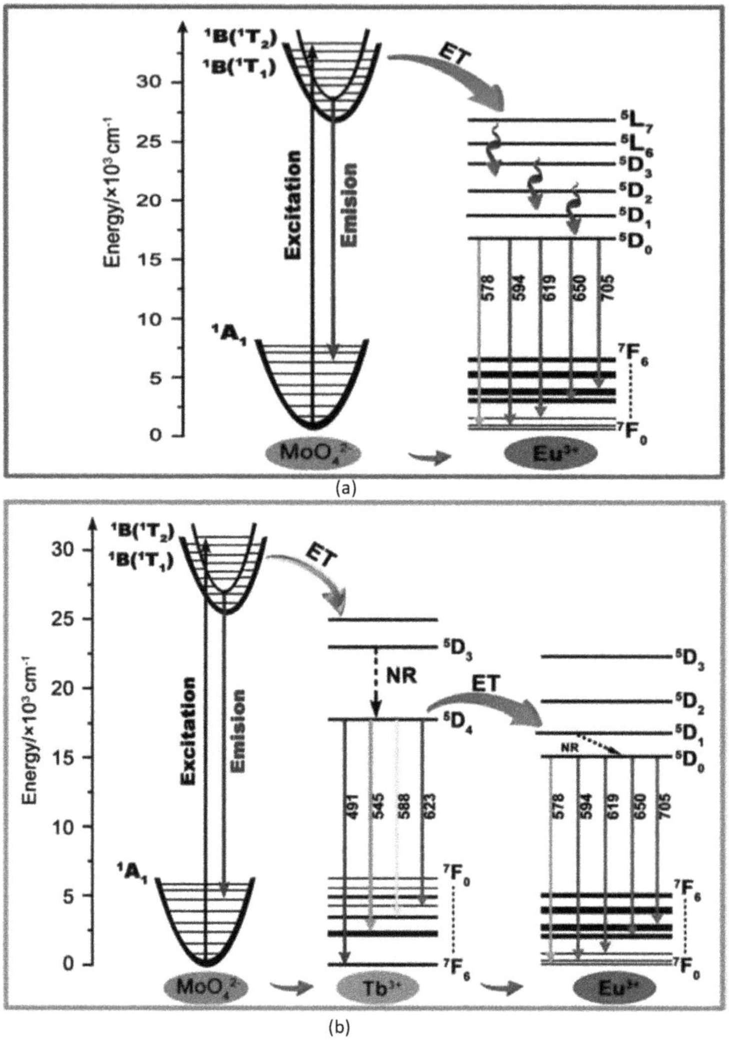

Figure 8.12 (a) The energy transfer process in the host lattice to Eu^{3+} and (b) to the Eu^{3+} through the Tb^{3+} ion. Reproduced from Ref. [83] with permission of The Royal Society of Chemistry.

8.4.5 White Light Generation in $LiGd(WO_4)_2$:RE^3

White-light emission was realized in the $LiGd(WO_4)_2$ lattice with different RE^{3+} (RE = Tm, Tb, Dy, Eu) concentrations. The emission spectra together show the characteristic emissions of RE^{3+} ions (Tm^{3+}, blue; Tb^{3+}, green; Dy^{3+}, yellow; and Eu^{3+}, red) under UV excitation. By just altering the activator Eu^{3+} concentration [84] in the $LiGd(WO_4)_2$:2% Tm^{3+}, 4% Tb^{3+}, x% Eu^{3+} phosphor, single-phase white emission was attained (adequate mixing of blue Tm^{3+}, green Tb^{3+}, and red Eu^{3+} emissions leads to white light); here the energy transfer from Tb^{3+} to Eu^{3+} ions was found to play a key role (Fig. 8.13). The PL properties of as-prepared materials indicate the single-host white-light-emitting phosphor could find potential applications in UV-excited WLEDs.

8.4.6 $M_5RE(BO_4)_4$ (M = Li, Na, K; RE = La, Eu, Y; B = W, Mo)

The compounds $Na_5RE(MoO_4)_4$:RE^{3+} = Y^{3+}, La^{3+} share the scheelite-related structure with RE^{3+} occupying the eight-coordinate sites [85]. The luminescence of $Na_5La(MoO_4)_4$:xEu^{3+}(x = 0–1, in steps of 0.1) was investigated and compared with that of the $NaEu(MoO_4)_2$ phosphor. Single red LEDs were fabricated by combining the red phosphors $Na_5Eu(MoO_4)_4$ and $NaEu(MoO_4)_2$ along with InGaN chips (~400 nm emitting) [86]. The phosphor exhibits higher emission under ~400 nm excitation with appropriate CIE coordinates (close to the NTSC standard values). Highly intense red emission was found in a Na-containing phosphor among the solid solutions of tungstate and molybdates $M_5Eu(WO_4)_{4-x}(MoO_4)_x$ (M = Li, Na, K)-based phosphors [87]. Compared to commercially available red phosphors using WLEDs, the color-rendering index (CRI) value for $Na_5Eu(WO_4)_2(MoO_4)_2$-incorporated WLEDs shows a better value (82.3) than that of La_2O_2S:Eu^{3+}-incorporated WLEDs (70.8). Recently, Dan Qin et al. also reported the synthesis of the $Na_5Ln(WO_4)_{4-z}(MoO4)_z$:$Tb^{3+}$,$Eu^{3+}$ (Ln = La, Y, Gd) phosphor by the sol-gel technique. The phosphor exhibits multicolor emission, which is required for WLEDs. The $Na_5Y_{1-x}Eu_x(MoO_4)_4$ phosphor shows the brightest red light among $Na_5Ln_{1-x}Eu_x(MoO_4)_4$ phosphors with a fixed Eu^{3+} concentration. Tb^{3+} to Eu^{3+} energy transfer studies were also investigated [88].

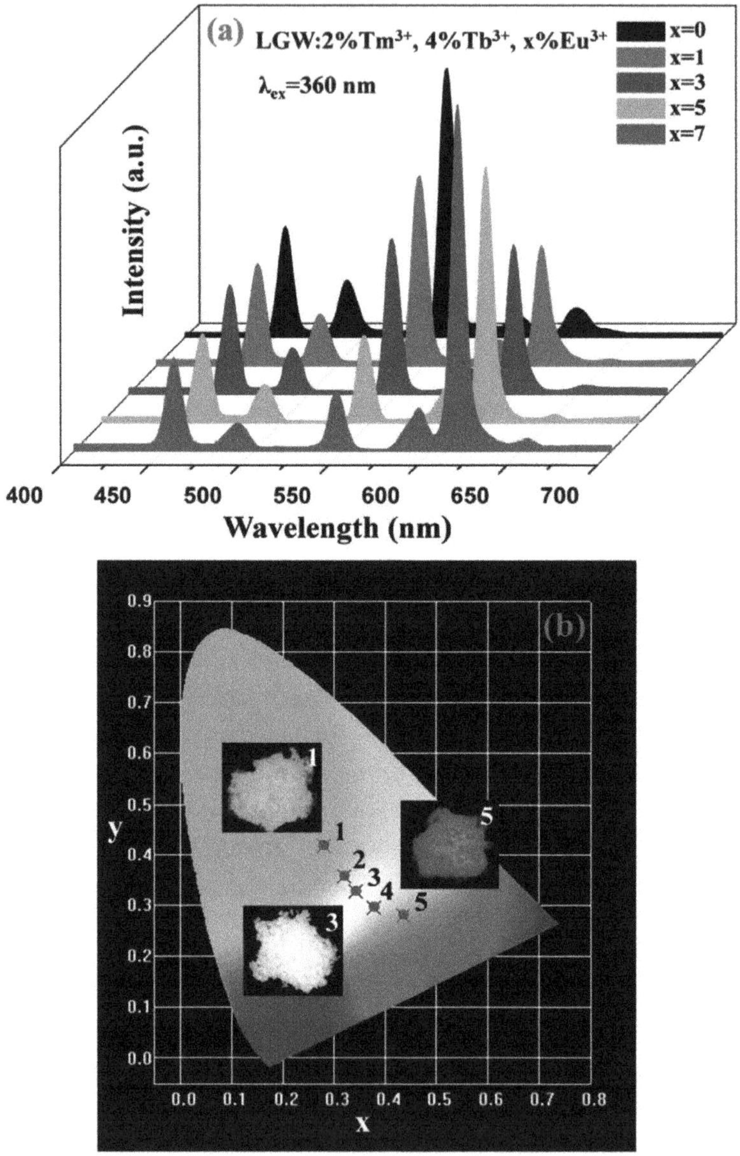

Figure 8.13 (a) PL emission spectra of $LiGd(WO_4)_2$:2% Tm^{3+}, 4% Tb^{3+}, *x*% Eu^{3+} samples with different Eu^{3+} concentrations ($0 \leq x \leq 7$). (b) CIE coordinate diagram of $LiGd(WO_4)_2$:2% Tm^{3+}, 4% Tb^{3+}, *x*% Eu^{3+} with different Eu^{3+} concentrations: (1) $x = 0$; (2) $x = 1$; (3) $x = 3$; (4) $x = 5$; and (5) $x = 7$. The insets are their corresponding digital photographs taken under 365 nm excitation. Reproduced from Ref. [84] with permission of The Royal Society of Chemistry.

8.4.7 $Li_{3.5}Ln_{1.5}(MoO_4)_4$ (Ln = Y, Eu)

The lithium-rich red-emitting $Li_{3.5}(Y_{1x}Eu_x)_{1.5}(MoO_4)_4$ phosphor with x = 0, 0.1, 0.25, 0.5, 0.75, and 1 was reported by Litterscheid et al. [89]. The compound $Li_{3.5}(Y_{1-x}Eu_x)_{1.5}(MoO_4)_4$ is isostructural with $Li_{3.5}Ho_{1.5}(MoO_4)_4$ [90]. The crystallographic study reveals that of the available three sites in the lattice, Li occupies two of them exclusively, whereas one is occupied by mixing of both Li and lanthanides (Fig. 8.14a). The CT band of $O^{2-}\rightarrow Mo^{6+}$ appeared at 354 nm. As the host lattice absorption is efficient, one can expect that the energy transfer host lattice to Eu^{3+} leads to intensification of red emission. The direct excitation Eu^{3+} levels yield a good QE of 44% (due to low vibrational quenching processes). The temperature-dependent PL study reveals that the emission intensity at 100°C is approximately 70% of that at RT, while it drops to approximately 10% at 375°C. Figure 8.14b shows the digital photograph of a crystal of $Li_{3.5}Eu_{1.5}(MoO_4)_4$ under daylight (left) and UV light (right).

8.4.8 Molybdates $M_2Gd_4(MoO_4)_7$ (M = Li, Na)

Molybdates $M_2Gd_4(MoO_4)_7$ (M = Li, Na) with a defect scheelite structure have been investigated for applications in solid-state laser materials [91–94] as well as host lattices for Eu^{3+} luminescence [95]. $Li_2Gd_4(MoO_4)_7$ has a defect scheelite (space group $I41/a$) structure with the formula $Li_{0.286}Gd_{0.571}\phi_{0.143}MoO_4$ (ϕ represents vacancy) [96], which is isostructural with the scheelite $CaWO_4$. However, Li^+ and Gd^{3+} ions are not fully occupied in the A-site (Ca^{2+}) in $Li_2Gd_4(MoO_4)_7$, and 14.3% of the lattice remains unoccupied. Furthermore, lacking of inversion symmetry was found in the site at Gd^{3+} (S_4). Thus, these materials could act as a host for Eu^{3+} luminescence (to obtain pure red emission). Eu^{3+}-doped $M_2Gd_4(MoO_4)_7$ (M = Li, Na) red phosphors were synthesised and their luminescence properties studied. The critical concentration of Eu^{3+} in the Li and Na hosts is 0.85 and 0.70, respectively. The high symmetry ratio of the two phosphors indicates that the Eu^{3+} ion occupies a low-symmetry site in the lattice and is two times larger than that of red-emitting phosphors $Y_2(MoO_4)_3$:Eu^{3+} and $Li_3Ba_2Gd_3(MoO_4)_8$:Eu^{3+}. The phosphors show high QE upto 23.1% superior to that of known commercial red phosphors (Fig. 8.15).

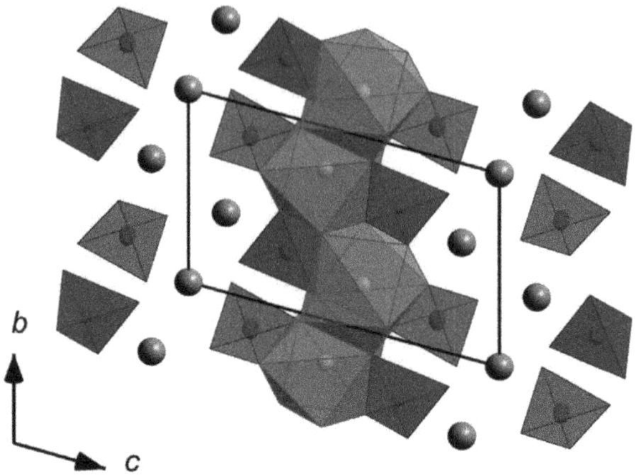

(a)

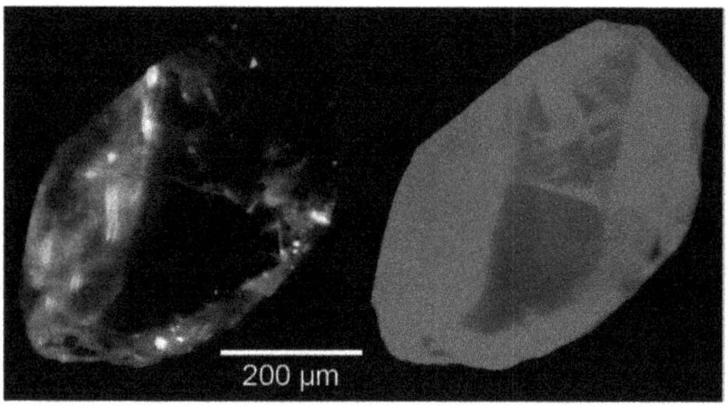

(b)

Figure 8.14 (a) Crystal structure of $Li_{3.5}(Y_{1-x}Eu_x)_{1.5}(MoO_4)_4$ (blue, MoO_4 tetrahedra; pink, MO_8 polyhedra, where M = 75% RE^+, 25% Li; RE: $Y_{1-x}Eu_x$; gray, Li). (b) Crystal of $Li_{3.5}Eu_{1.5}(MoO_4)_4$ under daylight (left) and UV light (right) [89] (Creative Commons Attribution-NonCommercial 3.0 Unported Licence).

Solid-solution $Li_2Y_{4-x}Eux(WO_4)_{7-y}(MoO_4)_y$ ($0 \leq x \leq 4$, $0 \leq y \leq 7$) phosphors were synthesized and their luminescent properties were studied. All compositions show red emission, the relative ratio of Mo/W was 7:0, the optimum doped concentration of Eu^{3+} was 2.8 mol, and the phosphor showed strong red emission lines at 615 nm corresponding to the forced ED $^5D_0 \rightarrow {}^7F_2$ transition of Eu^{3+}. Compared to $Na_2Y_2Eu_2(MoO_4)_7$ and $K_2Y_2Eu_2(MoO_4)_7$, the fluorescence intensity of the $Li_2Y_{1.2}Eu_{2.8}(MoO_4)_7$ phosphor was found to be the strongest [97]. A comparison of excitation and emission spectra is shown in Fig. 8.15.

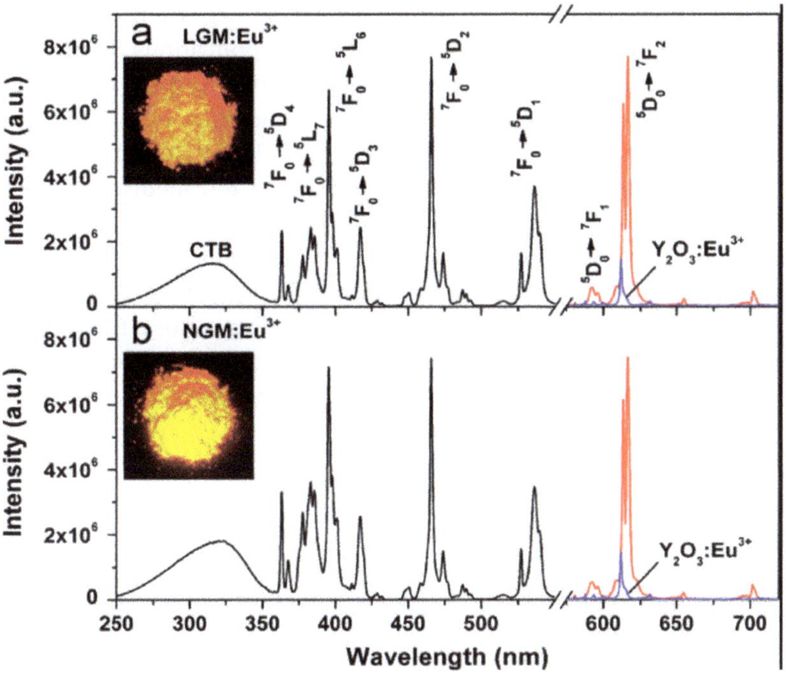

Figure 8.15 Excitation (λ_{em} = 617 nm) and emission (λ_{ex} = 466 nm) spectra of red phosphors (a) LGM:0.85Eu^{3+} and (b) NGM:0.70Eu^{3+}. Emission spectra for Eu^{3+}-doped $M_2Gd_4(MoO_4)_7$ (M = Li, Na) and Y_2O_3:Eu^{3+} are represented by red and blue lines, respectively. The insets show the images of the phosphors excited at 365 nm in a UV box. The emission spectrum (λ_{ex} = 468 nm) of $(Y_{0.95}Eu_{0.05})_2O_3$ is shown for comparison.

8.4.9 $Ca_4GdNbMo_4O_{20}$ with Powellite-Type Structure

The Eu^{3+}-activated $Ca_4LnNbMo_4O_{20}$(Ln = Gd, La) phosphor with a powellite-type structure was synthesized and its PL properties investigated in detail [98, 99]. This structure is considered to be a combination between two individual hosts, 4($CaMoO_4$) + GdNbO4, which bear a different crystal structure ($CaMoO_4$ has a powellite-type structure and $GdNbO_4$ has a fergusonite structure) [100]. The phosphor exhibits three different types of CT transitions (O^{2-} →Eu^{3+}, NbO_4^{3-}, and MoO_4^{2-}) and partial energy transfer occurs from the NbO_4^{3-} and MoO_4^{2-} complexes to Eu^{3+} ions. The emission from $CaLaNbMoO_8$:Eu^{3+} showed greater intensity values than the

$CaMoO_4:Eu^{3+}$ phosphor under NUV wavelength (394 nm). However, the compound is not suitable for NUV LED applications; hence the shifting of the absorption edge from 360 nm to soft UV excitation is attractive for future research. Matas Janulevicius et al. reported highly efficient $Y_2Mo_4O_{15}:Eu^{3+}$ phosphors, and they show a broad CT band and efficient red emission due to MoO_4 tetrahedra and Eu^{3+} ions, respectively. The quantum yield (QY) was found to be 85% for the 50% Eu^{3+}-substituted phosphors. Temperature-dependent PL studies indicate that the Eu^{3+}-rich phosphors show good thermal stability (show stable emission up to 400 K and lose half of the efficiency only at 515 K). $Y_2Mo_4O_{15}$:75%Eu^{3+} ceramic disks were made (thickness of 0.71 and 0.98 mm) and the fabricated LED (375 and 400 nm) shows red photons, whereas integrating with a 455 nm LED yields purple color [101].

8.4.10 $R_2Zr_3(MoO_4)_9:Eu^{3+}$ (R = La, Sm, Gd)

Red-emitting phosphors $RE_2Zr_3(MoO_4)_9:Eu^{3+}$ (RE = La, Sm, Gd) were prepared using the convenient solid-state reaction method and PL studies were performed. The compound $R_2Zr_3(MoO_4)_9$ crystallizes in a trigonal crystal system with the space group *R*-3*c* [102]. The 3D network is built by vertices shared between ZrO_6 octahedra and MoO_4 tetrahedra (Fig. 8.16). The RE ions are orderly distributed in the structure and well separated by ZrO_6 and MoO_4. This rigid host lattice could enhance the thermal stability of the RE-activated phosphor. All the compositions show red emission (ED transition), consistent with the local site symmetry, under NUV as well as CT band excitation. The intensity of red emission of the compositions increases in the following order of Sm→La→Gd. As mentioned earlier the Eu–O–M angle plays an important role in the QE of phosphors. In the $Gd_2Zr_3(MoO_4)_9$ structure the Mo–O–Eu angle is 165°, and it is apparent to provide an efficient energy transfer to the Eu^{3+} ion [103]. The investigated QEs and temperature-dependent PL studies of Eu^{3+}-activated $R_2Zr_3(MoO_4)_9$ (R = La, Sm, Gd) reveals that the Gd containing host lattice shows higher QEs and thermal stability than that of La and Sm- containing systems. The QE values are 58%, 45%, and 72% for the phosphors of RE = La, Sm, and Gd, respectively. The values are higher than other reported red-emitting phosphors such as $Gd_6MoO_{12}:0.25Eu^{3+}$ nanopowder (62%, λ_{ex} =

395 nm) [104], $SrY_{0.7}Eu_{1.4}(MoO_4)_4$ (48%, λ_{ex} = 395 nm) [105], and commercial $Y_2O_2S:Eu^{3+}$ (35%, λ_{ex} = 317 nm) [106]. These results indicate that Eu^{3+}-doped $R_2Zr_3(MoO_4)_9$ (R = La, Sm, Gd) compounds have better potential as red-emitting phosphors for WLEDs.

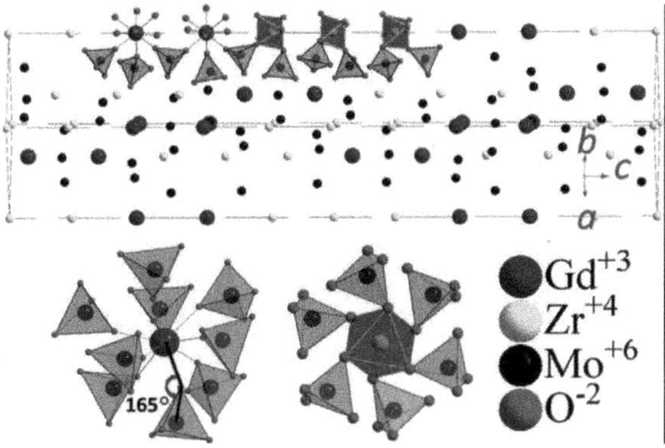

Figure 8.16 $Gd_2Zr_3(MoO_4)_9$ crystal structure along the [001] direction and the surrounding Gd and Zr polyhedron. The angle of Gd(Eu)–O–Mo is about 165°. From Ref. [103]. With permission of Springer.

8.4.11 $LaBWO_6:Eu^{3+}$

Huang et al. synthesized Eu^{3+}-doped borotungstate $La_{1-x}Eu_xBWO_6$ (x = 0.01–0.25) by using the Pechini method. The luminescence properties of these phosphors are not studied extensively so far. The PL excitation and emission spectra were recorded and compared with the commercial red-emitting phosphor $Y_2O_2S:Eu^{3+}$. $LaBWO_6$,Eu^{3+}'-activated phosphors have efficient absorption in the region of NUV wavelengths compared to the previous one, which is stronger than the absorption in the CT region. Typical Eu^{3+} emission from $^5D_0 \rightarrow {}^7F_J$ (J = 0–4) was observed. The noticeable observation was unusual high emission intensity of $^5D_0 \rightarrow {}^7F_4$ spectra for $LaBWO_6$, while for the other two samples, Eu^{3+}-doped La_3BWO_9 and $La_4B_2WO_{12}$, the $^5D_0 \rightarrow {}^7F_2$ ED transition predominated. The reason behind this increase of $^5D_0 \rightarrow {}^7F_4$ spectra can be ascribed to a highly polarizable chemical environment with local symmetry corresponding to a distorted coordination polyhedron. The concentration quenching phenomena was observation beyond an x = 0.2 doping concentration of the activator ion due to energy migration among Eu^{3+} ions in the lattice. The best observation was in Eu^{3+}-concentrated Eu_3BWO_9 and $Eu_4B_2WO_{12}$ systems, which show no concentration quenching in the lattice. The CIE chromaticity coordinates of $La_{1-x}Eu_xBWO_6$ (x = 0.01–0.25) are (x = 0.67, y = 0.33), which are closer to the NTSC standard [107].

8.4.12 $Gd_3B(W,Mo)O_9:Eu^{3+}$

Taking into consideration the crystal structure and luminescence properties, a series of highly efficient red-emitting phosphors ($Gd_3B(W,Mo)O_9:Eu^{3+}$) were designed and synthesized by Dong et al. using a traditional high-temperature solid-state reaction. This compound has the space group $P6_3$, with lattice parameters of $a = b$ = 8.5571 Å, c = 5.4079 Å, and cell volume = 342.93 Å. The phosphors consist of a single phase; insertion of the Eu^{3+} ion into the host lattice does not affect the lattice structure much as the ionic radius is similar to Gd^{3+} and Eu^{3+}, so it is expected to replace the existing Gd^{3+} ions in the lattice. The major peaks that appeared in the emission spectra are assigned to Eu^{3+} transitions. At any excitation wavelength the $^5D_0 \rightarrow {}^7F_2$ spectral line was found to be dominating, indicating the noncentrosymmetric site residence of Eu^{3+}. Concentration-dependent studies conclude that 0.08 is the optimum doping concentration for Eu^{3+} ions. To evaluate the application of the red phosphor $Gd_3B(W,Mo)O_9:Eu^{3+}$ in NUV WLEDs, the WLED was fabricated using the same red and yellow phosphors. A bright-white emission was observed from the device when a direct current of 20 mA was provided. The calculated CIE color coordinates are well in agreement with the standard for white color according to the NTSC (0.320, 0.349). The color correlation temperature (CCT) was found to be 6031 K, almost close to that of daylight. In conclusion, the $Gd_3B(W,Mo)O_9:Eu^{3+}$ red phosphor can be efficiently used in tricolor phosphor-coated NUV LEDs [108].

8.4.13 $La_3BW_{1-x}Mo_xO_9:Eu^{3+}$

A series of $La_3BW_{1-x}Mo_xO_9:Eu^{3+}$ (x = 0–0.4) polycrystalline powders were prepared by by Huang et al. using solid-state reactions. The RE-based borotungsates Ln_3BWO_9 can be used as luminescence materials when activated with Eu^{3+}. The phosphor crystallizes in a hexagonal crystal system (noncentrosymmetric structure) with the space group $P6_3$ in which the single sites are occupied by $[BO_3]$, $[WO_6]$, and Ln^{3+} ions. The replacement of WO_6 by MoO_6 is well known in solid-state chemistry because of similar properties and size effect, but sometimes it may lead to adverse effects and break down the host structure completely. So efforts have been made to make a solid solution between Mo/W by adjusting and maintaining their ratio, and changes in structural and luminescence properties were studied. By incorporating Mo^{6+}, the single-phase $La_3BW_{1-x}Mo_xO_9{:}Eu^{3+}$ can exist up to $x = 0.3$, while a further increase of Mo content impurity phases coexist with preliminary $LaBWO_6$. Another remarkable observation in the patterns was made that the diffraction lines started shifting to a higher 2θ angle along with an increment of Mo^{6+}. The excitation spectra of $La_{2.7}BW_{1-x}Mo_xO_9{:}0.3Eu^{3+}$ without doping of Mo^{6+} show typical absorption emission properties for Eu^{3+}-activated phosphors; when Mo^{6+} was incorporated, shifting of the CT band toward a higher wavelength was evidenced, which is an advantage for UV LED phosphors. In the hexagonal $La_{2.7}BW_{1-x}Mo_xO_9{:}0.3Eu^{3+}$, it was assumed that Eu^{3+} substitute the La^{3+} site, which has ninefold coordination with O atoms (C_1 point symmetry). Clearly, with an increase of the incorporation of Mo^{6+} amount, the emission derived from $^5D_0 \rightarrow ^7F_J$ was consecutively enhanced along with the asymmetric ratio (AR). The emission intensity of the $La_{2.7}BW_{0.7}Mo_{0.3}O_9{:}0.3Eu^{3+}$ phosphor is over six times stronger than that of $Y_2O_2S{:}Eu^{3+}$, and they can be efficiently used as bright-red phosphors for LED applications because of the isomorphism effect [109].

8.4.14 $Y_2MoO_6{:}Eu^{3+}$

The detailed investigation of structural properties of Y_2MoO_6 was initially reported by Alonso et al. in 2004 [110]. It revealed that the yttrium ion occupies three nonequivalent sites, 4e, 4e, and C_2 as a point site and 8f (with C_1 as a point site) and these are coordinated to eight oxygen atoms. By doping of Eu^{3+} into the Y_2MoO_6 host lattice, efficient narrow-band-emitting phosphors have been obtained. However, a detailed study of concentration quenching and energy transfer was performed by Han et al. [111]. A series of $Y_{2(1-x)}Eu_{2x}MoO_6$

phosphors ($0 \leq x \leq 1$) were synthesized by the conventional solid-state method, and dependence of luminescence and energy transfer was studied in detail. XRD measurement reveals the phase formation, and the replacement of smaller Y^{3+} ions by larger Eu^{3+} ion leads to shifting of diffraction peaks to a lower (2θ) angle. IR spectroscopy confirms the formation of an M–O bond from the different signals that appear below 1000 cm^{-1}. The major and strong signal at around 693 cm^{-1} is attributed to the MoO_4^{2-} group stretching frequency. The excitation and emission spectra of the samples are similar to the typical 4f–4f transition of the Eu^{3+} ion, which consists of a broad CT band originating due to CT from oxygen to molybdenum in the host lattice, and others belong to Eu^{3+} excitation peaks. The emission spectra consist of several sharp lines ranging from 550 to 700 nm, which belong to $^5D_0 \rightarrow ^7F_J$ transitions of Eu^{3+} (where J = 0–4). The Eu^{3+} ion is present at the noncentrosymmetric site in the lattice, resulting in the appearance of a $^5D_0 \rightarrow ^7F_2$ line at around 611 nm. The chromaticity color coordinate diagram well matches with the $Y_2O_2S:Eu^{3+}$ phosphor with appropriate color purity. The CT band edge at around 363 nm (~27548 cm^{-1}) is almost equal to the energy level of $^5L_8$ (27435 cm^{-1}) and $^5D_4$ (27641 cm^{-1}), so the energy absorbed is transferred to the Eu^{3+} nonradiatively. Finally, emission from $^5D_0$ to ^7F_J (J = 0, 1, 2, 3) occurs through the nonradiative relaxation process from $^5L_8$ and $^5D_4$ to $^5D_0$ levels. The luminescence intensity of phosphor materials varies with doping concentration of the Eu^{3+} ion. For this solid solution the optimum doping concentration was found to be 30 mol% (x = 0.3), beyond which concentration quenching occurs due to exchange interaction. In this particular host lattice the concentration quenching mechanism of Eu^{3+} could be due to dipole–dipole (d–d) or dipole–quadrupole (d–q) interaction in Y_2MoO_6 solid solutions. Keeping in view the absorption peak located at 363 nm and 465 nm match with the NUV (385–390 nm) and blue LED (460 nm) excitation wavelength, we conclude that they may be potential new-generation red phosphors for WLED applications.

8.4.15 $Lu_2MoO_6:Eu^{3+}$

Eu^{3+}-doped RE molybdates and tungstates $RE_2MO_6:Eu^{3+}$, where RE = Y, Gd, La, and Lu and M = Mo/W, have attracted the attention of many researchers owing to their strong absorption properties in the NUV region, followed by energy transfer of the same to the Eu^{3+} ion by nonradiatively boosting the red emission. Li et al. [112] previously reported the PL properties of Lu_2MoO_6:Eu^{3+}; the phosphor was synthesized by sol-gel methodology; however, no brief report has been made on phosphors synthesized by the solid-state reaction. Li et al. made efforts to develop an efficient red phosphor Lu_2MoO_6:Eu^{3+} and studied the effect of increasing activator ion concentration in the host lattice, along with luminescence properties and decay lifetime [113]. The crystal structure reveals that the phosphors crystallizes in a single monoclinic phase, where Lu^{3+} occupies three nonequivalent crystallographic sites: 4e, 4e with C_2 site symmetry, and 8f with C_1 site symmetry. And all of them are coordinated with eight oxygen atoms. The Mo atoms are coordinated with four oxygen atoms, of which two are at a short distance and the other two are at a long distance. The emission spectra of Lu_2MoO_6:Eu^{3+} consist of several sharp lines; however, the intensity of red emission at around 610 nm belongs to $^5D_0 \rightarrow ^7F_2$ transition of Eu^{3+} ion. In this host lattice, Eu^{3+} mainly occupies the nonequivalent crystallographic sites (both C_1 and C_2). The emission intensity of 610 nm gradually increases with increasing concentration of Eu^{3+} and reaches a maximum at 10% doping concentration of Eu^{3+}. The concentration quenching mechanism can be well explained through Dexter's theory [114, 115]: quenching occurs through the dipole–dipole mechanism. The comparison of the present phosphor was done with the commercial oxysulfide phosphor Y_2O_2S:Eu^{3+} and emission intensity found to be 2.2 times higher than that of the presently studied phosphor. The CIE color coordinates also well agree with the general requirement of a red phosphor for SSL.

8.4.16 Eu^{3+} Luminescence in $Y_6W_xMo_{(1-x)}O_{12}$

As mentioned earlier, the host lattice absorption can be enhanced by using a host containing octahedral tungsten (Fig. 8.17) or molybdenum groups rather than tetrahedral moieties. Eu^{3+} luminescence was studied in Y_6WO_{12} (showed intensive red light emission) [116–122]. Optical property tuning was carried out in the Y_6WO_{12} host lattice by replacing Mo as well as N ions and it was shown that the Y_6WO_{12} host lattice could absorb NUV light. And when W was gradually replaced by Mo, the absorbed light shifted from the NUV to the violet region [119]. The PL properties of $Y_2W_{1-x}Mo_xO_6$:Eu^{3+} and Y_2WO_6:Eu^{3+},Bi^{3+} are investigated [123]. The excitation band edge moves from 370 to 420 nm with the incorporation of Mo^{6+} ions. In addition, Bi^{3+}'-doped samples show energy transfer from Bi^{3+} to Eu^{3+} ions. Systematic investigation of Eu^{3+} luminescence was carried out in a $Y_6W_xMo_{(1-x)}O_{12}$ solid solution [124]. $R_6W(Mo)O_{12}$, in which R can be Y or RE elements having a small ionic radius [125], was reported to be isostructural with the RE oxide Tb_7O_{12} [126] or binary oxide Y_6UO_{12} [127] and contains WO_6 (or MoO_6) groups.

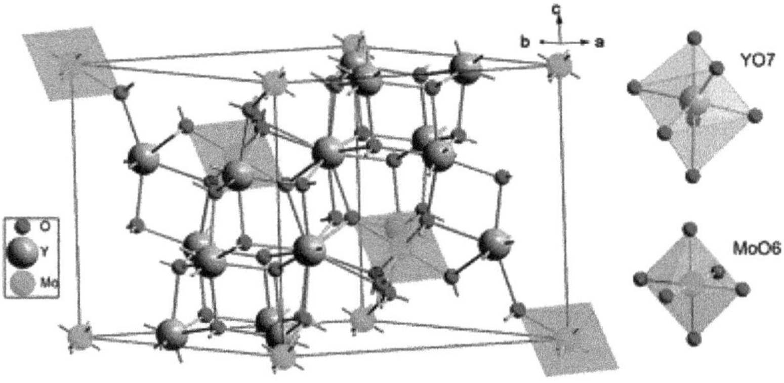

Figure 8.17 The hexagonal cell of Y_6MoO_{12}, as well as the coordination of Mo (green) and Y (gray) with O (red) in the cell.

By successful replacement of Mo by W in the $Y_6W_xMo_{(1-x)}O_{12}$ lattice the absorption edge shifted toward NUV (Fig. 8.18; host lattices can absorb UV or violet light efficiently), as well as energy transfer was observed from the MO_6 group to the activator Eu^{3+} ion, resulting in intensive red-light emission. The PL emission intensity strongly depends on the ratio of W to Mo. The MoO_6 groups quench the WO_6 excitation greatly, whereas the introduction of WO_6 groups not only significantly enhances the MoO_6 excitation but also increases the ratio of red emission to yellow emission. According to their PL properties, it is recommended that $Y_6W_xMo_{(1-x)}O_{12}$:Eu compounds can be candidates as red phosphors for WLEDs.

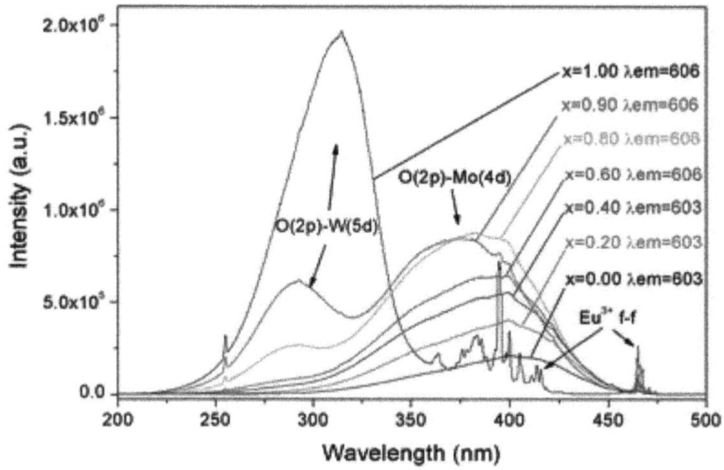

Figure 8.18 PL excitation spectra of as-prepared $Y_6W_xMo_{(1-x)}O_{12}$:Eu powders. Broad bands come from the excitation of electronic transition from O 2p to Mo 4d (or W 5d), and sharp peaks originate from the transition among the Eu^{3+} ion f–f energy levels. Reproduced from Ref. [124] with permission of The Royal Society of Chemistry.

Recently, Eu^{3+}-activated $Y_6(WMo)_{0.5}O_{12}$ phosphors were synthesised by the citrate complexation method and detailed investigation of the tenergy transfer mechanism has been executed in terms of the structure and crystallinity of the host lattice. The host lattice crystallized in a cubic phase at 800°C–1200°C and in a hexagonal phase at 1300°C. In general, the crystal structure plays an important role in the PL behavior, even though the phosphor compositions showed intensive absorption of NUV light in the range of 300–420 nm and of blue light around 467 nm. However, the energy transfer from the host lattices to Eu^{3+} ions in the hexagonal phase was found to be more competent than that of cubic phase because of less energy quenching. The outcome of the studies reveals that the hexagonal phase is suitable for soft UV-chip-based WLEDs, while the cubic phase is more apt for blue chip-based WLEDs. The compositions show (YWMO:Eu-800 and YWMO:Eu-1300) high color purity, and quenching concentrations were found to be 10 mol% and 15 mol%, respectively [128]. In addition, the energy transfer host lattice to Sm^{3+} has also been observed in Sm^{3+}-doped Y_6WO_{12} phosphors and shows orange-red emission [129].

8.5 Perovskite and Double-Perovskite

The structure of $(NH_4)_3FeF_6$ was first determined by Pauling and the structure is common to a series of compounds of the type A_3BX_6 [130]. $(NH_4)_3MoO_3F_3$ constitutes an intermediate line (between the complex fluoride such as $(NH_4)_3FeF_6$ and ternary oxides such as A_3BO_6), which has also been investigated by Pauling [130]. In 1951, Steward and Rooksby made a series of alkaline earth tungstates containing one or more alkaline earth metals with R_3WO_6 type [131]. The structure of these compounds is found to be analogous with $(NH_4)_3FeF_6$. However, some of the compounds are closely related to the $(NH_4)_3FeF_6$ structure and some of them do not belong to the cubic structure. In Na_3AlF_6, the structure is made by an AlF_6 octahedra framework, similar to that of FeF_6 groups in $(NH_4)_3FeF_6$, but the octahedra are slightly rotated or displaced out of the highest-symmetry orientation [132]. The true structure of Na_3AlF_6 therefore becomes monoclinic and falls into the space group $P2_1/m$. Similarly, the observed departures from the cubic system in R_3WO_6-type tungstates may be due to displacement from symmetrical orientation in WO_6 octahedra, as observed in the case of the Na_3AlF_6 structure. Indeed, the observed multiplicity in X-ray powder diffraction lines in Ca_3WO_6 closely resembles that of Na_3AlF_6 powder photographs.

Blasse has reported a series of mixed-metal oxides with perovskite-related structures. In the sublattice of the smaller cations, 1:1 and 1:2 long-range orders have been found in the perovskite as well as in the hexagonal barium titanate structure, and the Jahn–Teller distortions of these compounds have also been discussed [133]. New anion-deficient strontium and barium tantalates ($Sr_3TaO_{5.5}$ and $Ba_3TaO_{5.5}$) have been reported, and these compounds crystallize in the face-centered cubic structure of $(NH_4)_3FeF_6$ with the space group O_h [134]. Oxide ion conduction has been reported in $Ba_2BB'O_{5.5}$ (B = Li, Na; B′ = Mo, W, Te) with an anion-deficient double-perovskite structure [135]. A detail comprehensive examination of the factors that influence B-cation arrangement, that is, the charge, size, and electronic configuration of the B-cations and the A/B size ratio, has been reported [136]. A series of molybdenum-containing double-perovskites with the formula $A^{II}{}_2B^{II}MoO_6$ and $A^{II}{}_2B^{II}W_{0.5}Mo_{0.5}O_6$ (A^{II} = Ba, Sr, and Ca; B^{II} = Mg, Ni, Co, Cd, and Ca) have been reported. A new fitness factor has been proposed, which can discriminate the crystal system of the $A^{II}_2B^{II}B'^{VI}O_6$ double-perovskites more exactly than the well-known tolerance factor [137].

8.5.1 Spectral Properties of Double-Perovskites

The CT absorption of the tungstate group depends on the choice of A and B in ordered perovskites A_2BWO_6 [138]. From the reflection spectra the absorption edge of A_2BWO_6 compounds is only weakly influenced by the choice of A (Table 8.2); the larger the ionic radius of A [139], the longer the wavelength corresponding to the absorption edge. It is generally assumed that the absorption edge of tungstates is due to absorption in the tungstate group corresponding to a so-called *charge transfer transition* in which an electron is transferred from the oxygen ion (highest filled molecular orbital) to the tungsten ion (lowest empty molecular orbital, mainly 5d (t_{2g})).

Table 8.2 Position of the absorption edge of compounds A_2BWO_6 in dependence on the ionic radii of A (twelve-coordination) and B (six-coordination)

Compound	Absorption edge (nm)	Ionic radii (Å) [139]
Ba_2MgWO_6	314	1.60 (Ba)
Sr_2MgWO_6	304	1.40 (Sr)
Ca_2MgWO_6	300	1.35 (Ca)
Ba_2MgWO_6	314	0.72 (Mg)
Ba_2ZnWO_6	317	0.75 (Zn)
Ba_2CdWO_6	317	0.95 (Cd)
Ba_2CaWO_6	318	1.00 (Ca)
Ba_2SrWO_6	316	1.13 (Sr)
Ba_2BaWO_6	320	1.36 (Ba)

Source: From Ref. [139]. © International Union of Crystallography.

That the absorption edge depends on the choice of A and B in the formula A_2BWO_6 is surprising, because the B ions are the nearest neighbors of the tungstate group. The cation coordination around O^{2-} ions in A_2BWO_6 is shown in Fig. 8.19. The anion is linearly coordinated by one B^{2+} and one W^{6+} ion and at larger distances by a square of four A^{2+} ions, this square being perpendicular to the collinear array B^{2+}–O^{2-}–W^{6+}. The oxygen 2p orbitals are situated between the square (occupied) geometry and the t_{2g} orbitals of the W^{6+} ion (empty). The absorption corresponds to an electronic transition from the 2p orbitals to the t_{2g} orbitals. The spectral position of the CT band depends on the charge and radius of the cations surrounding the anion. The CT absorption is situated at higher energies for higher charges and smaller radius, since the anion valence electrons will be stabilized by the strong field due to the positive cations.

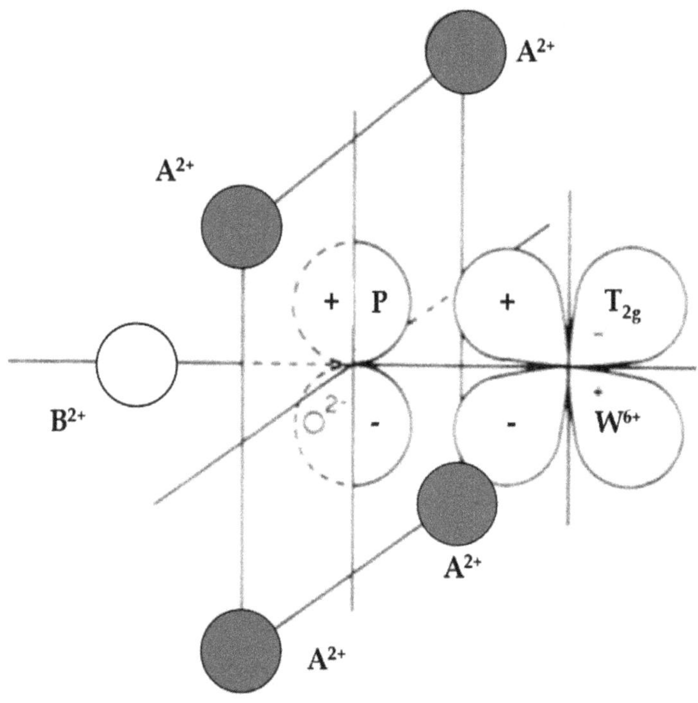

Figure 8.19 Anion coordination in ordered perovskite A_2BWO_6. Reprinted from Ref. [138], Copyright (1973), with permission from Elsevier.

Luminescence has been observed at 77 K for all compounds (Table 8.2) under UV excitation. The observed emission color is ranging from blue to yellow. Ba_2MgWO_6 shows reasonable luminescence at RT. Similar type of luminescence studies have been carried out in A_2BWO_6 with different A- and B-site substitution [140]. Pb^{2+} luminescence has been studied in $Sr_{1-x}Pb_xLaLiWO_6$ and $Ba_{2-x}Pb_xMgWO_6$ and the emission is observed from tungstate as well as Pb^{2+} [141]. Similarly, octahedral uranate group luminescence has also been studied in an ordered perovskite structure [142]. The influence of deviation from a 1:1 order perovskite structure in tungstates leads to clustering of WO_6 octahedra, and the luminescence of WO_6 clusters is different than that of an isolated corner-sharing WO_6 octahedron [143].

8.5.2 Eu^{3+} Luminescence in Perovskite Structure

The compound Ba_2GdNbO_6 crystallizes in an ordered perovskite structure. The Gd^{3+} ion and the Nb^{5+} ion occupy the sublattice of the smaller cations in an ordered way [144]. The regular octahedron of oxygen is occupied by the RE ion (Gd^{3+}). Even for the complete lattice (symmetry group O_h), the inversion symmetry remains valid for more remote neighbors. The emission spectrum of Eu^{3+}-activated Ba_2GdNbO_6 shows a sharp line at 595 nm (magnetic dipole [MD] transition) and weak and broad lines in the region 610–620 nm (due to ED transition). The ED transition is strictly forbidden, since the Eu^{3+} ion occupies a centrosymmetric site in this lattice. At most two lines can be expected; for a pure electronic transition within the Eu^{3+} ion on the basis of the number of levels available ($^5D_0$ is not split and $^7F_2$ is split into two levels by the crystal field of symmetry O_h). However, the spectra show six weak lines and these lines are therefore assigned to transitions involving the simultaneous excitation of noncentrosymmetric vibrational modes of the surroundings of the RE ion (vibronic transition). This type of transition has been reported in a Eu^{3+}-activated $SrTiO_3$ lattice [145]. In this case, the MD transition is allowed as such (Eu^{3+} occupies a centrosymmetric site). Here the local distortion of the crystal lattice at the site of the activator can be excluded, since the ionic radii of Gd^{3+} and Eu^{3+} are similar. If 50% of the Nb^{5+} ions were replaced by Sb^{5+} ions (similar charge and radius) in the Ba_2GdNbO_6 crystal structure, the observed emission is equal to that of Eu^{3+}-activated Ba_2GdNbO_6, except for a peak at 613 nm with a peak intensity about one-third of the 595 nm MD transition. The ED transition at 613 nm is due to the $^5D_0$–$^7F_2$ transition, which now is no longer strictly forbidden [146]. Eu^{3+}-substituted (A-site) distorted perovskites $LaAlO_3$ (rhombohedral) and $GdAlO_3$ (orthorhombic) show ED and MD transitions, which have equal emission intensity. This is because the aluminates are distorted perovskites and one can expect distortion in the A-site coordination, and hence the surroundings of the Eu^{3+} ion do not have inversion symmetry. Remarkable changes are observed in the emission intensity of the MD and ED transitions by varying the next-nearest neighbors of the RE ion. Eu^{3+}-doped $LaAl_{0.5}Ga_{0.5}O_3$ shows the same emission peaks as $LaAlO_3$: Eu^{3+}, but the peak intensity of ED transition is twice that of the MD transition [146].

Eu^{3+} luminescence has been studied in a layered perovskite structure, $SrRE_2Al_2O_7$, $BaRE_2Ti_3O_{10}$, $RETa_3O_9$, and $RbRETa_2O_7$ (RE = rare earth), with 2D arrangement of RE ions. The layered perovskite compound $RbRETa_2O_7$ [147], consisting of TaO_6 octahedral layers, is a new member of the Dion–Jacobson series. Concentration quenching of Eu^{3+} emission has been observed in layered perovskite compounds $SrRE_2Al_2O_7$, $BaRE_2Ti_3O_{10}$, $RETa_3O_9$, and $RbRETa_2O_7$ (RE = rare earth) with a 2D arrangement of RE ions [148]. To elucidate the concentration quenching of Eu^{3+} emission in these layered perovskites, a percolation model has been used. It has been established that concentration quenching occurs by multipole–multipole interaction among nearest-neighbor sites in the RE sublattice. Depending on the nature of the metal ions present in the B-site in this structure, one can expect high absorption in the UV-to-NUV region and energy transfer from the absorbing ion to the RE ion is also possible in this structure. Designing new and novel phosphor materials with a perovskite and double-perovskite structure is one of the attractive research tasks.

Double-perovskites received much research interest due to their emerging applications as dilute magnetic semiconductors, electrode materials for solid oxide fuel cells, etc. [149, 150]. It is known that hexavalent Mo and W are stabilized not in the primitive ABO_3 perovskite structure but in the ordered double-perovskite structure, $A_2^{II}B^{II}B^{VI}O_6$ ($A_2BB'O_6$), in which A^{II} is an alkaline earth ion, B^{II} is a divalent metal ion such as Mg, Ca, Ni, and Cu, and B^{VI} is a hexavalent Mo or W ion. It has been reported that if the charge difference between B and B′ is four, the compounds adopt an ordered perovskite structure with the rock salt arrangement of B and B' ions [136]. RE luminescence has been studied materials containing WO_4 and MoO_4 tetrahedra with scheelite and related structures. In the structures the efficiency of the CT band (due to MoO_4/WO_4 tetrahedra) is not good enough to give high emission intensity for NUV excitation. This inefficient CT band has opened the door to look at structures that contain MO_6 (M = W, Mo) octahedra with high absorption in the NUV region.

After the significant effort Sivakumar and Varadaraju identified in 2006 efficient energy transfer from the MO_6 group (M = W, Mo) of the host lattice to Eu^{3+} ions in doped A_2CaMoO_6 (A = Sr, Ba) under NUV excitation [151, 152]. The Sr/Ba ratio in this structure $(Sr_xBa_{1-x})_2CaW_yMo_{1-y}O_6$, mainly influences the emission through changes in the crystal structure, whereas the W/Mo ratio changes the nature of the absorbing group. Although WO_6^- groups do not absorb well in the NUV, it has been shown that inclusion of W can reduce cross-relaxation losses by blocking energy transfer [153]. After that numerous research was carried out in $A_2BB'O_6:Eu^{3+}$ (A = Ca, Sr, Ba; B = Mg, Ca; B' = W, Mo) phosphors to identify the perfect crystal composition by a crystal chemical approach, and all the compounds are listed in Table 8.3 with their respective references.

Table 8.3 Eu^{3+}-doped double-perovskites with crystal chemical substitution

S.No.	Compound	Reference
1	Eu^{3+}-doped A_2CaWO_6 (A = Sr, Ba)	[153]
2	$Ba_{2-z}Sr_zCaMo_{1-y}W_yO_6$:Eu,Li	[154]
3	$Sr_2Ca_{1-2x}Eu_xNa_xMoO_6$	[155]
4	$Sr_2ZnWO_6:Eu^{3+}$	[156]
5	$Ba_2MgW(Mo)O_6$:Eu	[157]
6	$Sr_3WO_6:K^+,Eu^{3+}$	[158]
7	$Ca_3WO_6:Eu^{3+}$	[159]
8	$(Sr_{0.98-m}Ba_mEu_{0.02})_2Ca(Mo_{1-n}W_n)O_6$	[160]
9	Mo co-doped in $Ba_2CaWO_6:Eu^{3+},Li^+$	[161]
10	Eu^{3+} Ca_3WO_6	[162]
11	$Ba_3WO_6:Eu^{3+}$	[163]
12	Eu^{3+}-, Sm^{3+}-, and Pr^{3+}-doped Ca_2ZnWO_6	[164]
13	$Mg_{0.6}Ca_{2.16}Mo_{0.2}W_{0.8}O_6:Eu_{0.12}^{3+}/Na_{0.12}^+$	[165]
14	$Sr_2CaW_xMo_{1-x}O_6:Eu^{3+},Li^+$:	[166]
15	$Ba_2ZnMoO_6:Eu^{3+}$	[167]
16	Eu^{3+}-substituted $(Sr_xBa_{1-x})_2CaW_yMo_{1-y}O_6$	[168]
17	$Sr_2ZnW_{1-x}Mo_xO_6:Eu^{3+},Li^+$	[169]

Even though efficient energy transfer occurs from the host lattice pseudocubic, orthorhombic activator ion, the observed color purity is not red enough since both MD and ED transitions (orange-red emission) exist in the emission spectrum. As mentioned earlier, to obtain pure-red emission, the Eu^{3+} ion should be located at a noncentrosymmetric site (a site without an inversion centre). The research devoted on Eu^{3+} in double-perovskite structures give leads to identify the distorted double-perovskite structure for Eu^{3+} luminescence, and among all, fortunately, the Ca_2MgMO_6:Eu^{3+} phosphor shows intense pure-red emission at around 615 nm due to distorted Eu^{3+} local site symmetry [170]. Figure 8.20 shows the structural dependence of ED and MD transition variation in a double-perovskite structure.

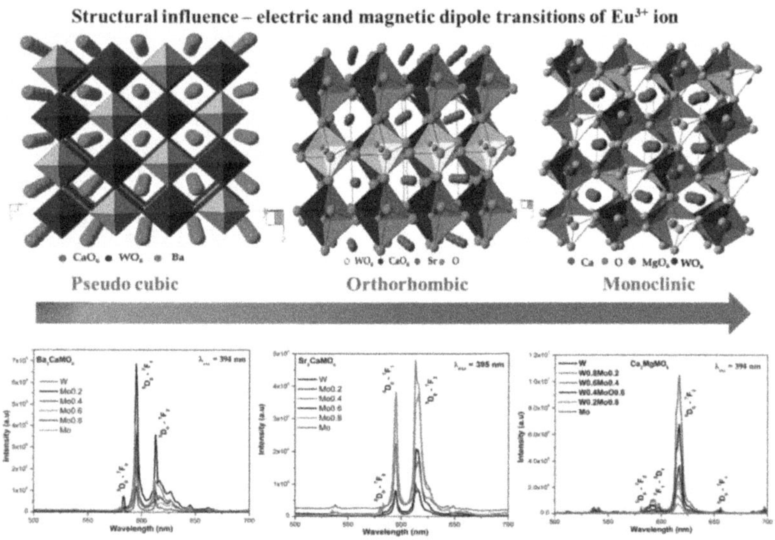

Figure 8.20 Structural influence on ED and MD transition of Eu^{3+} in double-perovskite structures.

8.5.3 $NaREMgWO_6$ (RE = La, Gd, Y)

In the double-perovskite structure, by implementing a combination of $A^+ + A^{3+}$ cations instead of $2A^{2+}$ we can induce lower site symmetry or the A-site in the lattice. Ultimately this facilitates a change within the dominating transition (improves color purity of red emission).

Simultaneously, ions with a smaller radius can be used to replace A-site ions in order to reduce symmetry. It is noteworthy that the double-perovskites $AA'BB'O_6$ (A = Na, K; A' = La, Gd; B = Mg; B' = W) have been found to be excellent phosphors due to their excellent energy transfer efficiency of the CTB, as well as superior luminescence intensity compared to that of commercially available red phosphors [153]. Additionally, it is intriguing that the structure of double-perovskites $AA'BB'O_6$ varies with the different A and A' cation pairs.

$AREMgMO_6$ (A = Li, Na; RE = La, Gd, Y; M = W, Mo) with a double-perovskite structure have been studied for the host lattice for RE doping [171, 172]. Recently, synthesis and PL properties of the $NaLaMgWO_6:RE^{3+}$ (RE = Eu, Sm, Tb) phosphor were studied for WLED applications. Effort has been made to replace the existing Na^+ ions by Li^+ ions expected to enhance the luminescence properties. Double-perovskite structure formation without any impurity was found. The noticeable observation was enriched luminescence properties, which was only possible by increase of radiative ED transition for Eu^{3+} due to the low crystal symmetry and particularly the shortened Eu–O bond length. This phenomenon is called A-site-induced energy transfer. Overall it can be concluded that the distorted symmetry was responsible for attending the reduced lifetime, amplified intensity ratio of red/orange emission, and anticipated red-light quality.

Perovskite-type $NaLaMgWO_6$ with both the A-site and B-site order has been the subject of several recent studies [173–175]. Instead of following the regular synthetic route a sol-gel technique was preferred for the phosphor ($NaLaMgWO_6:Eu^{3+}$) synthesis using citric acid and polyethylene glycol as complexing agents. $NaLaMgWO_6$ has a monoclinic double-perovskite structure with the space group $C2/m$ (No. 12) and the lattice parameters a = 7.8074(1) Å, b = 7.8158(1) Å, c = 7.8977(1) Å, and β = 90.136(1)8 [173]. It shows significant long-range ordering of the A-site cations (layered ordering), Na^+ and La^{3+}, and rock salt ordering of the B-site. In $NaLaMgWO_6$ perovskites layered ordering of A-site cations creates a bonding instability that is compensated by a second-order Jahn–Teller distortion of the B_0-cation. These two distortions are synergistic, and the removal of one leads to the disappearance of the other [173]. The structural parameters are highly responsible for concentration quenching; for the presently studied phosphors the quenching concentration was found to be high compared to other phosphors. The special structural characteristics may be the reason behind the high concentration quenching.

The tolerance factor (t) for perovskite $NaLaMgWO_6$ is $t = 0.952 < 1$. The value of t was found to be less than 1, which states that the size of the A-site cation is too small for the cubo-octahedral cavity formed by BO_6 units. This made the perovskite-based phosphor suitable for doping of activators.

Sharits et al. made further efforts to study the doubly ordered double-perovskite ($NaREMgWO_6$; RE = La, Gd, Y) structures synthesized through a general solid-state method [176]. The host substituted with RE = Y was found to be isostructural with the gadolinium analog with the space group $P2_1$ and a monoclinic crystal structure (Fig. 8.21).

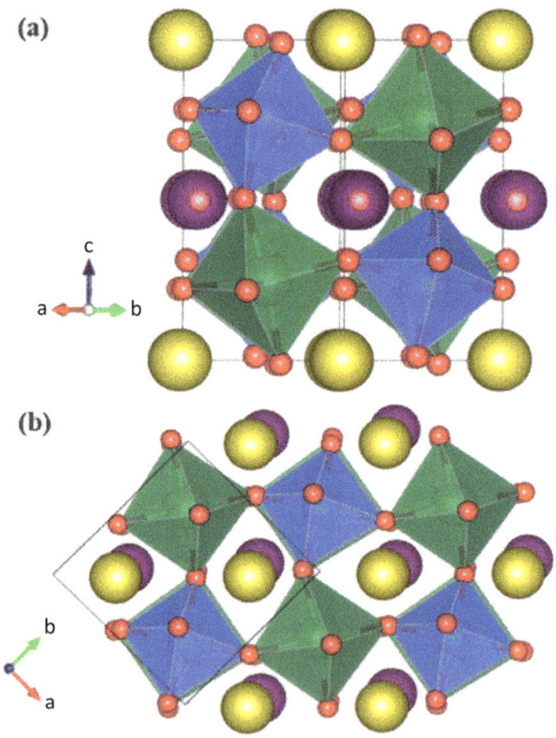

Figure 8.21 View of the crystal structure of $NaYMgWO_6$ as viewed along the (a) [110] and (b) [001] directions. The Na^+ and Y^{3+} ions are yellow and purple, respectively, while the Mg- and W-centered octahedra are green and blue, respectively. The black line represents the unit cell. Reprinted with permission from Ref. [176]. Copyright (2016) American Chemical Society.

The 4f–4f transition was not affected by host lattice absorption. The dominating $^5D_0 \rightarrow ^7F_2$ ED transition at ~615 nm is (with 394 nm excitation) found to the strongest emission for all three hosts, which is a most essential feature of a red-emitting phosphor. Further substitution of Eu^{3+} onto a more compacted RE site in the $NaY_{1-x}Eu_xMgWO_6$ and $NaGd_{1-x}Eu_xMgWO_6$ hosts results in a blue shift of the CT band and an increase in the intensity of the $^5D_0 \rightarrow ^7F_2$ transition related to $NaLa_{1-x}Eu_xMgWO_6$. Concentration quenching studies reveal that luminescence saturation for these types of phosphors occurs at doping concentration of the Eu^{3+} ion. The Na(Y, Gd)$MgWO_6$ hosts reveal brighter PL properties than $NaLaMgWO_6$, as well as behaving as a sensitizer. In conclusion, they are promising red phosphors when compared with Y_2O_3:Eu^{3+}, found to be better in each characteristic.

8.5.4 A_2LnMO_6 (A = Ca, Sr, Ba; Ln = La, Gd, Y; M = Sb, Nb, Ta)

Eu^{3+}-activated A_2LnMO_6 (A = alkaline earth metal = Ca, Sr, Ba; Ln = RE = La, Gd, Y; M = transition/d block metal = Sb, Nb, Ta) phosphors with a double-perovskite structure were reported [177]. The structural advantages of these types of structures are that they have an A-site with lower symmetry where necessary structural modification of the octahedron LnO_6 (AO_6) and MO_6 welcome more dopants into the host lattice cavity.

Powder XRD of Ca_2LaSbO_6 confirms this host crystallizes in a monoclinic structure with the space group $P2_1/n$ (cell constants of a = 5.6852(1) Å, b = 5.8781(1) Å, c = 8.1723(1) Å, and β = 89.99(4)°). The interesting structural observation was the presence of six different crystal sites (Ca_1, Ca_2, Sb, O_1, O_2, O_3), where 50% Ca and 50% La are present in the Ca_1 site. The general formula can be rewritten as $(Ca_{1/2}\ La_{1/2})_2CaSbO_6$, which is analogous with the classic double-perovskite structure $A_2B_1B_2O_6$ (A = $Ca_{1/2}La_{1/2}$; B_1 = Ca; B_2 = Sb). The emission properties were convincing as the increase of Eu^{3+} content was leading to an increase in emission intensity. The intensity reaches a maximum at doping concentration x = 0.4; further increasing the activator concentration leads to reduced luminescence due to an increase of optically active centers (energy migration among Eu^{3+} ions). The structure of Sr_2LaSbO_6 ($A_2B_1B_2O_6$: A = Sr; B_1 = La; B_2 =

Sb) is derived from the Ca_2LaSbO_6 structure and differs from the crystal structure of Ca_2LaSbO_6. The Sr atom is situated in the Ca_1 site of Ca_2LaSbO_6 to form a highly distorted polyhedron SrO_8, and the La atom is residing at the Ca_2 site to form LaO_6 as an alternative of LaO_8. The structural change is a result of different ionic radii of Ca^{2+} (1.12 Å), Sr^{2+} (1.26 Å), and La^{3+} (1.16 Å) [139]. The La atoms in Ca_2LaSbO_6 are mainly at the greater coordination site (Ca_1) instead of the Ca_2 site, and the La atoms in Sr_2LaSbO_6 should occupy the octahedral site. $(Sr,Ba)_2(La,Gd)SbO_6$ with the space group $P2_1/n$ is of the same structure as Sr_2LaSbO_6. But Sr_2YSbO_6 and Ba_2LnSbO_6 (Ln = Gd, Y) crystallize in the cubic system with the space group Fm-$3m$. The basic crystal structure and the coordination environments of La and Sb are alike to those of Sr_2LaSbO_6. In $A_2LnSbO_6:Eu^{3+}$ (A = Sr, Ba), the doping Eu^{3+} atoms choose to occupy the six-coordinated Ln-site (Fig. 8.22a).

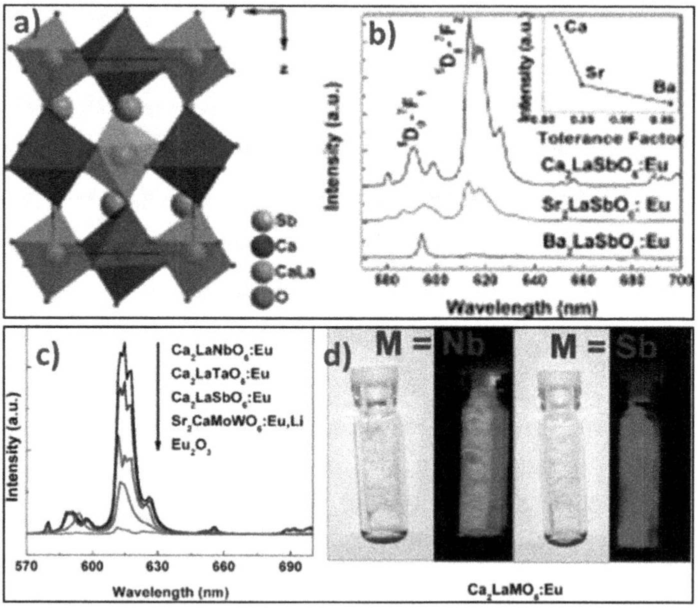

Figure 8.22 (a) Crystal structure of Ca_2LaSbO_6 viewed along [1 0 0], (b) PL emission intensity and tolerance factor, and (c) luminescent spectra of $Ca_2La_{0.6}MO_6:0.4Eu^{3+}$ (M = Sb, Nb, Ta), $Sr_2Ca_{0.8}Mo_{0.1}W_{0.9}O_6:0.1Eu^{3+}0.1Li^+$, and Eu_2O_3 excited at 465 nm. (d) Optical images of $Ca_2LaMO_6:0.4Eu^{3+}$ (M = Nb, Sb) irradiated under a UV lamp. Reprinted from Ref. [177], Copyright (2011), with permission from Elsevier.

To further establish the connection between emission intensity and crystal structure, the tolerance factors were calculated. There was a rise in emission intensity with a decline of the tolerance factor, and the brightest Ca_2LaSbO_6 has the lowest value (0.8165) among the three Sb samples. The lower tolerance factor imitates higher distortion of the crystal structure, which corresponds to the weaker energy transfer in the framework of B-site and oxygen atoms. A number of double-perovskites with similar structural properties were investigated and it was concluded that the structure and emission intensity are associated with each other. The higher the distortion in site symmetry, the greater the emission intensity. The phosphors (Eu^{3+}-doped Ca_2LaMO_6; M = Sb, Nb, Ta) possess excellent red luminescent emission under NUV, blue, and green light irradiation (Fig. 8.22) and are very capable materials for WLED and additional optical applications.

Jingshan Hou et al. reported a single-phase $CaLa_{1-x}MgM'O_6$: $x$$Eu^{3+}$ (M' = Nb, Ta; $0 < x < 61$) with perovskite red phosphors. Eu^{3+} doping does not affect the structure of the host lattice at all. The same was investigated for potential applications in WLEDs. All the compositions show excitation bands in the NUV (396) and blue (466 nm) regions with enhanced red emission intensity of the $Ca(La_{0.7}Eu_{0.3})MgTaO_6$ and $Ca(La_{0.6}Eu_{0.4})MgNbO_6$ phosphors. Comparative studies with commercial red phosphor $(Y_{0.95}Eu_{0.05})_2O_3$ reveal an enormous increase in (3 to 4 times) emission intensity under NUV light excitation, while under blue light it is 2 and 2.5 times, respectively. The good thermal stability of $CaLaMgM'O_6$:Eu^{3+} can be attributed to its distinctive layered ordering of the perovskite structure (can be a good replacement for existing sulfide-based phosphors with lesser stability). $Ca(La_{0.6}Eu_{0.4})MgNbO_6$ achieved a high quantum efficiency (QE) of 312% of Y_2O_3:Eu^{3+} and 98.4% of commercial red-emitting compound $Sr_2Si_5N_8$:Eu^{2+} excited by blue (466 nm) light [178].

A Eu^{3+}-substituted narrow-band emitting $SrLa_{1-x}MgTaO_6$:$x$$Eu^{3+}$ was reported by Yue Guo et al. [179], which belongs to a double-perovskite family. From the crystal structure investigation of $SrLaMgTaO_6$ rock salt ordering of A-cations (Mg^{2+} and Ta^{5+}) was found and Sr^{2+} or La^{3+} on the A-sites with coordination number 12. The smallest full width at half maximum (FWHM) value for red emission ($^5D_0 \rightarrow {}^7F_2$) 6.49 nm was obtained for the present phosphors. The high-purity and intense red emission (3 times higher) was detected for the double-perovskite phosphor compared to Y_2O_3:Eu^{3+}. The calculated CIE color coordinates from the emission data at $x = 0.66, y = 0.33$ and $x = 0.66, y = 0.34$ are close to the NTSC standard.

Jicheng Zhu et al. synthesised a series of Eu^{3+}-activated double-perovskite $Ca_{(2-x)}Ba_xLaNbO_6$:Eu^{3+} phosphors [180]. The advantages of Eu^{3+} substitution in the La^{3+} site of the $Ca_{(2-x)}Ba_xLaNbO^6$ phosphor is that structural changes can happen. In fact both compounds are compounds are crystallizing in a similar crystal structure (monoclinic); nevertheless their space groups are dissimilar (monoclinic $P2_1/n$ in Ca_2LaNbO_6 and monoclinic $C2/m$ in Ba_2LaNbO_6) (Figs. 8.23a and 8.23b). In a Ca-based crystal structure, there are two Ca sites available in the lattice and one site is occupied by the disordered 50% Ca and 50% La, whereas only one Ba and La site is present in Ba_2LaNbO_6 (in other words La^{3+} ions experience a different crystal coordinate environment in Ca_2LaNbO_6 and Ba_2LaNbO_6. The ED transition dominates $^5D_0$–$^7F_2$, peaking at 618 nm for Ca_2LaNbO_6:Eu^{3+} (Eu^{3+} ion located in highly distorted La site, LaO_8, in the Ca_2LaNbO_6 lattice), the intensity of the ED transition decreases, and simultaneously the MD transition also gains intensity with increasing Ba content in the lattice (Eu^{3+} ions located at the six-coordinated sites, EuO_6, of the Ba_2LaNbO_6 phase). The temperature-dependent PL emission intensity variation has also been correlated with the help of crystal structures. Selected compositions can find potential application as red phosphors for warm WLEDs.

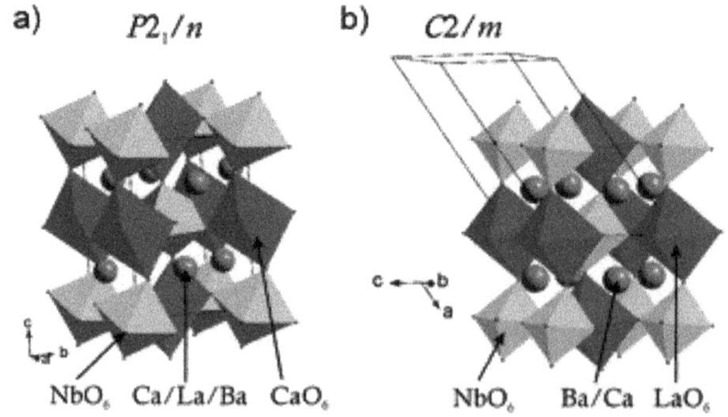

Figure 8.23 The crystal structure of (a) Ca_2LaNbO_6 and (b) Ba_2LaNbO_6. Reproduced from Ref. [180] with permission of The Royal Society of Chemistry.

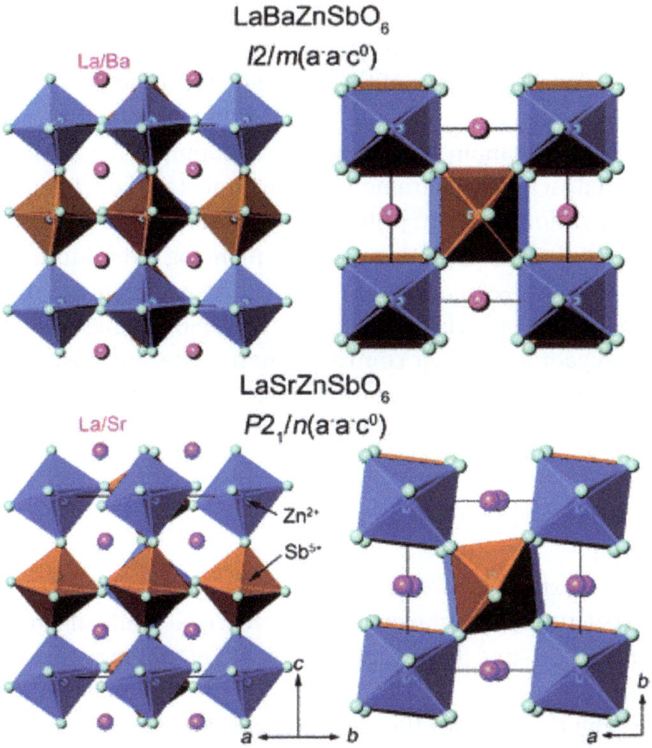

Figure 8.24 Crystal structure views along the [110] (left) and [001] (right) directions for the completely B-site-ordered perovskites $LaBaZnSbO_6$ and $LaSrZnSbO_6$. Blue and brown octahedra represent ZnO_6 and SbO_6, respectively. La/Ba/Sr cations are shown as purple spheres. Reproduced from Ref. [181] with permission of The Royal Society of Chemistry.

One can achieve complete B-site ordering by choosing a large size and charge difference by selecting Zn^{2+} (r = 0.74 Å, in six-coordination) and Sb^{5+} (r = 0.6 Å) in the double-perovskite structure ($LaBaZnSbO_6$) (Fig. 8.24). Powder XRD confirms the space group of the structure to be $I2/m$ (rarely been seen in double-perovskites). Compositionally, increasing the Sr^{2+} content found that the symmetry of $LaBa_{1-x}Sr_xZnSbO_6$ decreases further, which is a rare observation. The transition boundary is around a concentration of x = 0.4. Rietveld structural refinements well agreed with calculated data and confirmed the B-site ordering in the rock salt structure. The absorption and emission spectra have a very similar profile to other

Eu^{3+}-activated phosphors. The doping of Sr^{2+} ions is of significant importance; a blue shift in the CT band of around 25 nm was observed. The absorbance of the CT band excitation is much higher than those of the 4f–4f transitions, which increases with increasing Eu^{3+} content. The main advantage of Sr^{2+} doping is the achievement of improved structural parameters. The important conclusions were that A-site cation symmetry alteration was achieved via Sr^{2+} doping, attaining a lower symmetry and enhancing the luminescence properties of $La_{0.95}Eu_{0.05}Ba_{1-x}Sr_xZnSbO_6$. Also the choice of $Fm\bar{3}m$ or $I4/m$ is inappropriate for $LaBa_{1-x}Sr_xZnSbO_6$, due to cubic or tetragonal geometry of yjr centrosymmetric site. Moreover the Eu^{3+} is situated in a noncentrosymmetric site dominating the $^7F_0 \rightarrow ^5D_2$ transition [181].

8.6 Pyrochlore Structures

Pyrochlore oxides, expressed by the general formula $A_2^{3+}B_2^{4+}O_7$, have a structure derived from the fluorite. Two kinds of cations are ordered on the A- and B-sites, and one-eighth of the anions are missing with ordered vacancies. Disordering of the cation sites coupled with oxygen disorder on the anion vacancies within the pyrochlore structure results in a defective fluorite structure [182]. The crystal structure of $RE_2Ti_2O_7$ is shown in Fig. 8.25. Cation A can be a trivalent RE, a divalent alkaline earth, or a monovalent alkali ion. Cation B will be a 3d, 4d, or 5d transition metal ion of an appropriate oxidation state for charge neutrality. The A-site (16d) cations are eight-coordinated and are located within scalenohedra (distorted cubes) that contain six equal oxygens (O) and two additional axial oxygens (O’) at a slightly shorter distance from the central cations, whereas the smaller B-site (16c) cation resides in sixfold coordination, forming (BO_6) oxygen octahedra. The coordination geometry of A ions in the pyrochlore structure shows. As the fractional coordinate x of O’ oxygen decreases, the $(AO_8)^{n-}$ scalenohedra become highly depressed in shape. A number of Eu^{3+}-doped pyrochlores, $A_2B_2O_7$ (A = Y, La, Gd; B = Ti, Zr, Hf, Sn), have been reported as promising candidates for phosphors.

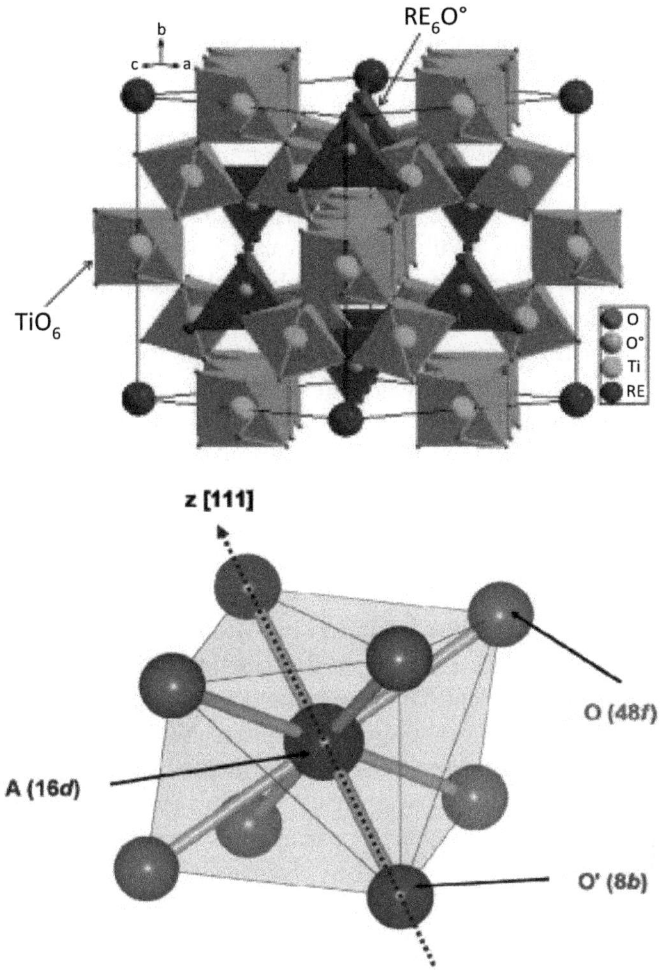

Figure 8.25 The AO_8 scalenohedra in a pyrochlore structure, $A_2B_2O_7$. The *z* direction corresponds to the [1 1 1] crystallographic direction of the pyrochlore. (Top) Reprinted from Ref. [183], Copyright (2016), with permission from Elsevier. (Bottom) Reprinted from Ref. [184], Copyright (2008), with permission from Elsevier.

As mentioned earlier pyrochlore and weberite, which generally suffer through anion deficiency, belong to the family of fluorite-related ($A_2B_2O_7$) structures and also maintain the closed-packed cation layers [183]. The weberite crystal structure with the space group I_{mma} (No. 74), of the general formula $A_2B_2X_7$ (A and B are cations, X is an anion, O or F), is a category of an anion-deficient fluorite superstructure (AX_2), which resembles the structure of pyrochlore or layered perovskite.

Pyrochlore ($A_2B_2O_7$) could be an attractive host for a different crystal site (A- and B-sites; 3+ and 4+ or 2+ and 5+) with adjustment of the ionic radius and charge neutrality criteria. The general formula can be rewritten as $A_2^{3+}B_2^{4+}O_7$ (3+, 4+), $A_2^{2+}B_2^{5+}O_7$ (2+, 5+). $A_2B_2O_7$ compounds exhibit a wide variety of interesting physical properties; as RE substitution offers high chemical stability, lattice stiffness, thermal stability, etc., the B atom can be any transition or posttransition material with different valances, and A can be an RE ion with an inert lone pair of electrons [184]. Eu^{3+}-doped pyrochlores, $A_2B_2O_7$ (A = Y, La, Gd; B = Ti, Zr, Hf, Sn), have been studied as potential orange-red-emitting phosphors by various researchers [183–185]. McCauly et al. reported Eu^{3+} luminescence and the splitting of the $^5D_0 \rightarrow ^7F_1$ transition by replacement of the larger B-cations with small-ionic-radii cations, resulting in relatively more diffuse emission spectra [186]. An oxide-based Eu^{3+}-doped pyrochlore $M_2Sn_2O_7$ (M = Gd and Y) has been successfully prepared by Li et al. through a simple and facile hydrothermal synthesis without using a catalyst. They performed a detailed investigation of the phosphor via different spectroscopic techniques. An enhanced PL property was observed due to structural advantages [187]. The energy difference between valence band (VB) conduction can be estimated by three different parameters: first the energy of CT from the top of the VB to the Eu^{3+} ion, second the red shift of the first 4f→5d transition, and third the energy of CT from the VB to the conduction band (CB) [187]. A synthesis of sol-gel-assisted thin films of pyrochlore-type yttrium stannate ($Y_2Sn_2O_7$) activated by Eu^{3+} was reported by Fujihara et al. Especially they discussed the luminescence behavior of Eu^{3+} ions introduced in a $Y_2Sn_2O_7$ host crystal that has a high site symmetry of Y^{3+}. Luminescence spectra specified the multiband orange-red luminescence at single-wavelength excitation due to the weak $^5D_0 \rightarrow ^7F_0$ transition and the doubly split $^5D_0 \rightarrow ^7F_1$ and $^5D_0 \rightarrow ^7F_2$ transitions [188].

A fresh zirconate-based pyrochlore-type red phosphor, $Nd_2Zr_2O_7$:Eu^{3+}, was synthesized by Santosh et al. using the gel combustion method. In $Nd_2Zr_2O_7$, Nd^{3+} obeys eightfold coordination, whereas Zr^{4+} tries to attain sixfold coordination and the site symmetry of Nd^{3+} is D_{3d}. There are rare reports of studying the photophysical properties of this phosphor ($Nd_2Zr_2O_7$) with varying Eu^{3+} concentrations. Formation of an absolute pyrochlore phase was confirmed from XRD studies. The red shift of the CT band with increasing Eu^{3+} ion doping concentration was evidenced. The emission spectra consist of typical Eu^{3+} emission along with host lattice absorption, which is an indirect confirmation of incomplete energy transfer due to oxygen-related defects at 0.1% of Eu^{3+} concentration. With growing doping concentration of Eu^{3+}, the intensity of $^5D_0 \rightarrow {}^7F_0$ was found to slowly decrease due to compensation of oxygen vacancies by the high level of europium doping [189].

A series of $(La_{1-x}Eu_x)_2M_2O_7$ (M = Zr, Hf, Sn) pyrochlores were synthesized using the high-temperature solid-state reaction mechanism [184]. To have a better understanding of structural relationships with luminescence properties, the pyrochlore structure is a good option. The most probable position for Eu^{3+} to occupy is the A-site because of similar ionic radii. The A-cations are eight-coordinated and are located within scalenohedra (distorted cubes) that contain six equal oxygens (O) and two additional axial oxygens (O_0) at a slightly shorter distance from the central cations. The $^5D_0 \rightarrow {}^7F_0$ transition of Eu^{3+} ions, which occupy only one site, is not split, because both its excited and ground states are nondegenerate in nature. In the nonsplitted $^5D_0 \rightarrow {}^7F_0$ transition states, the phosphor has a single emission center (Eu^{3+}). The Eu^{3+} sites in $(La_{0.95}Eu_{0.05})_2Zr_2O_7$ and $(La_{0.95}Eu_{0.05})_2Hf_2O_7$ heated at lower temperatures could have a local symmetry distorted from the inversion center, which results in enhanced intensity of electronic dipole transitions. So the distortion around the Eu^{3+} ion is directly connected with the crystallinity of the samples. So complete control on crystallinity can efficiently improve luminescence properties.

8.7 Eu^{3+}-Doped Ternary Rare Earth Antimonates R_3SBO_7 (R = LA, GD, Y)

The ternary RE compounds with the general formula R_3MO_7 (M = Sb, Ta, Re, Ru, etc.) could be used for the host lattice for Eu^{3+} luminescence due to several advantages like high chemical stability, lattice stiffness, and thermal stability. Furthermore, the structure can accommodate various crystal chemical substitutions at the R and M sites with mixed valences (3+ and 4+ or 2+ and 5+) along with oxygen vacancies. The smaller RE ions (Dy^{3+}→Lu^{3+}) together with yttrium ions can easily be substituted in the R-sites of the R_3SbO_7 lattices, inducing a fluorite-type superstructure with some defects. The R_3SbO_7 lattice is supposed to have a fluorite-like structure ($M_4{}^{4+}O_8$): three trivalent ions (R^{3+}) are accommodated in the place of four tetravalent metal ions (M^{4+}) and one pentavalent ion (Sb^{5+}) and one oxide vacancy, which leads to the formation of a perfluorite cell [190]. Ordering can be anticipated and found in both metal and anion sites due to substantial differences in ionic radii between the R^{3+} and M^{5+} ions [191]. Depending on the Ln^{3+} and M^{5+} ions there is a wide possibility of almost similar structures with different ordering and defect formation. The crystal structures have been investigated in detail for antimonite compounds R_3SbO_7 [192]. However, their PL properties with lanthanide ions not been explored. Recently, the Eu^{3+}-doped RE antimonates R_3SbO_7 (R = La, Gd, Y) were explored as red-emitting phosphors for WLEDs [193]. Ln_3SbO_7 compounds crystallize in an orthorhombic superstructure of cubic fluorite (space group *Cmcm* for Ln = La, Pr, Nd and *C*2221 for Ln = Nd–Lu). Yokogawa et al. earlier described the structure as a weberite type with the space group $C222_1$ [194]. The distinctive feature of the weberite structure is that MO_6 octahedra chains formed by MO_6 octahedra linking to each other in the corner with parallel chains of LnO_8 distorted cubes. Thus in a weberite structure there are equal numbers of six- and eight-coordinated cations in an ordered BO_6–AO_8 layer arrangement [195].

Jing Wang et al. [193] synthesized solid-state-assisted Eu^{3+}-doped Ln_3SbO_7, where Ln = La, Gd, and Y. Excitation and emission spectra show characteristic emission of Eu^{3+}. The excitation CT band is stronger than f–f transitions, which indicates that the phosphors have efficient excitation under UV light. The $^5D_0\rightarrow{}^7F_4$ emission is dominating in the emission spectra of La_3SbO_7, which is different from (Gd,Y)SbO_7 , where the $^5D_0\rightarrow{}^7F_1$ and $^5D_0\rightarrow{}^7F_0$ peak's are of highest intensity for Gd- and Y-based compounds, respectively. The unusually higher intensity of the $^5D_0\rightarrow{}^7F_4$ transition could not be explained clearly; however, it is expected that the host La_3SbO_7 is not much affected by incorporation of Eu^{3+}. The dominated $^5D_0\rightarrow{}^7F_0$

transition in yttrium-based host lattices can be explained by the lattices having enough space to accommodate an extra oxygen because the sample was prepared in air and the extra oxygen in the lattices could be responsible. The luminescence efficiency increases with the increase of Eu^{3+} doping and reaches a maximum at about $x = 0.4$; the above happens due to luminescence saturation within the host lattice (concentration quenching). For the La_3SbO_7 lattice theer is a sharp decrease in the luminescence value because of formation of a new phase and change of the space group. The luminescence QE of La_3SbO_7:$0.4Eu^{3+}$ can reach 63.8% under excitation of UV light at RT.

The luminescence properties of Eu^{3+}-activated Gd_3MO_7 (M = Nb, Sb, and Ta) and concentration quenching of Eu^{3+} were also studied in the Gd_3TaO_7 host lattice by Linda et al. [196]. Structural characterization (XRD, FTIR, and Raman) techniques showed that niobate and tantalate samples crystallize in the weberite-type structure, whereas the antimonate crystallize in the fluorite structure. In comparison with antimonate and tantalate systems, the niobate sample has a broader but less intense CTB with a maximum at 294 nm in the NUV region. The shifting of the CT band toward a longer wavelength (red shift) observed for the niobate sample can be ascribed to the difference in their lower ionization potential. Niobium (6.88 V) possesses a lower ionization potential value than tantalum (7.89 V). All three samples exhibit decent absorption in the UV region, signifying that they can be efficiently excited UV LED chips. In the niobate and tantalate samples the intensity of 4f–4f electronic transitions is more than the CT band. However, $Gd_{2.85}SbO_7$:$0.15Eu^{3+}$ shows very weak excitation spectra compared to the other two samples. The most convincing explanation for this can be the structural difference between the samples. The fluorite structure of the antimonate sample is a more disordered one than the weberite, along with the process of ET in case the SbO_8 moiety to Eu^{3+} is not efficient enough as compared to NbO_6 and TaO_6 groups. Thus the intensity of 4f–4f transitions in nabobium and tantalite is more. It is well known that the asymmetric ratio (AR) is a measure of structural disorder in the lattice; the higher the value of AR, the higher the disorder environment around Eu^{3+}. For the present samples, AR (2.8) of the tantalate sample is higher than that of the niobate sample (2.4), proving that $Eu3^{+}$ ions are in more noncentrosymmetric sites. To further examine the site distribution of Eu^{3+} ions, the $^5D_0 \rightarrow ^7F_0$ transition under 394 nm excitation was monitored in detail and no apparent split was detected, indicating the occupancy of Eu^{3+} ions in a single site (eight-coordinated site). The probability of ED transitions ($^5D_0 \rightarrow ^7F_2$) is rapidly increased with increasing doping concentration of Eu^{3+} ions, as they tend to occupy a highly distorted eight-coordinated site. The probability of ED transitions is higher (2.4 times) in the tantalate system, with Eu^{3+} being located at a more distorted site than in the niobate sample.

The concentration quenching mechanism was studied for $Gd_{3-x}TaO_7{:}xEu^{3+}$ (x = 0.05, 0.1, 0.15, and 0.2) with different concentration ranges (up to x = 0.15), and it was found that the luminescence intensity increases with doping content as multiphonon and cross-relaxation processes are the major reasons for electrons in the excited states to enter the $^5D_0$ level. By further increase in the doping concentration, energy transfer processes known as second-order interactions and spin relaxation occur easily between the coupled excited Eu^{3+} ions, resulting in the decrease in luminescence intensity.

Eu^{3+}-activated Ln_3TaO_7 (Ln = La, Gd, Y, Lu) red-emitting phosphors were synthesized. The Ln^{3+} substitution induces variation of the crystalline structure from a defect fluorite to a weberite type. These phosphors exhibit strong absorption at the NUV wavelength and emit an orange-red color. The Eu^{3+} luminescence in La_3TaO_7 represents structural variation. The characteristic $^5D_0 \rightarrow ^7F_J$ (J = 0–4) transition was observed. The structural changes in the host lattice were responsible for variation in luminescence properties of ternary lanthanide-tantalum oxides, Ln_3TaO_7, activated with Eu^{3+} [197].

The influence of the substitution of Eu^{3+} at the aliovalent site in the $Y_2Ce_2O_7$ lattice ($Y_{1.9}Ce_2O_7{:}0.1Eu^{3+}$, $Y_2Ce_{1.9}O_7{:}0.1Eu^{3+}$, and $Y_2Ce_{2-x}O_7{:}xEu^{3+}$; x = 0.05, 0.10, 0.15, 0.20, 0.25 and 0.50) on the structural and PL properties was studied. Induced structural changes are encountered from a defective fluorite structure to a C-type structure for the substitution of Eu^{3+} at the Ce^{4+} site, whereas substitution at the Y^{3+} site showed a biphasic nature of the defective fluorite and C-type structures. XRD and Raman studies indicate the formation of a C-type structure with excess oxygen; the Y^{3+} ion is in octahedral coordination, while the Ce^{4+} ion is in eightfold or less coordination. The substitution of Eu^{3+} at the Ce^{4+} site is confirmed. The increased luminescence intensity (due to allowed doping concentration up to 50 mol%) and lifetime are mainly attributed to the uniform distribution of Eu^{3+} ions in the lattice due to cation ordering linked with oxygen vacancy. The emission spectra of $Y_2Ce_{1.5}O_7$:0.50Eu^{3+} and the commercial red phosphor ($Y_2O_3:Eu^{3+}$) from Philips were compared using the same excitation wavelength of 466 nm (blue region). It is observed that the red emission intensity of $Y_2Ce_{1.5}O_7$:0.50Eu^{3+} is slightly larger than that of the commercial red phosphor [198].

8.8 Garnet

Garnet-type crystals with a general formula $\{C\}_3B_2(A)_3O_{12}$ are considered to be appropriate hosts for optical materials due to their stable lattice structure and large heat conductivity [199, 200]. In recent years, there have been many investigations on garnet phosphors doped with lanthanide ions and their applications in WLEDs and field emission displays (FEDs), including the most famous yellow phosphor $Y_3Al_5O_{12}:Ce^{3+}$, green phosphors $Lu_3Al_5O_{12}:Ce^{3+}$ and $Ca_3Sc_2Si_3O_{12}:Ce^{3+}$ for blue LED chips, and YAG:Tb^{3+} for FEDs. Currently, commercial WLEDs are mostly fabricated by combining a blue InGaN chip and the yellow-emitting color convertors $Y_3Al_5O_{12}:Ce^{3+}$ (YAG:Ce^{3+}) with a garnet structure dispersed in an organic resin or silicone. $Gd_3Ga_5O_{12}:Eu^{3+}$ red-emitting phosphors with a garnet structure have also been reported [201]. Recently, garnet-structured metal oxides with the general formula $Li_6MLa_2N_2O_{12}$ (M = Ca, Sr, Ba; N = Ta, Sb, Nb) have been extensively studied as lithium ion conductors because of their chemical and electrochemical stability [202–205]. Garnet-based red-emitting phosphors $Li_6MLa_2Nb_2O_{12}:Eu^{3+}$ (M = Ca, Sr, Ba) and $Li_6CaLa_2M_2O_{12}:Eu^{3+}$ (M = Ta, Sb) were studied for WLEDs and the luminescence properties correlated with the crystal structure of the lattice [206].

A series of garnet-based $Li_6CaLa_{2-2x}Eu_2xSb_2O_{12}$ (x = 0.1–1.0) red phosphors have been successfully synthesized via a high-temperature solid-state reaction and their luminescence properties for WLEDs investigated to resolve the major disadvantages associated with silicon-based phosphors: they are thermally unstable and exhibit a high color correlation temperature (CCT) [207]. $Li_6CaLa_2Sb_2O_{12}$ is isostructural with the parent compound $Li_5La_3Sb_2O_{12}$ [206]. Figure

8.26 shows the crystal structure of $Li_6CaLa_2Sb_2O_{12}$ viewed from the [001] direction. The Ca^{2+}/La^{3+} ions are surrounded by eight oxygen atoms and are located in the 24c position of the $Ia\bar{3}d$ crystal structure. The Sb^{5+} and Li^+ ions are surrounded by four and three oxygen atoms on the 16a site and 24d or 96h site, respectively.

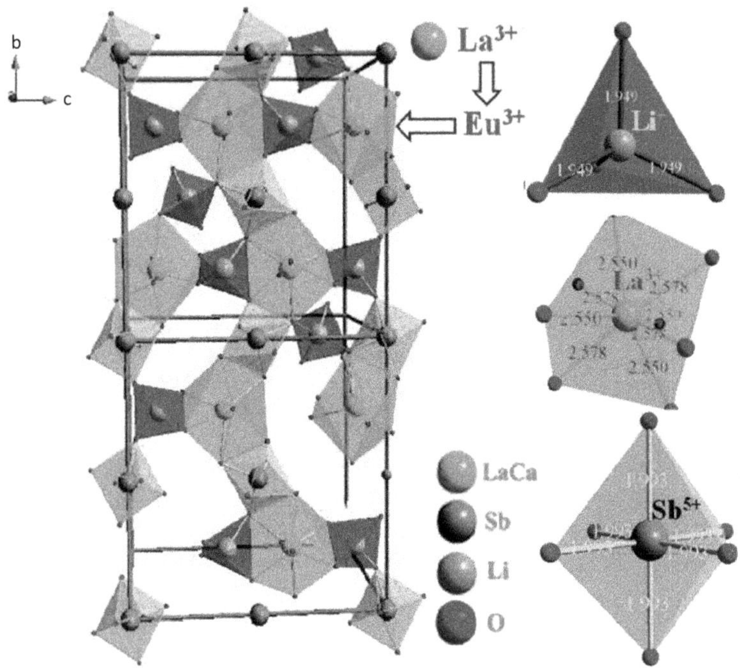

Figure 8.26 Simulated crystal structure of $Li_6CaLa_2Sb_2O_{12}$ double cell viewed in the [001] direction and the coordination environment of the $[LiO_4]$ tetrahedral $[Ca/LaO_8]$ dodecahedral and $[SbO_6]$ octahedral.

Eu^{3+}-doped $Li_6CaLa_2Sb_2O_{12}$ shows highly efficient red luminescence at 611 nm ($^5D_0 \rightarrow {}^7F_2$) ED transition of Eu^{3+}. Surprisingly there is no concentration quenching observed in this lattice (even 100 mol% of Eu^{3+} substation) and excellent thermal stability with only 9.7% emission loss at 553 K. The large Eu^{3+}–Eu^{3+} distance (7.048–7.105 Å) and subsequently the hindering of unwanted energy migration among them in the $Li_6CaLa_2Sb_2O_{12}$ crystalline lattice could be the reasons.

A successful attempt was made to assess the aptness of a red converter ($Li_6CaLa_2Sb_2O_{12}:Eu^{3+}$). Both $Y_3Al_5O_{12}:Ce^{3+}$ yellow and

$Li_6CaLa_2Sb_2O_{12}$:Eu^{3+} red garnet-based phosphor co-doped glass ceramics were fabricated by a low-temperature co-sintering technique. The quantum yield (QY) of the glass ceramic reached as high as 89.3%, since the adverse energy transfers between Ce^{3+} and Eu^{3+} were efficiently suppressed due to the spatial separation of Ce^{3+} in $Y_3Al_5O_{12}$ and Eu^{3+} in $Li_6CaLa_2Sb_2O_{12}$ in the crystal lattice. The constructed WLED exhibited an optimal luminous efficacy of 101 lm W^{-1}, a CCT of 5449 K, and a CRI of 73.7.

The electronic structure and luminescence properties of a garnet-based $Ca_2YZr_2Al_3O_{12}$:Eu^{3+} red-emitting phosphor for NUV LEDs and FEDs was studied [208]. The calculated bandgap based on density functional theory (DFT) is direct with the value of 4.39 eV and is consistent with the value obtained from diffuse reflectance spectra (DRS). The phosphor exhibits typical red emission peaking at 611 nm, and the brightness of the optimal sample is superior to that of the commercial Y_2O_3:Eu^{3+} phosphor. In the lattice the Eu^{3+} ion could replace the Y^{3+} ion and occupy the 24c symmetric position (due to the same valence, +3, and similar radii, when eight-coordinated, Eu^{3+}: 1.066 Å, Y^{3+}: 1.019 Å), and the absence of inversion symmetry for the Eu^{3+} site give rises to the dominant $^5D_0 \to {}^7F_2$ transition. The CYZA:Eu^{3+} phosphor was combined with the blue BAM:Eu^{2+} phosphor and the green LuAG:Ce^{3+} phosphor and then combined together with an NUV LED chip (λ_{max} = 395 nm) for WLED fabrication. The WLED lamp exhibits white light possessing a lower CCT (4745 K) and a higher CRI (86.72). The phosphor is highly stable with excellent color stability.

8.9 Eu^{3+}-Rich Phosphors without Concentration Quenching

Several possible excitation lines in the region of 390–405 nm can ultimately contribute to the absorption of (near) UV LEDs with emission centered on 395 nm as well as a strong absorption band at around 465 nm. Hence, the red phosphor is made attractive with a Eu-rich host lattice.

In recent years, researchers have focused on a $Li_3Ba_2Gd_{3-x}(MoO_4)_8$ host matrix for luminescent ions because of their excellent chemical and thermal stability and favorable luminescence features (no luminescence quenching at a high doping concentration of the Eu^{3+} ion due to structural benefits) related to the sulfide- and nitride-based materials. They possess several advantages over other host materials due to their unique structure, good chemical stability, low-temperature preparation, greater luminescence characteristics, and use of cost-effective raw materials. Recently, Kozhevnikova and Kopylova reported ternary molybdates with the general formula $Li_3Ba_2R_3(MoO_4)_8$ (R = La–Lu, Y) and $Li_3Ba_2R_3(MoO_4)_8$ (R = La–Lu, Y) with a stratified scheelite structure [209, 210].

8.9.1 $Li_3Ba_2La_3(MoO_4)_8$:Eu^{3+}

Katelnikovas et al. showed significant absorption of the emission of an NUV LED when using thick $Li_3Ba_2Eu_3(MoO_4)_8$ ceramic disks, combining the increased absorption due to the high dopant concentration and the long optical path length in the ceramic disk. A series of polycrystalline $Li_3Ba_2La_{3-x}Eu_x(MoO_4)_8$ samples were reported [211]. The synthesized samples crystallize in the monoclinic lattice with the space group $C_{2/c}$, where Z = 2. In the lattice, 66.7% of Li ions possess distorted octahedral coordination, whereas the rest share the same eightfold coordinated site with Ba^{2+} and Gd^{3+} ions. Due to the similar ionic radii of Eu^{3+} and Gd^{3+}, they form a solid solution between them. The color of the $Li_3Ba_2La_{3-x}Eu_x(MoO_4)_8$ sample is white to the naked eye, indicating no absorption in the visible range, while with increasing Eu^{3+} content, samples with the maximum content of Eu^{3+} (fully substituted) show moderate reddish appearance, suggesting some absorption in the blue–violet spectral range, causing Eu^{3+} luminescence. The absorption and emission properties were found to be interesting. The spectra exhibit typical 4f–4f intraconfigurational transitions of Eu^{3+}, where the broad band from (250–350 nm) attributed to the CT band and narrow bands 350–600 nm are assigned to $^7F_0 \rightarrow ^5D_J$ transitions of Eu^{3+} (where J = 0–4). A broadening and slight shifting of the CT band was observed (red shift) with increasing content of Eu^{3+}. The intensity of lines originating from $^5D_0 \rightarrow ^7F_1$, $^5D_0 \rightarrow ^7F_3$, and $^5D_0 \rightarrow ^7F_4$ is much weaker compared to the $^5D_0 \rightarrow ^7F_2$ transition, proving that Eu^{3+} resides at the noncentrosymmetric site as ED transition requires a lower symmetry on the respective Eu^{3+} sites (Fig. 8.27). The strong absorption is one of the top requirements to be used for practical applications (WLEDs), and the major shortcoming of lanthanide ions is they exhibit poor absorption due to parity and spin-forbidden nature of 4f–4f transitions. Different methods can be followed to overcome these problems: adding a sensitizer, increasing the activator concentration, or making ceramics (the absorption is increased due to a longer optical pathway of the incident photons). Katelnikovas et al. prepared $Li_3Ba_2La_3(MoO_4)_8$ ceramic disks and carried out different characterizations. The increase of ceramic thickness slightly increased the Eu^{3+} emission intensity compared to a blue LED.

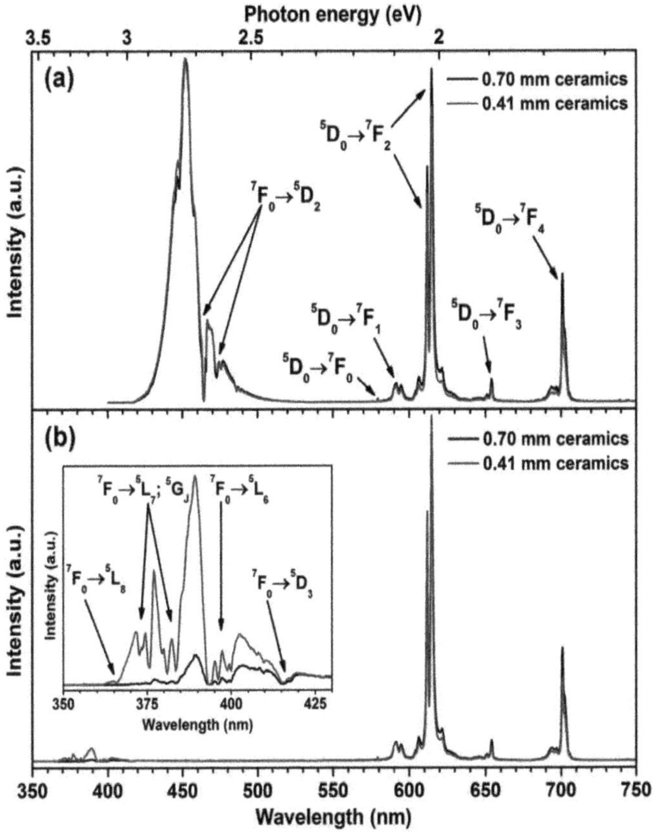

Figure 8.27 Emission spectra of (a) $Li_3Ba_2Eu_3(MoO_4)_8$ ceramics excited by a blue LED and (b) $Li_3Ba_2Eu_3(MoO_4)_8$ ceramics excited by an NUV LED. Inset shows an enlarged fragment in the NUV LED emission range. Reproduced from Ref. [211] with permission of The Royal Society of Chemistry.

$Li_3Ba_2Eu_3(MoO_4)_8$ ceramics can be used for red-emitting LEDs or as a red component in WLEDs pumped by an NUV-emitting semiconductor die.

8.9.2 Li3$Ba_2Gd_3(MoO_4)_8$:Eu^{3+}

Anthuban et al. reported the preparation of Eu^{3+}-doped $Li_3Ba_2Gd_3(MoO_4)_8$ phosphors by the mechanochemically assisted solid-state reaction route, along with the investigation of the red emission enhancement for an optimal Eu^{3+} doping concentration. Furthermore, the synthesized phosphors were compared with commercially available Y_2O_2S:Eu^{3+} in order to find out variations in the luminescence intensity. Along with excellent luminescence properties, they exhibit some superior properties, such as ferroelectricity, laser hosts, phosphors, and catalysis, due to their robust disordered structure [212, 213].

The structure and optical properties of phosphor materials are highly dependent on their synthetic condition. Except for conventional solid-state reactions, researchers have adopted many other synthetic routes based on wet-chemistry procedures. The mechanochemically assisted solid-state reaction is one of the techniques recently accepted by people for its simple, economical, effective, and convenient process in the preparation of nanoluminescent materials, in addition to providing several advantages over existing methods, like a lower synthesis temperature and shorter heating time. The powder XRD method confirms the formation of a single phase of $Li_3Ba_2Gd_3(MoO_4)_8$ phosphors with the monoclinic structure of the space group $C_{2/c}$. The sharp and intense patterns are a witness to the development of a highly crystalline phase via a mechanochemically assisted solid-state reaction at RT. The phase formation is again in accordance with Vegard's law, which states that to form a perfect solid solution between two compounds, the cations of the two compounds should have ionic radii within ±15 % of each other [212]. The incoming Eu^{3+} ions are believed to occupy the sites by substituting the current Gd^{3+} ions in the lattice because of their analogous ionic radii. Eu^{3+} (r = 1.066 Å and CN = 8, where CN is the coordination number of the metal ions) is adjoining Gd^{3+} (r = 1.053 Å and CN = 8) [139].

When the excitation spectrum was monitored at 616 nm, the spectrum showed typical absorption lines of Eu^{3+} absorption, but the intensity of the lines varies with increasing Eu^{3+} content. Among all the 4f–4f transitions of Eu^{3+}, the peaks at 394 and 464 nm excitation wavelengths (which are emission wavelengths of NUV and blue LED chips, respectively) are much stronger than the other transitions of Eu^{3+}. The reason behind the increase in the emission of Eu^{3+} is the increased absorption strength of $^7F_0 \rightarrow {}^5L_6$ and $^7F_0 \rightarrow {}^5D_2$ transitions. The photoluminescence emission (PLE) spectra contains sharp lines from 530 nm to 710 nm, which are related with the $^5D_0 \rightarrow {}^7F_J$ (J = 1, 2, 3, 4) transitions from the higher excited state of Eu^{3+} to the ground state. The stronger intensity of the $^5D_0 \rightarrow {}^7F_2$ transition over $^5D_0 \rightarrow {}^7F_1$ is evidence that Eu^{3+} is located in a distorted (or asymmetric) cation environment and indicates that Eu^{3+} ions occupy the sites of Gd. The concentration variation was performed to obtain the optimal doping concentration for Eu^{3+} in a $Li_3Ba_2Gd_3(MoO_4)_8$ host. The steady enhancement of PL intensity was observed with increasing doping concentration, which reaches a maximum at 0.07 mol content followed by a decrease with further increase in concentration. With different excitations (394 and 465 nm), it was observed that all the phosphors excited with irradiation of 394 nm possess stronger emission intensity compared to 465 nm, ultimately proving that they may be used as potential red-emitting phosphors in NUV-based WLEDs than blue GaN-based WLEDs.

Cheng et al. also studied the luminescence, concentration, and thermal quenching behavior of $Li_3Ba_2Gd_3(MoO_4)_8{:}Eu^{3+}$ and equated its emission intensity with that of commercial phosphors, ZnS:(Mn^{2+},Te^{2+}) [214]. The scanning electron microscopy (SEM) micrographs show aggregated particles with sizes from 3 to 5 μm, and the larger particles are attributed to the Li content in $Li_3Ba_2(Gd_{3-x}Eu_x)(MoO_4)_8$, which enhances grain growth and improves crystallinity at a low calcination temperature. In general, when the Eu^{3+} ions are located at a low-symmetry site, the $^5D_0 \rightarrow {}^7F_2$ emission transition often dominates in the emission spectrum. Similar kinds of observations are noted for the present phosphor samples. The CIE chromaticity coordinates of the $Li_3Ba_2Gd_{0.6}Eu_{2.4}(MoO_4)_8$ phosphor for red emission are (0.67, 0.33), which follows the NTSC system standard for red chromaticity. ZnS:(Mn^{2+},Te^{2+}) is used as the standard. The intensity of the emission of the $^5D_0 \rightarrow {}^7F_2$ transition for $Li_3Ba_2Gd_{0.6}(MoO_4)_8{:}Eu_{2.4}$ is 3.8 times higher than that of the commercial red sulfide phosphors. The interesting observations for the presently studied phosphors were the emission from $^5D_1 \rightarrow ^7F_1$ increases with increasing Eu^{3+} ion concentration and gradually decreases as the doping concentration becomes higher than $x > 0.1$. With increasing Eu^{3+} content the probability of energy transfer between two adjacent Eu^{3+} ions increases as the distance decreases; the higher-level ($^5D_1$) emission can be easily quenched via cross-relaxation and the $^5D_0$ emission becomes dominant. The lifetime (τ_D) of the $^5D_0 \rightarrow ^7F_2$ transition, efficiency (η), and distance between the Eu^{3+}–Eu^{3+} pair of $Li_3Ba_2Gd_3(MoO_4)_8$ phosphors were calculated. Thermal quenching studies of the same phosphors are highly stable at LED fabrication temperature. Figure 8.28 represent the thermal quenching of the $^5D_0$ level with the help of a charge transfer state (CTS) band.

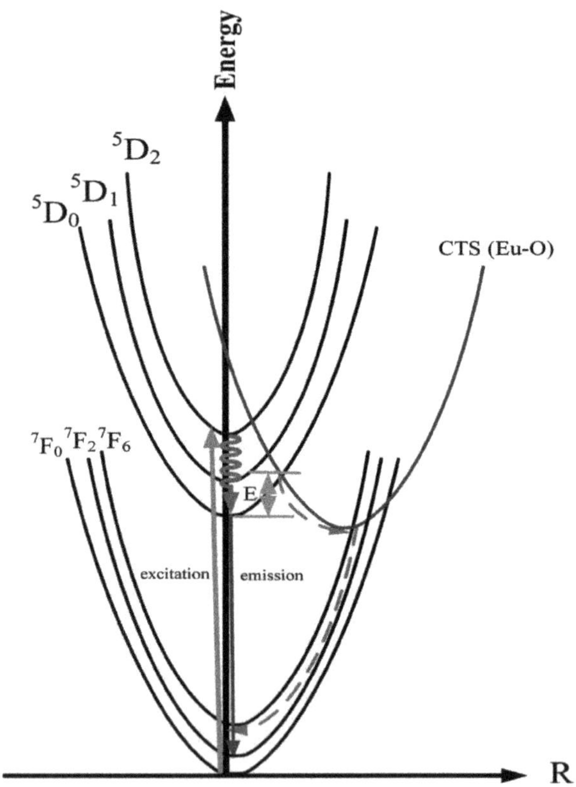

Figure 8.28 Plot of pathways for the thermal quenching of the $^5D_0$ state through a CTS band. Reprinted with permission from Ref. [214]. Copyright (2010) American Chemical Society.

8.9.3 $Li_3BaSrLn_3(WO_4)_8$:Eu^{3+}

Wei et al. reported fast synthesis of red-emitting $Li_3BaSrLa_3(WO_4)_8$:Eu^{3+} phosphors through a microwave-assisted solid-state reaction [215]. The conventional solid-state method is the most common technique used by researchers to synthesize phosphor materials. Similarly, microwave heating has been widely used to prepare a variety of organic compounds for more than 15 years [216]. The advantage of this method over others is that during a short interval of time it directly interacts with the reactant compounds and can lead to a sudden rise in the reaction temperature, while in the solid-state reaction, heating over a long period is a must, except that narrow size distribution, higher purity, and better crystallinity of phosphors can be achieved through this method. Their group succeeded in synthesizing a series of $Li_3BaSrLn_3(WO_4)_8$:Eu^{3+} phosphors through a microwave-assisted solid-state reaction and performed different studies on them. The powder XRD pattern agreed with the JCPDS card and confirmed the formation of a monoclinic phase. PL studies show that $Li_3BaSrLa_3(WO_4)_8$:Eu^{3+} could be a promising red phosphor for WLEDs under NUV excitation, where the Eu^{3+} is in sites without inversion symmetry, which helps red-emitting phosphors with better performances and improved quality. Variation of emission intensity was observed for different sintering temperatures; with increasing temperature the enhancement of intensity was found up to 900°C, and afterward it started decreasing, showing that 900°C is the best sintering temperature for these phosphors. Similarly the concentration variation profile suggested a rapid increase of emission intensity with increasing Eu^{3+} content up to a defined level of concentration (60%); afterward it increased furthermore to 90% but slowly, and the same is the optimum doping concentration. Further increase of Eu^{3+} content leads to concentration quenching as the distance between activator ions will be shorter and defects originating from a lattice mismatch will increase, leading to concentration quenching from cross-relaxation and nonradiative energy transfer, respectively. Comparatively the microwave-assisted method showed higher intensity over the solid-state method. The thermal stability of phosphors is one of the important technological parameters for evaluating a phosphor to be applied as a WLED [217]; with increasing temperature the emission intensity decreases, and a decay of 22% for the phosphor at 130°C

(403 K) is observed, indicating that the phosphor has relatively good thermal stability against the thermal quenching effect. The $Li_3BaSrLa_3(WO_4)_8$:90at%Eu^{3+} phosphor also exhibited impressive purity of red emission with coordinate values (x = 0.670, y = 0.330) from NTSC.

8.9.4 $Li_3BaSrGd_{3-x}Eu_x(MO_4)_8$ (M = W/Mo)

We also reported synthesis and complete investigation of structural and optical parameters of a series of new red-emitting $Li_3BaSrGd_{3-x}Eu_x(MO_4)_8$ (M = W, Mo) phosphors with a stratified scheelite structure by a conventional solid-state reaction [218]. Both the phosphors of molybdate and tungstate series belong to a monoclinic crystal structure with the space group $C_{2/c}$ (Fig. 8.29). The fully substituted Eu^{3+} (x = 3) phosphor shows the highest emission intensity compared to that other phosphor compositions, and no concentration quenching was found in the presently studied host lattice. Both the Eu^{3+}-substituted tungstate and molybdate host lattices were excited at 394 and 465 nm. All compositions show red emission at 612 nm, whatever the excitation wavelength (including CT band excitation). Another interesting observation was found for the presently studied phosphors: tungstate-based phosphors possess higher emission intensity compared to molybdate-based phosphors. The investigation reveals the significant information about the importance of choosing a host lattice for obtaining high red luminescence intensity of the Eu^{3+} ion.The Judd–Ofelt (J–O) theory is a well-known and imperative tool for exploring the spectral properties of lanthanide (Ln^{3+}) ions in host lattices and is specifically used by researchers to obtain appropriate information regarding 4f orbital transition intensities in their respective host materials [219, 220]. A higher value of Ω_2 than Ω_4 indicates the probability that the $^5D_0 \rightarrow ^7F_2$ transition is indeed dominating (Fig. 8.29). A decay of 50% for the phosphor at 150°C is observed, indicating that the phosphors have relatively good thermal stability against the thermal quenching effect (as compared to sulfide-based red phosphor). The currently synthesized phosphors were compared with commercially available phosphors such as Y_2O_3 and Y_2O_2S doped with Eu^{3+} to observe their potential. The QY calculation was performed for currently synthesized phosphors. The QY of $Li_3BaSrGd_{0.3}Eu_{2.7}(WO_4)_8$ was found to be 38.42 ± 0.09%. The QY of commercial Y_2O_3:Eu^{3+} can be 90% under 270 nm excitation but was found to be only ~9.6% under 394 nm excitation. The QY of our sample is found to be almost four times higher than that of commercial $Y_2O_3:Eu^{3+}$ at an excitation wavelength of 394 nm. The excitation of phosphors can also be done in the blue LED region along with UV LED excitation; for the above reason they may find potential applications in WLEDs.

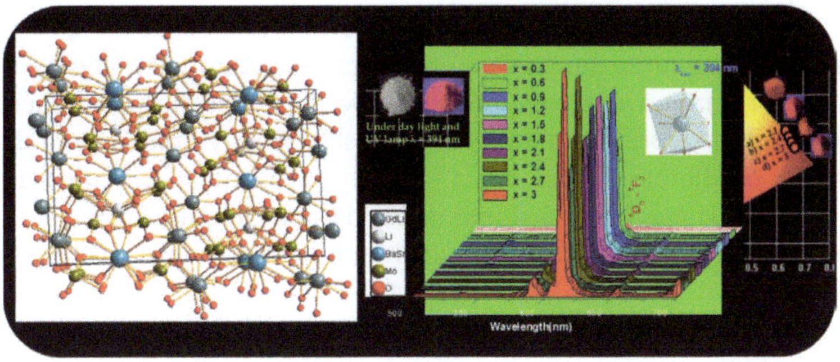

Figure 8.29 Crystal structure and PL emission at excitation of 394 nm for $Li_3BaSrGd_{3-x}Eu_x(MoO_4)_8$ phosphors along with their CIE color coordinates, which shows the color saturation with increasing Eu^{3+} content.

8.10 EU^{III} Complexes for White LEDs

Owing to their amazing intrinsic photophysical properties (narrow emission lines, large QYs, and high resistance to photobleaching), RE complexes have attracted increasing attention for a wide range of applications. In fact their use in potential practical applications is restricted due to their low absorption coefficients (low oscillator strength). It is essential to identify good harvesting (sensitizing) ligands based on β-diketonates, aromatic carboxylates, or heterocyclic compounds, capable of efficient ET to the lanthanide excited state, and make stable complexation toward lanthanide ions. For the past several decades, researchers have made numerous efforts toward understanding ET between ligands (ancillary or anionic ligand) and lanthanide ions with the aim of obtaining high-luminescence QYs. Indeed, high-photoluminescence quantum yields (PLQYs) and reasonable stability (photo- as well as thermal) are vital for the high-tech application of lanthanide complexes in the areas of energy conversion (luminescent dyes or solar concentrators) and devices such as LEDs [221, 222]. However, lanthanide complexes showing a quantitative ligand-to-metal energy transfer leading to luminescence QYs higher than 90% remain extremely rare [223–225].

[Eu(tta)$_3$(phen)] (where tta is 2-thenoyltrifluoroacetonate and phen is 1,10-phenanthroline) was recorded as one of the most prominent red-emitting europium-based molecular complex phosphors [226]. As mentioned earlier the intraconfigurational 4f→electronic transitions of Eu^{III} (excitation and emission spectra) are predominantly sensitive to the local structural site (local Eu^{3+} ion coordination environment). It is well known that the spacing between the vibrational levels in C–F oscillators is smaller than that in C–H oscillators. Hence, most often, the vibrational luminescence quenching is much less pronounced and can frequently be neglected. So, it is possible to afford significant enhancement in the QY by the replacement of the C–H bonds of a β-diketone ligand with lower-energy C–F oscillators; situated near the metal center, it could help decrease the nonradiative energy loss [227, 228]. The photophysical process in the Eu^{III} complex can be called a ligand-sensitized luminescence process (antenna effect) and is schematically represented in Fig. 8.30.

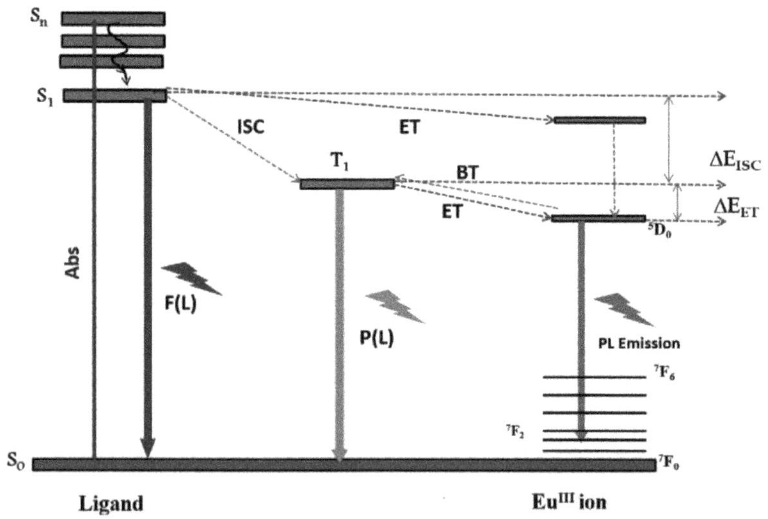

Figure 8.30 Schematic representation of the photophysical process of the Eu^{III} complex. Abs, absorption; F(L), fluorescence from ligand; P(L), phosphorescence from ligand; PL, photoluminescence; ISC, intersystem crossing; ET, energy transfer; BT, back energy transfer; ΔE_{ISC}, energy difference for ISC; ΔE_{ET}, energy difference for ET.

8.10.1 Significance of Energy Transfer from Ligand to Eu^{3+} ($^5D_0$) Level

The energy transfer or sensitization process usually involves a triplet excited state of a ligand, thus requiring efficient intersystem conversion in the ligand (facilitated by the heavy metal ion effect) as well as an appropriate alignment of the triplet state energy (ET) over that of the accepting lanthanide state(s) [229].

It is essential to understand the energy transfer between singlet ($^1\pi\pi^*$) and triplet ($^3\pi\pi^*$) energy levels of ligands and subsequent triplet- ($^3\pi\pi^*$) to $^5D_0$-level energy transfer to get a characteristic red emission from the Eu^{III} metal ion. It is well identified that the Eu^{III} metal ion excited the $^5D_0$ energy level located at 17,500 cm^{-1} [230, 231]. Latva et al. investigated the difference between ligand ($^3\pi\pi^*$) and the Eu excited state ($^5D_0$), which should be >2500 cm^{-1} [232]. In addition, the singlet state should be located higher than 25,000 cm^{-1} and the triplet state higher than 20,000 cm^{-1}. Therefore when designing a ligand for a Eu complex, the ligand should have a triplet energy level higher than that of the $^5D_0$ level of the Eu^{3+} ion.

Recently, the Eu complex gained much attention due to its emerging technological applications such as SSL (essential component as a red-emitting phosphor for NUV- or blue LED–based warm WLEDs). By combining the Eu complex with inorganic color convertors the white-light emission can be realized. For example, hybrid WLEDs could be fabricated by combining the phosphors in a polymer matrix; the same is placed on the top of an LED chip, and by integrating the light from the LED chip and emission from the phosphor white light could be produced (shown in Fig. 8.31).

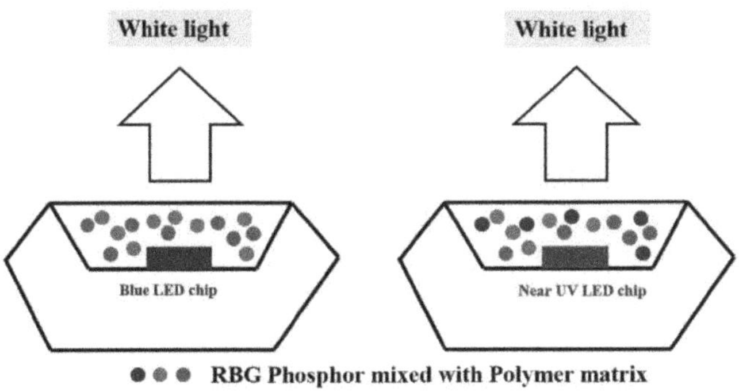

Figure 8.31 Structure of a WLED based on an NUV LED chip.

The molecular structure used in this section is summarized in Figs. 8.32, 8.34, 8.36, and 8.37.

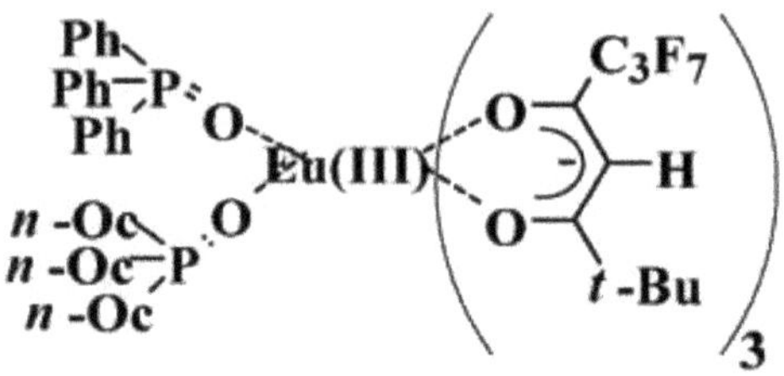

Figure 8.32 The molecular structure of a Eu complex.

Lee et al. fabricated a hybrid WLED by combining $SrAl_2O_4:Eu^{2+}$ (for green, 520 nm, emission) and $Sr_4Al_{14}O_{25}:Eu^{2+}$ (for blue, 490 nm, emission) and a new organic Eu–metal complex, $Eu(BTFA)_3phen$ (for red, 614 nm, emission, complex 1). The fabricated WLED electroluminescence spectrum shows typical RBG emission and the CIE values (0.321, 0.365) are better than that of the typical commercial Nichia WLED (0.297, 0.297). It is worth noting that the preliminary lifetime test indicated that the phosphor is robust, showing no degradation over a period of 100 h [233].

Several Eu-based complexes (complex 1–6 along with three phosphine-based Eu complexes) were taken and their PL properties for WLEDs studied. Among all the complexes, the phosphine oxide–based β-diketone Eu complex (given below) showed a better luminescence QY. The Eu complex (described below) integrated with the NUV LED gave intense red light with luminous intensity over 200 mcd (20 mA), and these types of complexes (fluorescence compounds) are promising for applications in warm WLEDs [234].

Xiang et al. designed and synthesized a novel europium (III) tetrabasic complex, Eu $(TPBDTFA)_2$ (TPATFA)phen (complex 7), for WLED applications. Complex 7 exhibits bright-red emission due to the $^5D_0 \rightarrow {}^7F_J$ transition of Eu^{3+} ions with appropriate CIE chromaticity coordinates (x = 0.64, y = 0.35) under λ_{exc} = 230–470 nm. The luminescence QY for complex 7 is 0.126 and the decomposition temperature is found to be high (383.4°C) from thermogravimetric analysis (TGA). An intense red-emitting LED was fabricated by combining the monophosphor (complex 7) with a ~395 nm emitting InGaN chip [235].

Pei He et al. reported an efficient red-emitting Eu^{III} organometallic complex Eu(ECTFBD)$_3$phen (complex 8), where ECTFBD is 1-(9-ethyl-9H-carbazol-3-yl)-4,4,4-trifluorobutane-1,3-dione and phen is 1,10-phenanthroline. Combined spectroscopic analysis and TGA reveals that complex 8 exhibits high thermal stability and excellent PL properties. Ligand-sensitized luminescence was observed due to the energy match between the triplet level of the ligand and the Eu excited ($^5D_0$) levels. A bright-red LED was fabricated based on an InGaN chip and the CIE coordinates (x = 0.65, y = 0.320) are close to the NTSC values for red with a high CRI value [236].

Similarly complex 9, Eu$_2$(BTFBDEC)$_3$(phen)$_2$ (BTFBDEC is 3,6-bis(4,4,4-trifluoro-1,3-Butyldiketone)-9-ethyl-carbazole, also possesses good thermal and PL properties with a good absorption NUV region. Hence it is possible use the same in NUV-based WLEDs. An intense-red-light-emitting diode was fabricated by combining complex 9 with a ~395 nm emitting InGaN chip [237].

White-light-emitting terbium and europium β-diketonate complexes with different doping ratios were made in a poly(*N*-vinylcarbazole) (PVK) matrix and their PL properties studied. The various energy transfer processes involved in these doping systems are well analyzed. Intermolecular energy transfer from PVK to Tb(DBM)$_3$phen(TbDP) and Eu(DBM)$_3$phen(EuDP) and from TbDP to EuDP exists in the present systems. By simply adjusting the dopant mixing ratio and the wavelength excitation, one can tune the emission color, and by further optimizing the dopant concentration, white light was realized (shown in Fig. 8.33) [238].

Liu et al. in 2009 reported a quadruple stranded dinuclear Eu^{III} complex Na$_2$[Eu$_2$(dbt)$_4$] (complex 10), where dbt is 2,8-bis(40,40,40-trifluoro-1',3'-dioxobutyl)-dibenzothiophene. The complex showed typical ED transitions of the Eu^{3+} ion with a QE of 23%. The LEDs were fabricated by amalgamation of a 395 nm emitting InGaN chip, with complex 10 as a red-emitting phosphor, while an original LED without a phosphor and an LED with Y$_2$O$_2$S:Eu^{3+} were also made for comparison. The emission spectra of the InGaN–complex 10 LEDs show red emission, whereas the original InGaN LED without a phosphor shows NUV emission under a 20 mA forward bias. The dependence of CIE chromaticity coordinates of the fabricated LEDs on the complex-to-silicone mass ratio has also been investigated. The results show that the complex-to-silicone mass ratios are 1:5, 1:10, and 1:20, and the CIE chromaticity coordinates of the fabricated LEDs are ($x = 0.6224, y = 0.3238$), ($x = 0.5322, y = 0.2701$), and ($x = 0.4997, y = 0.2471$), respectively. The CIE chromaticity coordinates of an LED of a 1:1 mass ratio of $Y_2O_2S:Eu^{3+}$ to silicone are ($x = 0.4033, y = 0.1871$). The efficiencies of the original LED without a phosphor or with $Y_2O_2S:Eu^{3+}$ were only 0.39 and 0.20 lm/W, though the efficiencies of the fabricated LEDs with complex 10 are 1.55, 1.86, and 2.18 lm/W [239].

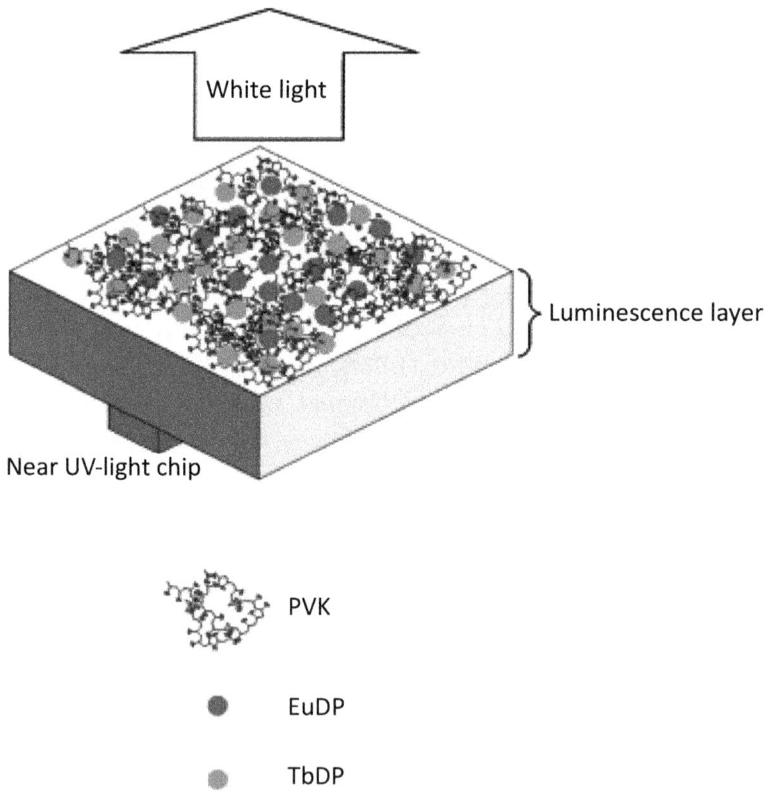

Figure 8.33 The model of UV-excited WLEDs.

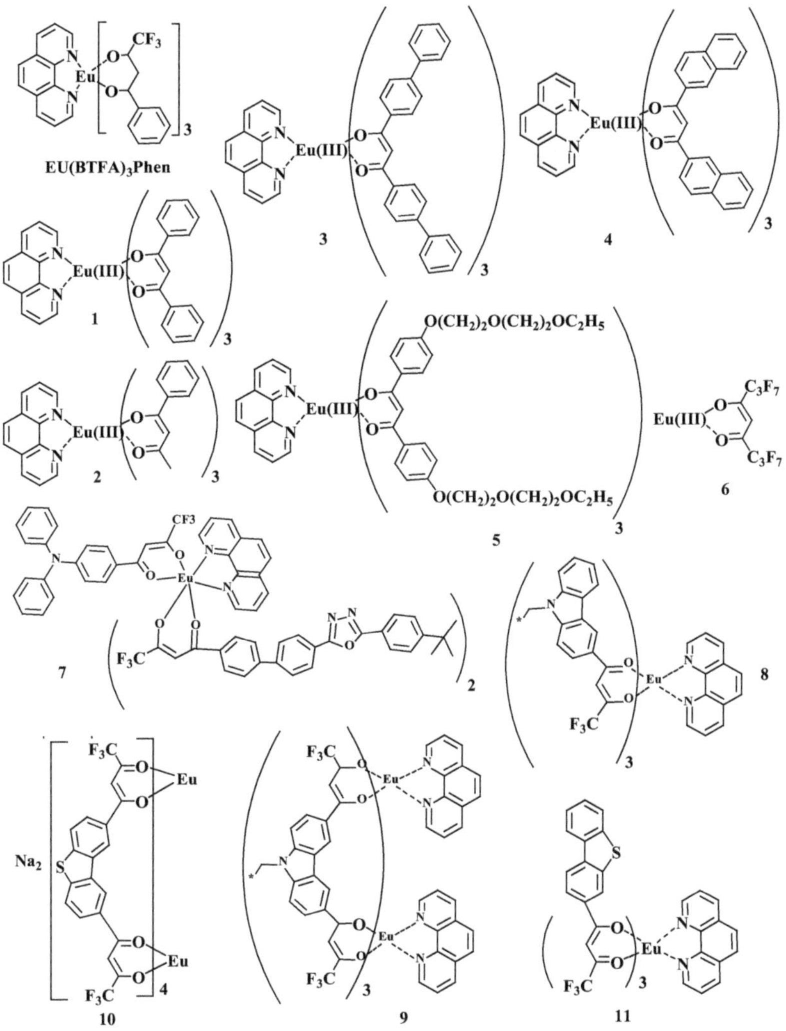

Figure 8.34 The molecular structure of complexes 1–11.

Due to their structural advantage the ligand, namely 2-(4′,4′,4′-trifluoro-1′,3′-dioxobutyl)-dibenzothiophene (Hdbt), was synthesized and used for the Eu complex (complex 11) and further evaluated for WLED applications. The ligand's structural advantages are (i) it contains a β-diketone unit, which could coordinate successfully to Eu^{3+} ions to form a stable Eu^{III} complex; (ii) the dibenzothiophene unit has a suitable conjugated system and could absorb the NUV energy effectively and then transfer it to the Eu^{3+} ion; and (iii) it has no absorption in the blue and green emission range. The complex efficiently emits red emission due to ED transition and absorption in the NUV region. The complex was used to fabricate a red LED (integrated with a ~395 nm emitting InGaN chip), and the LED's CIE chromaticity coordinates are (x = 0.5835, y = 0.2857) and the luminescence efficiency is 1.29 lm/w [240].

Carbazole-based β-diketonates with different alkyl substituents (CH_3, C_2H_5, C_3H_7, C_5H_{11}) at the *N*-position in the carbazole ring were synthesized and used for making ternary Eu^{III} complexes with phen (complexes similar to 12a but the *N*-1 position of the ligand should be different: CH_3, C_2H_5, C_3H_7, C_5H_{11}). The complexes show good thermal stability (decomposition temperature > –300°C), which is suitable for LED applications. Efficient energy transfer from the ligand to metal ions was observed due to the higher location of the triplet level of ligands than $Eu^{3+5}D_0$ levels. Even though all the complexes show red emission, the emission intensity decreases in the following order: $Eu(N\text{-}C2)_3phen$ > $Eu(N\text{-}C3)_3phen$ > $Eu(N\text{-}C5)_3phen$ > $Eu(N\text{-}C1)_3phen$. By using these complexes a bright-red LED was fabricated, and this study clearly indicates that the complexes can be used as a red component for fabrication of NUV-based WLEDs [241].

The functionalization in the 2- or 2,7-postiion in the carbazole ring, (2-(4'4'4'-trifluoro-1'3'-dioxobutyl)-carbazole (2-TFDBC) and 2,7-bis(4'4'4'-trifluoro-1'3'-dioxobutyl)-carbazole (2,7-BTFDBC)), was carried out in order to enhance the visible light absorption (with extended π conjugation). The same was used to metallate the Eu^{3+} ion and two Eu^{III} ternary complexes were made, $Eu(2\text{-}TFDBC)_3phen$ and $Eu_2(2,7\text{-}BTFDBC)_3(phen)_2$ (complex 12a), to use in the WLED based on a blue InGaN chip. The QY of the Eu complex mainly depends on the location of the triplet level of the ligands, and these ligands show higher triplet energy levels than the $Eu^{3+5}D_0$ levels. The complexes emit a bright-red color in the region of 615 nm with blue ray excitation and are potential candidates for WLEDs [242].

He et al. in 2010 synthesized a carbazole-functionalized β-diketone (TTA-based) ligand for the novel Eu^{III} ternary complex $Eu(EMOCTFBD)_3phen$ (complex 12b) (EMOCTFBD is 1-(9-ethyl-7-methoxyl-9*H*-carbazol-2-yl)-4,4,4-trifluorobutane-1,3-dione). The methoxy moiety introduction significantly extends to the blue region, intensifying the absorption intensity. The complex emits bright-red emission due to the antenna effect. The decomposition temperature of Eu(EMOCTFBD)$_3$phen is found to be 338.3°C, which is good enough for luminescence applications. The fabricated red LED using this complex integrated with a ~460 nm emitting InGaN chip emits a bright-red color (Fig. 8.35) [243].

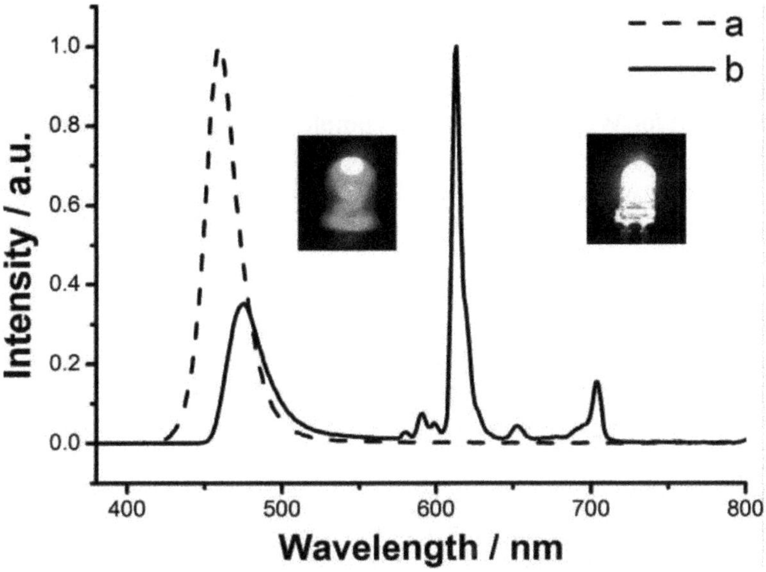

Figure 8.35 Emission spectra and photographs of the original InGaN LED without a phosphor (a, left) and the LED with Eu(EMOCTFBD)$_3$phen (b, right) under excitation of a 20 mA forward bias. Inert: photographs of lighting LEDs.

Huihui Wang et al. designed and synthesized two new multinuclear Eu^{III} complexes, [Eu(AFTFBD)$_3$]$_2$(1,4-bmb) and [Eu(AFTFBD)$_3$]$_3$(tmb) (complex 13), where AFTFBD is 2-acetylfluorene-4,4,4-trifluorobutane-1,3-dione, 1,4-bmb is 1,4-bis[2-(20-pyridyl)benzimidazolyl]benzene, and tmb is 1,3,5-tris[2-(20-pyridyl)benzimidazolyl] benzene. Red luminescence of the Eu^{3+} ion due to $^5D_0$–^7F_J (J = 0–4) transitions under 395 nm light excitation was observed for all the Eu complexes. The fabricated LED (coating the selected complex 13 onto ~395 nm emitting InGaN chips) shows bright-red light with appropriate CIE color coordinates. Complex 13 is found to be a promising candidate as a red component for application in fabrication of NUV-based WLEDs [244].

A trivalent europium complex (complex 14) was synthesized and its PL properties studied. Fortunately the complex has a broad excitation band in the NUV spectral range and emits bright-red emission due to the $^5D_0 \rightarrow ^7F_J$ transitions of Eu^{3+} ions. The luminescence QY for complex 14 is 0.17 and the complex shows high thermal stability (confirmed from a TGA decomposition temperature of 344°C). An intense-red-light-emitting diode was fabricated by combining complex 14 with a ~395 nm-emitting InGaN chip [245].

A β-diketone-based Eu complex (complex 15), $Eu(IDTFBD)_3phen$, where IDTFBD is 4-imidazol-4,4,4-trifluorobutane-1,3-dione, was designed for WLED applications. Complex 15 was characterized by IR, UV-visible, TGA, and PL spectroscopy. The complex exhibited a strong and wide excitation band between 320 and 420 nm (due to the formation of a big π-conjugated system). Complex 15 exhibits strong red emission of the Eu^{3+} ion due to the $^5D_0 \rightarrow ^7F_J$ (J = 0–4) transitions with NUV excitation, and similar types of observations were found for the corresponding Gd^{3+} complex. The triplet energy of the ligand was measured, and it is higher than that of Eu^{3+} excited levels. Finally, a bright-red LED was fabricated by coating complex 15 onto a ~395 nm emitting InGaN chip, and the efficiency of the fabricated LED was 0.531 lm/W, much higher than that of the original LED without a phosphor (0.33 lm/W). The LED exhibited appropriate CIE chromaticity coordinates in the red area [246].

The design and synthesis of a novel indane-based β-diketone with trifluorobutane in the contraposition, 5-acetylindane-4,4,4-trifluorobutane-1,3-dione (HAITFBD) and its Eu^{III} ternary complex, $Eu(AITFBD)_3phen$, were reported. Complex 16 was characterized by spectroscopic analysis and TGA and showed high thermal stability and wide and strong excitation bands from 300 nm to 425 nm when monitored at 611 nm (synchronize well with InGaN chips emission). The complex exhibited intense-red emission under excitation of NUV light with CIE chromaticity coordinates (x = 0.63, y = 0.34). An efficient antenna effect is expected due to the lowest triplet state energy of the primary ligand AITFBD (17,730 cm^{-1}), which is higher than that of the lowest excitation state energy level of the central Eu^{3+} ion, $^5D_0$. Finally, a bright-red LED was fabricated by combining complex 16 with a 380 nm emitting InGaN chip, and the efficiency of the fabricated LED was 0.51 lm/W, more than twice the original LED without a phosphor (0.23 lm/W). The results indicate that the red phosphor is a potential candidate for NUV-based WLEDs [247].

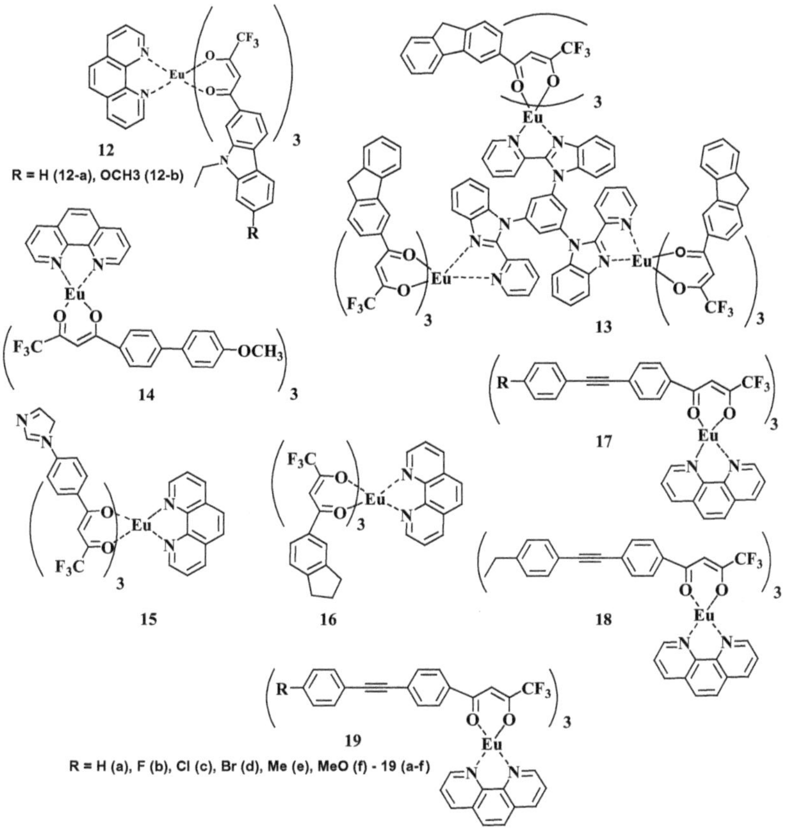

Figure 8.36 The molecular structure of complexes 12–19.

Wang et al. synthesized a novel indane-based β-diketone ligand for a ternary Eu^{III} complex (complex 17). The ligand, 5-acetylindane-4,4,4-trifluorobutane-1,3-dione (HAITFBD), was synthesized through the Claisen condensation reaction from the ketone and ester. A complex excitation band was observed in a wide spectral range (300–425 nm), and this could be attributed to the availability of a good and extended π-conjugated system in the complex. A strong and narrow Eu^{3+} red emission was observed at ~611 nm with 380 nm excitation; by exciting ligand absorption, the characteristic emission from Eu^{3+} ions was observed (no emission from the ligand). This clearly indicates that the effective intermolecular energy transfer from the coordinated ligands to the luminescent central Eu^{3+} ions in the complex occurs [248]. Integrating the $Eu(AITFBD)_3phen$

complex with a ~380 nm emitting InGaN chip shows a bright-red LED [249].

Guang Shao et al. reported a new Eu^{III} ternary complex based on a fluorinated β-diketonate ligand and 1,10-phenanthroline as an ancillary ligand and assessed it as a candidate for WLEDs. Complex 18 exhibits a high decomposition temperature (316°C) and good absorption in the NUV region. The complex exhibits an efficient energy transfer pathway from the ligands to the central Eu^{3+} ion due to energy level matching. An intense-red-light-emitting diode was obtained by coating complex 18 onto a 395 nm emitting InGaN chip. The results reveal that the energy transfer from the InGaN chip to the complex is effective, and it can be a suitable candidate for WLEDs [250].

Shao et al. synthesized six Eu^{III} ternary complexes (complex 19a–f) using different β-diketonate ligands (antenna) and 1,10-phenanthroline as an ancillary ligand. The ligand-to-metal CT band could be possible as the calculated triplet energy level values are above the $^5D_0$ level of the Eu^{3+} ion (lowest triplet state levels of ligands 2a–f could be calculated by the shortest wavelength of the phosphorescence peaks, which are about 19,267 cm^{-1} [519 nm], 18,975 cm^{-1} [527 nm], 19,230 cm^{-1} [520 nm], 20,366 cm^{-1} [491 nm], 18,248 cm^{-1} [548 nm], and 19,723 cm^{-1} [507 nm]). Bright-red-light-emitting diodes were then fabricated by coating the complexes onto 395 nm emitting InGaN chips. The light emission from the InGaN chips could be completely absorbed in the spectra of the LEDs [251].

The complex $Eu(DBM)_3(phen)$ is one of the renowned red-emitting Eu^{3+}-based molecular phosphors with a high QY. However, due to its low absorption in the visible region (blue region), its application as a red phosphor in WLEDs is restricted. Recently, taking all the advantages of the ligand DBM and phen, Zhengliang Wang et al. designed an additional ligand, 2,7-bis(4'4'4'-trifluoro-1,3-dioxobutyl)-(9-ethyl-9H-carbazole) (2,7-BTFDBC), with extended π conjugation (facilitates the blue absorption) for the Eu complex (complex 20). It is evident from the PL emission and excitation spectra of the synthesized complexes that all the complexes were efficiently excited with blue light and emitted bright-red emission (ED transition confirms the ligand-to-metal charge transfer [LMCT]). The $Eu_2(2,7\text{-}BTFDBC)(DBM)_4(phen)_2$ was to fabricate a single red LED combined with a ~460 nm emitting GaN chip [252].

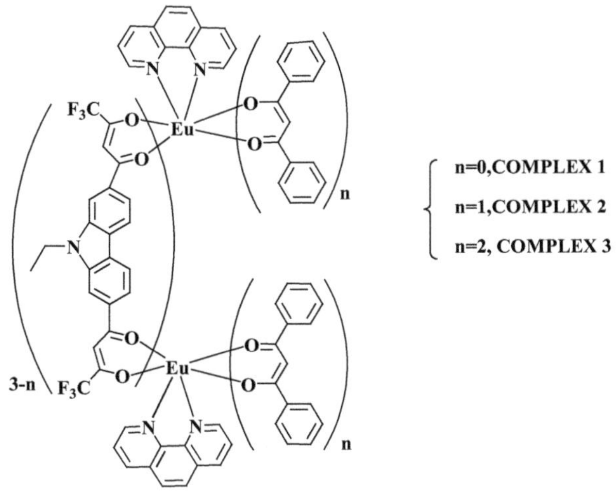

20

Figure 8.37 The molecular structure of complex 20.

8.11 Conclusions

This chapter was dedicated to inorganic color-converting oxide-based phosphors and their application in the field of SSL, including warm WLEDs. We tried to demonstrate the Eu^{3+} luminescence in a host lattice with different crystal structures, which includes scheelite-related, double-perovskite, pyrochlore, garnet, and other tungstates and/molybdates. In addition, we also made efforts to summarize the Eu^{III} complex, which has been used for WLED applications, with special emphasis on its spectral properties. The influence of structural parameters always plays a key role in changes in Eu^{3+} luminescence. Keeping this in mind, we reviewed recent developments in the red phosphor with its structure–composition–property correlation. Even though enormous research effort has been devoted for the past two decades in the field of synthesizing new narrow-band red-emitting phosphors, it still needs major advancements or breakthroughs in terms of PL efficiency and thermal stability. With a view to the future, it is clear that there exists great scope for further structural and vibrant investigations in order to increase the understanding of narrow-band red-emitting phosphors.

Acknowledgments

Financial support from the Naval Research Board (NRB) is gratefully acknowledged.

References

1. Pust, P., Schimdt, P. J. and Schnick, W. (2015). A revolution in lighting, *Nat. Mater.*, **14**, pp. 454–458.
2. Nakamura, S., Mukai, T. and Senoh, M. (1994). Candela-class high-brightness InGaN/AlGaN double-hetero structure blue-light-emitting diodes, *Appl. Phys. Lett.*, **64**, pp. 1687–1689.
3. Pimputkar, S., Speck, J. S., Den Baars, S. P. and Nakamura, S. (2009). Prospects for LED lighting, *Nat. Photonics*, **3**, pp. 180–182.
4. Blasse, G. and Grabmaier, B. C. (1994). *Luminescent Materials* (Springer-Verlag, Berlin).
5. Jüstel, T., Nikol, H. and Ronda, C. (1998). New developments in the field of luminescent materials for lighting and displays, *Angew. Chem. Int. Ed.*, **37**, pp. 3084–3103.
6. Huang, X. (2014). Solid-state lighting: red phosphor converts white LEDs, *Nat. Photonics*, **8**, pp. 748–749.
7. Crawford, M. H. (2009). LEds for solid-state lighting: performance challenges and recent advances, *IEEE J. Sel. Top. Quantum Electron.*, **15**, pp. 1028–1040.
8. Hoppe, H. A. (2009). Recent developments in the field of inorganic phosphor, *Angew. Chem. Int. Ed.*, **48**, pp. 3572–3582.
9. Shang, M., Li, C. and Lin, J. (2014). How to produce white light in a single-phase host?, *Chem. Soc. Rev.*, **43**, pp. 1372–1386.
10. George, N. C., Denault, K. A. and Seshadri, R. (2013). Phosphors for solid-state white, *Annu. Rev. Mater. Res.*, **43**, pp. 481–501.
11. Jørgensen, C. K. (1962). Electron transfer spectra Sm(III), Eu (III), Tm (III), Yb (III), *Mol. Phys.*, **5**, p. 271.
12. Reisfeld, R. and Jørgensen, C. K. (1977). *Lasers and Excited States of Rare Earth* (Springer, Berlin).
13. Jørgensen, C. K. (1970). Electron transfer spectra, *Prog. Inorg. Chem.*, **12**, p. 101.
14. Dimitrov, V. and Komatsu, T. (2012). Correlation among electronegativity, cation polarizability, optical basicity and single bond strength of simple oxides, *J. Solid State Chem.*, **196**, pp. 574–578.

15. Gao, X. L., Wang, Y. H., Wang, D. and Liu, B. T. (2009). Luminescent properties of $KGd_{1-x}(WO_4)_2:Eu_x^{3+}$ and $KGd_{1-x}(WO_4)_{2-y}(MoO_4)_y:Eu_x^{3+}$ phosphors in UV–VU regions, *J. Lumin.*, **129**, pp. 840–843.

16. Dorenbos, P. (2005). The Eu^{3+} charge transfer energy and the relation with the band gap of compounds, *J. Lumin.*, **111**, pp. 89–104.

17. Blasse, G. (1966). On the Eu^{3+} florescence of mixed metal oxides. IV. The photoluminescence efficiency of Eu^{3+}-activated oxides, *J. Chem. Phys.*, **45**, pp. 2356–2360.

18. El'yashevich, M. A. (1961). Spectra of the rare earths, *U.S. Atomic Energy Commission Transl. No. 4403*, New York.

19. Van Uitert, L. G. and Soden, R. R. (1962). Influence of ion size upon the intensity of Eu^{3+} fluorescence in the tungstates, *J. Chem. Phys.*, **32**, pp. 517–519.

20. Jayaraman, A., Kourouklis, G. A., Vanuitert, L. G., Grodkiewicz, W. H. and Maines, R. G. (1989). A high-pressure Raman study of $KReO_4$, $RbReO_4$ and $CsReO_4$ to 25 GPa and pressure-induced phase transitions, *Physica A*, **156**, pp. 325–340.

21. Birbenbach, L. and Buschendorf, F. (1936). *Zeitschrift für Physikalische Chemie*, **16**, pp. 102–112.

22. Jayaraman, A., Wang, S. W. and Sharma, S. K. (1995). High-pressure Raman investigation on $CdMoO_4$ and pressure-induced phase transformations, *Phys. Rev. B: Condens. Matter*, **52**, pp. 9886–9889.

23. Jayaraman, A., Wang, S. W. and Sharma, S. K. (1995). New pressure induces phase changes in $CdWO_4$ from raman spectroscopic and optical microscopic studies, *Curr. Sci.*, **69**, pp. 44–48.

24. Jayaraman, A., Batlogg, B. and Vanuitert, L. G. (1985). High-pressure Raman study of alkaline-earth tungstates and a new pressure-induced phase transition in $BaWO_4$, *Phys. Rev. B: Condens. Matter*, **31**, pp. 5423–5472.

25. Wang, X., Loa, I., Syassen, K., Hanfland, M. and Ferrand, B. (2004). Structural properties of the zircon- and scheelite-type phases of YVO_4 at high pressure, *Phys. Rev. B: Condens. Matter*, **70**, p. 064109.

26. Trunov, U. K. and Efremov, V. A. (1981). Characteristic features of the crystal chemistry of lanthanide molybdates and tungstates, *Kristallografiya*, **26**, pp. 67–71.

27. Hirano, M. and Morikawa, H. (2003). Hydrothermal synthesis and phase stability of new zircon- and scheelite-type $ZrGeO_4$, *Chem. Mater.*, **15**, pp. 2561–2566.

28. Dickinson, R. G. (1920). The crystal structures of wulfenite and scheelite, *J. Am. Chem. Soc.*, **42**(1), pp. 85–93.
29. Lia, H., Zhou, S. and Zhang, S. (2007). The relationship between the thermal expansions and structures of ABO_4 oxides, *J. Solid State Chem.*, **180**, p. 589.
30. Brazdil, J. F. (2015). Scheelite: a versatile structural template for selective alkene oxidation catalysts, *Catal. Sci. Technol.*, **5**, pp. 3452–3458.
31. Hua, Y., Zhuanga, W., Yeb, H., Wanga, D., Zhanga, S. and Huanga, X. (2005). A novel red phosphor for white light emitting diodes, *J. Alloys Compd.*, **390**, pp. 226–229.
32. Yen, W. H., Shionoya, S. and Yamamoto, H. (2006). *Fundamentals of Phosphors* (CRC Press, Boca Raton).
33. Blasse, G. and Bril, A. (1970). Luminescence phenomena in compoundswith fergusonite structure, *J. Lumin.*, **3**, pp. 109–131.
34. Van Uitert, L. G., Soden, R. R. and Linares, R. C. (1962). Enhancement of rare-earth ion fluorescence by lattice processes in oxides, *J. Chem. Phys.*, **36**, pp. 1793–1796.
35. Golub, A. M. and Maksin, V. I. (1977). Solid-phase preparation of double tungstates of certain rare-earth elements with silver or thallium, *Russ. J. Inorg. Chem.*, **22**, pp. 566–568.
36. Su, Y., Li, L. and Li, G. (2008). Synthesis and optimum luminescence of $CaWO_4$-based red phosphors with codoping of Eu^{3+} and Na^+, *Chem. Mater.*, **20**(19), pp. 6060–6067.
37. Zhang, W., Long, J., Fan, A. and Li, J. (2012). Effect of replacement of Ca by Ln (Ln = Y, Gd) on the structural and luminescence properties of $CaWO_4:Eu^{3+}$ red phosphors prepared via co-precipitation, *Mater. Res. Bull.*, **47**, pp. 3479–3483.
38. Singh, B. P., Parchur, A. K., Ningthoujam, R. S., Ansari, A. A., Singh, P. and Rai, S. B. (2014). Enhanced photoluminescence in $CaMoO_4:Eu^{3+}$ by Gd^{3+} co-doping, *Dalton Trans.*, **43**, p. 4779.
39. Yan, S., Zhang, J., Zhang, X., Lu, S., Ren, X., Nie, Z. and Wang, X. (2007). Enhanced red emission in $CaMoO_4:Bi^{3+}$, Eu^{3+}, *J. Phys. Chem. C*, **111**, pp. 13256–13260.
40. Cao, F. (2012). Sol–gel synthesis of red-emitting phosphor Ca–Sr–Mo–W–O– (Eu^{3+}, La^{3+}) powders and luminescence properties, *Ind. Eng. Chem. Res.*, **51**, pp. 3569–3574.
41. Shen, X., Li, L., He, F., Meng, X. and Song, F. (2012). Effects of doped-Li^+ and -Eu^{3+} ions content on structure and luminescent properties of

$Li_xSr_{1-2x}(MoO_4)$:$Eu^{3+}{}_x$ red-emitting phosphors for white LEDs, *Mater. Chem. Phys.*, **132**, pp. 471–475.

42. Zhou, F., Wei, J., Wu, J., Gong, F., Yi, L. and Huang, J. (2009). Potential red-emitting phosphor for white LED solid-state lighting, *J. Alloys Compd.*, **476**, pp. 390–392.

43. Wei, S., Yun, L., Li, F., Sun, J. and Li, S. (2015). Photoluminescent properties of Eu^{3+}-doped alkaline earth metal molybdates red phosphors with high quenching concentration, *Ceram. Int.*, **41**, pp. 1093–1100.

44. Shi, Y., Liu, B., Liu, B., Li, C. and Wang, Z. (2015). Enhanced red emission in $Ca_{0.5}La(MoO_4)_2$:Eu^{3+} phosphors for UV LEDs, *RSC Adv.*, **5**, pp. 95953–95959.

45. Cao, F., Tian, Y., Chen, Y., Xiao, L. and Wu, Q. (2009). Novel red phosphors for solid-state lighting: $Ca_{0.54}Sr_{0.34-1.5x}Eu_{0.08}La_x$ $(MoO_4)_y(WO_4)_{1-y}$, *J. Alloys Compd.*, **475**, pp. 387–390.

46. Guzik, M., Tomaszewicz, E., Kaczmarek, S. M., Cybińska, J. and Fuks, H. (2010). Spectroscopic investigations of $Cd_{0.25}Gd_{0.50}\Upsilon_{0.25}WO_4$:$Eu^{3+}$: a new promising red phosphor, *J. Non-Cryst. Solids*, **356**, pp. 1902–1907.

47. Wang, L., Moon, B. K., Park, S. H., Kim, J. H., Shi, J., Kim, K. H. and Jeong, J. H. (2016). Synthesis and photoluminescence of Bi^{3+}, Eu^{3+} doped $CdWO_4$ phosphors: application of energy level rules of Bi^{3+} ions, *New J. Chem.*, **40**, p. 3552.

48. Jiang, P., Gao, W., Cong, R. and Yang, T. (2015). Structural investigation of the A-site vacancy in scheelites and the luminescence behaviour of two continuous solid solutions $A_{1-1.5x}Eu_x\gamma_{0.5x}$ WO_4 and $A_{0.64-0.5y}Eu_{0.24}Li_y\gamma_{0.12-0.5y}WO_4$ (A = Ca, Sr; Υ = vacancy), *Dalton Trans.*, **44**, p. 6175.

49. Guzik, M., Tomaszewicz, E., Guyot, B., Legendziewicza, J. and Boulonc, G. (2015). Eu^{3+} luminescence from different sites in a scheelite-type cadmium molybdate red phosphor with vacancies, *J. Mater. Chem. C*, **3**, p. 8583.

50. Van Uitert, L. G. (1963). A comparison of the intensities of emission of Eu^{3+} and Tb^{3+} in tungstates and molybdates. *J. Electrochem. Soc.*, **110**(1), pp. 46–51.

51. Toradi, C. C., Page, C., Brixner, L. H., Blasse, G. and Dirksen, G. J. (1987). Structure and luminescence of some $CsLnW_2O_8$ compounds, *J. Solid State. Chem.*, **69**, pp. 171–178.

52. Van Vliet, J. P. M., Blasse, G. and Brixner, L. H. (1988). Luminescence properties of alkali europium double tungstates and molybdates $AEuM_2O_8$, *J. Solid State Chem.*, **76**, pp. 160–166.

53. Jeitschko, W. and Sleight, A. W. (1973). The crystal structure of $HgMoO_4$ and related compounds, *Acta Cryst.*, **B29**, pp. 869–875.

54. Blasse, G. (1975). Energy transfer in mercury-doped calcium tungstate and molybdate, *J. Solid State. Chem.*, **13**, pp. 339–344.

55. Blasse, G. and Vanden Henuvel, G. P. M. (1974). Luminescence of mercuric molybdate and tungstate, *J. Lumin.*, **9**, pp. 74–78.

56. Neeraj, S., Kijima, N. and Cheetham, A. K. (2004). Novel red phosphors for solid-state lighting: the system $NaM(WO_4)_{2-x}(MoO_4)_x$:Eu^{3+} (M = Gd, Y, Bi), *Chem. Phys. Lett.*, **387**, pp. 2–6.

57. Wang, Z., Liang, H., Gong, M. and Su, Q. (2005). Potential red-emitting phosphor for LED solid-state lighting, *Electrochem. Solid-State Lett.*, **8**, pp. H33–H35.

58. Wang, Z., Liang, H., Zhou, L., Wu, H., Gong, M. and Su, Q. (2005). Luminescence of $(Li_{0.333}Na_{0.334}K_{0.333})$ $Eu(MoO_4)_2$ and its application in near UV InGaN-based light-emitting diode, *Chem. Phys. Lett.*, **412**, pp. 313–316.

59. Wang, J., Jing, X., Yan, C., Lin, J. and Liao, F. (2006). Influence of fluoride on f–f transitions of Eu^{3+} in $LiEuM_2O_8$ (M=Mo, W), *J. Lumin.*, **121**, pp. 57–61.

60. Sivakumar, V. and Varadaraju, U. V. (2005). Intense red-emitting phosphors for white LEDs, *J. Electrochem. Soc.*, **152**(10), pp. H168–H169.

61. Shi, F., Meng, J. and Ren, Y. (1997). Structure and luminescent properties of some new $AgLnW_2O_8$ compounds, *J. Mater. Chem.*, **7**, pp. 773–776.

62. Pujol, M. C., Solé, R., Massons, J., Gavaldà, J., Solans, X., Zaldo, C., Díaz, F. and Aguiló, M. (2001). Structural study of monoclinic $KGd(WO_4)_2$ and effects of lanthanide substitution, *J. Appl. Cryst.*, **34**(1), pp. 1–6.

63. Macalik, L., Hanuza, J. and Kaminskii, A. A. (2000). Polarized Raman spectra of the oriented $NaY(WO_4)_2$ and $KY(WO_4)_2$ single crystals, *J. Mol. Struct.*, **555**, pp. 289–297.

64. Shi, F., Meng, J. and Ren, Y. (1996). Structure and luminescent properties of three new silver lanthanide molybdates, *J. Solid State Chem.*, **121**, pp. 236–239.

65. Sivakumar, V. and Varadaraju, U. V. (2006). Intense red phosphor for white LED based on blue GaN LEDs, *J. Electrochem. Soc.*, **153**(3), pp. H54–H57.

66. Morozov, A. V., Bertha, A., Meert, K. W., Rompaey, S. V., Batuk, D., Martinez, G. T., Aert, S. V., Smet, P. F., Raskina, M. V., Poelman, D., Abakumov, A. M. and Hadermann, J. (2013). Incommensurate modulation and

luminescence in the $CaGd_{2(1-x)}Eu_{2x}(MoO_4)_{4(1-y)}(WO_4)_{4y}$ ($0 \leq x \leq 1$, $0 \leq y \leq 1$) red phosphors, *Chem. Mater.*, **25**(21), pp. 4387–4395.

67. Guo, C., Yang, H. K. and Jeong, J.-H. (2010). Preparation and luminescent properties of phosphor $MGd_2(MoO_4)_4$:Eu^{3+} (M = Ca, Sr and Ba), *J. Lumin.*, **130**, p. 1390.

68. Haque, M. M. and Kim, D. K. (2009). Luminescent properties of Eu^{3+} activated $MLa_2(MoO_4)_4$ based (M = Ba, Sr and Ca) novel red-emitting phosphors, *Mater. Lett.*, **63**, p. 793.

69. Li, H., Yang, H. K., Moon, B. K., Choi, B. C., Jeong, J. H., Jang, K., Lee, H. S. and Yi, S. S. (2012). Structure dependence of the photoluminescence properties of Eu^{3+} ion-activated $R_2Mo_4O_{15}$ (R = Y, La, and Gd), *J. Electrochem. Soc.*,**159**, p. J61.

70. Abakumov, A. M., Morozov, V. A., Tsirlin, A. A., Verbeeck, J. and Hadermann, J. (2014). Cation ordering and flexibility of the BO_4 2– tetrahedra in incommensurately modulated $CaEu_2(BO_4)_4$ (B = Mo, W) scheelites, *Inorg. Chem.*, **53**(17), pp. 9407–9415.

71. Klevtsova, R. F. and Borisov, S. V. (1967). X-ray diffraction studies of $KY(MoO_4)_2$ double molybdate, *Dokl. Acad. Nauk SSSR*, **177**, pp. 1333–1336 (in Russian).

72. Powder Diffraction File, Cards 32-0804, 33-1035 and 51-1846 JCPDS; International Centre for Diffraction Data, 1601 Park Lane, Swarthmore, PA 19081.

73. Cavalli, E., Meschini, C., Toncelli, A., Tonelli, M. and Bettinelli, M. (1997). Optical spectroscopy of Tm^{3+} doped in $KLa(MoO_4)_2$ crystals, *J. Phys. Chem. Solids*, **58**, pp. 587–595.

74. Potapova, O. G., Protasova, V. I. and Kharchenko, L. Yu. (1987). X-ray diffraction study of $KLa(MoO_4)_2$ crystals prepared by hydrothermal method, *Zh. Neorg. Khim.*, **32**, pp. 2933–2938 (in Russian); *Russ. J. Inorg. Chem. (Transl. Zh. Neorg. Khim.)*, **32**, p. 1703 (in English).

75. Klevtsov, P. V., Kozeeva, L. P. and Pavlyuk, A. A. (1975). Polymorphism and crystallization of potassium-rare earth element molybdates $KLn(MoO_4)_2$ (Ln = La, Ce, Pr and Nd), *Crystallogr. Rep.*, **20**, pp. 1216–1220 (in Russian).

76. Morozov, V. A., Arakcheeva, A. V., Chapuis, G., Guiblin, N., Rossell, M. D. and Van Tendeloo, G. (2006). $KNd(MoO_4)_2$: a new incommensurate modulated structure in the scheelite family, *Chem. Mater.*, **18**, pp. 4075–4082.

77. Arakcheeva, A., Pattison, P., Chapuis, G., Rossell, M., Filaretov, A., Morozov, V. and Van Tendeloo, G. (2008). $KSm(MoO_4)_2$, an

incommensurately modulated and partially disordered scheelite-like structure, *Acta Crystallogr., Sect. B: Struct. Sci.*, **B64**, pp. 160–171.

78. Klevtsova, R. F., Kozeeva, L. P. and Klevtsov, P. V. (1974). Preparation and structure of the crystals of potassium-europium molybdate $KEu(MoO_4)_2$. *Crystallogr. Rep.*, **19**, pp. 89–94 (in Russian).

79. Yi, L., Zhou, L., Wang, Z., Sun, J., Gong, F., Wan, W. and Wang, W. (2010). $KGd(MoO_4)_2:Eu^{3+}$ as a promising red phosphor for light-emitting diode application, *Curr. Appl. Phys.*, **10**, pp. 208–213.

80. Morozov, V. A, Arakcheeva, A. V., Pattison, P., Meert, K. W., Smet, P. F., Poelman, D., Gauquelin, N., Verbeeck, J., Abakumov, A. M. and Hadermann, J. (2015). Energy transfer in Eu^{3+} doped silicates: use as thermographic phosphor, *Chem. Mater.*, **27**(16), pp. 5519–5530.

81. Cheng, F., Xia, Z., Molokeev, M. S. and Jing, X. (2015). Effects of composition modulation on the luminescence properties of Eu^{3+} doped $Li_{1-x}Ag_xLu(MoO_4)_2$ solid-solution phosphors, *Dalton Trans.*, **44**, p. 18078.

82. Yang, Z., Weitao, G. and Guiling, N. (2016). Novel red-emitting $LiGd(WO_4)_2:Eu^{3+}$ phosphor with high thermal stability and high color purity for application in white light-emitting diodes, *New J. Chem.*, **40**, pp. 10136–10143.

83. Lei, Z., Shanshan, H., Xianju, Z., Jianfeng, T. and Jun, Y. (2016). One-step surfactant-free synthesis of Eu^{3+}-activated $NaTb(MoO_4)_2$ microcrystals with controllable shape and their multi color luminescence properties, *Cryst. Eng.*, **18**, pp. 7590–7600.

84. Zhang, Y., Gong, W., Yu, J., Lin, Y. and Ning, G. (2015). Tunable white-light emission via energy transfer in single-phase $LiGd(WO_4)_2:Re^{3+}$ (Re = Tm, Tb, Dy, Eu) phosphors for UV-excited WLEDs, *RSC Adv.*, **5**, pp. 96272–96280.

85. Stedman, N. J., Cheetham, A. K. and Battle, P. D. (1994). Crystal structures of two sodium yttrium molybdates: $NaY(MoO_4)_2$and $Na_5Y(MoO_4)_4$, *J. Mater. Chem.*, **4**, p. 707.

86. Wang, Z., Liang, H., Wang, J., Gong, M. and Su, Q. (2006). Red-light-emitting diodes fabricated by near-ultraviolet InGaN chips with molybdate phosphors, *Appl. Phys. Lett.*, **89**, p. 071921.

87. Chiu, C.-H., Liu, C.-H., Huang, S.-B. and Chena, T.-M. (2008). Synthesis and luminescence properties of intensely red-emitting $M_5Eu(WO_4)_{4-x}(MoO_4)_x$(M = Li , Na , K) phosphors, *J. Electrochem. Soc.*, **155**(3), pp. J71–J78.

88. Qin, D. and Tang, W. (2016). Energy transfer and multicolour emission in single-phase $Na_5Ln(WO_4)_{4-z}(MoO_4)_z$:$Tb^{3+}$,$Eu^{3+}$ (Ln = La, Y, Gd) phosphors, *RSC Adv.*, **6**, p. 45376.

89. Litterscheid, C., Kruger, S., Euler, M., Dreizler, A., Wickleder, C. and Albert, B. (2016). Solid solution between lithium-rich yttrium and europium molybdate as new efficient red-emitting phosphors, *J. Mater. Chem. C*, **4**, p. 596.

90. Ipatova, E. N., Klevtsova, R. F., Solov'eva, L. P. and Klevtsov, P. V. (1982). Thermal stability and crystal structure of the double molybdate $Li_7Ho_3(MoO_4)_8$, *J. Struct. Chem.*, **23**, pp. 418–422.

91. Zhu, H. M., Lin, Y. F., Chen, Y. J., Gong, X. H., Tan, Q. G., Luo, Z. D. and Huang, Y. D. (2007). Polarized spectral characterization and laser demonstration of Nd^{3+}: $Li_2Gd_4(MoO_4)_7$ crystal, *J. Appl. Phys.*, **102**, p. 063104.

92. Zhu, H. M., Chen, Y. J., Lin, Y. F., Gong, X. H., Liao, J. S., Chen, X. Y., Luo, Z. D. and Huang, Y. D. (2007). Polarized spectral properties of Yb^{3+}: $Li_2Gd_4(MoO_4)_7$ crystal: a candidate for tunable and ultrashort pulse lasers, *J. Phys. D: Appl. Phys.*, **40**, pp. 6936–6941.

93. Zhao, W., Lin, Z. B., Zhang, L. Z. and Wang, G. F. (2011). Growth and spectroscopic properties of Nd^{3+}-doped $Na_2Gd_4(MoO_4)_7$ crystal, *J. Alloys Compd.*, **509**, pp. 2815–2818.

94. Zhao, W., Zhang, L. Z. and Wang, G. F. (2009). Growth, thermal and spectral characterization of Er^{3+}-doped $Li_2Gd_4(MoO_4)_7$crystal, *J. Cryst. Growth*, **311**, p. 2336.

95. Zhao, C., Yin, X., Huang, F. and Hang, Y. (2011). Synthesis and photoluminescence properties of the high-brightness Eu^{3+}-doped $M_2Gd_4(MoO_4)_7$ (M=Li, Na) red phosphors, *J. Solid State Chem.*, **184**, pp. 3190–3194.

96. Brixner, L. H. (1975). $Li_{0.286}Gd_{0.571\ 0.143}MoO_4$: a scheelite type solid solution, *J. Phys. Soc. Jpn.*, **38**, p. 1218.

97. Jingjing, R., Feiyun, G. and Jianzhong, C. (2013). Synthesis and luminescence characteristics of $Li_2Y_{4-x}Eu_x(WO_4)_{7-y}(MoO4)_y$ red-emitting phosphor for white LED, *J. Rare Earths*, **31**(11), p. 1059.

98. Thomas, M., Rao, P. P., Deepa, M., Chandran, M. R. and Koshy, P. (2009). Novel powellite-based red-emitting phosphors: $CaLa_{1-x}NbMoO_8$:xEu^{3+} for white light emitting diodes, *J. Solid State Chem.*, **182**, pp. 203–207.

99. Geng, J., Yu, C., Wu, H., Wu, B. and Tian, L. (2013). Photoluminescence characteristics of red-emitting Eu^{3+}-activated $Ca_4GdNbMo_4O_{20}$ phosphor, *J. Lumin.*, **140**, pp. 71–73.

100. Nair, K. R., Rao, P. P., Sameera, S., Mohan, V. S., Chandran, M. R. and Koshy, R. (2008). New powellite type oxides in Ca-R-Nb-Mo-0 system (R=Y, La, Nd, Sm or Bi): their synthesis, structure and dielectric properties, *Mater. Lett.*, **62**, p. 2868.

101. Janulevicius, M., Marmokas, P., Misevicius, M., Grigorjevaite, J., Mikoliunaite, L., Sakirzanovas, S. and Katelnikovas, A. (2016). Luminescence and luminescence quenching of highly efficient $Y_2Mo_4O_{15}:Eu^{3+}$ phosphors and ceramics, *Sci. Rep.*, **6**, p. 26098.

102. Klevtsova, R. F., Solodovnikov, S. F., Tushinova, Y. L., Bazarov, B. G., Glinskaya, L. A. and Bazarova, Z. G. (2000). A new type of mixed framework in the crystal structure of binary molybdate $Nd_2Zr_3(MoO_4)_9$, *J. Struct. Chem.*, **41**, p. 280.

103. Qi, S., Huang, Y., Cheng, H. and Seo, H. J. (2016). Luminescence and application of red-emitting phosphors of Eu^{3+}-activated $R_2Zr_3(MoO_4)_9$ (R = La, Sm, Gd), *Electron. Mater. Lett.*, **12**, pp. 171–177.

104. Qin, L., Wei, D., Huang, Y., Kim, S. I., Yu, Y. M. and Seo, H. J. (2013). Efficient and thermally stable red luminescence from nano-sized phosphor of $Gd_6MoO_{12}:Eu^{3+}$, *J. Nanopart. Res.*, **15**, pp. 1940–1948.

105. Qin, L., Huang, Y. L., Tsuboi, T. and Seo, H. J. (2012). The red-emitting phosphors of Eu^{3+}-activated $MR_2(MoO_4)_4$ (M = Ba, Sr, Ca; R = La^{3+}, Gd^{3+}, Y^{3+}) for light emitting diodes, *Mater. Res. Bull.*, **47**, pp. 4498–4502.

106. Huang, Y. L., Nakai, Y., Tsuboi, T. and Seo, H. J. (2011). The new red-emitting phosphor of oxyfluoride $Ca_2RF_4PO_4:Eu^{3+}$ (R=Gd, Y) for solid state lighting applications, *Opt. Mater. Express*, **19**, pp. 6303–6311.

107. Yanlin, H. and Hyo, J. S. (2012). A novel red-emitting nano-phosphor of Eu^{3+}-doped $LaBWO_6$, *Mater. Lett.*, **84**, pp. 107–109.

108. Dong, S., Ye, S., Wang, L., Chen, X., Yang, S., Zhao, Y., Wang, J., Jing, X. and Zhang, Q. (2014). $Gd_3B(W,Mo)O_9$: Eu^{3+} red phosphor: from structure design to photoluminescence behaviour and near-UV white-LEDs performance, *J. Alloys Compd.*, **610**, pp. 402–408.

109. Huang, J., Hou, B., Ling, H., Liu, J. and Yu, X. (2014). Crystal structure, electronic structure, and photoluminescence properties of $La_3BW_{1-x}Mo_xO_9:Eu^{3+}$ red phosphor, *Inorg. Chem.*, **53**, pp. 9541–9547.

110. Alonso, J. A., Rivillas, F., Martinez-Lope, M. J. and Pomjakushin, V. (2004). Preparation and structural study from neutron diffraction data of R_2MoO_6 (R=Dy, Ho, Er, Tm, Yb, Y), *J. Solid State Chem.*, **177**, pp. 2470–2476.

111. Han, B., Zhang, J., Wang, Z., Liu, Y. and Shi, H. (2014). Investigation on the concentration quenching and energy transfer of red-light-emitting phosphor $Y_2MoO_6:Eu^{3+}$, *J. Lumin.*, **149**, pp. 150–154.

112. Li, H., Yang, H. K., Jeong, J. H., Jang, K., Lee, H. S. and Yi, S. S. (2011). Sol–gel synthesis, structure and photoluminescence properties of nanocrystalline Lu_2MoO_6:Eu, *Mater. Res. Bull.*, **46**, pp. 1352–1358.

113. Li, L., Shen, J., Zhou, X.-J., Pan, X., Chang, W.-X., He, Q. W. and Wei, X.-T. (2016). Luminescent properties of $Lu_2MoO_6:Eu^{3+}$ red phosphor for solid state lighting, *Nonferrous Met. Soc. China*, **26**, pp. 1670–1675.

114. Dexter, D. L. (2009). A theory of sensitized luminescence in solids, *J. Chem. Phys.*, **17**, pp. 7138–7144.

115. Dexter, D. L. and James, H. S. (1954). Theory of concentration quenching in inorganic phosphor, *J. Chem. Phys.*, **22**, pp. 1063–1070.

116. Beaury, O., Faucher, M. and Caro, P. (1978). Crystal structure and fluorescence spectrum of $3Y_2O_3$, $WO_3:Eu^{3+}$, *Mater. Res. Bull.*, **13**, pp. 175–185.

117. Ma, J., Yoshimura, M., Kakihana, M. and Yashima, M. (1998). Synthesis of ZrO_2-Y_6WO_{12} solid solution powders by a polymerized complex method, *J. Mater. Res.*, **13**, pp. 939–943.

118. Escolástico, S., Vert, V. B. and Serra, J. M. (2009). Preparation and characterization of nanocrystalline mixed proton-electronic conducting materials based on the system Ln_6WO_{12}, *Chem. Mater.*, **21**, pp. 3079–3089.

119. Chevire, F., Clabau, F., Larcher, O., Orhan, E., Tessier, F. and Marchand, R. (2009). Tunability of the optical properties in the $Y_6(W,Mo)(O,N)_{12}$ system, *Solid State Sci.*, **11**, pp. 533–536.

120. Xue, B. and Sun, J. (2013). Synthesis and tuning orange to green up-conversion color in Y_6WO_{12}: Er/Yb phosphor, *Opt. Mater.*, **36**, p. 278.

121. Koh, J.-H., Sorge, E., Wen, T.-C. and Shetty, D. K. (2013). Thermal expansion behaviors of yttrium tungstates in the WO_3-Y_2O_3 system, *Ceram. Int.*, **39**, pp. 8421–8427.

122. Chevire, F., Munoz, F., Baker, C. F., Tessier, F., Larcher, O., Boujday, S., Colbeau-Justin, C. and Marchand, R. (2006). UV absorption properties of ceria-modified compositions within the fluorite-type solid solution CeO_2–Y_6WO_{12}, *J. Solid State Chem.*, **179**, pp. 3184–3190.

123. Tian, L., Yang, P., Wu, H. and Li, F. (2010). Luminescence properties of $Y_2WO_6:Eu^{3+}$ incorporated with Mo^{6+} or Bi^{3+} ions as red phosphors for light-emitting diode applications, *J. Lumin.*, **130**, pp. 717–721.

124. Li, H., Yang, H. K., Moon, B. K., Choi, B. C., Jeong, J. H., Jang, K., Lee, H. S. and Yi, S. S. (2011). Investigation of the structure and photoluminescence properties of Eu^{3+}ion-activated $Y_6W_xMo_{1-x}O_{12}$, *J. Mater. Chem.*, **21**, pp. 4531–4537.

125. Aitken, E. A., Bartram, S. F. and Juenke, E. F. (1964). Crystal chemistry of the rhombohedral $MO_3.3R_2O_3$ compounds, *Inorg. Chem.*, **3**, pp. 949–954.

126. Zhang, J., Von Dreele, R. B. and Eyring, L. (1993). The structures of Tb_7O_{12} and $Tb_{11}O_{20}$, *J. Solid State Chem.*, **104**, pp. 21–32.

127. Bartram, S. F. (1966). Crystal structure of the rhombohedral $MO_3.3R_2O_3$ compounds (M = U, W, or Mo) and their relation to ordered R_7O_{12} phases, *Inorg. Chem.*, **5**, pp. 749–754.

128. Li, H. Y., Noh, H. M., Moon, B. K., Choi, B. C., Jeong, J. H., Jang, K., Lee, H. S. and Yi, S. S. (2013). Wide-band excited $Y_6(WMo)_{0.5}O_{12}$: Eu red phosphor for white light emitting diode: structure evolution, photoluminescence properties, and energy transfer mechanisms involved, *Inorg. Chem.*, **52**, pp. 11210–11217.

129. Yu, R., Guo, Y., Wang, L., Mi Noh, H., Kee Moon, B., Chun Choi, B. and Hyun Jeong, J. (2014). Characterizations and optical properties of orange–red emitting Sm^{3+}-doped Y_6WO_{12} phosphors, *J. Lumin.*, **155**, pp. 317–321

130. Pauling, L. (1924). The crystal structures of ammonium fluoferrate, fluo-aluminate and oxyfluomolybdate, *J. Am. Chem. Soc.*, **46**, pp. 2738–2751.

131. Steward, E. G. and Rooksby, H. P. (1951). Pseudo-cubic alkaline-earth tungstates and molybdates of the R_3MX_6 type, *Acta. Cryst.*, **4**, pp. 503–507.

132. Naray-Szabo, S. and Sasvari, K. (1938). The structure of cryolite Na_3AlF_6, *Z. Kristallogr.*, **99**, pp. 27–31.

133. Blasse, G. (1965). Magnetic properties of mixed metal oxides containing trivalent cobalt, *J. Appl. Phys.*, **36**, p. 879.

134. Brixner, L. H. (1958). Preparation and structure of the strontium and barium tantalates $Sr_3TaO_{5.5}$ and $Ba_3TaO_{5.5}$, *J. Am. Chem. Soc.*, **80**, pp. 3214–3215.

135. Thangadurai, V., Shukla, A. K. and Gopalakrishnan, J. (1997). Oxide-ion conduction in anion-deficient double perovskites, $Ba_2BB'O_{5.5}$ (B = Li, Na; B′ = Mo, W, Te), *Solid State Ionics*, **104**, pp. 277–283.

136. Anderson, M. T., Greenwood, K. B., Taylor, G. A. and Poeppelmeier, K. R. (1993). B-cation arrangements in double perovskites, *Prog. Solid State Chem.*, **22**(3), pp. 197–233.

137. Teraoka, Y., Wei, M. and Kagawa, S. (1998). Double perovskites containing hexavalent molybdenum and tungsten: synthesis,

structural investigation and proposal of a fitness factor to discriminate the crystal symmetry, *J. Mater. Chem.*, **8**(11), pp. 2323–2325.

138. Blasse, G. and Corsmit, A. F. (1973). Electronic and vibrational spectra of ordered perovskites, *J. Solid State Chem.*, **6**, pp. 513–518.

139. Shannon, R. D. and Prewitt, C. T. (1969). Effective ionic radii in oxides and fluorides, *Acta Cryst.*, **B25**, pp. 925–945.

140. Bode, J. H. G. and Van Oosterhout, A. B. (1975). Defect luminescence of ordered perovskites A_2BWO_6, *J. Lumin.*, **10**, pp. 237–242.

141. Bleijenberg, K. C. and Blasse, G. (1975/76). Luminescence of lead-containing tungstates with perovskite structure, *J. Lumin.*, **11**, pp. 279–284.

142. de Hair, J. Th. W. and Blasse, G. (1976). The luminescence properties of the octahedral urinate group in oxides with perovskite structure, *J. Solid State Chem.*, **19**, p. 263.

143. Bleijenberg, K. C. and Blasse, G. (1976). The influence of deviations from 1:1 order in perovskite on optical properties, *J. Solid State Chem.*, **17**, pp. 71–74.

144. Galasso, F. and Darby, W. (1962). Preparation, structure, and properties of K_2NbO_3, *J. Phys. Chem.*, **66**, pp. 1318–1320.

145. Weber, M. J. and Scehaufele, R. F. (1965). Vibronic spectrum of Eu3+ in strontium titanate, *Phys. Rev.*, **138**, p. A1544.

146. Blasse, G. (1966). Compounds with α-PbO_2 structure, *Z. Anorg. Allg. Chem.*, **345**, pp. 222–224.

147. Toda, K., Honma, T. and Sato, M. (1997). Unusual concentration of europium luminescence in new layered perovskite compound, $RbLa_{1-x}Eu_xTa_2O_7$ ($0 \leq x \leq 1$), *J. Lumin.*, **71**, pp. 71–77.

148. Honma, T., Toda, K., Ye, Z.-G. and Sato, M. (1998). Concentration quenching of the Eu^{3+}-activated luminescence in some layered perovskites with two-dimensional arrangement, *J. Phys. Chem. Solids*, **59**, pp. 1187–1193.

149. Kobayashi, K.-I., Kimura, T., Sawada, H., Terakura, K. and Tokura, Y. (1998). Room-temperature magnetoresistance in an oxide material with an ordered double-perovskite structure, *Nature*, **395**, pp. 677–680.

150. Huang, Y., Dass, R. I., Xing, Z. and Goodenough, J. B. (2006). Double perovskites as anode materials for solid-oxide fuel cells, *Science*, **312**, pp. 254–257.

151. Sivakumar, V. and Varadaraju, U. V. (2006). A promising orange-red phosphor under near UV excitation, *Electrochem. Solid-State Lett.*, **9**, pp. H35–H38.

152. Sivakumar, V. and Varadaraju, U. V. (2007). An orange-red phosphor under near-UV excitation for white light emitting diodes, *J. Electrochem. Soc.*, **154**, pp. J28–J31.

153. Sivakumar, V. and Varadaraju, U. V. (2008). Synthesis, phase transition and photoluminescence studies on Eu^{3+}-substituted double perovskites—a novel orange-red phosphor for solid-state lighting, *J. Solid State Chem.*, **181**(12), pp. 3344–3351.

154. Ye, S., Wang, C. H., Liu, Z. S., Lu, J. and Jing, X.-P. (2008). Photoluminescence and energy transfer of phosphor series $Ba_{2-z}Sr_zCaMo_{1-y}W_yO_6$: Eu, Li for white light UV-LED applications, *Appl. Phys. B*, **91**, pp. 551–557.

155. Xia, Z., Sun, J., Du, H., Chen, D. and Sun, J. (2010). Luminescence properties of double-perovskite $Sr_2Ca_{1-2x}Eu_xNa_xMoO_6$ red-emitting phosphors prepared by the citric acid-assisted sol-gel method, *J. Mater. Sci.*, **45**, pp. 1553–1559.

156. Zhang, X., Li, Z., Zhang, H., Ouyang, S. and Zou, Z. (2009). Luminescence properties of Sr_2ZnWO_6:Eu^{3+} phosphors, *J. Alloys Compd.*, **469**, pp. L6–L9.

157. Li, H., Yang, H. K., Moon, B. K., Choi, B. C., Jeong, J. H., Jang, K., Lee, H. S. and Yi, S. S. (2011). Color-conversion and photoluminescence properties of $Ba_2MgW(Mo)O_6$:Eu phosphor, *J. Alloys Compd.*, **509**, pp. 8788–8793.

158. Zhao, X., Ding, Y., Yu, T. and Zou, Z. (2013). An efficient charge compensated red phosphor Sr_3WO_6:K^+,Eu^{3+}- For white LEDs, *J. Alloys Compd.*, **553**, pp. 221–224.

159. Zhang, S., Hu, Y., Chen, L., Wang, X., Ju, G. and Fan, Y. (2013). Photoluminescence properties of Ca_3WO_6:Eu^{3+} red phosphor, *J. Lumin.*, **142**, pp. 116–121.

160. Zhang, L., Han, P., Han, Y., Lu, Z., Yang, H., Wang, L. and Zhang, Q. (2013). $(Sr_{0.98-m}Ba_mEu_{0.02})_2Ca(Mo_{1-n}W_n)O_6$ phosphor with ultraviolet excitation for white LEDs, *J. Alloys Compd.*, **558**, pp. 229–235.

161. Sun, X., Hao, Z., Li, C., He, X., Qi, H., Yu, L., Luo, Y., Zhang, J., Gao, J. and Zhong, R. (2013). Enhanced orange-red emission by using Mo codoped in Ba_2CaWO_6:Eu^{3+},Li^+ phosphor under near UV excitation, *J. Lumin.*, **134**, pp. 191–194.

162. Zhao, X., Wang, J., Fan, L., Ding, Y., Li, Z., Yu, T. and Zou, Z. (2013). Efficient red phosphor double-perovskite Ca_3WO_6 with A-site substitution of Eu^{3+}, *Dalton Trans.*, **42**, pp. 13502–13508.

163. Li, Y. and Liu, X. (2014). Structure and luminescence properties of $Ba_3WO_6:Eu^{3+}$ nanowire phosphors obtained by conventional solid-state reaction method, *Opt. Mater.*, **38**, pp. 211–216.

164. Dabre, K. V. and Dhoble, S. J. (2014). Synthesis and photoluminescence properties of Eu^{3+},Sm^{3+} and Pr^{3+} doped Ca_2ZnWO_6 phosphors for phosphor converted LED, *J. Lumin.*, **150**, pp. 55–58.

165. Khanna, A. and Dutta, P. S. (2015). Near-UV and blue wavelength excitable $Mg_{0.6}Ca_{2.16}Mo_{0.2}W_{0.8}O_6:Eu_{0.12}{}^{3+}/Na_{0.12}{}^{+}$ high efficiency red phosphors, *J. Solid State Chem.*, **225**, pp. 120–134.

166. Xiao, N., Shen, J., Xiao, T., Wu, B., Luo, X., Li, L., Wang, Z. and Zhou, X. (2015). $Sr_2CaW_xMo_{1-x}O_6:Eu^{3+},Li^{+}$: an emission tunable phosphor through site symmetry and excitation wavelength, *Mater. Res. Bull.*, **70**, pp. 684–690.

167. Li, Y. and Liu, X. (2015). Sol-gel synthesis, structure and luminescence properties of $Ba_2ZnMoO_6:Eu^{3+}$phosphors, *Mater. Res. Bull.*, **64**, pp. 88–92.

168. Sletnes, M., Lindgren, M., Valmalette, J. C., Wagner, N. P., Grande, T. and Einasrud, M.-A. (2016). Photoluminescence of A- and B-site Eu^{3+}-substituted $(Sr_xBa_{1-x})_2CaW_yMo_{1-y}O_6$ phosphors, *J. Solid State Chem.*, **237**, pp. 72–80.

169. Li, L., Pan, Y., Zhou, X., Zhao, C., Wang, Y., Jiang, S., Suchocki, A. and Brik, M. G. (2016). Luminescence enhancement in the $Sr_2ZnW_{1-x}Mo_xO_6:Eu^{3+}$, Li^{+} phosphor for near ultraviolet based solid state lighting, *J. Alloys Compd.*, **685**, pp. 917–926.

170. Sivakumar, V. (2007). Investigation of lanthanide luminescence in mixed metal oxides and silicates, PhD Thesis, IIT Madras.

171. Liu, Q., Shen, J., Xu, T., Wang, L., Zhang, J., Zhang, Q. and Zhang, L. (2016). Enhanced luminescence of a Eu^{3+}-activated double perovskite (Na, Li) $LaMgWO_6$ phosphor based on a site inducing energy transfer, *Ceram. Int.*, **42**, pp. 13855–13862.

172. Li, L., Chang, W., Chen, W., Feng, Z., Zhao, C., Jiang, P., Wang, Y., Zhou, X. and Suchocki, A. (2016). Double perovskite $LiLaMgWO_6:Eu^{3+}$ novel red-emitting phosphors for solid-state lighting: synthesis, structure and photoluminescent properties, *Ceram. Int.*, **43**, pp. 2720–2729.

173. Knapp, M. C. and Woodward, P. M. (2006). A-site cation ordering in AA′BB′O6 perovskites, *J. Solid State Chem.*, **179**, pp. 1076–1085.

174. King, G. and Woodward, P. M. (2010). Cation ordering in perovskites, *J. Mater. Chem.*, **20**, pp. 5785–5796.

175. Garcia-Martin, S., Urones-Garrote, E., Knapp, M. C., King, G. and Woodward, P. M. (2008). Transmission electron microscopy studies of $NaLaMgWO_6$: spontaneous formation of compositionally modulated stripes, *J. Am. Chem. Soc.*, **130**, pp. 15028–15037.

176. Sharits, A. R., Khoury, J. F. and Woodward, P. M. (2016). Evaluating $NaREMgWO_6$ (RE = La, Gd, Y) doubly ordered double perovskites as Eu^{3+} phosphor hosts, *Inorg. Chem.*, **55**(23), pp. 12383–12390.

177. Yin, X., Wang, Y., Huang, F., Xia, Y., Wan, D. and Yao, J. (2011). Excellent red phosphors of double perovskite Ca_2LaMO_6:Eu (M=Sb,Nb,Ta) with distorted coordination environment, *J. Solid State Chem.*, **184**, pp. 3324–3328.

178. Hou, J., Yin, X., Fang, Y. Z., Huang, F. and Jiang, W. (2012). Novel red-emitting perovskite-type phosphor $CaLa_{1-x}MgMoO_6$: xEu^{3+} (M0 = Nb, Ta) for white LED application, *Opt. Mater.*, **34**, pp. 1394–1397.

179. Guo, Y., Moon, B. K., Park, S. H., Jeong, J. H., Kim, J. H., Jang, K. and Yu, R. (2015). A red-emitting perovskite-type $SrLa_{(1-x)}MgTaO_6$:xEu^{3+} for white LED application, *J. Lumin.*, **167**, pp. 381–385.

180. Zhu, J., Xia, Z., Zhang, Y., Molokeev, M. S. and Liu, Q. (2015). Structural phase transitions and photoluminescence properties of Eu^{3+} doped $Ca_{(2-x)}Ba_xLaNbO_6$ phosphors, *Dalton Trans.*, **44**, pp. 18536–18543.

181. Jiang, P., Zhou, Z., Gao, W., Cong, R. and Tao, Y. (2016). B-site ordered double perovskite $LaBa_{1-x}Sr_xZnSbO_6$ ($0 \leq x \leq 1$): Sr^{2+}-doping-induced symmetry evolution and structure–luminescence correlations, *Dalton Trans.*, **45**, pp. 3949–3957.

182. Subramanian, M. A., Aravamudan, G. and Subba Rao, G. V. (1983). Oxide pyrochlores: a review, *Prog. Solid State Chem.*, **15**, pp. 55–143.

183. Garbout,A., Kallel-Kchaou, N. and Férid, M. (2016). Relationship between the structural characteristics and photoluminescent properties of $LnEuTi_2O_7$ (Ln=Gd and Y) pyrochlores, *J. Lumin.*, **169**, pp. 359–366.

184. Hirayama, M., Sonoyama, K., Yamada, A. and Kanno, R. (2008). Relationship between structural characteristics and photoluminescent properties of $(La_{1-x}Eu_x)_2M_2O_7$(M = Zr, Hf, Sn) pyrochlores, *J. Lumin.*, **128**, pp. 1819–1825.

185. Cai, L. and Nino, J. C. (2009). Complex ceramic structures. I. Weberites, *Acta Crystallogr., Sect. B: Struct. Sci.*, **65**, p. 269.

186. McCauley, R. A. and Hummel, F. A. (1973). Luminescence as an indication of 4+ distortion in $A_2^{3+}B_2^{4+}O_7$ type pyrochlores, *J. Lumin.*, **6**, p. 105.

187. Li, K., Li, H., Zhang, H., Yu, R., Wang, H. and Yan, H. (2006). Hydrothermal synthesis of Eu^{3+}-doped $Y_2Sn_2O_7$ nanocrystals, *Mater. Res. Bull.*, **41**, pp. 191–197.

188. Fujihara, S. and Tokumo, K. (2005). Multiband orange-red luminescence of Eu3+ ions based on the pyrochlore-structured host crystal, *Chem. Mater.*, **17**, pp. 5587–5593.

189. Gupta, S. K., Reghukumar, C. and Kadam, R. M. (2016). Eu^{3+} local site analysis and emission characteristics of novel $Nd_2Zr_2O_7$:Eu phosphor: insight into the effect of europium concentration on its photoluminescence properties, *RSC Adv.*, **6**, pp. 53614–53624.

190. Kahn-Harari, A., Mazerrolles, L., Michel, D. and Robert, F. (1995). Structural description of La_3NbO_7, *J. Solid State Chem.*, **116**, pp. 103–106.

191. Hinatsu, Y., Ebisawa, H. and Doi, Y. (2009). Magnetic properties of orthorhombic fluorite-related oxides Ln_3SbO_7 (Ln = rare earths), *J. Solid State Chem.*, **182**, pp. 1694–1699.

192. Fennell, T., Bramwell, S. T. and Green, M. A. (2001). Structural and magnetic characterization of Ho_3SbO_7 and Dy_3SbO_7, *Can. J. Phys.*, **79**, p. 1415.

193. Wang, J., Cheng, Y., Huang, Y., Cai, P., Kim, S. and Seo, H. J. (2014). Structural and luminescent properties of red-emitting Eu^{3+}-doped ternary rare earth antimonates R_3SbO_7 (R = La, Gd, Y), *J. Mater. Chem. C*, **2**, pp. 5559–5569.

194. Yokogawa, Y., Yoshimura, M. and Somiya, S. (1989). Lattice energy and polymorphism of fluorite-related rare earth-tantalum double oxides, *Solid State Ionics*, **35**, pp. 275–279.

195. Cai, L., Denev, S., Gopalan, V. and Nino, J. C. (2010). Phase transition in weberite-type Gd_3NbO_7, *J. Am. Ceram. Soc.*, **93**, pp. 875–880.

196. Francis, T. L., Rao, P. P., Thomas, M., Mahesh, S. K., Reshmi, V. R. and Sreena, T. S. (2014). Structural influence on the photoluminescence properties of Eu^{3+} doped Gd_3MO_7 (M = Nb, Sb, Ta) red phosphors, *Phys. Chem. Chem. Phys.*, **16**, pp. 17108–17115.

197. Francis, T. L., Rao, P. P., Mahesh, S. K., Sreena, T. S. and Parvathi Babu, S. (2016). Effect of host structure on the photoluminescence properties of Ln_3TaO_7:Eu^{3+} red phosphors, *Opt. Mater.*, **52**, pp. 134–143.

198. Raj, A. K. V., Rao, P. P., Sreena, T. S., Sameera, S., James, V. and Renju, U. A. (2014). Remarkable changes in the photoluminescent properties of $Y_2Ce_2O_7$:Eu^{3+} red phosphors through modification of the cerium oxidation states and oxygen vacancy ordering, *Phys. Chem. Chem. Phys.*, **16**, pp. 23699–23710.

199. Murugan, R., Thangadurai, V. and Weppner, W. (2007). Fast lithium ion conduction in garnet-type $Li_7La_3Zr_2O_{12}$, *Angew. Chem. Int. Ed.*, **46**, pp. 7778–7781.

200. Katelnikovas, A., Sakirzanovas, S., Dutczak, D., Plewa, J., Enseling, D., Winkler, H., Kareiva, A. and Jüstel, T. (2013). Synthesis and optical properties of yellow emitting garnet phosphors for pcLEDs, *J. Lumin.*, **136**, pp. 17–25.

201. Sharma, S. K., Som, S., Jain, R. and Kunti, A. K. (2015). Spectral and CIE parameters of red emitting $Gd_3Ga_5O_{12}$:Eu^{3+}phosphor, *J. Lumin.*, **159**, pp. 317–324.

202. Murugan, R., Weppner, W., Schmid-Beurmann, P. and Thangadurai, V. (2008). Structure and lithium ion conductivity of garnet-like $Li_5La_3Sb_2O_{12}$ and $Li_6SrLa_2Sb_2O_{12}$, *Mater. Res. Bull.*, **43**, pp. 2579–2591.

203. Liu, F. X., Fang, Y. Z., Hou, J. S., Zhang, N. and Ma, Z. F. (2014). Garnet-based red emitting phosphors $Li_6MLa_2Nb_2O_{12}$: Eu^{3+}(M=Ca, Sr, Ba): photoluminescence improvement by changing crystal lattice, *Ceram. Int.*, **40**, pp. 3237–3241.

204. Bernuy-Lopez, C., Manalastas, W. J., Amo, J. M. L., Aguadero, A., Aguesse, F. and Kilner, J. A. (2014). Atmosphere controlled processing of Ga-substituted garnets for high Li-ion conductivity ceramics, *Chem. Mater.*, **26**, pp. 3610–3617.

205. Zeier, W. G., Zhou, S. L., Lopez-Bermudez, B., Page, K. and Melot, B. C. (2014). Dependence of the Li-ion conductivity and activation energies on the crystal structure and ionic radii in $Li_6MLa_2Ta_2O_{12}$, *ACS Appl. Mater. Interfaces*, **6**, pp. 10900–10907.

206. Liu, F., Fang, Y., Zhang, N., Hou, J., Ma, Z. and Zhao, G. (2014). Blue light excited $Li_6CaLa_2M_2O_{12}$:Eu^{3+} (*M*=Ta, Sb) red-emitting phosphors: structure and photoluminescence properties, *Ceram. Int.*, **40**, pp. 14781–14786.

207. Zhong, J., Chen, D., Zhao, W., Zhou, Y., Yu, H., Chen, L. and Ji, Z. (2015). Garnet-based $Li_6CaLa_2Sb_2O_{12}$:Eu^{3+} red phosphors: a potential color-converting material for warm white light-emitting diodes, *J. Mater. Chem. C*, **3**, pp. 4500–4510.

208. Wang, X., Zhao, Z., Wu, Q., Li, Y. and Wang, Y. (2016). A garnet-based $Ca_2YZr_2Al_3O_{12}$:Eu^{3+} red-emitting phosphor for n-UV light emitting diodes and field emission displays: electronic structure and luminescence properties, *Inorg. Chem.*, **55**, pp. 11072–11077.

209. Kozhevnikova, N. M. and Kopylova, O. A. (2011). Synthesis and X-ray diffraction and IR spectroscopy studies of ternary molybdates $Li_3Ba_2R_3(MoO_4)_8$ (R = La–Lu, Y), *Russ. J. Inorg. Chem.*, **56**, pp. 935–938.

210. Kozhevnikova, N. M. and Kopylova, O. A. (2011). Synthesis and study of $Li_3BaSrR_3(MoO_4)_8$ (R = La–Lu, Y) with stratified scheelite structure, *Russ. J. Inorg. Chem.*, **84**, pp. 1498–1501.

211. Katelnikovas, A., Plewa, J., Sakirzanovas, S., Dutczak, D., Enseling, D., Baur, F., Winkler, H., Kareiva, A. and Justel, T. (2012). Synthesis and optical properties of $Li_3Ba_2La_3(MoO_4)_8$:Eu^{3+} powders and ceramics for pcLEDs, *J. Mater. Chem.*, **22**, pp. 22126–22134.

212. Rico, M., Han, X., Cascales, C., Esteban-Betegon, F. and Zaldo, C. (2011). Efficient mid-infrared laser operation of $Li_3Lu_{3-x}Tm_xBa_2(MoO_4)_8$ disordered crystal, *Opt. Express*, **19**, pp. 7640–7645.

213. Song, M., Zhang, L. and Wang, G. (2009). Growth and spectral properties of Nd3+-doped $Li_3Ba_2Ln_3(MoO_4)$ (Ln = La, Gd) crystals, *J. Alloys Compd.*, **480**, p. 839.

214. Chang, Y. C., Liang, C. H., Yan, S. A. and Chang, Y. S. (2010). Synthesis and photoluminescence characteristics of high color purity and brightness $Li_3Ba_2Gd_3(MoO_4)_8$:Eu^{3+} red phosphors, *J. Phys. Chem. C*, **114**, pp. 3645–3652.

215. Wei, B., Liu, Z., Xie, C., Yang, S., Tang, W., Gu, A., Wong, W.-T. and Wong, K.-L. (2015). Fast synthesis of red $Li_3BaSrLn_3(WO_4)_8$:Eu^{3+}phosphors for white LEDs under near-UV excitation by a microwave-assisted solid state reaction method and photoluminescence studies, *J. Mater. Chem. C*, **3**, pp. 12322–12327.

216. Xie, R.-J., Hirosaki, N., Li, Y. and Takeda, T. (2010). Rare-earth activated nitride phosphors: synthesis, luminescence and applications, *Materials*, **3**, pp. 3777–3793.

217. Xin, M., Tu, D., Zhu, H., Luo, W., Liu, Z., Huang, P., Li, R., Cao, Y. and Chen, X. (2015). Single-composition white-emitting $NaSrBO_3$:Ce^{3+},Sm^{3+},Tb^{3+} phosphors for NUV light-emitting diodes, *J. Mater. Chem. C*, **3**, pp. 7286–7293.

218. Singh, K. and Vaidyanathan, S. (2016). Synthesis and optical properties of new red emitting phosphor $Li_3BaSrGd_{3-x}Eu_x(MO_4)_8$ for white LEDs, *ChemistrySelect*, **1**, pp. 5448–5462.

219. Oflet, G. S. (1962). Intensities of crystal spectra of rare-earth ions, *J. Chem. Phys.*, **3**, p. 37.

220. Judd, B. R. (1962). Optical absorption intensities of rare-earth ions, *Phys. Rev.*, **3**, p. 127.

221. Bettencourt-Dias, A. (2007). Lanthanide-based emitting materials in light-emitting diodes, *Dalton Trans.*, pp. 2229–2241.

222. Bunzli, J.-C. G. and Eliseeva, S. V. (2013). Intriguing aspects of lanthanide luminescence, *Chem. Sci.*, **4**, pp. 1939–1949.

223. Starck, M., Kadjane, P., Bois, E., Darbouret, B., Incamps, A., Ziessel, R. and Charbonnière, L. J. (2011). Towards libraries of luminescent lanthanide complexes and labels from generic synthons, *Chem. Eur. J.*, **17**, p. 9164.

224. Brunet, E., Juanes, O., Sedano, R. and Rodriguez-Ubis, J.-C. (2002). Luminescence of lanthanides in covalently pillared zirconium phosphate, *Photochem. Photobiol. Sci.*, **1**, pp. 613–618.

225. De, G. F., Malta, O. L., Doneg, C. D., Simas, A. M., Longo, R. L., Santa-Cruz, P. A. and da Silva, E. F. (2000). Spectroscopic properties and design of highly luminescent lanthanide coordination complexes, *Coord. Chem. Rev.*, **196**, pp. 165–195.

226. Binnemans, K. (2005). Rare-earth beta-diketonates, in *Handbook on the Physics and Chemistry of Rare Earths*, eds. Gschneidner Jr., K. A., Bünzli, J.-C. G. and Pecharsky, V. K. (Elsevier: Amsterdam), Volume **35**, Chapter 225, pp. 107–272.

227. Yang, C., Fu, L. M., Wang, Y., Zhang, J. P., Wong, W. T., Ai, X. C., Qiao, Y. F., Zou, B. S. and Gui, L. L. (2004). A highly luminescent europium complex showing visible-light-sensitizedred emission: direct observation of the singlet pathway, *Angew. Chem. Int. Ed.*, **43**, pp. 5010–5013.

228. Dossing, A. (2005). Luminescence from lanthanide(3+) ions in solution, *Eur. J. Inorg. Chem.*, pp. 1425–1434.

229. Yang, C., Fu, L. M., Wang, Y., Zhang, J. P., Wong, W. T., Ai, X. C., Qiao, Y. F., Zou, B. S. and Gui, L. L. (2004). A highly luminescent europium complex showing visible-light-sensitized red emission: direct observation of the singlet pathway, *Angew. Chem. Int. Ed.*, **43**, pp. 5010–5013.

230. Shi, M., Li, F., Yi, T., Zhang, D., Hu, H. and Huang, C. (2005). Tuning the triplet energy levels of pyrazolone ligands to match the $^5D_0$ level of europium(III), *Inorg. Chem.*, **44**, pp. 8929–8936.

231. Li, T., Shang, W., Zhang, F., Mao, L., Tang, C., Song, M., Du, C. and Wu, Y. (2011). Luminescent properties of europium complexes with different long chains in Langmuir-Blodgett (LB) films, *Engineering*, **3**, pp. 301–311.

232. Latva, M., Takalo, H., Mukkala, V. M., Matachescu, C., Rodriguez-Ubis, J. C. and Kankare, J. (1999). Correlation between the lowest triplet state energy level of the ligand and lanthanide(III) luminescence quantum yield, *J. Lumin.*, **75**, pp. 149–169.

233. Lee, K. M., Cheah, K. W., An, B. L., Gong, M. L. and Liu, Y. L. (2005). Emission characteristics of inorganic/organic hybrid white-light phosphor, *Appl. Phys. A*, **80**, p. 337.

234. Iwanaga, H., Amano, A., Aiga, F., Harada, K. and Oguchi, M. (2006). Development of ultraviolet LED devices containing europium (III) complexes in fluorescence layer, *J. Alloys Compd.*, **408–412**, pp. 921–925.

235. Xiang, N. J., Leung, L. M., So, S. K., Wang, J., Su, Q. and Gong, M. L. (2006). Red InGaN-based light-emitting diodes with a novel europium (III) tetrabasic complex as mono-phosphor, *Mater. Lett.*, **60**(23), pp. 2909–2913.

236. He, P., Wang, H., Liu, S., Hu, W., Shi, J., Wang, G. and Gong, M. (2009). An efficient europium(III) organic complex as red phosphor applied in LED, *J. Electrochem. Soc.*, **156**(2), pp. E46–E49.

237. He, P., Wang, H., Liu, S., Shi, J., Wang, G. and Gong, M. (2009). Luminescent dinuclear Eu(III) organic complex as a red-emitting phosphor for fabrication of LEDs, *Electrochem. Solid-State Lett.*, **12**(5), pp. B61–B64.

238. Luo, Y., Yan, Q., Zhang, Z., Yu, X., Wu, W., Su, W. and Zhang, Q. (2009). White LED based on poly(N-vinylcarbazole) and lanthanide complexes ternary co-doping system, *J. Photochem. Photobiol., A*, **206**, pp. 102–108.

239. Liu, S., He, P., Wang, H., Shi, J. and Gong, M. (2009). A luminescent quadruple stranded dinuclear Eu(III) complex based on 2,8-bis(40,40,40-trifluoro-10,30-dioxobutyl)-dibenzothiophene for light-emitting diodes, *Inorg. Chem. Commun.*, **12**, pp. 506–508.

240. Liu, S.-G., Gong, M.-L., Wang, S. and Tan, X.-M. (2009). A luminescent Eu(III) complex based on 2-(4′, 4′, 4′-trifluoro-1′, 3′-dioxobutyl)-dibenzothiophene for light-emitting diodes, *Spectrochim. Acta, Part A*, **74**, pp. 731–734.

241. He, P., Wang, H., Liu, S., Shi, J., Wang, G. and Gong, M. (2009). Effect of different alkyl groups at the N-position on the luminescence of carbazole-based diketonate europium(III) complexes, *J. Phys. Chem. A*, **113**, pp. 12885–12890.

242. Freeman, A. W., Urvory, M. and Criswell, M. E. (2009). Triphenyl phosphine-mediated reductive cyclization of 2-nitrobiphenyls: a practical and convenient synthesis of carbazoles, *Inorg. Chem.*, **48**, pp. 11382–11387.

243. He, P., Wang, H. H., Yan, H. G., Hu, W., Shi, J. X. and Gong, M. L. (2010). A strong red-emitting carbazole based europium(III) complex excited by blue light, *Dalton Trans.*, **39**, pp. 8919–8924.

244. Wang, H., He, P., Liu, S., Shi, J. and Gong, M. (2010). New multinuclear europium(III) complexes as phosphors applied in fabrication of near UV-based light-emitting diodes, *Inorg. Chem. Commun.*, **13**, pp. 145–148.

245. Wang, Z., Xiang, N., Wang, Q., Zhang, G. and Gong, M. (2010). A red europium (III) ternary complex for InGaN-based light-emitting diode, *J. Lumin.*, **130**, pp. 35–37.

246. Wang, H., He, P., Yan, H., Shi, J. and Gong, M. (2011). A novel europium(III)–imidazol–diketonate–phenanthroline complex as a red phosphor applied in LED, *Inorg. Chem. Commun.*, **14**, pp. 1183–1185.

247. Wang, H., Hea, P., Yana, H. and Gong, M. (2011). Synthesis, characteristics and luminescent properties of a new europium(III)organic complex applied in near UV LED, *Sens. Actuators, B*, **156**, pp. 6–11.

248. Sabbatini, N., Guardingli, M. and Lehn, J. M. (1993). Luminescent lanthanide complexes as photochemical supramolecular devices, *Coord. Chem. Rev.*, **123**, pp. 201–228.

249. Shao, G., Li, Y., Feng, K., Gan, F. and Gong, M. (2012). Diphenylethyne based β-diketonate europium (III) complexes as red phosphors applied in LED, *Sens. Actuators, B*, **173**, pp. 692–697.

250. Shao, G., Zhang, N., Lin, D., Feng, K., Cao, R. and Gong, M. (2013). A new europium(III)-β-diketonate complex based on diphenylethyne as red phosphors applied in LED, *J. Lumin.*, **138**, pp. 195–200.

251. Shao, G., Yu, H., Zhang, N., He, Y., Feng, K., Yang, X., Cao, R. and Gong, M. (2014). Synthesis and photophysical properties of europium(III)–β-diketonate complexes applied in LEDs, *Phys. Chem. Chem. Phys.*, **16**, pp. 695–702.

252. Wang, Z., Yang, H., He, P., He, Y., Zhao, J. and Tang, H. (2016). A highly-efficient blue-light excitable red phosphor: intramolecular π-stacking interactions in one dinuclear europium(III) complex, *Dalton Trans.*, **45**, p. 2839.

Chapter 9

Molecular Designing of Luminescent Europium–Metal Complexes for OLEDs: An Overview

V. Sivakumar and B. Rajamouli
Department of Chemistry, National Institute of Technology Rourkela, Rourkela 769008, Odisha, India
vsiva@nitrkl.ac.in

This chapter deals with lanthanide-based molecular complexes that gained interest due to their emerging technological applications. Among all, trivalent europium complexes exhibit efficient red emission in the region around 614 nm due to electric dipole (ED) transition ($^5D_0$–$^7F_2$). For obtaining efficient red emission from a molecular Eu complex, one needs to choose suitable a multidentate ligand, which is capable of harvesting the photons to the Eu excited levels ($^5D_0$). For practical applications (such as OLEDs), the designed molecules need to be doped in the desired host materials (to match the energy levels) and with a suitable device structure. The internal quantum efficiency (IQE) has the probability to reach 100% (100% IQE complexes are rare), but in the case of external quantum

Phosphors: Synthesis and Applications
Edited by Sanjay J. Dhoble, B. Deva Prasad Raju, and Vijay Singh

ISBN 978-981-4774-49-9 (Hardcover), 978-042-9460-99-9 (eBook)
www.panstanford.com

efficiency (EQE) it is not that much satisfactory. To improve the same, a lot of research effort was directed for the past two decades in designing new functional molecular complexes to use in OELDs. This chapter intends to discuss the recent developments in designing new functional Eu molecular complexes and their use in OLEDs. The energy-efficient lighting and display devices will certainly upgrade human living as well as be ecofriendly.

9.1 Introduction: Basic Approach and the Nitty-Gritty

There are numerous processes that are available to get the radiative light emission from different sources. Among all the processes fluorescence and phosphorescence are major phenomena, which majorly depend on the timescale [1–3]. A short lifetime is fluorescence (K_f 10^{-7} to 10^{-9} s^{-1}) and a long lifetime is phosphorescence. It is worth to note that the fluorescence is described as without spin change, and phosphorescence has a change in spin transitions according to IUPAC rules (e.g., fluorescence, Yb [$^2F_{5/2}$–$^2F_{7/2}$], and phosphorescence, Eu [$^5D_4$–^7F_J]) [4]. However, among all the elements in the periodic table, the lanthanide series has the best and specific light-emitting features. The emission spectra of transition metal complexes and organic dyes are asymmetric and wide ranging with a huge full width at half maximum (FWHM), more than 3000 cm^{-1}, due to d orbital evolutions [5, 6]. But in the case of lanthanides (Ln's), it consists of 4f–4f electronic transition (Ln contraction), which leads to efficient monochromatic characteristic emission (luminescence); no ligand field interactions are involved in the 4f orbitals of Ln's, and hence the emission and excitation lines are sharp. The Ln series, with 15 elements, including lanthanum (La, G.S. $^1S_0$), consists of different valence states (common +3; +2 is the lowest and +4 is the highest valence) [7]. Ln reacts with another element via compound formation, and it usually loses three of its outer electrons. It leads tripositive ions and occasionally develops less stable +2 or +4 ions also [8, 9]. Ln's can be described in two ways: (i) those that are fluorescent, for example, praseodymium (Pr, G.S. $^3H_4$), neodymium (Nd, G.S. $^4I_{9/2}$), holmium (Ho, G.S. $^5I_8$), erbium (Er, $^4I_{15/2}$), ytterbium (Yb, near-infrared [NIR] region, G.S. $^2F_{7/2}$), and (ii) those that have orange phosphorescence, for example, samarium (Sm^{III}, G.S. $^6H_{5/2}$), red europium (Eu^{III}, G.S. $^7F_0$), ultraviolet gadolinium (Gd^{III}, G.S. $^8S_{7/2}$), green terbium (Tb^{III}, G.S. $^7F_6$), yellow dysprosium (Dy^{III}, $^6H_{15/2}$), and blue thulium (Tm^{III}, G.S. $^3H_6$) [10–12]. Figure 9.1 indicates the energy levels of the furthermost operational selected Ln metal ions (Ln = Tm, Tb, Dy, Sm, and Eu) according to their emission color.

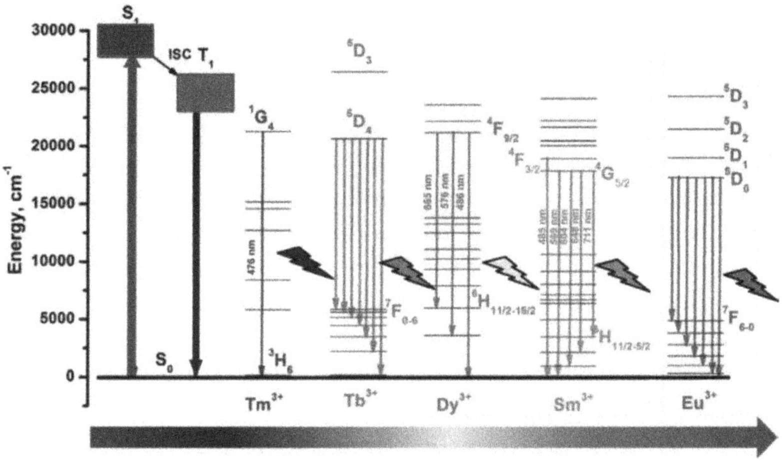

Figure 9.1 The energy levels of the most functioning selected Ln (Ln = Tm, Tb, Dy, Sm, and Eu) metal ions according to their emission color.

9.1.1 Orientation of Applications

The Ln series elements consist of specific emission colors; however, in some cases multicolor emissions are observed based on the temperature and oxidation states. For example, Tb emits green emission at room temperature but in the case of low temperature shows blue emission. In inorganic crystals, the low concentration of Tb shows blue emission, whereas high Tb concentration shows green emission [13]. Ln-based complexes are attractive due to their own merits such as unique optical properties (4f–4f and 4f–5d electronic transitions), high quantum yield (QY), long luminescent lifetime (ranging from microseconds to milliseconds), low long-term toxicity, and structural versatility [14–16]. Numerous reports have been documented on luminescent Ln complexes for different applications, such as tunable lasers, plastic optical fiber amplifiers

(POFAs), multicolor displays, solar energy convertors [17, 18] cathode ray tubes [19], aircraft and space shuttles, fluoroimmune assays, biomedical analyses, optical sensor cells [20], sensing reagents of chiral biological substrates and photonic crystals, tissue imaging, medical diagnosis, and protein labeling [21–25].

Eu^{3+}-based complexes are extensively attracting attention for medical diagnostics [27], biomarkers [28], electroluminescent materials [29], white light-emitting diode (environment friendly) [30] materials, owing to their simple energy levels of Eu^{3+} as well as efficient line-like or narrow-band red emission. Eu(III)- and Tb(III)-based complexes exhibit numerous appropriate characteristics when associated with those of conventional organic fluorophores, which makes them prominent candidates for biomedical imaging applications such as long excited-state lifetimes (usually in the millisecond range), a large energy shift between absorbed and emitted radiations (in the case of ligand sensitization), and very narrow emission bands; these two effects allow the separation of Ln(III) luminescence and the short-lived background fluorescence.

9.1.2 Assets and Historical Development

In recent days Ln's can found almost everywhere, from lighters to TV screens and from colored glass to control rods in nuclear reactors. The major source of Ln is monazite, a heavy, dark sand. These Ln elements forms low-solubility oxalates and fluorides, which are favored for separation of Ln's through the ion exchange process [31]. Among the Ln's, Ce was discovered quite a long time ago, in 1787, from a mineral called ytterite. The remaining Ln's were extracted from the same mineral by Swedish chemist Carl Gustar Mosander (1797–1858). The absorption spectra of Eu^{3+} by sharp lines were first portrayed in 1900 by Demarcay in solution. Urbain wrote about the red radiance of Eu_2O_3 (diluted in lime) in 1906. However, Prandtl confirmed Demarcay's observations and distributed a photo of an assimilation spectrum of Eu^{3+} for the first time in 1920 [32]. Weissman first reported the light emission of β-diketonates with Ln materials in 1942 [33, 34]. In 1955, Bernanose first observed electroluminescence (EL) from organic material such as acridine orange [35, 36]. Lempicki et al. first reported laser action in β-ketonated Eu(III) complexes in 1963 [37]. In the same year, Pop et al. observed organic fluorescence on anthracene crystal emission under vacuum [38]. In 1965, Helfrich and Schneider also introduced a single-layer device [39]. In 1967 Crosby introduced homoleptic materials [40, 41]. In 1980 Donlad reported hyperbranched molecules (dendrimers) [42, 43]. After 1987 Tang and VanSlyke [44–46] introduced a high-performance two-layer organic light-emitting diode (OLED) by vacuum deposition of organic layers with low bias voltage; the dynamic work moved on to OLEDs with different ligands based on Eu(III) complexes. In 1990, Burroughes et al. observed EL from an organic polymer and made OLEDs [47]. In the same year Friend synthesized a conducting polymer [48]. Ligand excitation energy is transferred through its triplet states to the central metal ion [49]. In general, centered metal Eu(III) complexes exhibit red emission. Frequently three, blue, green, and red, colors are essential to produce white-light emission. A mix of these colors produces white light, and it depends on the host, hole transfer (HT), and electron transfer (ET) materials with their appropriate voltages [50–52]. The use of binuclear molecular materials in multilayer devices produces different colors [53, 54]. Kido et al. used Tb^{3+} and Eu^{3+}, to produce white light [55, 56]. In 1998 Huang introduced a multiquantum well (MQW) into organic EL [57]. In 2002, Huang et al. reported p-type doped hole-transporting layer (HTL) and n-type doped electron-transporting layer (ETL) PIN OLEDs [58]. These p-i-n structure devices give high performance. Ln metal ions (Eu^{3+}) almost exhibit sharp emission bands because of the 4f shell. The Eu^{3+} metal ion gives 580, 592, 612, 652, and 703 nm corresponding f–f transitions of $^5D_0$–$^7F_0$, $^5D_0$–$^7F_1$, $^5D_0$–$^7F_2$, $^5D_0$–$^7F_3$, and $^5D_0$–$^7F_4$ [59, 60].

9.1.3 Classical Energy Transfer Excited States and Design Strategy

It is well documented that Eu(III)-based complexes are used as ratiometric optical sensors by monitoring two major emission bands, most commonly the magnetic dipole (MD) transition $^5D_0$–$^7F_1$ ($\Delta J = 1$) and the electric dipole (ED) transition $^5D_0$–$^7F_2$ ($\Delta J = 2$) bands [61, 62]. The MD-induced $\Delta J = 1$ transition is relatively insensitive to changes in coordination, whereas the $\Delta J = 2$ and $\Delta J = 4$ transitions are hypersensitive and generally change substantially upon formation of a new complex. Due to their parity-forbidden intraconfigurational 4f–4f transitions, the Eu(III) ions feature low absorption coefficients or oscillator strength, and hence their direct photoexcitation is inefficient. It is necessary to have a sensitizer (organic chromophoric ligand) that can indirectly populate the higher excited level via energy transfer to boost the emission intensity of the Eu(III) ion [63]. The energy transfer or sensitization process usually involves a triplet excited state of a ligand, thus requiring efficient intersystem conversion (ISC) in the ligand (facilitated by the heavy metal ion effect) as well as an appropriate alignment of the triplet state energy (T) over that of the accepting Ln state(s) (quantum efficiency ≈ 100, theoretical) [64]. In addition, ligand-to-metal (L-to-M) energy transfers embrace the L-to-M charge transfer (LMCT) and are analyzed by recent theoretical computations (combined density functional theory [DFT] and time-dependent density functional theory [TD-DFT] analysis). It is arising because of radiative and nonradiative competitive transitions. The intramolecular energy transfer process by Weissman [33, 65], Dexter [66], Crosby [67] et al. was already described in three major ways for typical binary arrangements involving anionic ligands and Ln ions. The first mechanism is ^{1}S to higher-energy or $^5D_0$ levels through internal conversion (IC), the second mechanism is energy transfer to Eu(III) from the triplet (^{3}T) by ^{1}S through ISC, and the third mechanism is direct energy transfer from the singlet via mixing of ^{1}S and ^{3}T to higher excited Eu(III) levels. In the case of Eu^{3+}, an optimal energy gap (ΔE) should be in the range of ~2500–3000 cm^{-1}. Improper alignment of triplet energy to the Eu excited levels (lower ΔE values) would enable back-transfer of energy and thus reduce sensitization efficiency or the possibility of obtaining multiple emissions from the ligand as well as the metal center. The exact supportive results were investigated by Latva et al. [64]. It is well identified that the Eu(III) metal ion excited $^5D_0$ energy level is located at 17250 cm^{-1}, [68]. In addition to that the singlet state should be located higher than 24,750 cm^{-1} and the triplet state higher than 19,750 cm^{-1}, leading to highly efficient emission and irreversible energy transfer. It was also being supported by Reinhoudt's empirical rule [69]. The energy transfer process from the ligand to the center Eu(III) metal ion through the ISC process is shown in Fig. 9.2.

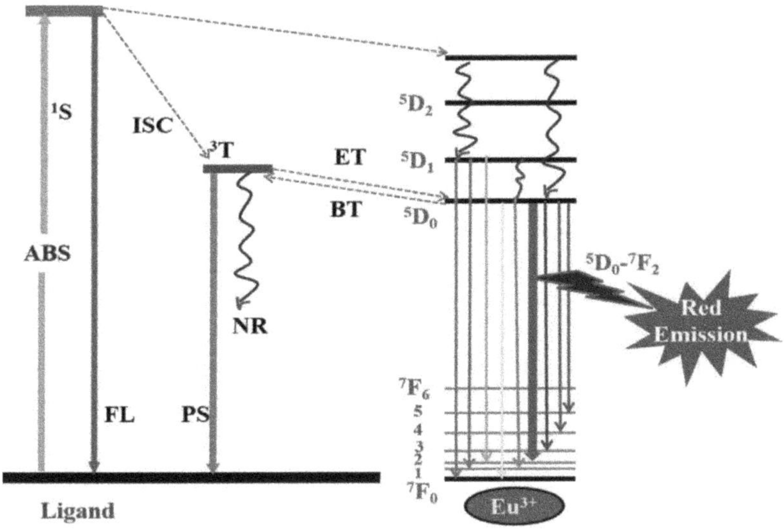

Figure 9.2 Energy transfer process from the ligand to the center Eu(III) metal ion through the ISC process. ^1S = singlet state; ^3T = triplet state; $^5D_{R\ (0,\ 1,\ 2,\ \ldots)}$ = excited state; FL = fluorescence; PL = phosphorescence; NR = nonradiative pathways; ISC = intersystem crossing; ET = energy transfer; BT = energy back transfer.

9.1.4 Electroluminescence

An OLED device is simple, where the emissive organic molecule is sandwiched between two electrodes. When the device is forward-biased, the electrons injected from the cathode (low work function) into the lowest unoccupied molecular orbital (LUMO) electronic state and holes injected from the anode into the neighboring highest occupied molecular orbital (HOMO) recombine in the middle of the device (emissive layer), leading to the formation of excitons. Finally, it produces light, the so-called electroluminescence (Fig. 9.3) [70, 71]. From the reported articles, mobility of holes is more in order compared to electron mobility [72]. In this case if there is no charge balance, the recombination moves toward the anode or cathode with respective voltages, resulting in a decrease in the luminescence from the emissive layer and unwanted emission by charge combination in the Alq_3 layer [73, 74]. Aziz et al. confirmed that the injection of holes into the ETL (Alq_3) is the main factor responsible for deficiency of the device [75]. Koji Itano et al. also observed exciplex formation at the interface of Alq_3 and lower transport potentials of HT materials [76]. For increasing recombination through charge balance in the emissive layer, they introduced hole-blocking layers (HBLs) [77] or electron-blocking layers (EBLs) [78]. In certain devices the HBL acts as both the ETL and hole-blocking properties [79]. The position of the branched groups' arrangement [80, 81], nature of electron-withdrawing groups, sublimation, distance between the ligand and central metal ion (Eu^{3+}) within the complex, recrystallization tendencies, phase separation, energy levels of the host and guest molecules (HOMO, LUMO levels), concentration of organic materials and size of their layers, addition of layers (thin film), character and nature of Eu^{3+} metal ion complexes (rigidity), and resistant oxidation influencing the carrier charge transportation result in manipulating efficiency [82–87]. To explore the optical and EL properties one needs to confirm the study in thin films. Smooth and uniform amorphous thin films can be generated by vacuum deposition. Spin-coated (inkjet, dry rolling, and screen printing) [88–90] amorphous materials are superior. The thickness of the ITO layer and the reactive index are responsible for the external coupling efficiency. For efficient carrier injection, hole and electron layers with a low HOMO or a high LUMO level are suitable [91]. The absorption energy of the guest is less than that of the emission energy of the host material and leads to an increase in the emission efficiency of the guest. Most of the devices are made by vacuum thermal evaporation, and it is satisfactory by an excellent ambipolar 4,4-bis(9-carbazolyl)-2,2-biphenyl (CBP) host matrix, and trimethylphenyl diamine (TPD) and n,n′-di(1-naphthyl)-n,n′-diphenyl-(1,1′-biphenyl)-4,4′-diamine (NPB) are hole-transporting moieties. Alq_3 is a good electron-transporting moiety. Polyvinylcarbazole (PVK) and poly(9,9-dioctylfluorene) (PFO) are spin-coated host mediums (including 2-(4-biphenylyl)-5-(4-*tert*-butylphenyl-1,3,4-oxadia) [PBD]); poly(3,4-ethylenedioxythiophene):poly(styrene sulfonate) (PEDOT:PSS) (hole injecting) and PVK (HTL) are hole-related moieties. 2,9-Dimethyl-4,7-diphenyl-1,10-phenanthroline (BCP) acts as a bifunctional emission layer in the device. The formation of excitons in the device is 25% singlet and 75% triplet (by theoretical prediction), and harvesting the triplet excitons to the ground state is an attractive research task. It is necessary to have heavy metal ions, as the triplet emission to the ground state is restricted, according to the spin selection rule. Ln-containing molecular complexes are able to give 100% internal quantum efficiency (IQE) by harvesting the unwanted excited triplet excitons to the ground state, and the same is shown Fig. 9.3 (right). The IQE can be calculated by the following equation. The QY (Φ) is the ratio of emitted to absorbed photons and reflects the efficiency of the emission process [92]. The EL efficiency $\eta_{el} = \eta_{int} \times \eta_{ex}$. The IQE ($\eta_{int}$) of an electroluminescent device is the ratio of formed photons to injected charges, which combine to form excitons. $\eta_{int} = \chi\Phi_{fl}\eta_r$. Here, η_r = fraction of injected charge carriers that form excitons, χ = fraction of charge carrier recombination that form singlet excitons, Φ_{fl} = photoluminescence (PL) efficiency of the dye, and η_{ex} = external quantum efficiency (EQE).

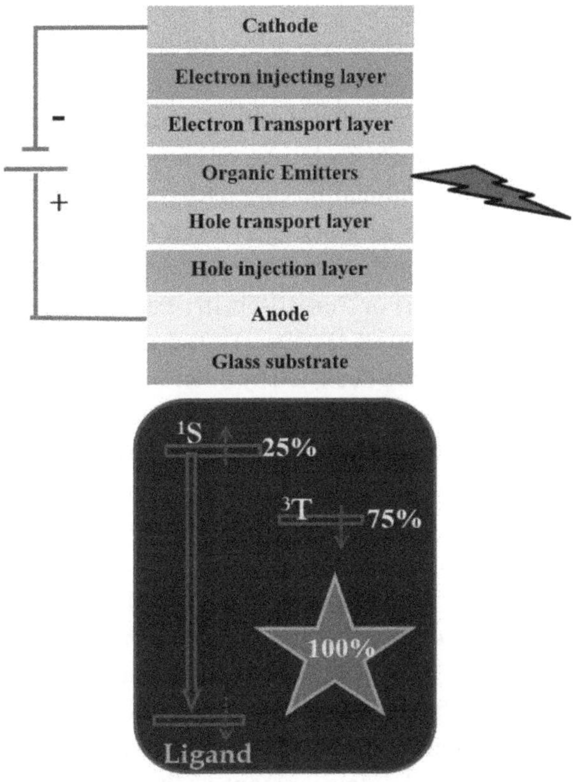

Figure 9.3 Electroluminescence emission from the OLED device (left) and the utilization of ^{1}S and ^{3}T for 100% IQE (right).

9.1.5 Theme of the Present Study

There are many review articles and book chapters available for Ln chemistry. We tried here to give an overall view of emissive Eu complexes with functional mono- to multidentate ligands. It is also necessary to understand the proactive design of the ligand that consists of β-diketone, N-N, N-N-N, N-N-N-N construction, and/or P = O moiety's with/without imidazole or bipyridine or spiro- or benzoimidazole. Functionalization of these molecules in the complex certainly gives the direction for energy transfer (fully/partial) and which is possible to get the tunable emission color from the Ln-metal-complex Utilizing these materials and obtaining red emissive Eu(III) complexes for OLEDs, with special emphasis on mono-, bi-, tri-, and multinuclear connection with anionic as well as neutral ligands in the complex is an attractive research task (Fig. 9.4). In general, as mentioned earlier, the Eu(III) complexes are expected to give red emission. However, on the other hand, it is also capable of producing white emission from single-molecular or multimolecular complexes. In general, different combinations are necessary to produce a white emissive complex, such as a blue-green-red (BGR) or a red-yellow combination by utilizing Eu(III) with multichromophores or Ir(III)-Eu(III) or Tb(III)-Eu(III) or Tm(III)-Tb(III)-Eu(III) ($^1G_4$–$^3H_6$, blue at ~430; $^5D_4$–$^7F_5$, green at ~545; $^5D_0$–$^7F_2$, red at ~612 nm) [93–99]. The Eu(III) complexes are designed by different sites of modification of β-diketone; dibenzoylmethane (DBM), 2-thenoyltrifluoroacetone (TTA), 1,3-di(2-thienyl)propane-1,3-dione (DTP), 2(4'-methoxy-benzoyl) benzoic acid (MeOBB), flha, 1,3-dicyano-1,3-propanedione (DCNP), dipivalomethane (DPM), 2,2,6,6-tetramethylheptane-3,5-dione), 4,4,5,5,6,6,6-heptafluoro-1-(2-naphthyl)hexane-1,3-dione (HFNH), and dinaphthoyl methane (DNM) as an anion ligand and 1,10-phenanthroline (phen) are utilized for OLEDs. Among all, TTA (^{3}T, 20,400 cm^{-1}), DBM (^{3}T, 20,300 cm^{-1}), phen (^{3}T, 21,480 cm^{-1}), and triphenylphosphine oxide (TPPO) (^{3}T, 18954 cm^{-1}) are found to be major contributions to PL and EL. The most widespread materials, which were used for the development of the device performance, are given below and their energy levels are also interpreted in Table 9.1. On the basis of the above discussion, the present discussion was made from Ln's to specific core emissive materials. The same was interpreted in Fig. 9.5.

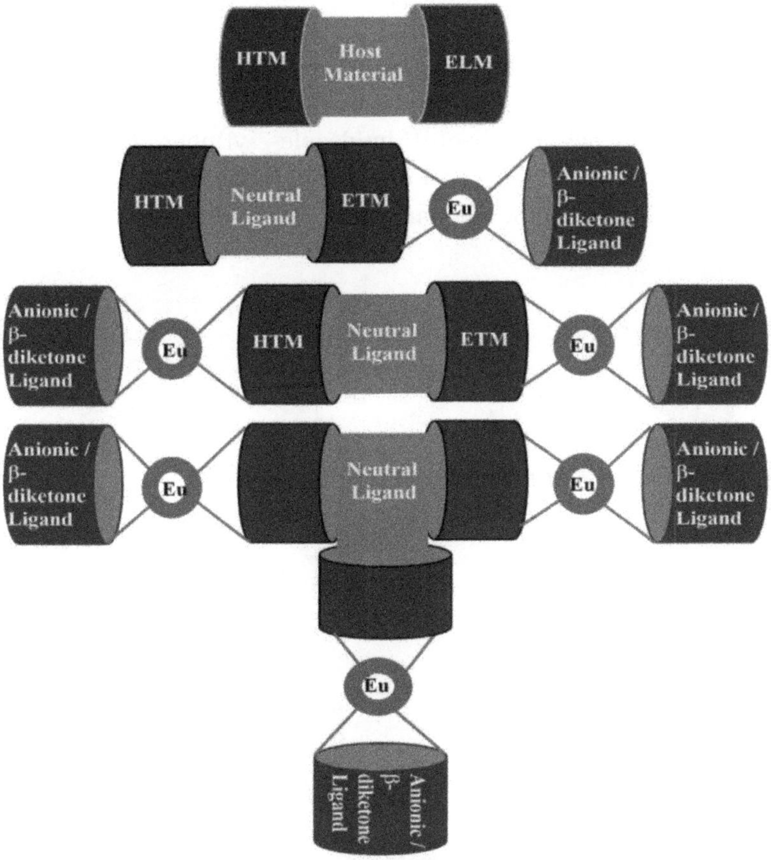

Figure 9.4 General schematic diagram of the connectivity nature of Eu(III) complexes as mono-, bi-, and trinuclear sites and corresponding general host materials. HTM, hole-transporting material; ETM, electron-transporting material.

In Fig. 9.6 is shown the electroluminescent device–associated layer materials from the emissive layer to the anode side, the cathode side, and their host materials. The materials that are used in the device are as follows: *anode material*: indium tin oxide (ITO, 4.7 eV); *cathode material*: Al (4.3 eV), LiF, Mg (3.8 eV), Ca (2.9 eV), Au (5.1 eV), In (4.1 eV), and their alloys (like Mg:Ag, 3.7

eV); *insulating layers*: LiF, CsF, AlOx; *host materials*: CBP, poly[2-(6′-cyano-6′-methylheptyloxy)-1,4-phenylene (CNPPP), PFO, PEDOT:PSS, PBD (electron transport also), PMPS, and poly(methyl methacrylate) (PMMA); *hole-transporting materials (HTMs)*: PVK, TPD, 4,4′-bis[N,N′-(3-tolyl)amino]-3,3′-dimethylbiphenyl (HMTPD), and NPB; and *electron-transporting materials (ETMs)*: Alq_3, Alq_2Oph, 1,3,5-tri(1-phenyl-1H-benzo[d] imidazole-2-yl)phen (TPBI), and 1,5,7-triazabicyclo[4,4,0]dec-5-ene (TBD). Figure 9.7 reflects the HOMO and LUMO of the host materials.

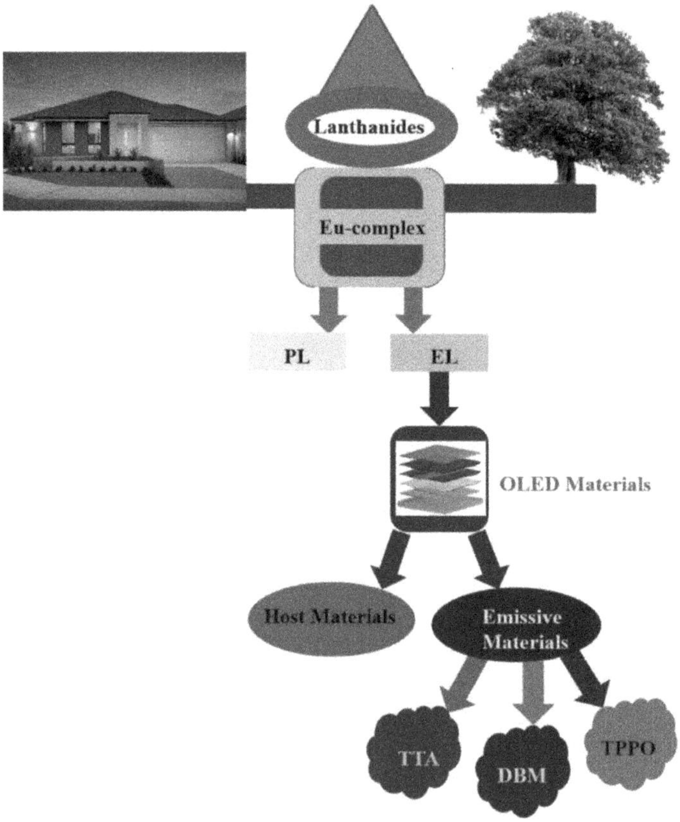

Figure 9.5 Flowchart for lanthanides to specific luminescent core materials. The greenery, new light architecture, and promising stores will all be a nice upgrade for the environment.

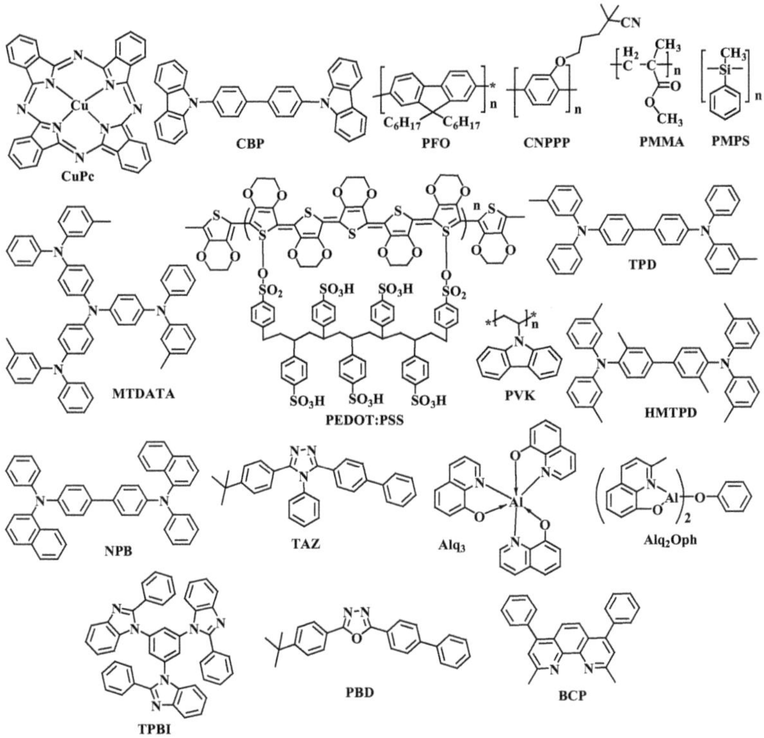

Figure 9.6 Different types of host, HT, and ET materials that are used in OLED devices.

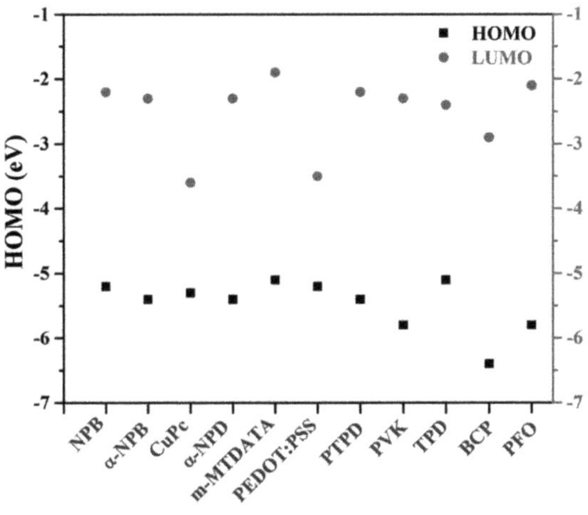

Figure 9.7 The HOMO and LUMO of the device host materials.

Table 9.1 Energy levels of the foremost device materials (HOMO and LUMO values in the negative scale)

S. No.	Name of the material	HOMO (eV)	LUMO (eV)	Ref.
1.	NPB	−5.2	−2.2	[100]
2.	α-NPB	5.4	2.3	[101–103]
3.	CuPc	5.3	3.6	[104]
4.	α-NPD	5.4	2.3	[105]
5.	m-MTDATA	5.1	1.9	[106, 107]
6.	PEDOT:PSS	5.2	3.5	[108, 109]
7.	PTPD	5.4	2.2	[110]
8.	PVK	5.8	2.3	[111]
9.	TPD	5.1	2.4	[112, 113]
10.	BCP	−6.4	−2.9	[114, 115]
11.	PFO	5.8	2.1	[116, 117]

9.2 Luminescent Eu(III) Complexes

The two or three (frequently) β-diketone bidentate anionic ligands occupy the coordination site of the Eu(III) ion; in addition, one or two neutral ligands (bi-, tri-, or tetra-) occupy the coordination sphere of the complex (coordination number [CN] ranging from 8 to 10). The first binary Eu(III) complex made was $Eu(TTA)_3$, and after that $Eu(DBM)_3$ was reported. On the basis of the popular anionic antennas, we designed/arranged Eu complexes with antennas in different ways to better understand Eu luminescence and its use in OLEDs. The β-diketone bidentate ligands were separated according to TTA and DBM, including the TPPO core resources. Figure 9.8 presents the TTA-based antennas for OLED applications.

9.3 TTA as an Anionic Ligand

Kido et al. and Okamoto introduced the Eu metal ion with a TTA-based EL device. The Eu complex AA has a nonvolatile nature. So the PMPS and Eu are spin-coated and PBD is made by vacuum deposition. The device configuration is glass substrate/ITO/PMPS doped with (20 wt%) AA (600 Å)/PBD (600 Å)/Mg (100 Å)/Ag

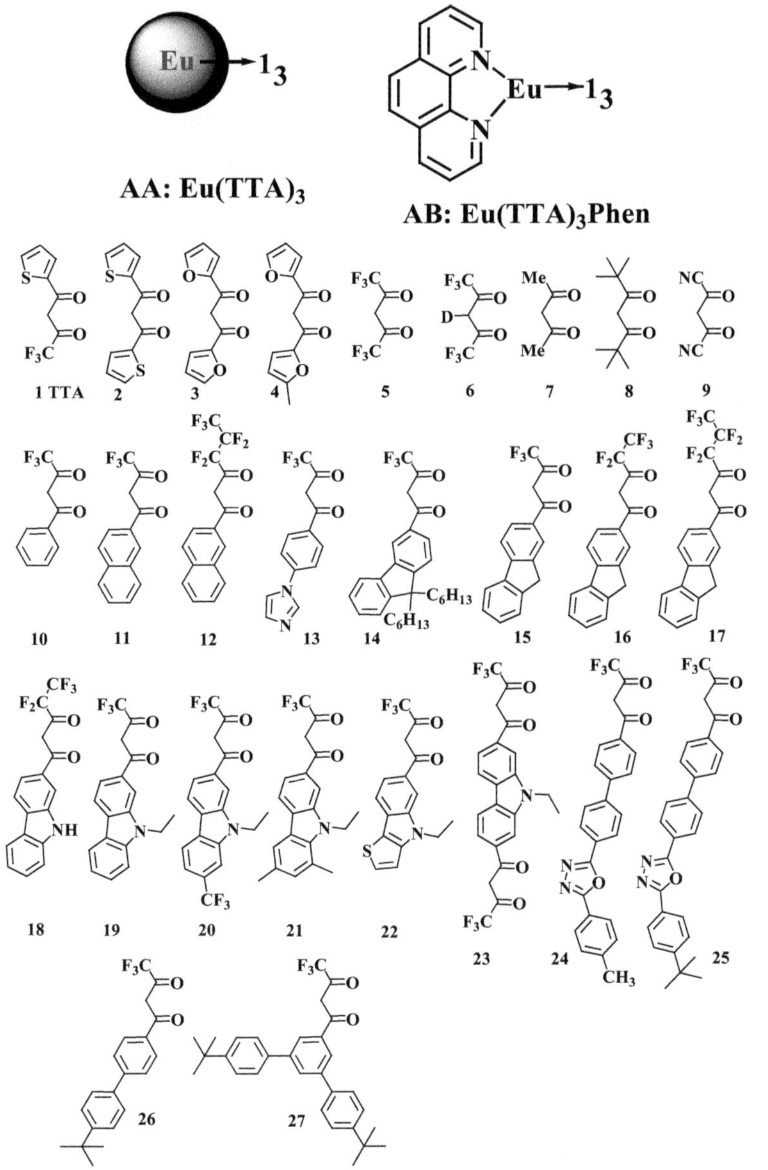

Figure 9.8 The β-diketonate (TTA-based) anionic antennas for Eu(III) complexes.

(1500 Å) and observed red emission with a turn-on voltage (TOV) of 12 V and a maximum intensity of 0.3 cd/m^2 at 18 V [118, 119]. After its invention, the same complex with connected phen was taken in different device structures and improvements made toward high efficiency. Takada et al. designed a microcavity structure using SiO_2/TiO_2 bilayers. The device structure is SiO_2/TiO_2 bilayers/ITO/TPD/AB/Alq_3/Mg:Ag; a planer microcavity is constituted by a dielectric reflector and a Mg:Ag metal electrode. When operated under a DC drive voltage, sharply directed emission from the europium complex is observed [120].

Sano and coworkers [121] studied the same Eu complex to improve EL using the co-deposition method for a complex doped in a host material (1AZM-Hex) in the emission layer. From the device with only the HTL or ETL, no emission was observed because of the poor transporting property in the solid state, but in the case of both layers, the device shows a maximum luminescence of 7 cd/m^2 (16 V, 125 mA/cm^2).

Introducing a host material in the device with only the HTL in different wt% of the Eu complex, 5 wt% shows red emission, and it is better than that in the two-layer device. The device structure is ITO/TPD/1AZM-Hex:AB/Mg:In (10:1, 2000 Å) and a maximum luminance of 137 cd/m^2 at 200 mA/cm^2.

Sano et al. [122] also studied the EL properties of AB. The synthesized AB has an emitting layer with red emission in the El OLEDs. The organic layer is vacuum-deposited, and the device structure is glass substrate/ITO/TPD/AB/Mg:In. The Eu complex's PL wavelength is 614 nm. This device's PL originates in the transition between energy levels of the Eu(III) ion and does not show emission effectively because of lower EL efficiency. The observed luminescence is poor at 100 cd/m^2. Adachi et al. [123] reported the energy transfer process in the device. The host material (CBP) was doped with a Eu complex (AB) in different concentrations. The device structure is ITO/TPD (50 nm)/CBP:AB (1%) (20 nm)/BCP (10 nm)/Alq_3 (30 nm)/Mg:Ag (10:1, 100 nm) and achieves a maximum external EL quantum efficiency (η) of 1.4% at a current density of 0.4 mA/cm^2 and a maximum luminescence of 505 cd/m^2 at 12 V and 100 $mAcm^{-2}$. A significant decrease of QE was observed with increasing current along with increase in the CBP host emission due to triplet–triplet annihilation on CBP molecules following back transfer from TTA because of near resonance of the TTA and CBP triplet state. From this, the conclusion was made that the direct trapping of electrons and holes and the subsequent formation of the excitons occur on the dopant, leading to high QE at low current densities. It was further analyzed by Ohmoria et al. [124] by fabricating a single-layer-based ITO/AB:PVK/Mg:In device. The emission peak was perceived from the Eu compound and PVK at 610 nm and 420 nm, respectively. The ratio of emission peak intensity decreases with increasing Eu concentration, and at >0.01 mol% of concentration, red emission was observed. From the energy band diagram, the excitons formed are transferred successfully to the Eu complex from the PVK host.

9.3.1 TTA Modification

In 1998, Keizouokada et al. [125] introduced a novel complex AC as an emitting layer between the HTL and the cathode (sandwich type) in an organic EL device, and the device structure is glass substrate/ITO (sheet resistance 10 Ω/square/TPD (600 Å)/AC (600 Å)/Al:Li (99:1) (2000 Å). Bright red emission was observed with a maximum luminescence of 34.3 cd/m^2 at 12 V and a current density of 25.7 mA/cm^2 and the organic layer was vacuum-deposited. The identical PL spectra and EL spectra suggested no interaction between the TPD layer and the Eu complex and recombination at the Eu complex layer. In the same year Okada et al. [126] substituted the methyl group in the complex AC (Eu(MDFP)$_3$(phen), AD) (MDFP = 5-methyl-l,3-di-(2-furyl)-l,3-propanedione; phen = 1,10-phenanthroline). The device structure is glass substrate/ITO/TPD (600 Å)/AD (600 Å)/Al: Li (99:1) (2000 Å), and it showed a maximum luminescence of 32.0 cd/m^2 at 9 V and 25.7 mA/cm^2.

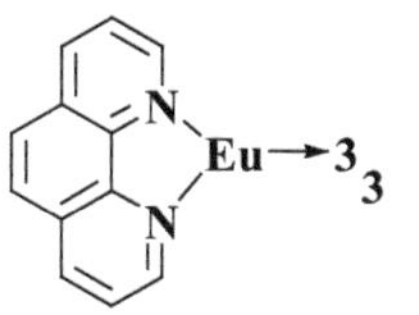

AC: Eu(DFP)$_3$(Phen)

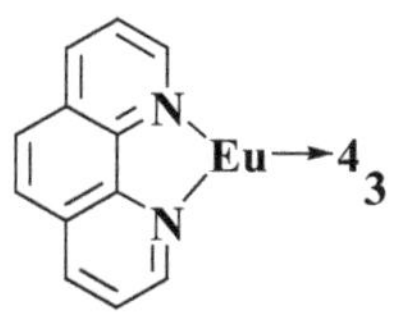

AD: Eu(MDFP)$_3$(Phen)

Zhang et al. [127] introduced a Eu complex layer with a PVK and PBD (30 wt%) blend in the device at different doping concentrations (0.2%–8%) to achieve equal holes and electrons in the device at the active (emissive) layer. The emission originates from the PVK:PBD host and decreases with increasing dopant concentration. Around

8% completely disappears; it means high efficiency energy transfer from the complex AG to the host system. Due to the large site isolation effect, the dendritic phenylene(polyphenylene)-functionalized Eu complex shows enhanced EL efficiency. This polyphenylene groups decrease self-quenching of the Eu complex compared to the other two, and it is observed by photoluminescence quantum efficiency (PLQE). The general configuration of the device of Eu complexes (X = AE, AF, and AG) is ITO/PEDOT:PSS (150 nm)/PVK:30 wt% PBD:X Eu complex (75 nm)/Ba (4 nm)/Al (200 nm). The charge trapping in the doping system is observed; it was confirmed by increasing the TOV with increasing doping concentration. The highest EQE and maximum luminescence observed for complexes, AE, AF, and AG are 0.22%, 0.34%, and 0.42% and 166, 272, and 277 cd/m^2, respectively, at 0.2 wt% doping concentration with red emission. By introducing TBPI as an HBL in the device with a Eu complex (0.5%), the observed EQE, luminous efficiency, and maximum luminescence are 0.72, 0.79, and 1.90; 0.78, 1.93, and 2.01; and 332, 640, and 319, respectively.

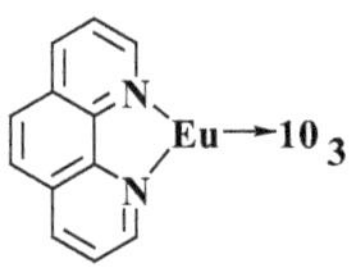

AE: Eu(TPD)$_3$Phen

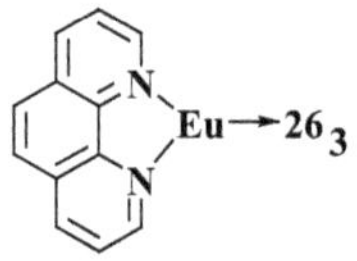

AF: Eu(Ph-TPD)$_3$Phen

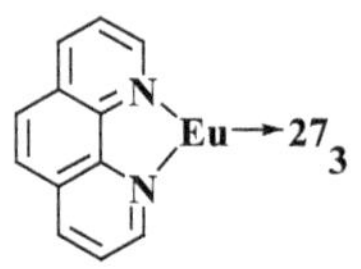

AG: Eu(G1-Ph-TPD)$_3$Phen

The absorption energy of the guest (G) is lesser than the emission energy of the host (H) material. On the basis of this Uekawn et al. (1997) [128] synthesized different groups and introduced AB, AH, and AI. PL emission was greater for AI (370 nm) than for AB (343 nm) and AH (351 nm) in the case of comparing the spectral overlap between AI and TPD emissions. The absorption peak of the AI complex shows lower energy with respect to the emission energy of the host (TPD), and it leads to energy transfer to the AI complex from TPD (HTL) higher compared to that of AB and AH complexes.

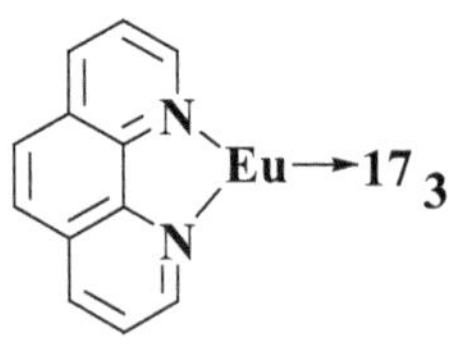

AH: Eu(FIHA)$_3$Phen

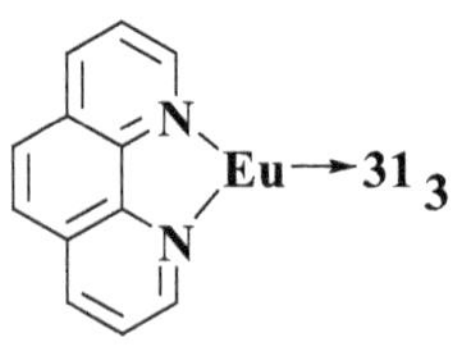

AI: Eu(DNM)$_3$Phen

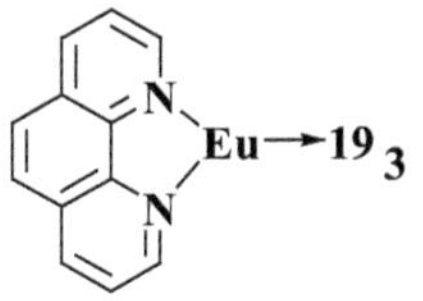

AJ: Eu(2-TFDBC)$_3$Phen

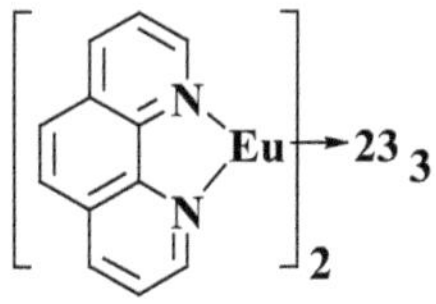

AK: Eu$_2$(2,7-BTFDBC)$_3$(Phen)$_2$

Gong et al. [129] introduced two novel carbazole-based Eu(III) complexes, AL and AM, (2-(4′ 4′ 4′-trifluoro-1′3′ - dioxobutyl)-carbazole (2-TFDBC) and 2,7-bis(4′ 4′ 4′-trifluoro-1′3′ -dioxobutyl)-carbazole (2,7-BTFDBC)). Both complexes have the best decomposition temperature (>361°C). The expanded π-conjugation leads a red shift in the complex excitation band (toward the visible region). The obtained high color purity almost reached NTSC standards with Commission internationale de L'Eclairage (CIE) coordinates of (0.68, 0.32). AL shows a relatively high PLQE (28%) compared to the rest of the complex (10%), which is due to the reduced Eu–Eu distance in dinulear structures.

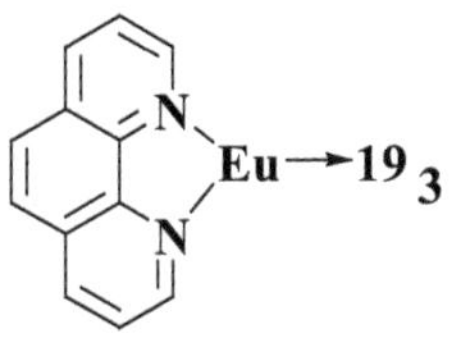

AL: Eu(2-TFDBC)$_3$Phen

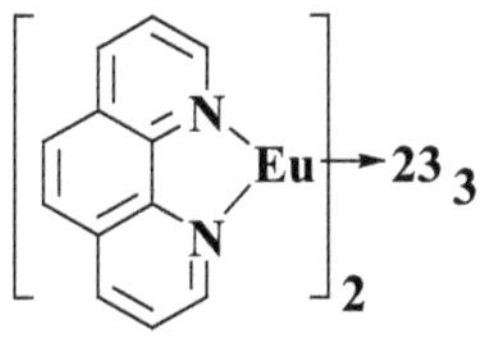

AM: Eu$_2$(2,7-BTFDBC)$_3$(Phen)$_2$

Fluorination is a feasible way of increasing emission efficiency via reduction of vibrational deactivation from associated high-energy oscillators. It also can be explained as curbed C-H vibration by fluorination, leading to increasing emission efficiency. In the case of the complex, AO and AN have shown low device efficiencies. However, the complex AP showed enhanced photoluminescence quantum yield (PLQY), which almost reached 34% [130–132].

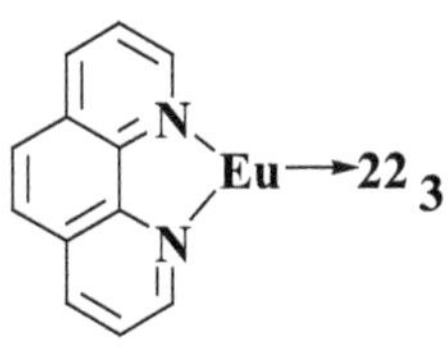

AN: Eu(22)$_3$Phen

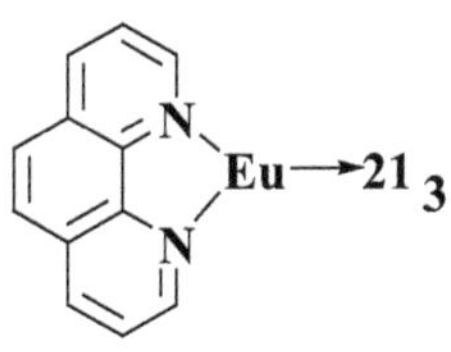

AO: Eu(21)$_3$Phen

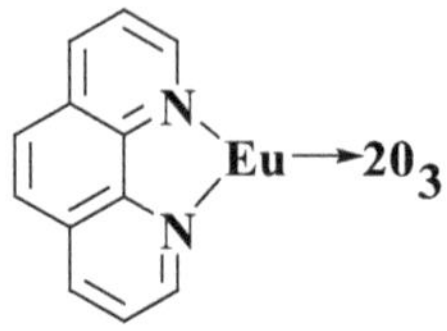

AP: Eu(ETFMCTFBD)$_3$Phen

As we know, pyridine, imidazole, and 1,3,4-oxadiazole (OXD) are strong electron-transporting-capable moieties. Utilizing these materials for Eu(III) complexes to modify 4,4,4-trifluoro-1-phenylbutane-1,3-dione (BTA) to improve sensitizability of the ligands as well as thermal and luminescent enactment, Gong and coworkers proposed similar type of complexes AQ, AR, and AS [133–135]. Introducing conjugate extension heterocyclic groups showed a bathochromic shift in the absorption spectra. These Eu(III) complexes have a PLQY of 14%–18%. A Eu ternary complex coated onto a ~395 nm emitting InGaN chip achieved red emission with appropriate CIE chromaticity coordinates. The complex AQ exhibited a ligand-sensitized luminescence process (antenna effect) and an LED with an improved power efficiency (PE) of 0.53 lm/W. AQ, AR, and AS show CIE coordinates (0.66, 0.33). In addition to achieving good efficiency and stability for the complexes, Eu(TPBDTFA)$_2$(TPATFA) phen, Eu(MEP)$_3$phen, and Eu(MEP)$_3$ are synthesized by mixing different ratios of two β-diketonates. The complex AT showed good thermal stability; however, it is less compared to that obtained by mixing different ratios of two β-diketonate complexes. Introducing the carbazole moiety in the ligand enlarged the π-conjugated system of the ligand as well as enhanced the Eu(III) complexes' luminescent intensity.

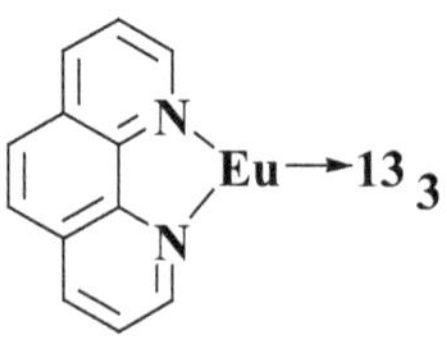

AQ:Eu(IDTFBD)$_3$ Phen

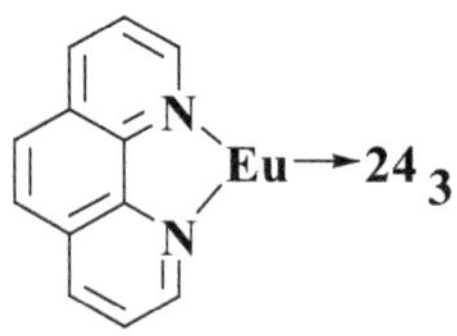

AR:Eu(MPBDTFA)$_3$ Phen

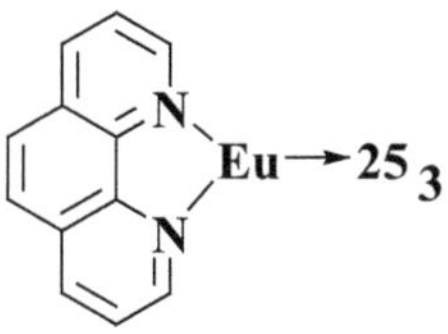

AS:Eu(TPBDTFA)$_3$ Phen

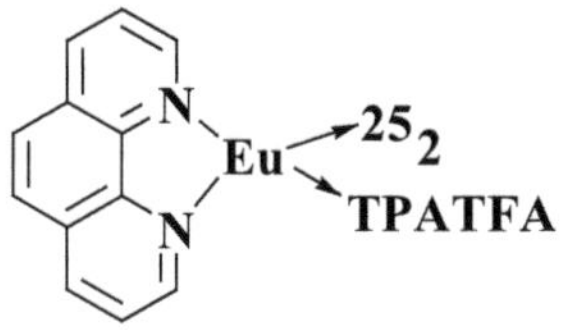

AT:Eu(TPBDTFA)$_2$(TPATFA)Phen

9.3.2 Modification of Phen

Pei Sun et al. introduced pyrazino[2,3-f][1,10]phenanthroline (PyPhen), methylpyrazino[2,3-f][1,10]phenan-throline (MPP), 4,5,9,14-tetraazabenzotriphenylene (DPPz), 11-methyldipyrido[3,2-a:2′,3′-c]phenazine (MDPz), 1,12-di-methyldipyrido[3,2-a:2′,3′-c]phenazine (DDPz), and benzo[i-]dipyrido[3,2-a:2,3-c]phenazine (BDPz) ligands with Eu(TTA)$_3$ moiety (AU, AV, AW, AX, AY, and AZ) [136–138], which are developed for high sensitizability and electroactivity for luminescent Eu(III) complexes. Li et al. worked additionally on the same by computational resources (DFT) and analyzed that the small-size ligands can undergo ligand-to-metal energy transfer (LMET) efficiently. It leads to higher ^{1}S and ^{3}T levels. It is strictly followed by AB, AU, and AV complexes through the empirical rule that the excited energy transfer (EET) must be larger than 0.43 eV and the excited intersystem crossing (EISC) must be larger than 0.62 eV. The rest of the complexes show low efficiency due to mismatching of the energy levels. It is also found that the contribution of fused pyrazine rings considerably reduced the LUMOs of the complexes (−2.8 to −3.0 eV). The device's general structure is ITO/TPD (or) NPB (50 nm)/Eu:CBP (1.7%–7.6%) (30 nm)/BCP (20–30 nm)/Alq$_3$ (25–35 nm)/Mg:Ag. Compared to all Eu complexes, AZ exhibits a 550 nm strong emission and a 612 nm weak emission from the center metal Eu ion, and the strong emission is closed to free BDPz. It is proposed that intramolecular energy transfer from the ligand to the Eu metal ion seems incomplete in the complex. AU:CBP at 2.6% exhibits luminescence (L) of 1309 cd/m^2 with less EQE (η_{ext}) but at 7.6% exhibits 757 cd/m^2 with EQE 1.15% and current efficiency (CE) 2.16 cd/A. AV:CBP at 1.7% exhibits 1256 cd/m^2 with EQE 1%. AW:CBP at 2% exhibits 2046 cd/m^2 with less

EQE but at 4.5% exhibits 1670 cd/m^2 with EQE 2.05%. AY:CBP at 2% gives 1528 cd/m^2 with less EQE and at 5% gives 1248 cd/m^2 with an improved EQE of 1.24%. The author observed that by improving the concentration of the dopant, AV and EQE are changing in different ways (unmatchable).

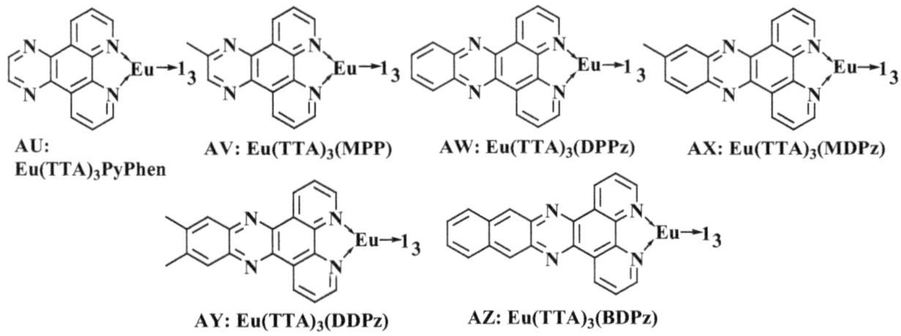

The relative HOMO and LUMO energy levels of the materials that are used in the device structure [NPB/Eu:CBP/BCP/Alq] are shown in Fig. 9.9.

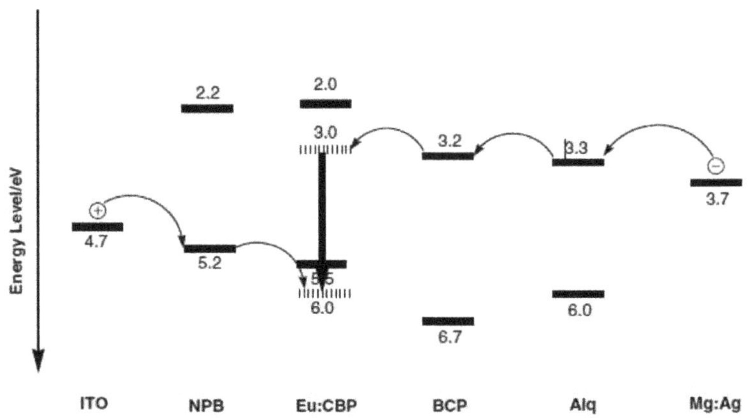

Figure 9.9 Relative HOMO/LUMO energy levels of the device structure [NPB/Eu:CBP/BCP/Alq] [135].

The above structure-related Eu(III) complexes were introduced by utilizing an indole-fused PyPhen-based ligand BB [139]. The complex consisted of a bipolar structure that combines electron-donating indole and electron-withdrawing pyrazine moieties. Addition to the same indole-fused ligand moiety was replaced by thiadiazole for the Eu(III) complex ($Eu(TTA)_3(TDAP)$, BC) [140] (shrink the conjugation by thiadiazole-connected phen) introduced by Pereira et al. in 2011. An electron-deficient five-membered thiadiazole ring was incorporated to enhance the electron injection ability of the ligand. The fabricated bilayer device is ITO/NPB/CBP:BC/Al. The HOMO and LUMO of BC were located at −5.7 and −2.6 eV, respectively.

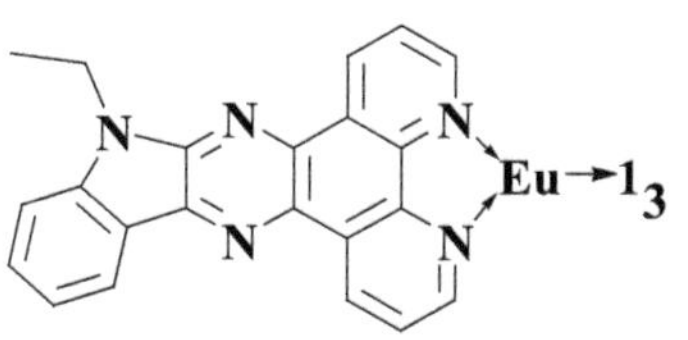

BB: $Eu(TTA)_3$DIQ-Et

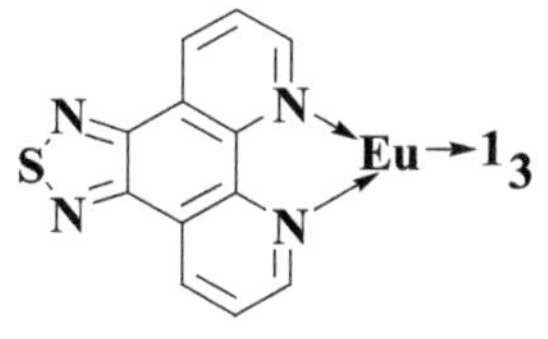

BC: $Eu(TTA)_3$TDAP

Phen was substituted with different types of electron-withdrawing and electron-donating groups. The overview of a chlorine-substituted group showed improved EL performance compared to that of a nitryl substitution. The EL efficiency increased significantly by reduction of the energy loss initiated by light hydrogen atom vibration. The devices' structures are ITO/TPD/CBP:AB/BCP/Alq_3/LiF/Al, η_c: 1.73 cd/A; ITO/TPD/CBP:BD/BCP/Alq_3/LiF/Al, η_c: 2.34 cd/A; and ITO/TPD/CBP:BE/BCP/Alq_3/LiF/Al, η_c: 0.98 cd/A [141]. Fang and Ma also introduced a methyl-substituted phen, trivalent Eu(III) complex BG into an ITO/TPD (40 nm)/Eu complex:CBP (1%, 30 nm)/BCP (20 nm)/Alq_3 (30 nm)/LiF (1 nm)/Al (100 nm) device. A maximum luminance up to 800 cd/m^2 at a drive voltage of 24.5 V, an EQE of 4.3%, a CE of 4.7 cd/A, and a PE of 1.6 lm/W were observed, and at a brightness of 100 cd/m^2, the quantum efficiency reaches 2.2% (2.3 cd/A). The EL QE decreases with increasing current; a recombination of long-lived triplet excitons causes luminescence. The same is explained with the help of triplet–triplet annihilation on CBP molecules, following back transfer from TTA due to the neighboring resonance [142].

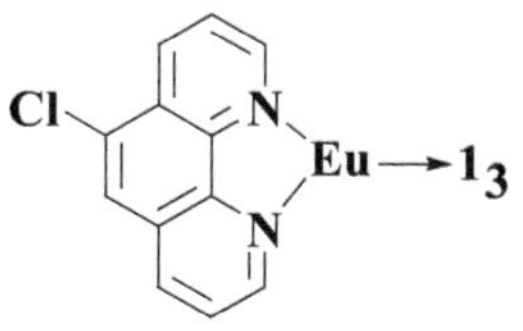

BD: $Eu(TTA)_3$ClPhen

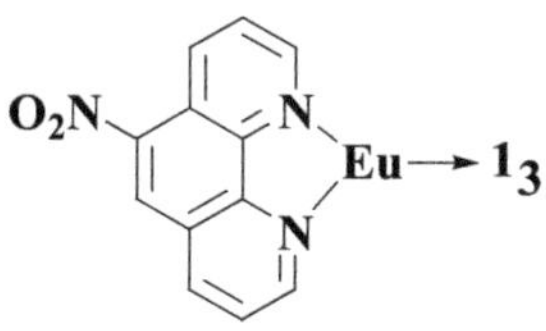

BE: $Eu(TTA)_3$NPhen

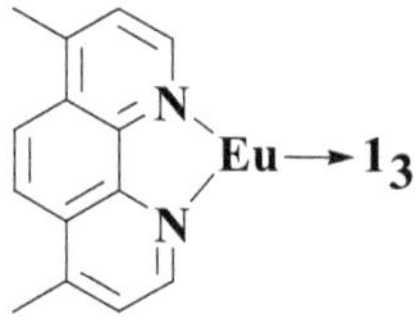

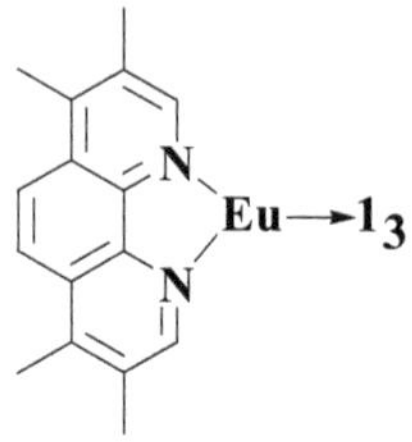

BF: Eu(TTA)$_3$DmPhen **BG: Eu(TTA)$_3$TmPhen**

The 4,7-position-substituted phen by carbazole moieties was introduced by Zucchi and coworkers [143] for the Eu^{3+} complex BH. The energy transfer from the ^3T level of the ligand to the Eu(III) metal ion was well suited for the LMET process. It is also supported by the ligand-broadened absorption in the range of 300–400 nm and does not lead to a change of the ^3T state (21,000 cm^{-1}). The bulky nature of the neutral ligand was providing the protection for excited Eu^{3+} ions from quenching as well as improving the long emission lifetimes (0.66 and 0.80 ms in acetonitrile solutions and films, respectively). It leads to high PLQE (80%) in PMMA.

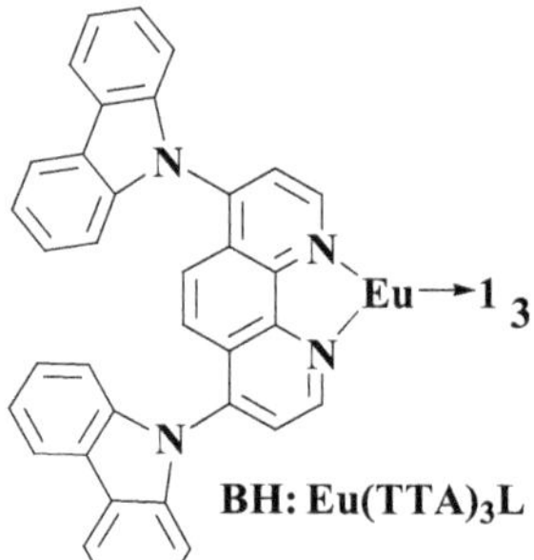

BH: Eu(TTA)$_3$L

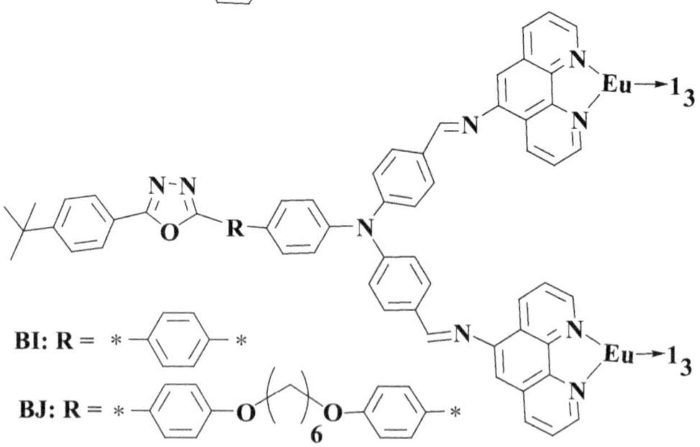

Utilizing both hole-transporting triphenylamine (TPA) and electron-transporting oxadiazole (OXD), two novel bipolar-transporting dinuclear Eu(III) complexes were reported (BI and BJ) [144]. Both are separated by a spacer moiety (hexyl linker) to arrest the conjugation. A fully conjugated complex shows high thermal stability with a decomposition temperature of 304°C (higher than that of the nonconjugated complex). The structure of single-layer polymeric light-emitting diodes (PLEDs) fabricated is ITO/ PEDOT:PSS(50 nm)/PVK-PBD (30 wt%):Eu(III) complex (70 nm)/LiF (0.5 nm)/Al (150 nm). The emitting layer consists of a host matrix of the PVK-PBD blend; the weight ratio of PBD is 30% and the doping weight varies from 1 to 2, 4, and 8 wt. for Eu(III) complexes' concentration. BI, the molecular complex, has good carrier injection and transporting abilities, and the fabricated device shows luminance up to 296 cd/m^2 at a low TOV of 8.5 V.

9.3.3 Both TTA and Phen Modification

In 1999, Keizouokada et al. synthesized a novel Eu-based BK as an essential component layer in double- and triple-layer EL devices by vacuum deposition. A triple-layer EL device has higher luminescence than that of a double-layer device due to the introduction of the ETL (Alq_3). Device fabrication was done in the following manner: glass substrate/ITO/TPD (600 Å)/BK (500 Å)/Alq_3 (200 Å)/Al:Li (99:1) (2000 Å). The TOV is <5, the maximum luminescence is 450 cd/m^2 at 15 V, and current density is 200 mA/cm^2. A decrease in luminescence was observed >16 V. This is because of the existence of dark sports created between the Eu layer and the Al–Li interface due to humidity and the presence of oxygen. With the decreasing thickness (300 Å) of the emissive layer in the device appears a broad peak at 535 nm coming from Alq_3 (500 Å), which shows that the energy transfers from the Eu layer to Alq_3. It shows better improvement when compared to that of the oxygen-containing antenna complex (AC).

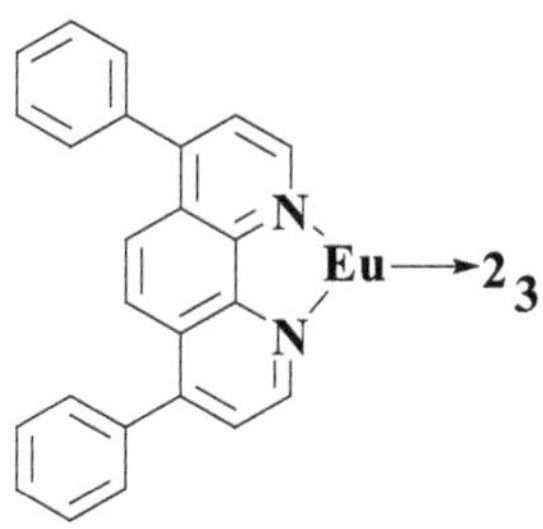

BK: Eu (DTP)$_3$(Dipphen)

9.3.4 Phosphine Oxide (P=O)-Based Complexes

The single-coordinate phosphine oxide ligands were reported with TPPO as a core, (4-diphenylaminophenyl)diphenylphosphineoxide (TAPO), (4-naphthalen-1-yl-phenylaminophenyl)diphenylphosphineoxide (NADAPO), and 9-[4-(diphenylphosphinoyl) phenyl]-9H-carbazole (CPPO). The corresponding Eu(III) complexes were BL, BM, and BO. The direct connection of the phosphine oxide ligand leads to proper ^1S and ^3T energy levels. It improves the intramolecular energy transfer in Eu(III) complexes [146, 147]. Huang and coworkers (2007) reported the bonding of electron-donating arylamine groups consisting of diphenphosphine oxide (DPPO) to produce donor-π-acceptor-type (D-π-A) monodentate Eu^{3+} complexes BL, BM, and BO, which showed high emission efficiencies (~36%–40%). The single-layer device, ITO/Eu^{3+} complex/Mg:Ag, showed improved luminance of 35, 59, and 21 cd/m^2, respectively. A similar way was followed for the four-layer devices and showed that both complexes (BL and BM) emit red emission and a maximum brightness of >1000 cdm^2 corresponding to TOVs lower than 5 V. Briefly, for the BL, shown for four-layer devices, maximum efficiencies were 5.07 cd/A, 3.62 lm/W, and PLQY 3.2%. Their remaining observations are at 100 cd/m^2 and were 2.08 cd/A, 0.63 lm/W, and 1.31%. The excellent spectral stability with a maximum EQY is more than 3%. BM showed a maximum luminance of 1158 cd/m^2 and maximum efficiencies of 5.88 cd/A, PE 3.69 lm/W, and EQY 3.71%. Xu et al. composed a D-A-type P=O ligand CPO [148] BN and arranged it in a complex for examination with its D-π-A simple CPPO and compared it with BO. Indeed, the BN-based four-layer implement is ITO/NPB/complex BN/BCP/Alq_3/Mg:Ag, which showed an improved luminance of 399 cd/m^2 and maximum efficiencies of 2.6 A^{-1}, 0.99 lm/W, and 1.68%. This focus obviously shows the prevalence of the D-π-A structure in the ligands.

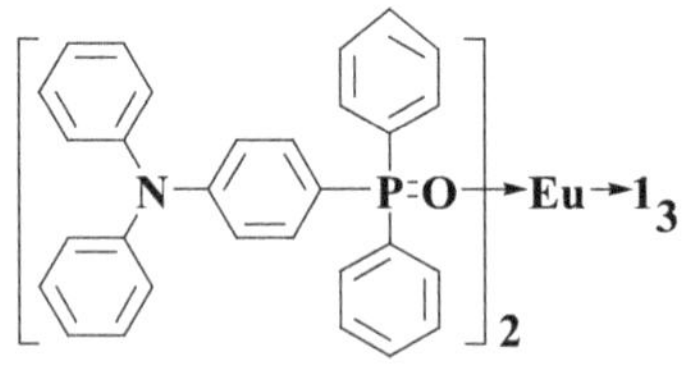

BL: $Eu(TTA)_3(TAPO)_2$

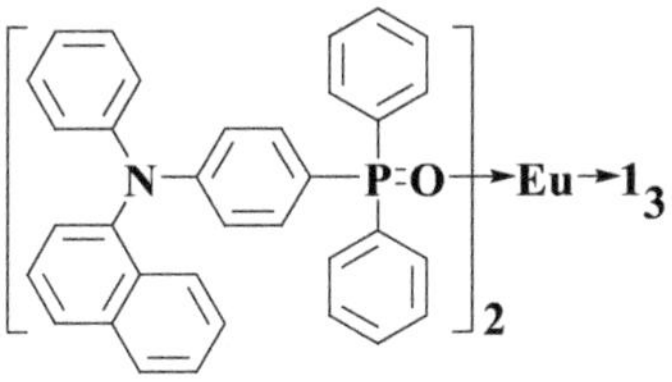

BM: Eu(TTA)$_3$(NADAPO)$_2$

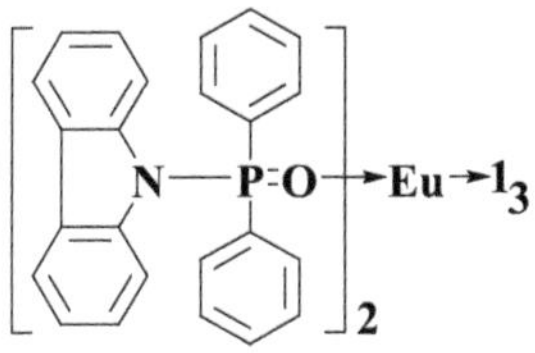

BN: Eu(TTA)$_3$(CPO)$_2$

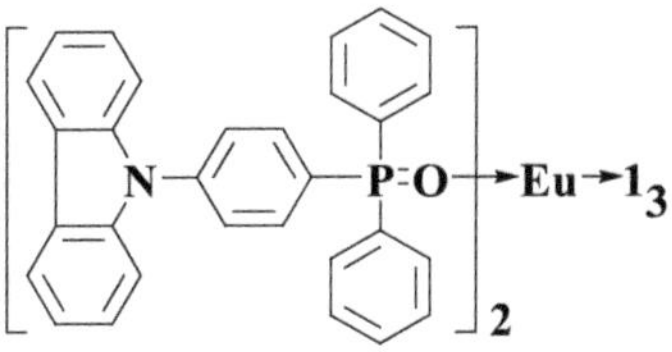

BO: Eu(TTA)$_3$(CPPO)$_2$

Intramolecular energy transfer processes of 1,8-bis(diphenylphosphino) naphthalene oxide (NAPO) and its Eu(III) complex, BP, were studied. The bidentate nature of NAPO defeats solvent-induced quenching by means of ligand–ligand interaction and the rigidity of the complex. In other words, NAPO has the ^{3}T energy level between the ^{1}S and ^{3}T energy levels of the TTA molecule. It leads to proper energy transfer to NAPO to TTA, and this would not be observed in the complex with TPPO (Eu(TTA)$_3$TPPO) [149, 150].

Bis(2-(diphenylphosphino)phenyl) ether oxide (DPEPO) was used to formulate rigid-nature BQ. The chelate structure of DPEPO leads to a glass transition temperature (T_g) of 93°C, a higher decomposition temperature of 317°C, and a melting temperature (T_m) of 167°C. The PLQY was found to be 55.3%. The improvement might be because DPEPO in BQ has the mezzo first triplet excited energy level (^{1}T) because it is located between the first singlet excited energy level (^{1}S) and ^{3}T of TTA. Compared to Eu(TTA)$_3$(TPPO)$_2$, BQ shows improved thermal and morphological stability. The device with four-layer ITO/NPB/CBP:BQ/BCP/Alq$_3$/LiF/Al shows a maximum luminance of 632 cd/m^2 and maximum efficiencies of 4.58 cd/A, 2.05 lm/W, and 2.89%. Reddy et al. also reported similar types of complexes by utilizing 4,4,5,5,5-pentafluoro-1-(9H-fluoren-2-yl)-1,3-pentanedione (Hpffpd) as a ligand (BR, BS). The complex BR shows a PLQE of 28% in thin-film form. The same group reported, on the basis of the same results, a xanthene-based ligand 4,5-bis(diphenylphosphino)-9,9-dimethylxanthene oxide (DDXPO). It was designed to improve the rigidity and condense structural relaxation-induced energy loss and showed an improved solid-state PLQE of 48%. Utilizing the same DDXPO ligand for BT exhibited a high PLQE of 47% (chloroform solution). Enhancement observed from the doping method (complex in a PMMA film) is almost 80% [151].

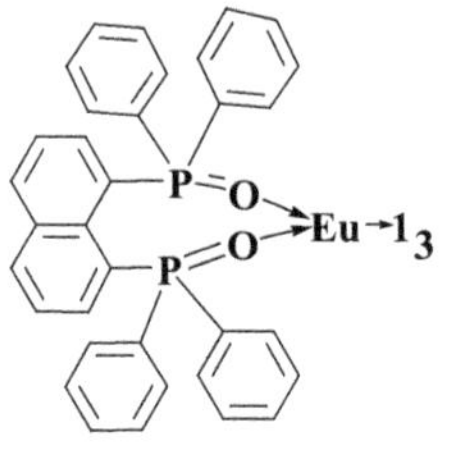

BP: Eu(TTA)$_3$NAPO

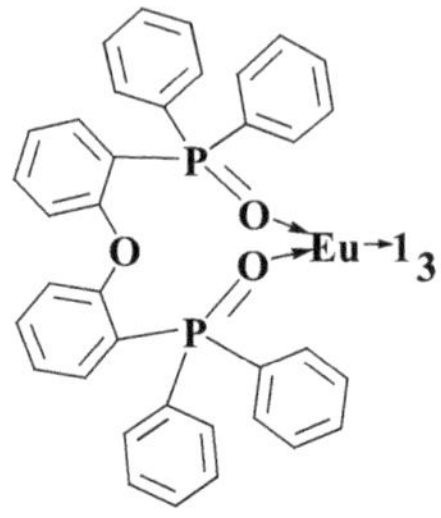

BQ: Eu(TTA)$_3$DPEPO

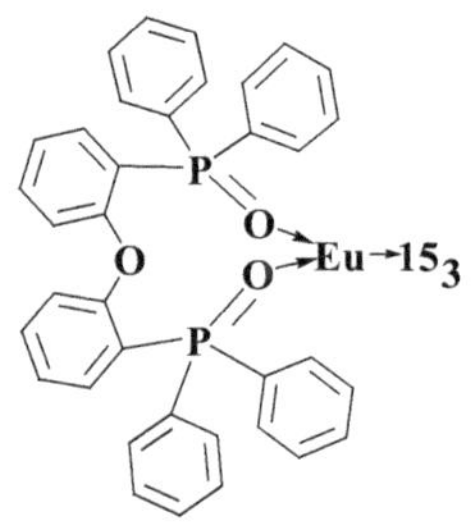

BR: Eu(pffpd)$_3$DPEPO

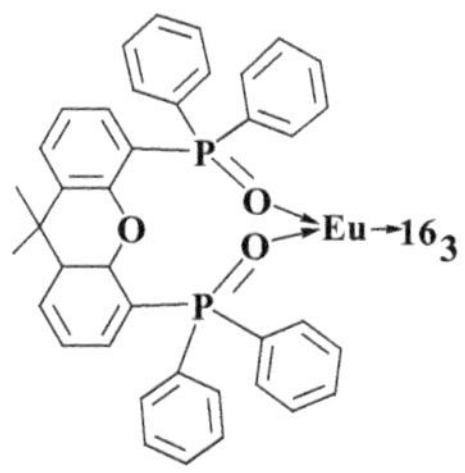

BS: Eu(pffpd)$_3$DDXPO

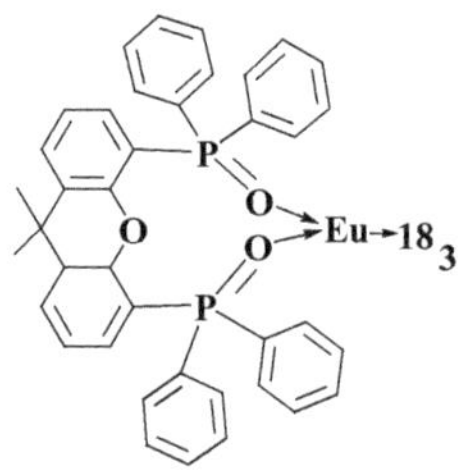

BT: Eu(CPFHP)$_3$DDXPO

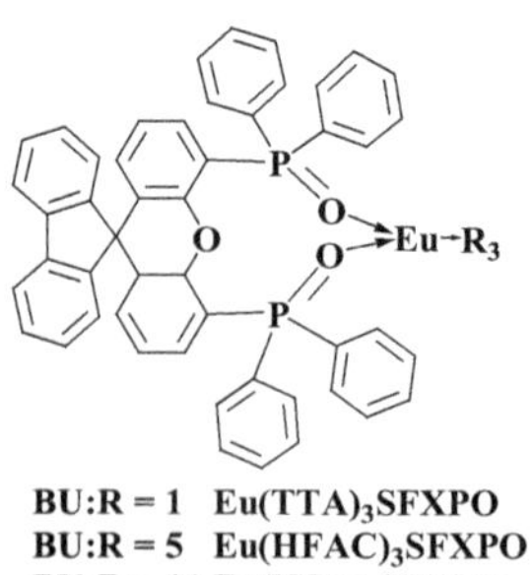

BU:R = 1 Eu(TTA)$_3$SFXPO
BU:R = 5 Eu(HFAC)$_3$SFXPO
BU:R = 11 Eu(NTA)$_3$SFXPO

The Eu(III) complexes were made based on the spiro-fluorene-xanthene-bis(diphenylphosphine) dioxide moieties with TPPO as a core for the SFXPO ligand (BU-R_1, BU-R_5, and BU-R_{11}) [152]. BU-R_1 and BU-R_{11} show pure-red emission with a peak at 612 nm with appropriate CIE coordinates (0.65, 0.32) and (0.66, 0.33), respectively. The solution-processed device with BU-R_1 shows an EQE of 2.4%, a PE of 0.6 lm/W, and a CE of 3.2 cd/A at 100 cd/m^2 brightness. However, in the case of BU-R_{11} improved EQE is 3.7% and PE 1.0 lm/W and CE 5.1 cd/A. These observations supporting the fact that the ternary complex structure is more encapsulated and protects the Eu(III) cation from vibrational quenching. It leads to increased light absorption from the antenna. In addition, the spiro-molecules improve the carrier injection and transport properties in the device.

Xu et al. [153–156] recently developed an effective postfunctionalization strategy to construct a series of PO–Eu(III) chelates with self-hosted features, ambipolar characteristics, and enhanced solubility. These ligands were used along with TTA and DBM to prepare light-emitting Eu complexes (BV-L_1 to L_{10} and BW-L_1 to L_{10}). It was observed that the solubility of the complexes was greatly enhanced, which can be attributed to the increase in the ratio of organic moieties in 4,4' positions, which also enhanced the antenna effect, allowing better charge transfer at both singlet and triplet states. DFT simulations and electrochemical analysis further verified the contributions of DPEPOAr$_n$ to the carrier-injecting/carrier-transporting ability of the complexes. With a simple single-layer spin-coated device structure of ITO/PEDOT:PSS/PVK:PBD:Eu(III) complex/Ca/Al, pure-red emissions were demonstrated with remarkably reduced driving voltages, such as a low TOV of 6 V. The first four devices had a high PLQE of 86%. BW-L_8 (DPNA = diphenylnaphthylamine) with the most enhanced electrical properties and suitable frontier molecular orbital (FMO) and triplet state locations endowed its devices with the biggest maximum luminance of >90 cd/m^2 and the highest EL efficiencies. This work verified the potential of small molecular EL Eu(III) complexes for solution-processed OLEDs through rational function integrations.

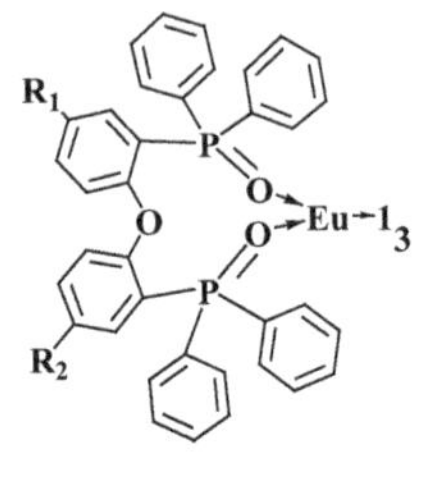

BV	
L1: R_1 = H, R_2 = CZ	Eu(TTA)$_3$DPEPOCz
L2: R_1 = R_2 = CZ	Eu(TTA)$_3$DPEPOCz2
L3: R_1 = H, R_2 = PhCz	Eu(TTA)$_3$DPEPOPhCz
L4: R_1 = R_2 = PhCz	Eu(TTA)3DPEPOPhCz2
L5: R_1 = H, R_2 = TPA	Eu(TTA)$_3$DPEPOTPA
L6: R_1 = R_2 = TPA	Eu(TTA)$_3$DPEPOTPA2
L7: R_1 = H, R_2 = DPNA	Eu(TTA)$_3$DPEPODPNA
L8: R_1 = R_2 = DPNA	Eu(TTA)$_3$DPEPODPNA2
L9: R_1 = H, R_2 = DEF	Eu(TTA)$_3$DPEPODEF
L10: R_1 = R_2 = DEF	Eu(TTA)$_3$DPEPODEF2

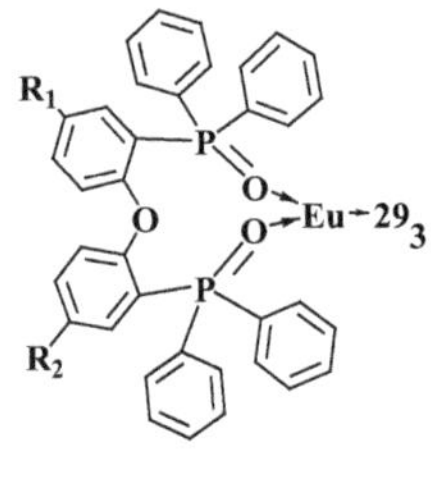

BW	
L1: R_1 = H, R_2 = CZ	Eu(DBM)$_3$DPEPOCz
L2: R_1 = R_2 = CZ	Eu(DBM)$_3$DPEPOCz2
L3: R_1 = H, R_2 = PhCz	Eu(DBM)$_3$DPEPOPhCz
L4: R_1 = R_2 = PhCz	Eu(DBM)3DPEPOPhCz2
L5: R_1 = H, R_2 = TPA	Eu(DBM)$_3$DPEPOTPA
L6: R_1 = R_2 = TPA	Eu(DBM)$_3$DPEPOTPA2
L7: R_1 = H, R_2 = DPNA	Eu(DBM)$_3$DPEPODPNA
L8: R_1 = R_2 = DPNA	Eu(DBM)$_3$DPEPODPNA2
L9: R_1 = H, R_2 = DEF	Eu(DBM)$_3$DPEPODEF
L10: R_1 = R_2 = DEF	Eu(DBM)$_3$DPEPODEF2

9.3.5 Pyridine Based Complexes

Zheng et al. [157] showed that the introduction of methoxy groups in the bipyridine ligands can enhance their EL properties (BX, BY). They fabricated an OLED device with the general device configuration ITO/NPB/Eu:CBP/BCP/Alq_3/LiF/Al and showed proper practical results. The complex BX realized a maximum brightness of 877 cd/m^2, and its maximum CE of 1.19 cd/A was achieved at a luminance of 3 cd/m^2. The device with a TTA complex had a maximum brightness of 181 cd/m^2 and a peak CE of 0.32 cd/A.

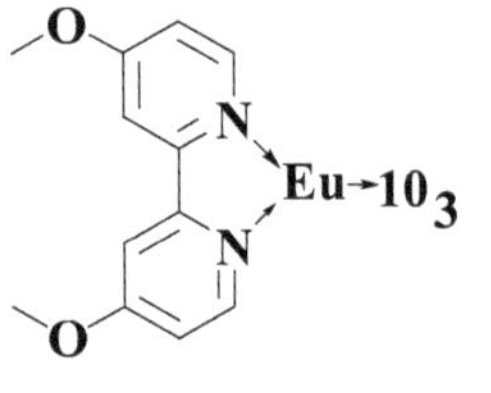

BX:Eu(BTA)$_3$dmbipy

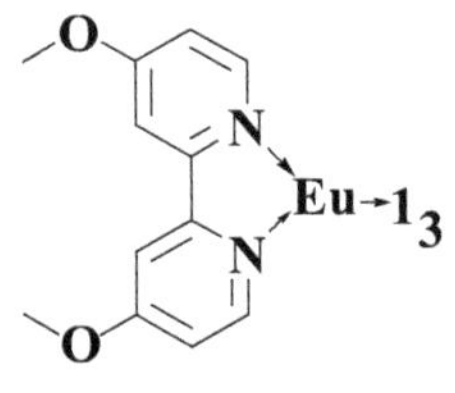

BY:Eu(TTA)$_3$dmbipy

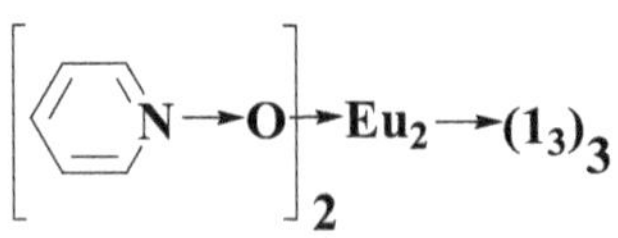

BZ:(TTA)$_3$Eu(PyO)$_2$Eu(TTA)$_3$

Similarly, You et al. [158] synthesized a pure-red-emitting binuclear Eu complex containing pyridine *N*-oxide as a ligand. The usual quenching effects thus observed due to the presence of binuclear complex BZ were decreased due to the presence of oxygen atoms in the same plane, encapsulating the Eu(III) ions. The devices showed pure-red emission at a peak wavelength of 612 nm with an FWHM of 3 nm, which is a characteristic emission from the Eu^{3+} ion based on electric dipole (ED)$^5D_0 \rightarrow {}^7F_2$ transition. The maximum brightness and EL efficiency reached 340 cd/m^2 at a driving voltage of 19 V and 2.4 cd/A (0.78 lm/W) at a current density of 0.14 mA/cm^2, respectively. Also it was seen that the transitions at 652 nm ($^5D_0$–$^7F_3$) and 702 nm ($^5D_0$–$^7F_4$) increased in intensity with the applied bias. This may be attributed to the increased probability for the ED transition, which is caused by the increase of the electric field surrounding the microenvironment.

9.3.6 Multi-Dentate-Nitrogen Beard Heterocyclic Availed Complexes

To develop β-diketone, 2-acetylfluorene-4,4,4-trifluorobutane-1,3-dione (AFTFBD)-based di- and trinuclear molecular Eu(III) complexes were introduced by Gong and coworkers. Two and three PyBM units were used to get the final construction of Eu molecular structures CC [159] and CD [160]. The obtained high decomposition temperature (340°C) indicated the durable coordinating ability between the PyBM moiety and Eu(III). In addition, both complexes showed unlike stoichiometry. PLQEs were found to be 15.4% and 16.2%, respectively. A PE of 0.3 lm/W was obtained when implanted with InGaN chips.

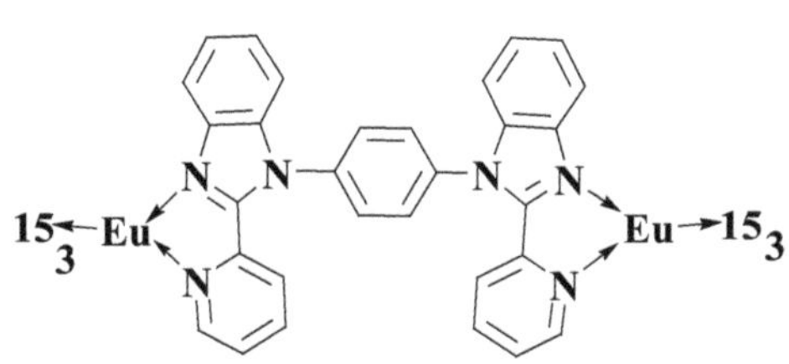

CC: $[Eu(AFTFBD)_3]_2$(1,4-bmb)

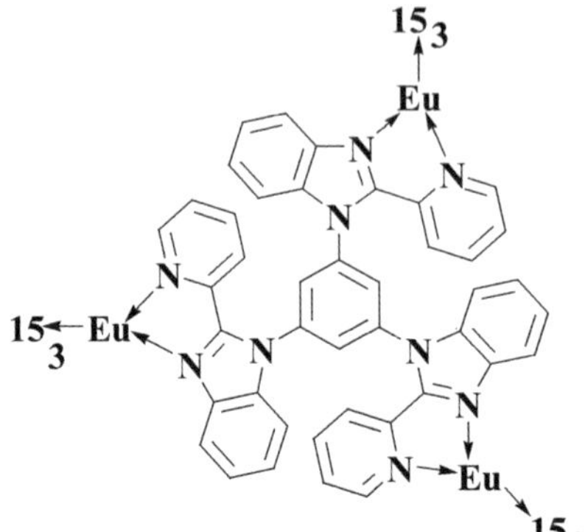

CD: $[Eu(AFTFBD)_3]_3(tmb)$

Chen and group in the same year prepared several derivatives of phen (PhoR: CF, CG, and CH), 2,2'-bipyridine (BpoR: CI, CJ, and CK) [161], and their corresponding Eu complexes along with TTA and studied their photophysical and electroluminescent properties. It is known that CE (Php = 2-(pyridyl)-1,10-phenanthroline) acts as a tridentate ligand, hence conferring the complexes with extra stability and making them more suitable for vacuum deposition. In addition, the complexes have a coordination number (CN) of 9, which leads to increased distortion of the complex, which is partially responsible for the bright red emission at 612 nm. It was inferred from the experiments that these complexes had high PLQYs due to suitable triplet energy levels of the ligands, between 19,724 and 22,472 cm^{-1}, for the sensitization of Eu(III) ($^5D_0$: 17,500 cm^{-1}). The phenanthroline-substituted complexes had a PLQE value almost twice that of their bipyridine counterparts because of their comparatively rigid structures. A four-layered device was prepared using the complex CG (PhoB = 2-(5-phenyl-1,3,4-oxadiazol-2-yl)-1,10-phenanthroline) as a dopant emitter in CBP, with a structure of TPD (30 nm)/CG:CBP (7.5%, 20 nm)/ BCP (2,9-dimethyl-4,7-diphenyl-1,10-phenanthroline, 20 nm)/Alq_3 (tris(8-hydroxyquinoline), 30 nm), and exhibited a pure emission from Eu^{3+} ions. The highest efficiency obtained was 5.5 lm/W, 8.7 cd/A and the maximum brightness achieved was 1086 cd/m^2. At a practical brightness of 100 cd/m^2, the efficiency remained above 2.0 cd/A.

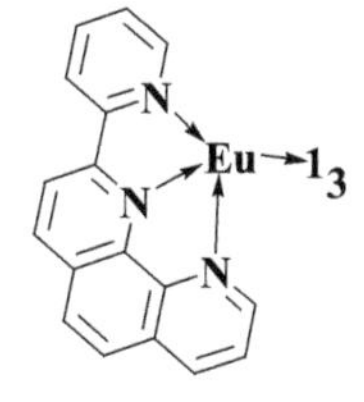

CE: Eu(TTA)$_3$PhP

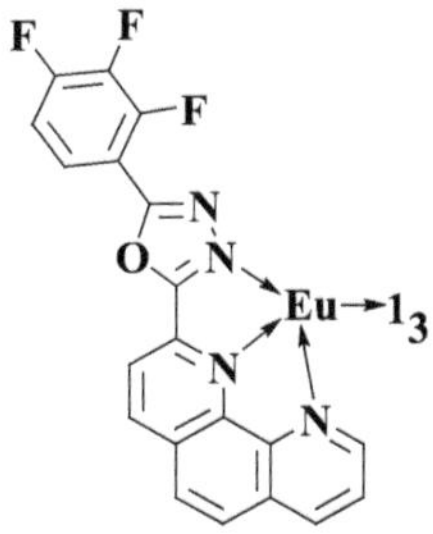

CF: Eu(TTA)$_3$PhoA

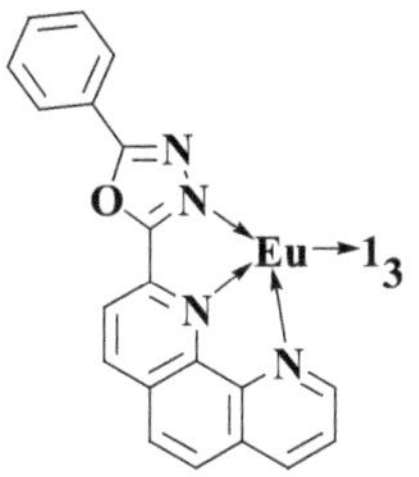

CG: Eu(TTA)$_3$PhoB

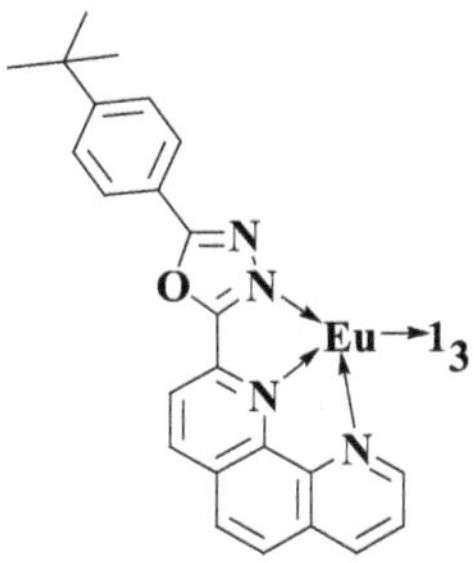

CH: Eu(TTA)$_3$PhoC

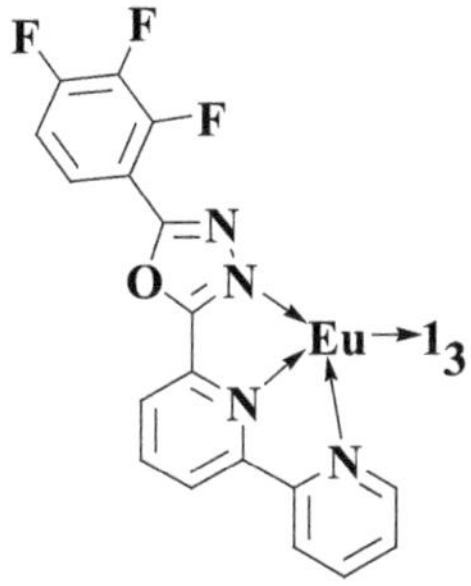

CI: Eu(TTA)$_3$BpoA

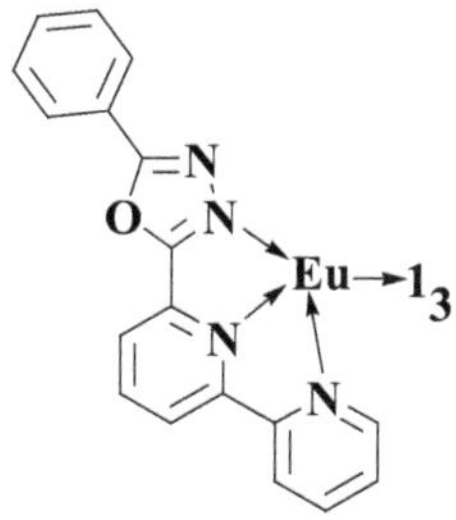

CJ: Eu(TTA)$_3$BpoB

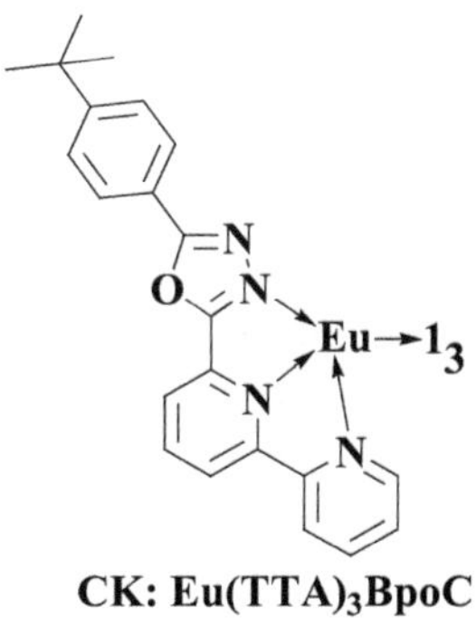

CK: Eu(TTA)$_3$BpoC

Silva et al. [157, 162, 163] produced a series of complexes (CL, CM, and CN) containing 2,4,6-tri(2-pyridyl)-1,3,5-triazine (TPTZ). It was observed that the presence of heterocyclic triazines, which themselves have a high electron affinity, facilitates better performance of the Eu layer in the devices. The complexes were turned into a device of the following sequential arrangement: ITO/NPB/CBP:Eu^{3+} complex/BCP/Alq_3/LiF/Al. All the devices showed characteristic red emission of Eu(III) along with a weak exciplex emission in the 500–600 nm region at very high voltages. The CM complex had a CE of 4.76 cd/A at 100 cd/m^2 and a maximum brightness of 781 cd/m^2, whereas the highest QY amongst the compounds was 69.7% in chloroform.

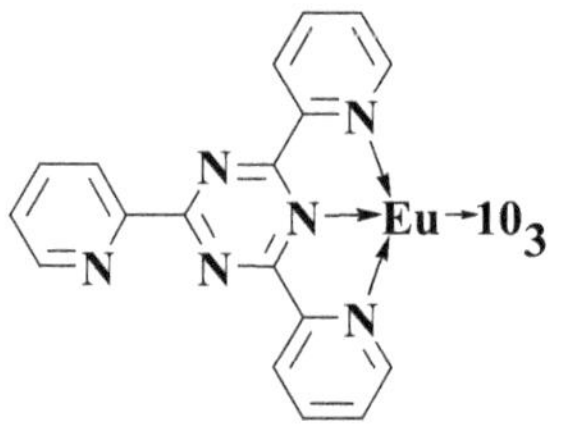

CL: $Eu(BTA)_3TPTZ$

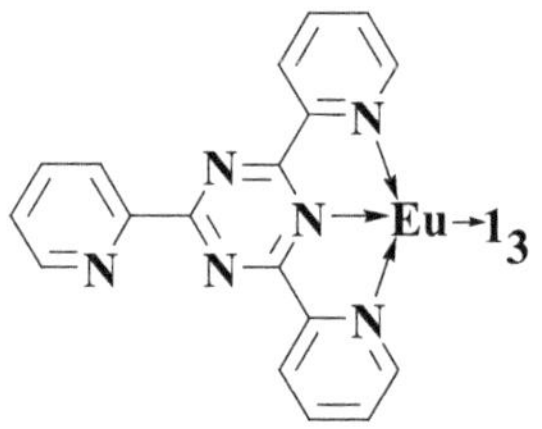

CM: $Eu(TTA)_3TPTZ$

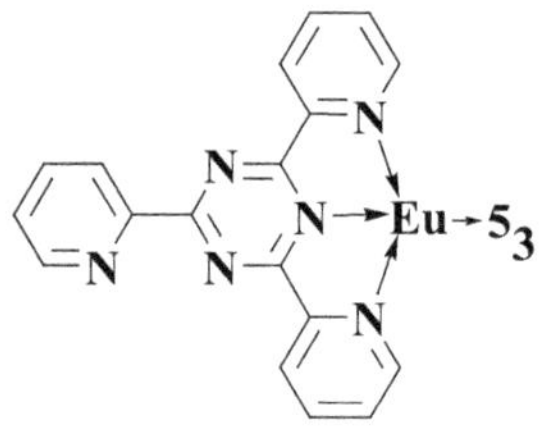

CN: $Eu(HFAC)_3TPTZ$

Law et al., in 2009, synthesized a white-light-emitting complex utilizing a neutral ligand consisting of pyrazole and triazine rings along with a diethylamine substituted at the *para* position of the triazine ring to enhance the photophysical properties of the complex. The ligand had a LUMO of ~2.9 eV, thus making it proficient in efficiently transferring and injecting electrons. As a whole the complex contained a double-emission center of bluish-green and red, combined to give a pure-white emission. The device was fabricated with the structure of ITO/NPB/CBP:CO/ BCP/Alq_3/LiF/Al, which gave red color emission at 8 V and white emission at 16 V with color coordinates of (0.34, 0.35) and a maximum brightness reaching up to 350 cd/m^2. The obtained results clearly indicated that the influence of the transporting capacity leads to a color change in the device [164].

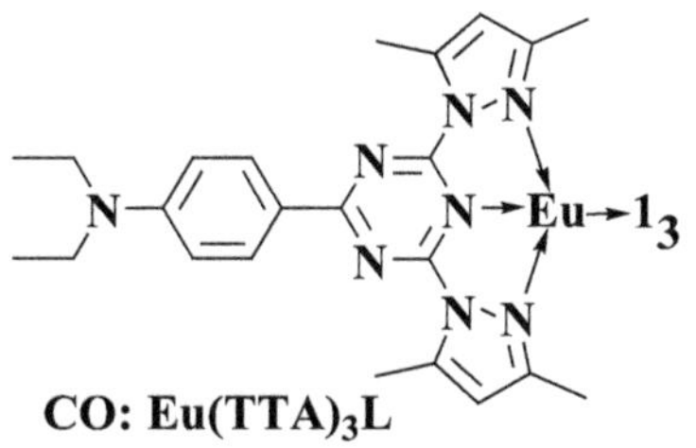

CO: $Eu(TTA)_3L$

9.4 DBM as an Anionic Ligand

The β-diketone-based complexes, which belong to the phenyl assembly (DBM), were mentioned in Fig. 9.10. In the report from Kido et al., in 1994, [165] based on a ternary Eu^{3+} complex, (CP) OLEDs, tremendous efforts have been devoted in the field of OLEDs using Ln complexes. This trivalent Eu complex was introduced into the device as an emissive material by doping method. The device structure is as follows: ITO/TPD (400 Å)/CP:PBD (1:3 mole ratio) (300 Å)/Alq_3 (300 Å)/Mg:Ag, TOV 6 V, and a maximum brightness of 460 cd/m^2 at driving voltage 16 V. The device shows a red emission peak at 614 nm. Co-deposition of PBD prevents the self-quenching, and addition of phenanthroline gives stability, creates a volatility nature, and influences fluorescence intensity. By decreasing the thickness of the emissive layer (300–200/100 Å), a broad peak is observed at 520 nm due to the migration of the emitting layer and the color (red) of the device changes.

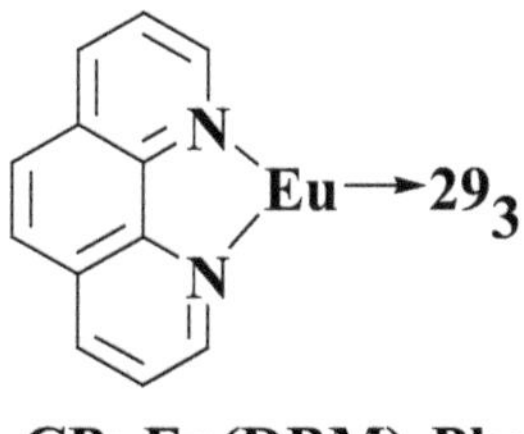

CP: $Eu(DBM)_3Phen$

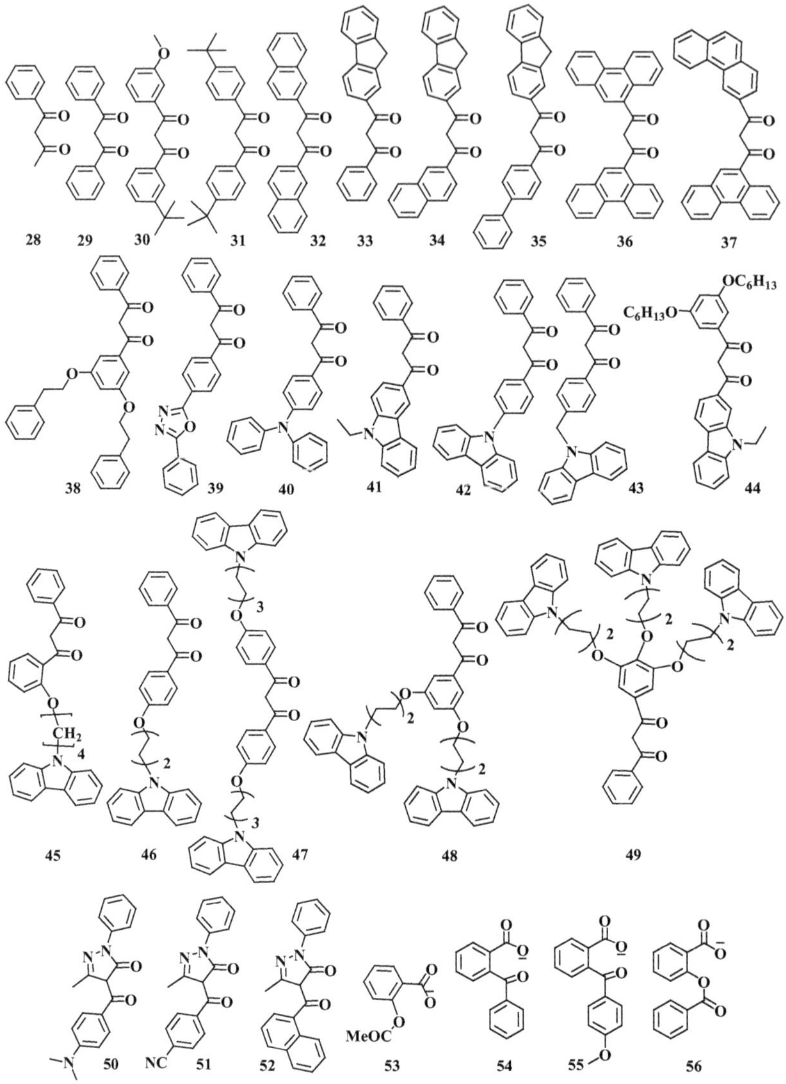

Figure 9.10 The β-diketonate (DBM-based) anionic antennas for Eu(III) complexes.

Heil and coworkers [166] introduced the complex CP in a device with the structure ITO/CuPC (5 nm)/α-NPD (*N,N′*-bis(naphthalen-1-yl)-*N,N′*-bis(phenyl)benzidine) (35 nm)/CP (20 nm)/BCP (15 nm)/Alq (25 nm)/LiF/Al, and a maximum brightness of 50 cd/m^2 was observed at 15 V. One more device was introduced with NPB instead of an NPD layer to given more brightness. The fabricated device is ITO/CuPC (5 nm)/α-NPB (35 nm)/NPB:CP (20:80 %) (20 nm)/ BCP (15 nm)/Alq (25 nm)/LiF/Al, and the maximum brightness observed is 200 cd/m^2 at 15 V. They demonstrated on the Eu(III) complex CP, how a dye molecule can be effectively integrated and luminesce in an organic EL layer structure, even if a simple energy transfer from excited host molecules to the dye is not possible due to lack of energy overlap. A suitable blocking layer's incorporation is a successful concept and is applicable to balanced transport, leading to high luminance efficiency. The complex CP was introduced into the device, which consists of a derivative of 1,2,4-triazole as an HBL and poly(*N*-vinylcarbazole) (PVK) as the host. The device, ITO/ PVK:CP:TPD:PBD/TAZ/Alq_{3}/MgAg, shows a brightness of 500 cd/ m^2 (18 V, 230 mA/cm^2).

The complex CP also doped with a PVK host matrix. Devices fabricated using a simple solution-based approach (spin coating) are ITO/PEDOT:PSS/PVK:CP/Ca/Al; the obtained brightness is 130 cd/ m^2 (25 V, 400 mA/cm^2), CE η_c is 0.063 cd/A, and EQE η_{ext} is 0.031% [167, 168].

9.4.1 Modification of DBM

Bazanet et al. incorporated carbazole and hexyloxybenzoyl groups with europium. The resistance to crystallization, solubility in organic solvents, and exceptionally low self-quenching prompted them to examine EL properties of this complex. They proposed a single-layer device ITO/CQ (600 Å)/Ca/Al [169]. The results demonstrated that the ligand design was successful in facilitating charge (electron and hole injection) and transport into and across the emissive (Eu complex) layer. The TOV is 5.3 V. At 15 mA/cm^2, luminescence is 9 cd/m^2 with an external EL quantum efficiency of 0.08%. Compared to the single-layer device, the TOV is higher for the second device, 8.7 V. The device is ITO/PVK (30 nm)/CQ (700 Å)/Ca/Al with an EQE of 0.3% at a luminescence of 50 cd/m^2 and a current density of 15 mA/cm^2.

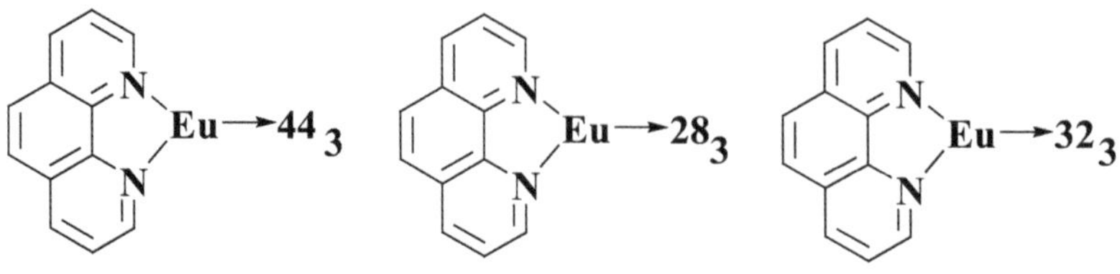

CQ: Eu(44)$_3$Phen CR: Eu(mppd)$_3$Phen CS: Eu(dnm)$_3$Phen

Bazan and coworkers [170] reported phenanthroline with a β-diketo-skeleton-based Eu complexes: CR, CP, and CS, and 5% of these Eu(III) complexes are introduced in the device with CNPPP as a blended material, leading to PL efficiencies of 0.17%, 0.26%, and 0.27%, respectively. The device configuration is ITO/PVK/2 wt% complex-CNPPP/cathode, showing a TOV of ~20 V and EL efficiencies are 0.1%, 0.7%, and 1.1%, respectively. The β-diketonate and phenanthroline ligands can prevent concentration quenching by sterically shielding neighboring Eu(III) ions from each other. The authors observed that the complex π–π* gap is matched with the CNPPP gap, showing the highest EL efficiency.

Tian et al. introduced three novel dendritic Eu complexes containing grafted carbazole units (CT, CU, CV, and CW). CU, CV, and CW [171, 172] are incorporated into a device with ITO/NPB/complex:CBP (2.5 wt%)/BCP /Mg:Ag. The relative luminescence QE is 3.3, 7.9, and 4.5, respectively, in the film state. The inserted single Eu complex, CW, in the device shows white emission. It is thought that blue and green emission is by forming an exciplex and an electroplex with CBP and β-diketonate, respectively. The maximum EQE is 1.1%, maximum brightness 229 cd/m^2, and current density 194 mA/cm^2, achieved at 20.5 V. Among three complexes, the cabazole-functionalized Eu complex CW shows weaker luminescence than the dicarbazole complex. The reason is that the special hindrance of three carbazole units is incorporated into one end of β-diketonate. This leads to less energy transfer between carbazole and β-diketonate. However, compared to the CT complex, these complexes show brighter red light. The results suggested that modifying ligands not only tunes the carrier transportation properties but also provides a meaningful result due to the use of electroplexes or exciplexes to realize broad or even white EL.

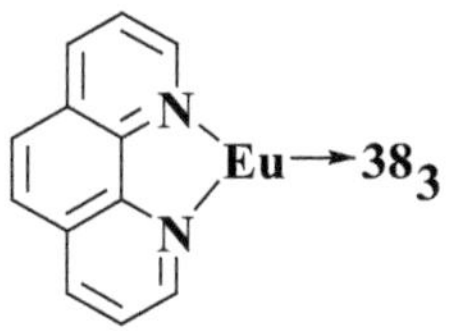

CT: Eu(BPPD)$_3$Phen

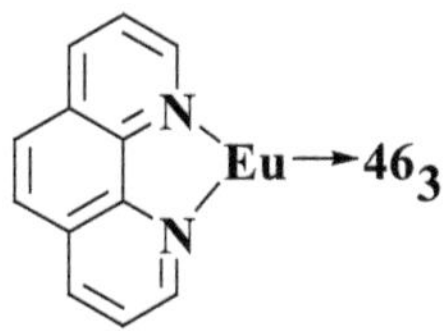

CU: Eu(MCPD)$_3$Phen

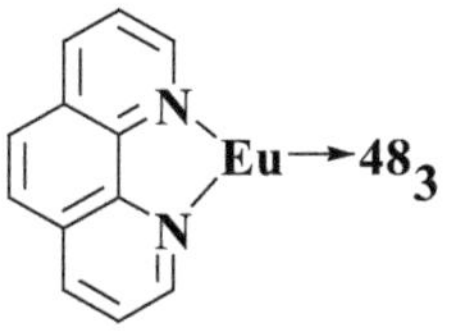

CV: Eu(BCPD)$_3$Phen

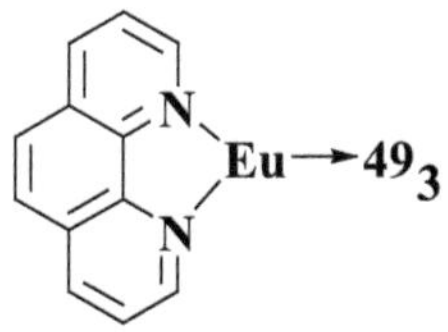

CW: Eu(TCPD)$_3$Phen

Bazan et al. made better investigations than earlier and reported a bulky t-butyl group–based Eu complex. In a comparison of the PL of CP and CX [173], CP has higher PL due to energy transfer hindered by bulky t-butyl groups. Loading at 0.19 mol% of the Eu complex, voltage increases to 6 V, and at 0.56 mol% for the device with CP and CX it is 8 V and 14 V, respectively. However, the presence of t-butyl groups in the complex reduces self-quenching and phase separation. It causes a longer operating lifetime and lower TOVs, and it increases power efficiencies.

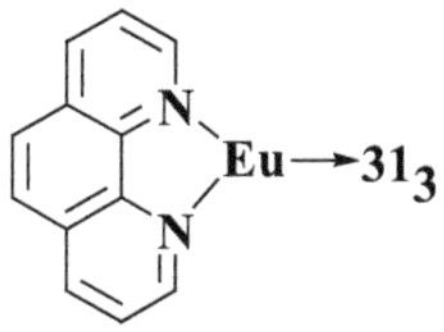

CX: Eu(t-t)$_3$Phen

9.4.2 Replacement of DBM Equivalents by Other Ligands

Noto and coworkers established a new Eu complex with 2,4-dioxopentanedinitrile (1,3-dicyano-1,3-propanedione (DCNP), DBM, and phen (1:2:1): CY. The Eu complex [174] was introduced into two devises with different host materials. The first device ITO/TPD (50 nm)/CY:PBD (10 wt%, 10 nm)/TAZ (25 nm)/Alq (30 nm)/Mg:Ag (10:1, 200 nm) exhibits an external efficiency of 3.5%, luminescence of 10 cd/m^2, and a PE of 2.0 lm/W at a current density of 0.17 mA/cm^2. Another device ITO/TPD (50 nm)/CY:CBP (10 wt%, 20 nm)/BCP (15 nm)/Alq (30 nm)/Mg:Ag (10:1, 200 nm) exhibits an external efficiency of 1.6%, luminescence of 84 cd/m^2, and a PE of 0.69 lm/W at a current density of 2.9 mA/cm^2. The first device has better balance charge-transporting properties than the rest of the second device.

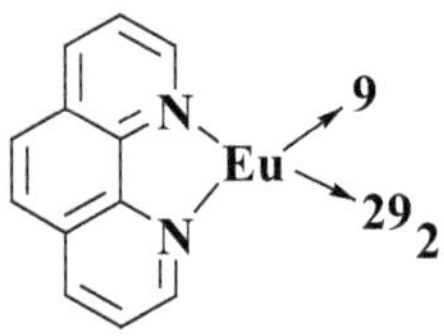

CY: Eu(DCNP) (DBM)$_2$ (Phen)

9.4.3 Modification of Phen

Lin et al. established various types of devices using complexes CP and CZ [175]. Both show a sharp emission peak at 614 nm. The device had TPD as the HTL and OXD-7 as the ETL and all organic layers of 50 nm, and the aluminum (Al) cathode was 200 nm. Two-layer structures showed poor carrier/injection capability. TPD/CZ/OXD-7 has more efficiency than the device with TPD only. It suggested poor hole-blocking and injection/transport capacity. TPD/CZ/OXD-7 has higher efficiency than TPD/CP/OXD-7 at a fixed voltage. The results signify that the device with Bath (CZ) has a better carrier transport tendency than other Eu complex devices. The presence of additional phenyl rings in the complex leads to lowering of radiationless transition.

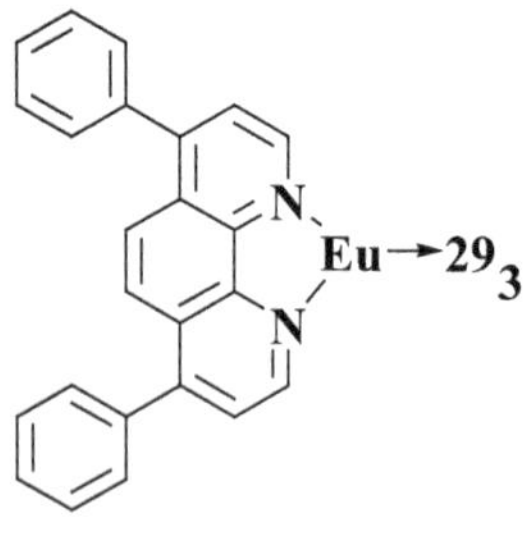

CZ: Eu (DBM)$_3$(Bath)

Liang and coworkers used the complex CZ [176] as the electron transport–emitting layer in ITO/PVK or NPB (50–100 nm)/CZ (50 nm)/Mg:Ag. The device showed bright-red EL emission with PVK. When the PVK is 100 nm, the TOV is 4 V and luminescence 400 cd/m^2 at 15 V. But in the case of NPB (50 nm), different results are observed with color change (white). The same complex was further developed by the same group in the same year and CZ was used as the electron transport–emitting layer in the ITO/TPD (30 nm)/CZ:TPD (30 nm)/CZ (50 nm)/Mg:Ag (0.9:0.1) device. The device has a TOV of 3 V, the observed luminescence is 820 cd/m^2 at 18 V, the EQE is 1%, and luminescence efficiency at a brightness of 100 cd/m^2 is 0.40 lm/W. Without a doping device emission is orange-red instead of pure red because exciplex forms an interface between Eu and the TPD layer. In this case only a luminescence of 150 cd/m^2 is observed at 18 V.

Hong and coworkers [177] introduced CZ into devices as a mixed layer for reducing the quenching of luminescence. Device fabrication was followed in the following order: ITO/TPD (30 nm)/TPD:CZ (1:1,1:2, and 1:3 ratios) (30 nm)/CZ (60 nm)/Ag:Mg (0.1:0.9). The holes are transferred from TPD to the Eu complex in bilayer devices, but addition of a mixed layer influences the carrier transport property. Mixed-layer devices showed higher efficiency than bilayer devices. The device with 1:2 ratios shows more efficiency than the rest of the devices. The EQE is 4.6% and the current density 0.01 mA/cm^2. The same group [178] also synthesized another complex using oxidazole substituted in phen at the fourth position. The LUMO level of the ligand was found to be −2.72 eV, thus facilitating proper electron transportation. A device was prepared with the structure ITO/PEDOT:PSS/PVK/PFO:PBD:DD/Ba/Al. The results included a maximum luminance of 568 cd/m^2 and a peak EQE of 1.26%.

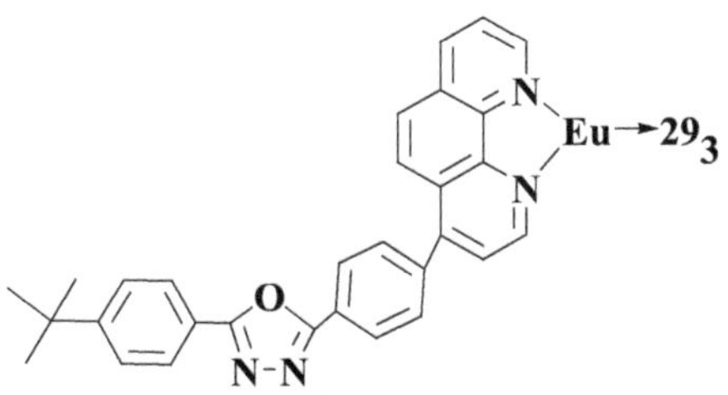

DD: Eu(DBM)$_3$(BuOXD-Phen)

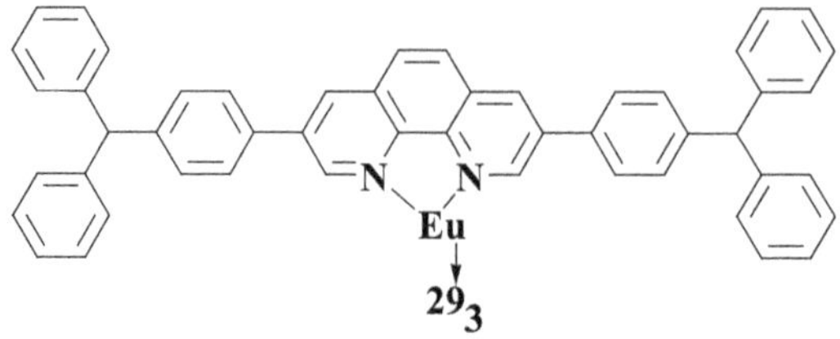

DE: Eu(DBM)$_3$(DTPA-Phen)

Liu et al. synthesized a di-triphenylamine-substituted phenanthroline and its corresponding Eu complex with dibenzoylmethanato ligands (DE) [179, 180]. The solid-state sample showed an absorption band up to 500 nm, which overlaps with that of the host matrices. thus allowing better charge transfer through Förster resonance energy transfer (FRET). The blend of the material with the host was used to create a bilayer device, which gave a narrow red emission at 612 nm with a maximum FWHM of 10 nm as the dopant concentration was varied between 1% and 8% wt./wt. The highest internal quantum efficiency (IQE) of 1.8% photons per electron at a current density of 1.2 mA/cm^2, with a maximum brightness of 1333 cd/m^2 at 173.2 mA/cm^2, was achieved from the device at 1% wt./wt. dopant concentration.

On the basis of a simple bilayer structure in the form of ITO/PEDOT:PSS/PVK/PFO:PBD:DF/Ba/Al [181], a peak luminance of 1783 cd/m^2 (at 2 wt% dopant concentration), and maximum efficiencies of 2.5% for EQE and 3.8 cd/A (at 1 wt% dopant concentration) for CE were realized. Fluorination on PyPhen forms fluorophenyls to further enhance conjugation and electron affinity and the site isolation effect. Efficient energy transfer indicates a high PLQE of 55%. The four-layer device, ITO/TPD/CBP:DG/Bphen/Alq$_3$/LiF/Al, has a maximum luminance of 1766 cd/m^2 and peak efficiencies of 4.6 cd/A and 2.27% [182].

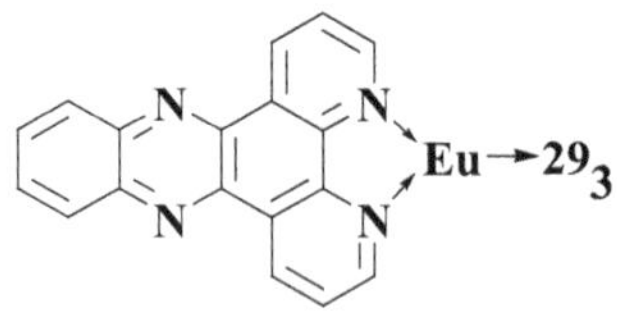

DF: Eu(DBM)$_3$(DPPz)

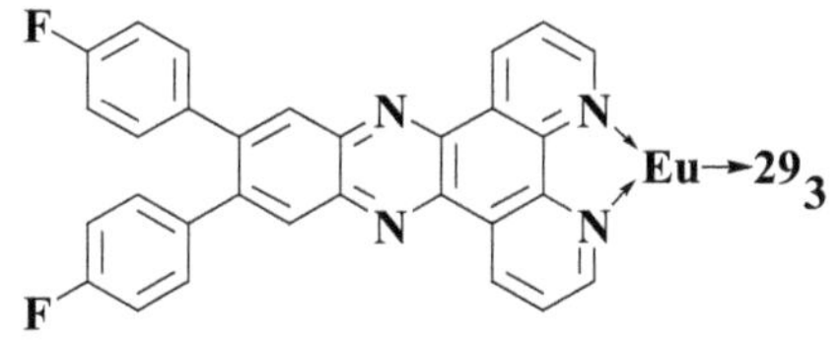

DG: Eu(DBM)$_3$(BFPP)

Min Sun and coworkers [183] synthesized a tris(dibenzoylmethanato)(2,4'-triphenylamino) imidazo[4,5-f]-1,10-phenanthroline) Eu(III) complex DH. It exhibits pure-red emission. TPA and phen are the hole-transporting material (HTM) and electron-transporting material (ETM), respectively. The Eu complex with single-layer device fabrication is ITO/DH (60 nm)/ Mg:Ag (0.9:0.1)/Ag, which exhibits a TOV of 8 V and a brightness of 19 cd/m^2 at 13.5 V and 280 mA/cm^2. A four-layer device ITO/ TPD (20 nm)/DH (40 nm)/BCP (20 nm)/AlQ (40 nm)/$Mg_{0.9}$ $Ag_{0.1}$/ Ag, with a driving voltage of 6 V, gives a maximum luminescence of 1305 cd/m^2 at 16 V and 225 mA/cm^2. The maximum EQY is 0.85% at 7.5 V and luminous yield 1.44 lm/W at 0.25 mA/cm^2.

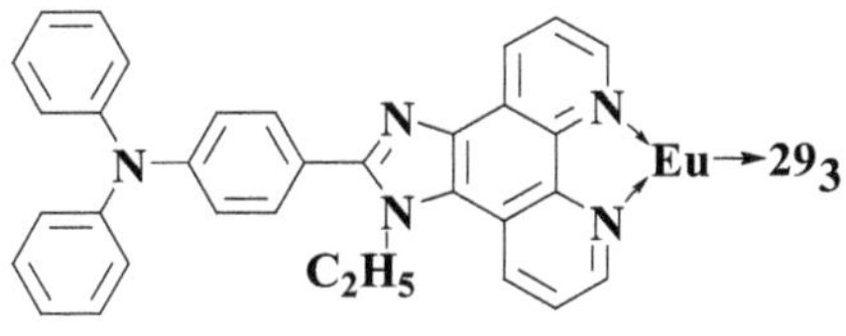

DH: Eu(DBM)$_3$(TPIP)

Li Yingkui [184] synthesized an OXD-containing Eu complex (DI, DJ). Insertion of an OXD-linked complex in the device followed, ITO/ m-TDATA/NPB/CBP:DJ/BPhen/Alq$_3$/LiF/Al, which had a device luminance that reached up to 610 cd/m^2 and a peak CE of 3.67 cd/A. Also the compound had a lifetime of 0.164 ms and a PLQE of 41%.

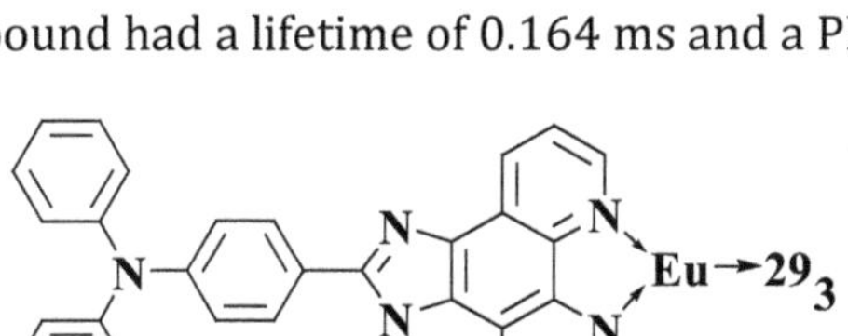

DI: Eu(DBM)$_3$(IPD)

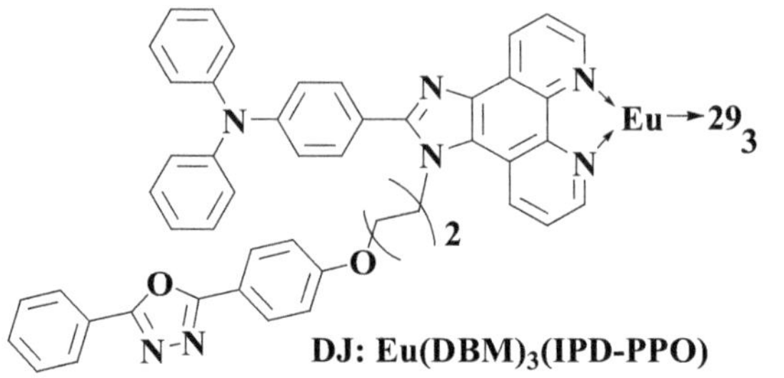

DJ: Eu(DBM)$_3$(IPD-PPO)

Gong and coworkers prepared a ligand combining TPA, OXD, and IP and formed a donor–acceptor–acceptor structure (DK). The resulting complex had a thermal decomposition temperature of 431°C and T_g of 153°C, respectively. The complex gave a blue-dominated emission in tetrahydrofuran (THF) [185].

Zhang and coworkers [186] made a Eu(III) complex based on the OXD moiety as a neutral as well as a substituted carbazole group as an antenna (DL). The carrier injection and transportation probability was proper in the complex, and it is made to avoid steric hindrance. Red emission was observed by an emission spectrum and showed LMET. The trilayer (undoped) device is ITO/NPB/DL/Alq_3/LiF/Al, which showed a maximum luminance of 199 cd/m^2 and a CE of 0.69 cd/A. The CBP-doped device showed a maximum luminance of 1845 cd/m^2 and a CE of 2.62 cd/A with a reduced TOV (5.5 V), indicating that the host material directed the efficiency toward a high value.

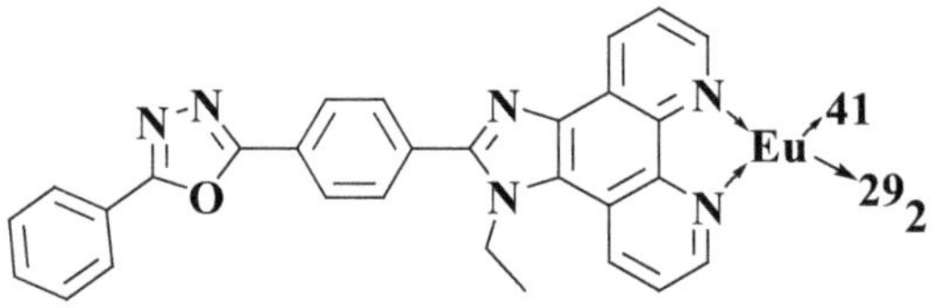

DL: Eu(DBM)$_2$(41)

Liu and coworkers utilized bipolar transporting Eu complexes by incorporating both hole- and electron-transporting moieties (carbazole and oxadiazole) units into a phenanthroline ligand [187]. The measured emission QY values of CP, DM-L_{1-5} were 0.78%, 9.9%, 10.5%, 10.1%, 9.5%, and 10.3%, respectively. These values aroundare 12-fold higher than those of the Eu(DBM)$_3$(phen) emission QY. DM-L_2 showed the highest QY; it indicates that a bipolar transporting unit into the ligand of phenanthroline is facilitating enhancement of emission QY. It is known that fluorination leads to reduction of the LUMO level. The same is also observed for the DM-L_1 complex.

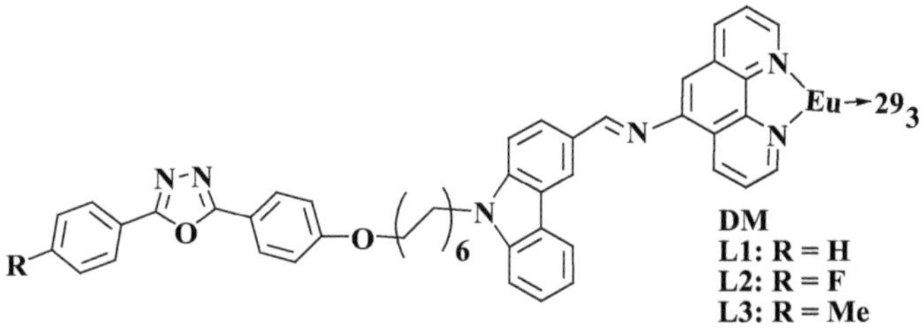

DM
L1: R = H
L2: R = F
L3: R = Me
L4: R = OMe
L5: R = t-But

$Eu(DBM)_3RPhOXD_6Cz$-Phen

The reported binuclear Eu(III) complex involving carbazole and oxadizaole fragments shows red EL [188]. Two intense UV-Vis absorption bands are observed at ~282 nm and 352 nm and show pure-red emission at 612 nm. The ligand's emission intensities were proportional to the dimensions of the terminal groups at OXD. The incomplete LMET process was probably caused by steric hindrance. Introducition into single-emitting-layer polymer light-emitting devices revealed saturated red emission at dopant concentrations from 1 wt% to 8 wt%. PLQEs of the mono- and binuclear complexes were almost similar. The focus on the influence of terminal groups was made less. The fluorine-based complex made a device by spin coating, and the configuration is ITO/PEDOT:PSS/PVK:PBD:DN-L_1/LiF/Al with observed brightness of 48.5 cd/m^2. This reduced the brightness expected due to the formation of exciplex and the pathetic electro-activity of the intricate long alkyl chain.

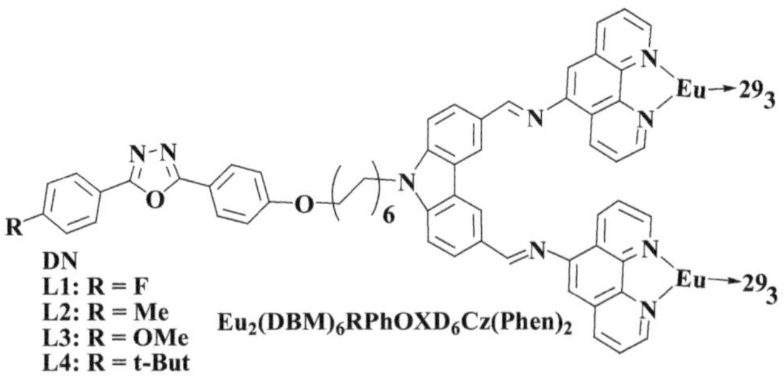

DN
L1: R = F
L2: R = Me
L3: R = OMe
L4: R = t-But

$Eu_2(DBM)_6RPhOXD_6Cz(Phen)_2$

Xin et at. formed tris(dibenzoylmethanato){1-ethyl-2-(*N*-ethyl-carbazole-yl-4) imidazo[4,5-f]-1,10-phenanthroline}europium(III),

DO, introduced into single- and multilayer devices as an emissive layer. In the single-layer device the observed TOV was 7 V and the higher brightness 20 cd/m^2 at 16 V. The multilayer device ITO/TPD (20 nm)/DO (40 nm)/BCP (20 nm)/Alq (40 nm)/$Mg_{0.9}$:$Ag_{0.1}$ (200 nm)/Ag (80 nm) shows better results than the device with Alq_3 (30 nm). The observed high PE is 2.7 lm/W at 5 V and 0.5 cd/m^2. Below 12 V, the characteristic emission from the centered Eu^{3+} ion is 350 cd/m^2 and 0.41 lm/W. By increasing the voltage below 16 V, saturated red emission (1500 cd/m^2) is observed. The high brightness is 2000 cd/m^2 at 20 V with orange emission was observed, and it is due to increasing of the emission from Alq_3 [189, 190].

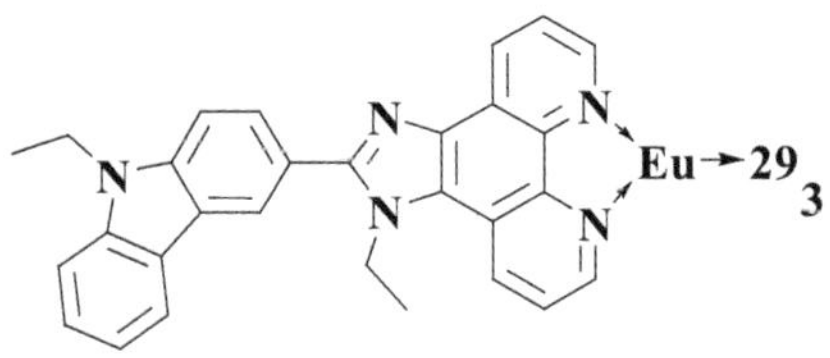

DO: $Eu(DBM)_3$(Phencarz)

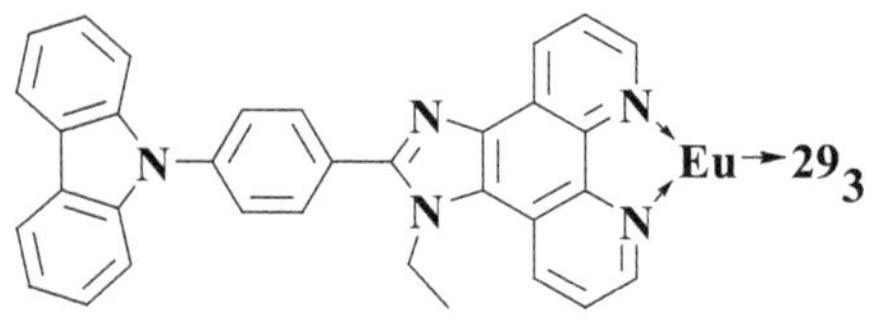

DP: $EU(DBM)_3$CPIP

Guan and coworkers synthesized a Eu complex with tris(dibenzoylmethanato)(1-ethyl-2-(4'-carbazole-9-yl) phenyl imidazo [4,5-f]-1,10-phenanthroline) Eu(III), DP [191]. The structure consists of a connectivity change as well as an increment of the conjugation. It possesses virtuous stability and hole transport property. The Eu complex was introduced into different devices as an emitting layer. Double-layer device fabrication is ITO/TPD (50 nm)/DP (60 nm)/$Mg_{0.9}Ag_{0.1}$ (200 nm)/Ag (80 nm) and exhibits Eu^{3+}-based pure-red emission with a TOV of 4 V, a brightness of 537 cd/m^2 at 13 V, and a PE of 0.089 lm/W at 6 V, 33 cd/m^2. A triple-layer device with the configuration of ITO/TPD (40 nm)/DP (30 nm)/CZ (50 nm)/$Mg_{0.9}Ag_{0.1}$ (200 nm)/Ag (80 nm) shows a TOV of 6 V, a highest brightness of 1050 cd/m^2 at 16 V, and the highest power efficiency of 0.50 lm/W at 8 V, 8 cd/m^2. A four-layer device with the configuration of ITO/TPD (40 nm)/DP (40 nm)/BCP (20 nm)/Alq_3 (10 nm)/$Mg_{0.9}Ag_{0.1}$ (200 nm)/Ag (80 nm) shows a TOV of 4 V and a highest power efficiency of 0.54 lm/W at 7 V, 56 cd/m^2. At a voltage greater than 12 V, emission originates from the central Eu ion and brightness is 375 cd/m^2, PE 0.19 lm/W. The observed a highest brightness of 1460 cd/m^2 at 14 V is seen in the form of orange emission because of the increased emission from Alq_3 not observed from TPD. It is suggested that although the Alq_3 layer is introduced, there is lower electron mobility in the Alq_3 layer.

Bian and coworkers synthesized two ligands, 2-(4'-dimethylaminophenyl)imidazo[4,5-f]-1,10-phenanthroline and 2-(4'-dimethylaminophenyl)imidazo[4,5-f]-1,10-phenanthroline, and chelated with Eu(III) complexes with DBM to form DQ and DR [192, 193]. Single-crystal data of DR showed a bond length and angles within the range of normal values; it is not coplanar due to ethyl groups. The absorption spectra showed that the low-energy side band was absent in DR due to less conjugation than DQ. The luminescence intensity was more for DR than DQ due to removal of high-energy N–H vibrations in DR after addition of an ethyl group and energy transfer from the ligand to Eu is more. From time-resolved spectra and highly resolved luminescence spectra at 77 K, irreproducible luminescence was observed for DR due to chemical instability under laser irradiation.

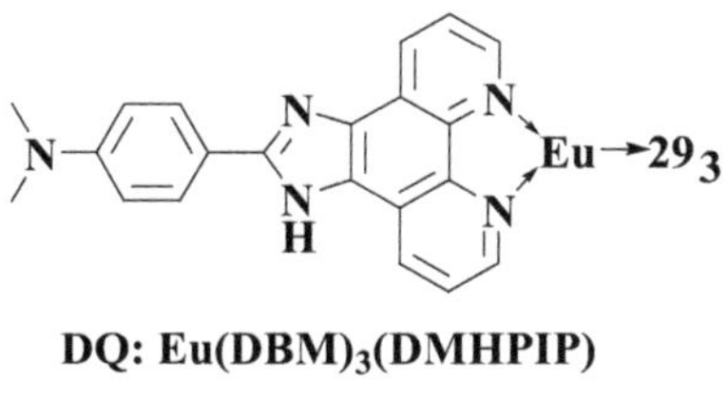

DQ: Eu(DBM)$_3$(DMHPIP)

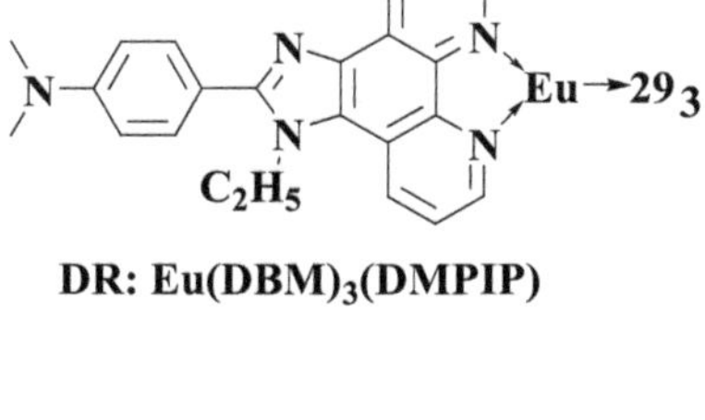

DR: Eu(DBM)$_3$(DMPIP)

Bian and coworkers [194] investigated further by changing the N1-functionalization of the phenanthroimidazole moiety (changing the substituent site) of the complex. They compared three ligands based on the $Eu(DBM)_3$ moiety in the device ITO/TPD (50 nm)/(DS/DT/DU) (50 nm)/Mg:Ag (200 nm)/Ag (100 nm). The device with DS exhibits a maximum luminescence of 3.7 cd/m^2 at 17 V, and its thermal decomposition temperature is 358°C. DT shows 197 cd/m^2 at 20 V with thermal decomposition 381°C, and DU exhibits 561 cd/m^2 at 16 V with decomposition temperature 408°C. From the above data DU exhibits more luminescence than DS because of increasing hole transport capability by using a carbzole unit, and also the thermal decomposition temperature increases with addition of carbazole. The PLQY and maximum EL for (DS/DT/DU) are 9.98, 10.98, 11.02, and 3.7 (17 V), 197 (20 V), and 561 (16 V), respectively. The configuration of the device ITO/TPD (50 nm)/DU (5 nm):BCP (5 nm)$]_4$/BCP (20 nm)/Alq (10 nm)/Mg:Ag (0.9:0.1)(110 nm)/Ag (100 nm) shows high, efficient pure-red emission with a luminescence of 1419 cd/m^2 (18 V) and high PE of 0.88 lm/W.

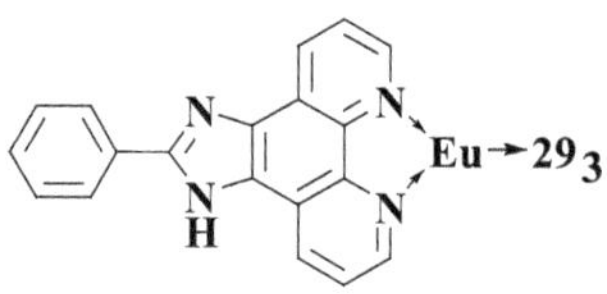

DS: $Eu(DBM)_3$(HPIP)

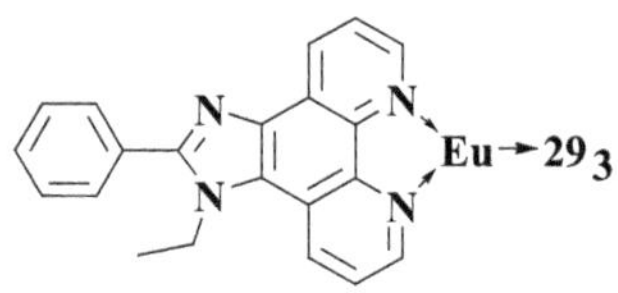

DT: $Eu(DBM)_3$(EPIP)

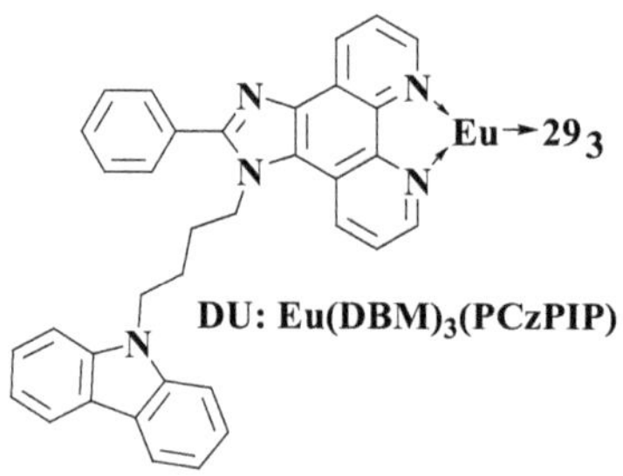

DU: $Eu(DBM)_3$(PCzPIP)

Lam et al. [195] synthesized a Eu complex with a dibenzoylmethane group as the primary and imidazole with X = H, Cl, CH_3 as the secondary ligand (DV, DW, and DX). The complex was introduced into a device ITO/CuPc/TPD/Eu complex/cathode.

The observed sharp emission of Eu complexes under investigation must be ET, and it favors light emitting. This emission (612 nm) and low FWHM are attributed to $^5D_0$–$^7F_2$ optical transition in the central metal ion. Introducing Eu complexes in the device only led to red emission at all applied voltages. From these three Eu complexes, the unsubstituted one (DV) shows good results compared to the other two because the H-Etpbip moiety is the least polar among the others. It also shows a TOV of 2 V, a maximum luminescence of 193 cd/m^2, and a luminescence efficiency of 0.438 cd/A. Also the mobility of the DV complex value is in the range of 10^{-5} cm^2/V/S. The local dipolar movement of the primary Etpbip ligand integrated the difference in the device performance and the inferior electron mobility also in comparison to the unsubstituted Eu complex.

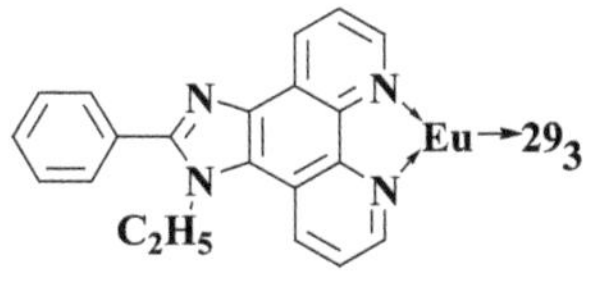

DV: Eu(DBM)$_3$(H-Etpbib)

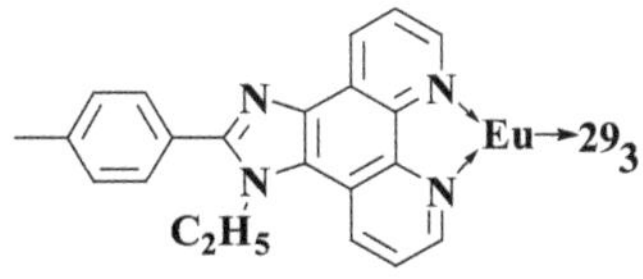

DW: Eu(DBM)$_3$(Me-Etpbib)

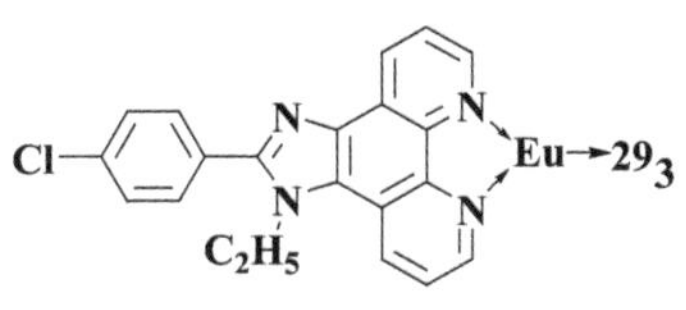

DX: Eu(DBM)$_3$(Cl-Etpbib)

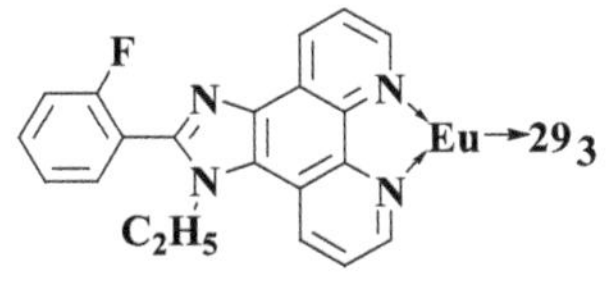

DY: Eu(DBM)$_3$(EFPIP)

Zhao and coworkers, in 2009, used a DY complex with a mono-fluorophenyl-substituted ligand, EFPIP. The fabrication of the device follows ITO/m-MTDATA /NPB/CBP:DY/BPhen/Alq$_3$/LiF/Al. Green emission was observed from the Alq$_3$ layer in the device without an ETL. It is clearly that the Eu^{3+} complexes have low electrical performance. A pure-red emission at 612 nm was achieved with CIE coordinates (0.64, 0.33). The device performance shows a maximum luminance of 465.2 cd/m^2 and a CE of 3.1 cd/A [196].

9.4.4 Both DBM and Phen Modification

The modification of phen as well as the antenna molecule was done by Li and coworkers, who introduced the Eu^{3+} complex DZ with a 4,5,9,14-tetraazabenzotriphenylene (DPPZ) ligand [181]. The obtained PLQE was 36% with a very short emission lifetime of 30 μs. The overturning triplet quenching-induced efficiency reduction was observed due to the short lifetime. The device ITO/m-MTDATA/NPB/CBP:DZ:Bphen/Alq_3/LiF/Al showed a maximum luminance of 2910 cd/m^2 and a CE of 3.0 cd/A, 13%, at a current density of 10 mA/cm^2.

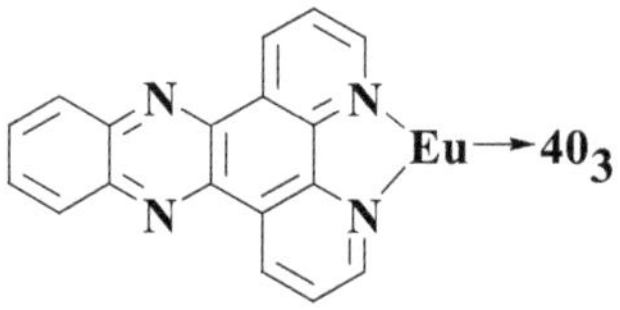

DZ: Eu(DPDBM)$_3$DPPZ

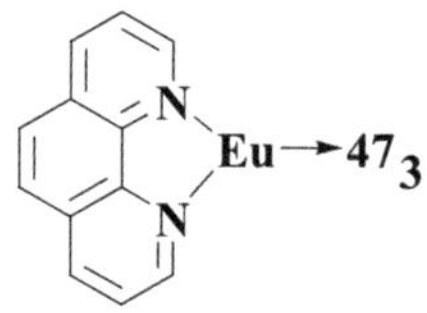

EE: Eu(DBM-Carb)$_3$Phen

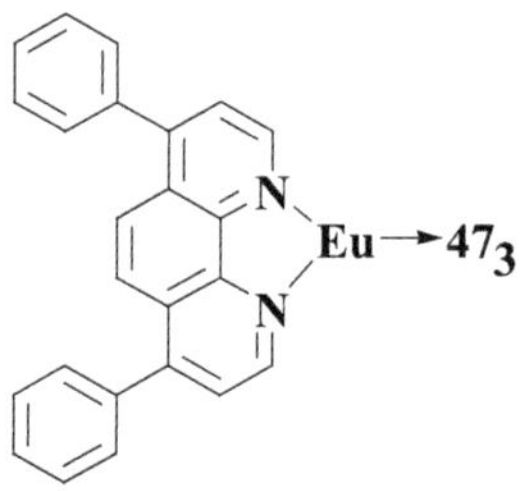

EF: Eu(DBM-Carb)$_3$BPhen

Zheng et al. introduced carbazole-based Eu(III) complexes EE and EF [197, 198]. Both complexes have DBM-Carb as a core symmetrically arranged diketone at the *para*-position of the phenyl rings and can reduce steric hindrance in the complex (anionic side). The PLQE of the Eu-phen and Eu-Bphen was 18% and 14%, respectively.

9.4.5 Spiro Complexes

In 2011, Wu et al. fabricated a device with the EG (sbf = 4,5-diaza-9,9'-spirobifluorene) complex as the emitter [199]. Due to good electron affinity of sbf, a trilayer device of ITO/TPD/EG:CBP/Alq_3/LiF/Al can be efficiently confined in the recombination zone without the need for an HBL. The doping concentration was limited to 8% to optimize emission. The device showed a remarkable PLQE of 41.23% and a long lifetime of 1.21 ms, and pure-red emission was obtained at 100 cd/m^2. The red emission could still be observed at an applied voltage of 19.8 V. The device also had a high yield, including a maximum luminance of 1365 cd/m^2 and maximum efficiencies of 5.21 cd/A, 1.6 lm/W, and 2.91%. The results indicated that the spiro complex without substitution enhances efficiency.

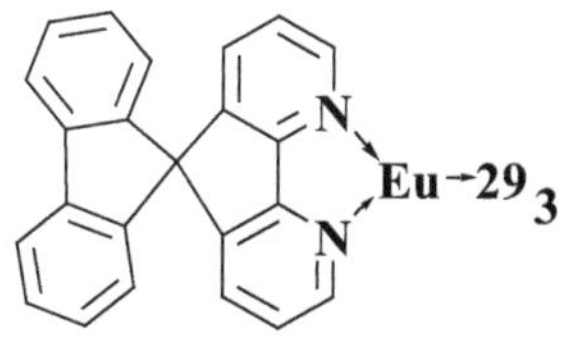

EG: Eu(DBM)$_3$(sbf)

Compared to the bipyridine and bipyridine oxide ligands 4,5-diazafluorene (DF) has a more rigid and planar configuration and allows much improved metal coordination and high electron injection. The researchers synthesized two secondary ligands, 9,9-di-(4-methoxyphenyl)-9-H-4,5-diazafluorene (OMe-Spiro-DF) and 9,9-di-(2-(4-(4-butyloxy)phenyl)-5-phenyl-1,3,4-oxadiazolyl)-phenyl-9-H-4,5-diazafluorene (OXD-Spiro-DF), and their corresponding Eu(DBM)$_3$ complexes (EH, EI) [200]. Compared to EH, EI had additional absorption bands at 310 nm, which may be attributed to the π–π* transitions. The bilayer undoped device of the configuration ITO/TPD/Eu(III)/Mg:Ag fabricated with the EI complex achieved a maximum luminance of 154 cd/m^2 and a peak CE of 0.055 cd/A, which is approximately four times the one observed with the EH compound. Integration of an oxadiazole unit into the ligand led to the recombination of charge carriers in the Eu(III) complex by utilizing balancing injection as well as transports.

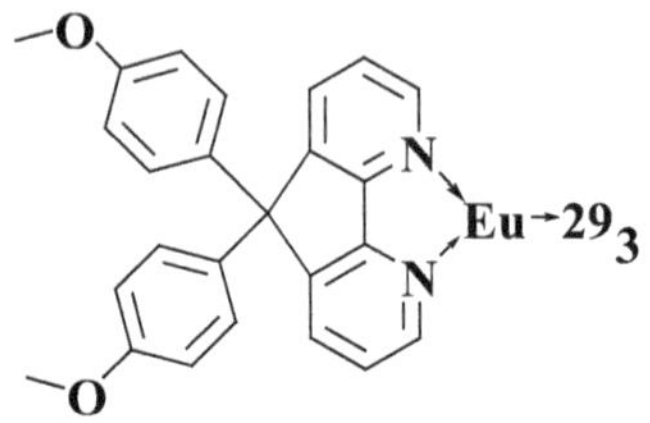

EH:Eu(DBM)$_3$OMe-Spiro-DF

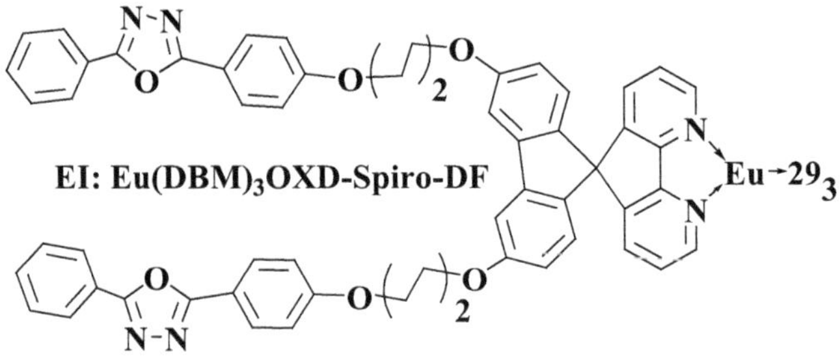

EI: Eu(DBM)$_3$OXD-Spiro-DF

9.4.6 Phosphine Oxide (P=O)-Based Complexes

Anionic ligands exert a crucial influence on optoelectronic properties of their Eu(III) complexes. Shi et al. used three pyrazolone-based ligands, namely 1-phenyl-3-methyl-4-(1-naphthoyl)-5-pyrazolone (52), 1-phenyl-3-methyl-4-(4-dimethylaminobenzoyl)-5-pyrazolone (50), and 1-phenyl-3-methyl-4-(4-cyanobenzoyl)-5-pyrazolone (51), and their corresponding Eu(III) complexes of the form EJ, EK, and EL [201]. All the three complexes emit the characteristic wavelength of 614 nm of Eu(III). The overall QY of EJ, EK, and EL complexes were 9.2×10^3, 3.1×10^3, and 0.4×10^3, respectively. The energy transfer efficiencies, probability of Ln emission for EJ and EK, were estimated to be 35.1%, 2.6% and 2.6%, 1.9% respectively. The PL and QY studies indicated that the modification of ligands tunes the triplet energy levels of three pyrazolone-based ligands to match the $^5D_0$ energy level of Eu^{3+} properly and improves the energy transfer efficiency from the antenna to Eu^{3+}, therefore, enhancing the Eu(III) emission intensity. The pure-red emission from the device ITO/NPB/CBP:EJ/BCP/Alq_3/Mg:Ag was predicted, and luminance is around 247 cd/m^2.

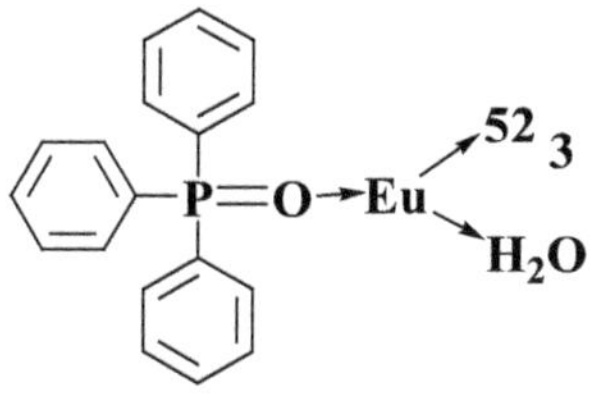

EJ: $Eu(TPPO)(52)_3(H_2O)$

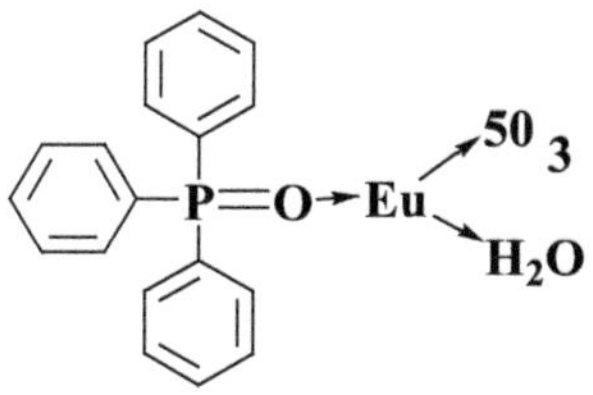

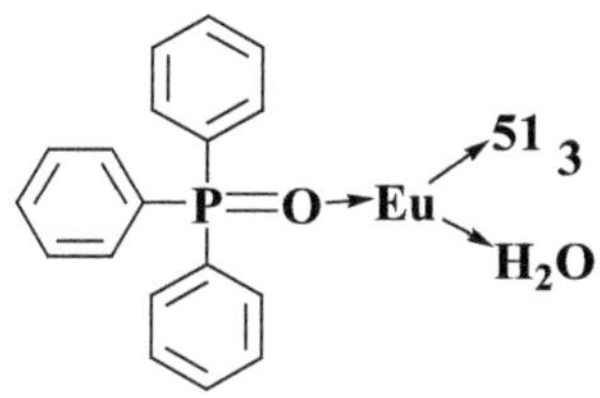

EK: $Eu(TPPO)(50)_3(H_2O)$ EL: $Eu(TPPO)(51)_3(H_2O)$

Hu et al. introduced a TPPO-based Eu complex EM. It showed transparent and homogeneous thin films are deposited by vacuum sublimation. The device configuration is ITO/TPD/EM/BCP/Alq_3/cathode (Mg:Ag or Al:Li), with bright and pure-red light. When they replaced Mg:Ag with a Li:Al cathode, they obtained pure-red emission with luminescence of 320 cd/m^2 at 14.5 V. They are also found the importance of insertion of an HBL (BCP) layer in the device. To get the pure-red emission, insertion of BCP is needed because EM possesses hole-transporting ability. It still has certain hole conductivity, and it shows low efficiency. To overcome these, in the same year the authors introduced multiple-stacked Eu complex/HB units. The EM lifetime was estimated to be longer than organic compounds'; it causes quenching by carriers (holes mainly). They fabricated the device ITO/TPD (50 nm)/{BCP (2.5 nm) /EM (2.5 nm)}$_n$ (n = 1, 2, 4 or 6)/BCP (20 nm)/Alq_3 (40 nm)/Mg:Ag to reduce the confrontation between the EM excited state and holes in the EM layer. When increasing n, I–V curves shifted toward higher voltage. It is one of the disadvantages of an EL device. However, luminescence efficiency improved with n and achieved a maximum luminescence of 380 cd/m^2 [202–204].

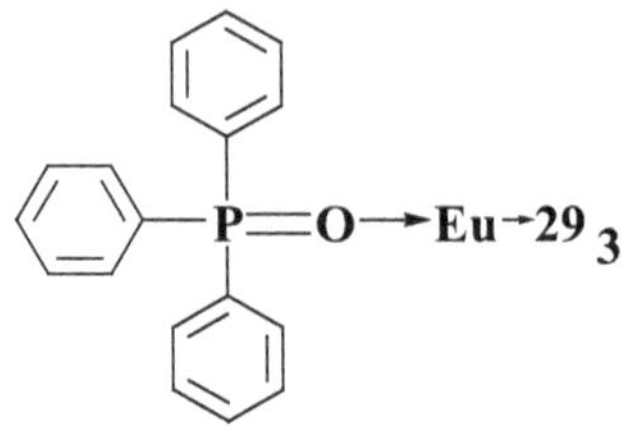

EM: Eu(DBM)$_3$TPPO

They modified the β-diketone and made a complex with TPPO as a neutral ligand, which are (57), (58), and (58). The HOMO and LUMO energy levels were found to be around −5.6 and −2.6 eV, respectively. The bilayer device, ITO/TPD/59/Al, showed a maximum luminance of 220 cd/m^2 and a peak CE of 0.1 cd/A.

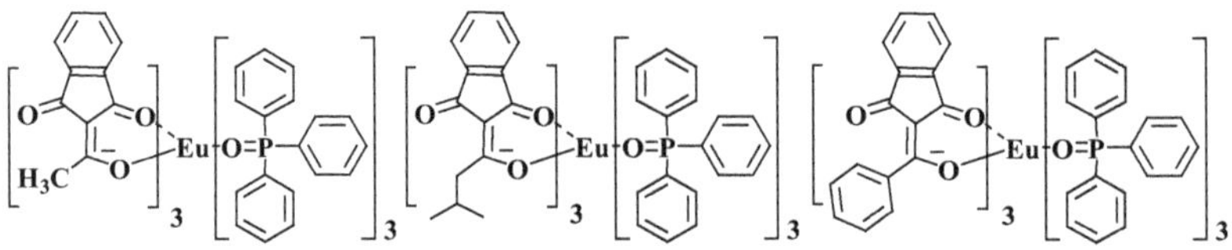

57 Eu(AIND)$_3$(TPPO)$_2$ 58 Eu(ISOAIND)$_3$(TPPO)$_2$ 59 Eu(BIND)$_3$(TPPO)$_2$

9.4.7 Pyridine-Based Complexes

Pyridine *N*-oxide shows a high binding affinity for Ln ions, and its strong electron-withdrawing ability helps improve the process of electron injection and transport. These properties were utilized by Ma et al., who prepared a neutral ligand and constructed a TTA-based Eu(III) complex (EN) [205]. The asymmetric configurations of both 2,2-bipyridine mono *N*-oxide (Obpy) and DBM facilitated the parity violation, thereby enhancing the $^5D_0$–$^7F_2$ transition probability of Eu(III) for intense pure-red emission at 612 nm. With a regular four-layered device structure, ITO/TPD/CBP:EN/BCP/Alq$_3$/LiF/Al), the complex exhibited a maximum luminance of 531 cd/m^2 and peak efficiencies of 4.6 cd/A, 2.33 lm/W, and 2.6%. These results are poor compared to the fluorination complex.

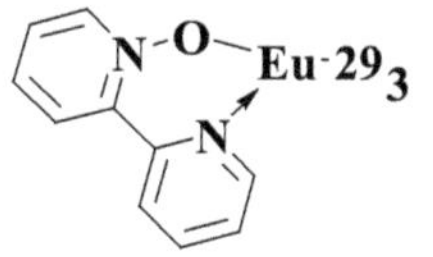

EN: Eu(DBM)$_3$Obpy

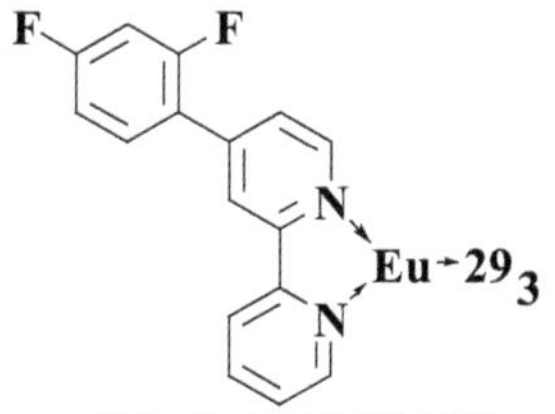

EO: Eu(DBM)$_3$DFbpy

Similarly, Huang and coworkers [206, 207] synthesized the complex of (2Fphen-bpy)Eu(DBM)$_3$ (2Fphenbpy = 4-(2,4-difluorophenylene)-2,2′-bipyridine, or DFpby) and incorporated them into the doped electroluminescent device of the structure ITO/TPD (40 nm)/EO:CBP (30 nm, 3%)/BCP (20 nm)/ Alq$_3$ (30 nm)/LiF (1 nm)/Al (100 nm). The device had a relatively low onset voltage of ~5.5 V at a luminance of 1 cd/m^2 and showed the characteristic emission of Eu(III) complexes at 612 nm with the observed maximum brightness of 491 cd/m^2 at ~17.4 V. The current and PE were observed to be 2.53 cd/A and 0.79 lm/W, respectively, at a brightness of 100 cd/m^2. The properties of the device could be optimized further by incorporating an orange fluorescence dye, 4-(dicyanomethylene)-2-t-butyl-6(1,1,7,7-tetramethyljulolidyl-9-enyl)-4H-pyran (DCJTB), as an auxiliary dopant (0.2 wt%) to form an EML, with a device of the form CBP:DCJTB:EO, having a maximum luminance of 1200 cd/m^2 as well as peak efficiencies of 7.3 cd/A and 2.0 lm/W; the increment of luminescence was observed compared to EN. It indicates that fluorination leads to high efficiency.

Hung et al. introduced two double- and triple-layer devices with HPBM and EPBM based on the Eu(DBM)$_3$ moiety [208, 209]. The double-layer device, ITO/TPD (40 nm)/EP or EQ (40 nm)/Al (100 nm), shows maximum luminescence, voltage, and current densities: 1.14 cd/m^2 at 23 V and 36 mA/cm^2; 29 cd/m^2 at 11 V and 210 mA/cm^2, respectively. From this device, exciplex formation was observed at the interface of the Eu complex and the TPD layer. In two Eu complexes EQ shows more luminescence. The same is introduced into the triple-layer device ITO/TPD (40 nm)/EQ (40 nm)/Alq (40 nm)/Al (100 nm), which showed increasing luminescence: a maximum luminescence of 180 cd/m^2 at 18 V. The spectral features are sensitive to the thickness of emitting and ET layers. Reducing the thickness of the emitting layer, a broad band appears from Alq_3. The complex ER showed better film formability, and it might be due to the presence of alkyl chain extension.

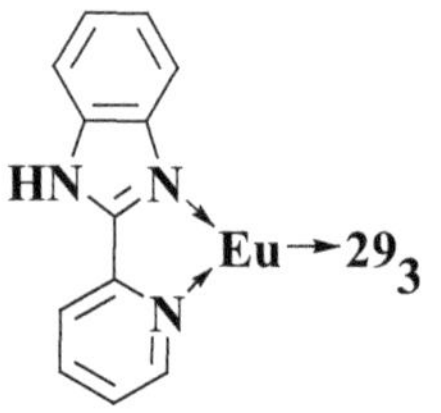

EP: $Eu(DBM)_3HPBM$

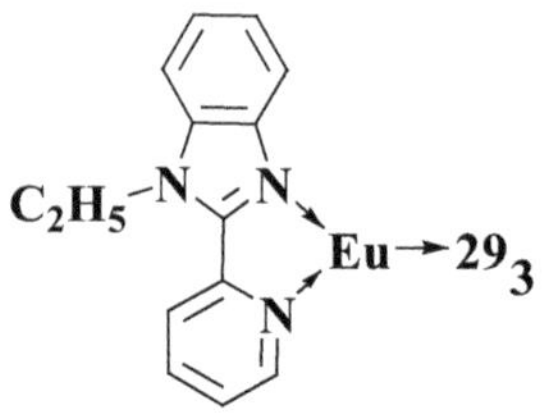

EQ: $Eu(DBM)_3EPBM$

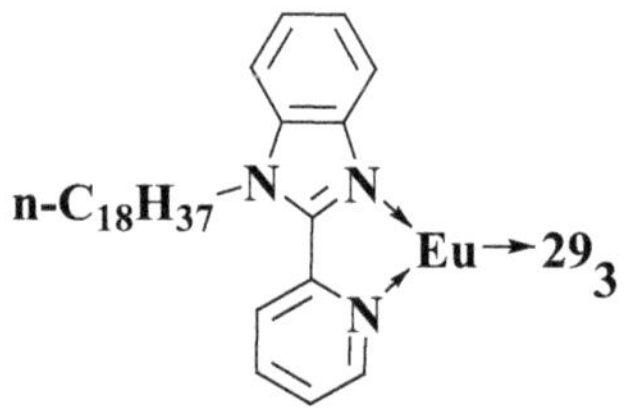

ER: $Eu(DBM)_3C_{18}PBM$

Zhang et al. [210] introduced a Eu(III) complex by incorporating the carbazole fragment in ES. Comparison of the device based on the unfunctional Eu(III) complex [$Eu(DBM)_3HPyBM$] and ES showed that the ES complex performed well due to improvement of hole transport ability. The device structure is ITO/ TPD (50 nm)/ES (30 nm)/1,3,5-tri(1-phenyl-1H-benzo[d] imidazole-2-yl)phen (TPBI), 20 nm)/LiF (1.5 nm)/Al, and it showed a maximum efficiency of 4.2 cd/A and the luminescence of red emission is 200 cd/m^2. TBPI was used as the ETL to confine the charge recombination effectively. EL and PL spectra suggested that the sharp red emission comes from the Eu(III) ion with a half width at half maximum (HWHM) of 10 nm and CIE color coordinates (0.67, 0.32). Up to ~17 V, the emission from TPD is negligible. When the increasing of voltage up to 19 V, there is a weak emission observed around 473 nm. It does not affect the purity of red color. These results all suggest that carbazole is helpful in balancing the charge carrier in the Eu(III) complex.

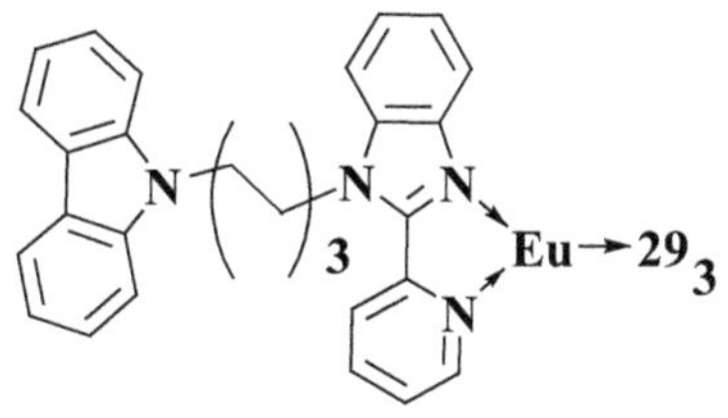

ES: Eu(DBM)$_3$CAR-PyBM

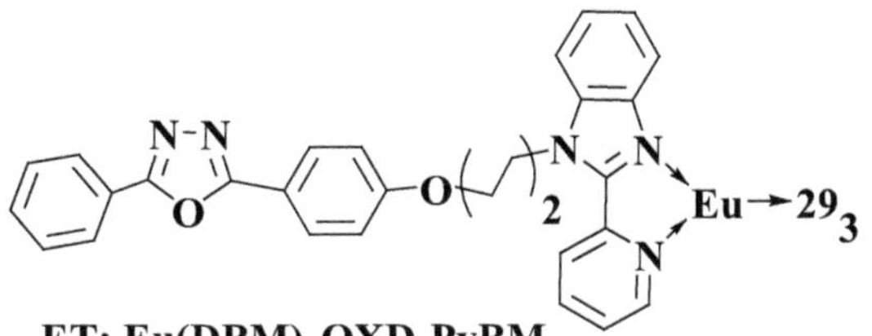

ET: Eu(DBM)$_3$OXD-PyBM

Liang et al. [209], to improve the EQE, constructed an oxadiazole-functionalized Eu complex–based device ITO/TPD (40 nm)/ET (50 nm)/LiF (1 nm)/Al (200 nm). PyBM (2-(2-pyridyl) benzimidazole) enhances the electron-transporting property and solubility. The device shows a TOV of 7.8 V, a luminous efficiency of 1.9 Cd/A, and an EQE of 1.7% with brightness 57 cd/m^2 at 13.8 V. The maximum luminescence is 322 cd/m^2 at 21 V. The bias voltage ranges from 8 to 21 V, and no exciplex was observed. This suggestes balanced charge carriers and thus effective recombination.

The organic benzimidazole, benzoxazole, and benzthiazole molecules have a planar nature, as well as high electron affinity to form bidentate neutral ligands, such as poly(benzobisoxazo1e) (PBO) and poly(benzobisthiazole) (PBT). The PBO- and PBT-based Eu(III) complexes (EU, EV) show pure-red emission [211]. The red emission of the complexes is also observed in the two devices. Device 1 is ITO/TPD (40 nm) Eu complexes (40 nm)/ALq$_3$ (40 nm)/Mg:Ag (200 nm)/Ag (80 nm). The emission is measured from low voltage to high voltage and reveals the emission color as red to yellow-orange. It clearly indicated the two Eu(III) complexes have hole-transporting abilities from Alq$_3$ emission. Device two is ITO/TPO (20 nm)/EV (40 nm)/BCP (10 nm) Alq$_3$ (40 nm)/Mg:Ag (200 nm)/Ag (80 nm). The maximum brightness reached 148 cd/m^2 at 19 V, PE 0.021 m/w, and

CE 0.08 cd/A. But the EU complex showed poor results: maximum brightness 43 cd/m^2 and CE 0.08 cd/A. From that we can conclude that EV has higher efficiency than EU.

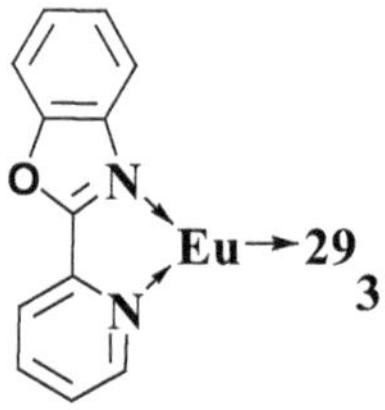

EU: Eu(DBM)$_3$PBO

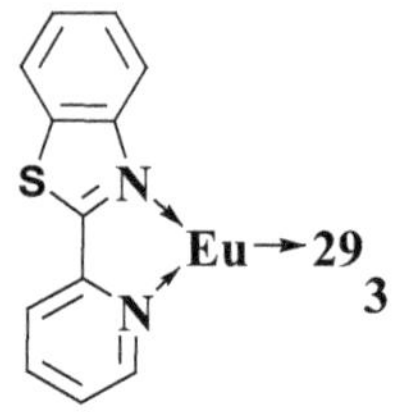

EV: Eu(DBM)$_3$PBT

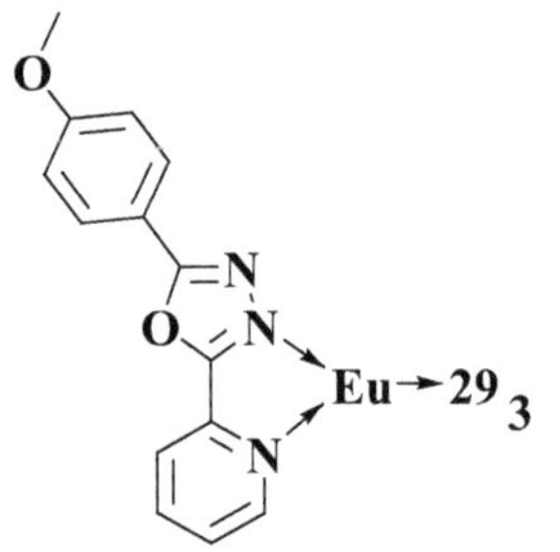

EW: Eu(DBM)$_3$MeO-OXD

Liu et al., in 2009, used 2-(4-methoxy)phenyl-5-(2-pyridyl)-1,3,4-oxadiazole (MeO-OXD) as a secondary bidentate ligand (EW) [212]. The UV results indicated that the sensitizing ability of the chelating ligands was comparable to phen, thus proving it efficient enough for further tests. The complex had an UV absorption maximum at 350 nm and a sharp emission peak at 612 nm. When incorporated into a device of the structure ITO/PEDOT:PSS/PVK:PBD:EW/Ba/Al, the device emitted pure-red emission characteristic to europium at 613 nm with a maximum light intensity of 12 cd/m^2 at 24 V.

Gusev et al. prepared mono- and bitriazole ligands to be used in ternary complexes with dbm as the primary ligand (EX, EY, and EZ). The triazoles are used as efficient ETMs in these complexes. In the excitation spectra, the complexes showed a broad absorption in the range of 280–400 nm, common in a lot of Eu(III) complexes. The complexes showed narrow emission at 612–613 nm, which can be attributed to the hypersensitive $^5D_0$–$^7F_2$ transition of Eu^{3+}. The PLQE of the complex containing the first ligand was only 2.34%, and that for the third was 29.1%, which also had the maximum luminous intensity amongst all the complexes produced. The PLQE of the latter was impressive as it is around four times that of the complex CP.

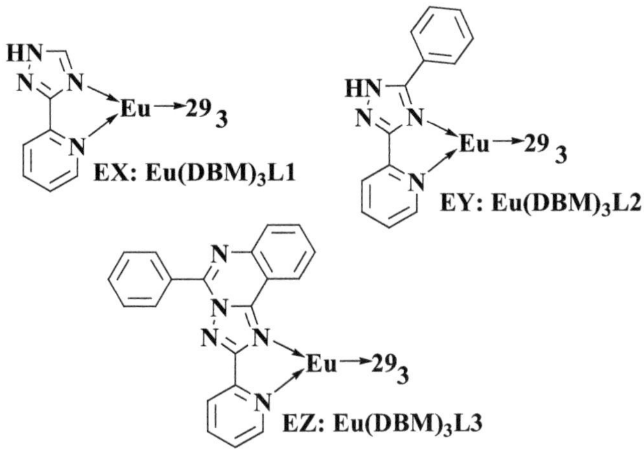

9.4.8 Multinitrogen Beard Heterocyclic Availed Complexes

In 2012, Li et al. [217, 218] introduced a trinuclear dendritic Eu(III) complex based on (1,3,5-tris[2-(2'-pyridyl)benzimidazoly]methyl-benzene) (TMMB)-europium(III) (FF), and the same was also improved by Wei et al. The devices were made by using a host as well as without a host. Device A was ITO/NPB (30 nm)/FF (60 nm)/LiF (1 nm)/Al(100 nm); device B was ITO/NPB (30 nm)/FF (40 nm)/BCP (30 nm)/LiF (1 nm)/Al(100 nm); and device C was ITO/NPB(30 nm)/CBP:1% FF (40 nm)/BCP(30 nm)/LiF(1 nm)/Al(100 nm). The TMMB ligand highlighted an extended Eu(III)–Eu(III) distance. The interrupted β-conjugation in the TMMB ligand leads to a rise in singlet and triplet energy levels. It also assists in achieving efficient energy transfer from the ligand to DBM and subsequently to Eu^{3+}. The fabricated device is ITO/NPB/CBP:FF/BCP/LiF/Al: a maximum luminance of 109 cd/m^2 and a maximum CE of 0.7 cd/A with pure-red emission are observed. The CIE coordinates are (0.64, 0.33). The results of the EL become pitiable for the trinuclear complex, which is related to high molecular weight and low volatility.

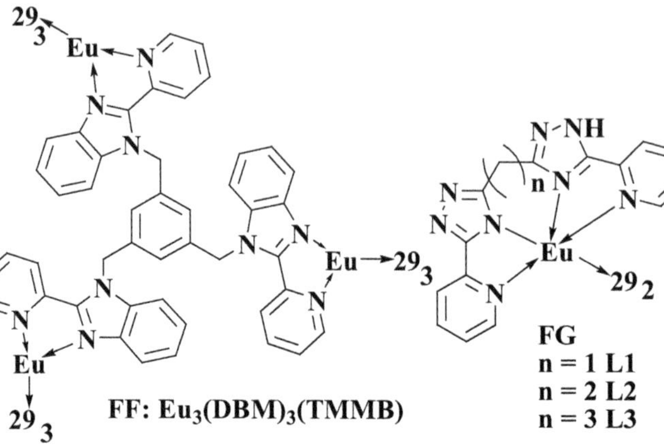

The team also created photoluminescent complexes with bis(5-(pyridine-2-yl)-1,2,4-triazol- 3-yl)methane (L1), 1,3-bis(5-pyridin-2-yl-1,2,4-triazol-3-yl)propane (L2), and 1,4-bis(5-pyridin-2-yl-1,2,4-triazol-3-yl)butane (L3) as ancillary ligands [213–216]. All three complexes showed broad bands in the excitation spectra between 220 and 400 nm. The solid-state QY was maximum for the first complex at 48.1%, whereas that for the other two was in the range of 30%. For the first complex (FG-L_1) a PVK-doped device was prepared with the complex as the active part. The device emitted red color at 8.8 V and 11.2 V, and its brightness reached a maximum at 19 V. For the other two complexes, double-layered devices of the structure ITO/NPB(40 nm)/FG-L_1/FG-L_2 (40 nm)/LiF (1 nm)/Al (100 nm) were prepared. Among them the second device gave the maximum brightness of 455 cd/m^2 at 19.2 V.

9.5 Conclusion

The design of new europium molecular complexes with a high intrinsic QY of europium-centered emission remains the most challenging research task for OLEDs, since these features are highly reliable on coordination spheres of the complex as well as efficient antennas (coordinated ligands). In this chapter we tried to correlate the linked antennas (capable of transporting and injecting charge in the device) in Eu(III) complexes and their photophysical, EL properties. Even though substantial research contributions have already been devoted in the arena of Ln-based complexes for OLED applications, their practical applications still remain unsatisfactory. To push Ln chemistry further for the improvement of OLED devices, it is necessary that the chemist should think creatively for the rational design of highly efficient luminescent Ln complexes.

Acknowledgments

The authors gratefully acknowledge the Board of Research in Nuclear Sciences (BRNS), Department of Atomic Energy (DAE), and Inspire faculty award INSPIRE (Innovation in Science Pursuit for Inspired Research), Department of Science and Technology (DST), India, for financial support.

References

1. Li, W. X., Xin, X. D., Feng, S. Y., Liu, Y., Zhang, J., Yang, B. A., and Li, Y. J. (2014). Fluorescence enhancement of europium (III) perchlorate by 1,10-phenanthroline on the 1-(naphthalen-2-yl)-2-(phenylsulthio) ethanone complex and luminescence mechanism, *Luminescence*, **29**, pp. 810–817.
2. Geist, F., Jackel, A., and Winter F. R. (2015). Dual ligand-based fluorescence and phosphorescence emission at room temperature from platinum thioxanthonyl complexes, *Dalton Trans.*, **44**, pp. 3974–3987.
3. Geist, F., Jackel, A., and Winter, F, R. (2015). Ligand based dual fluorescence and phosphorescence emission from BODIPY platinum complexes and its application to ratiometric singlet oxygen detection, *Inorg. Chem.*, **54**, pp. 10946–10957.
4. Siraj, N., El-Zahab, B., Hamdan, S., E. Karam, T., Haber, L. H., Li, M., Fakayode, S. O., Das, S., Valle, B., Strongin, R. M., Patonay, G., Sintim, H. O., Baker, G. A., Powe, A., Lowry, M., Karolin, J. O., DChris, G., and Isiah, M. W. (2016). Fluorescence, phosphorescence, and chemiluminescence, *Anal. Chem.*, **88**, pp. 170–202.
5. Duan, J. P., Sun, P. P., and Cheng, C. H. (2004). Europium complexes having an aminophenanthroline ligand as red dopants in electroluminescent devices, *Adv. Technol. Mater. Mater. Proc.*, **6**, pp. 95–102.
6. Vogler, A., and Kunkely, H. (2006). Excited state properties of lanthanide complexes: beyond f-f states, *Inorg. Chim. Acta*, **359**, pp. 4130–4138.

7. Morss Lester, R. (1984). Lanthanide and actinide oxidation states in complex oxides and halides, *Inorg. Chim. Acta*, **94**, p. 31.
8. Centennial, W. (1967). Coordination chemistry of the lanthanide elements–one hundred years of development and understanding, *Adv. Chem.*, **62**, Chapter 21, pp. 306–317.
9. Parker, D., Dickins, R. S., Puschmann, H., Crossland, C., and Howard, J. A. K. (2002). Being excited by lanthanide coordination complexes: aqua species, chirality, excited-state chemistry, and exchange dynamics, *Chem. Rev.*, **102**, pp. 1977–2010.
10. Deng, R., Yu, J., Zhang, H., Zhou, L., Peng, Z., Li, Z., and Guo, Z. (2007). Photoluminescence and electroluminescence properties of a samarium complex $Sm(TTA)_3phen$, *Chem. Phys. Lett.*, **443**, pp. 258–263.
11. Petoud, S., Muller, G., Moore Evan, G., Xu, J., Sokolnicki, J., Riehl, J. P., N. Le, U., M. Cohen, S., and N. Raymond, K. (2007). Brilliant Sm, Eu, Tb, and Dy chiral lanthanide complexes with strong circularly polarized luminescence, *J. Am. Chem. Soc.*, **129**, pp. 77–83.
12. Hu, M., Ma, D., Liu, C., Wang, J., Zhang, Z., and Meng, L. (2016). Intense white emission from a single-upconversion nanoparticle and tunable emission colour with laser power, *J. Mater. Chem. C*, **4**, pp. 6975–6981.
13. Blasse, G., and Grabmaier, B.C. (1994). *Luminescent Materials* (Springer-Verlag, Berlin).
14. Bunzli, J. C. G., and Eliseeva, S. V. (2012). Lanthanide NIR luminescence for telecommunications, bioanalyses and solar energy conversion, *J. Rare Earth.*, **28**, p. 824.
15. Bunzli, J. C. G., and Chauvin, A. S. (2014). In *Handbook on the Physics and Chemistry of Rare Earths*, eds. Bunzli, J. C. G., and Pecharski, V. K. (Elsevier Science, Amsterdam), **44**, pp. 169–281.
16. Huang, X., Han, S., Huang, W., and Liu, X. (2013). Enhancing solar cell efficiency: the search for luminescent materials as spectral converters, *Chem. Soc. Rev.*, **42**, pp. 173–201.
17. Liang, H., Zhang, Q., Zheng, Z., Ming, H., Li, Z., Xu, J., Chen, B., and Zhao, Hui. (2004). Optical amplification of Eu $(DBM)_3$Phen-doped polymer optical fiber, *Opt. Lett.*, **29**, pp. 477–479.
18. Hwan Oh, J., Song, H. M., Eom, Y. K., Ryu, J. H., Ju, M. J., and Kim, H. K. (2011). Wavelength conversion lanthanide (iii)-cored complex for highly efficient dye-sensitized solar cells, *Bull. Korean Chem. Soc.*, **32**, p. 2743.
19. Atwood, D. A. (2013). *The Rare Earth Elements: Fundamentals and Applications* (John Wiley & Sons, Chichester), p. 696.

20. Piccinelli, F., Bettinelli, M., Melchior, A., Grazioli, C., and Tolazzi, M. (2015). Structural, optical and sensing properties of novel Eu(III) complexes with furan- and pyridine-based ligands, *Dalton Trans.*, **44**, p. 182.

21. Zhou, X., Zhao, X., Wang, Y., Wu, B., Shen, J., Li, L., and Li, Q. (2014). Eu(III) and Tb(III) complexes with the nonsteroidal anti-inflammatory drug carprofen: synthesis, crystal structure, and photophysical properties, *Inorg. Chem.*, **53**, pp. 12275–12282.

22. Feng, J., and Zhang, H. (2013). Hybrid materials based on lanthanide organic complexes: a review, *Chem. Soc. Rev.*, **42**, p. 387.

23. Yablonovitch, E. (2001). Photonic crystals: semiconductors of light, *Sci. Am.*, pp. 47–55.

24. Peters, J. A., Vijverberg, C. A. M., Kieboom, A. P. G., and van Bekkum, H. (1983). Lanthanide (III) salts of (S)-carboxymethyloxysuccinic acid: chiral lanthanide shift reagents for aqueous solution, *Tetrahedron Lett.*, **24**, pp. 3141–3144.

25. Molander, G. A. (1992). Application of lanthanide reagents in organic synthesis, *Chem. Rev.*, **92**, pp. 29–68.

26. Momani, W. M., Taha, Z. A., Ajlouni, A. M., Abu Shaqra, Q. M., and Zouby, M. A. (2013). A study of in vitro antibacterial activity of lanthanides complexes with a tetradentate Schiff base ligand, *Asian Pac. J. Trop. Biomed.*, **3**, pp. 367–370.

27. Han, G., Deng, Y., Sun, J., Ling, J., and Shen, Z. (2015). Research into europium complexes as magnetic resonance imaging contrast agents (review), *Exp. Ther. Med.*, **9**, pp. 1561–1566.

28. Yang, C., Xu, J., Zhang, Y., Li, Y., Zheng, J., Liang, L., and Lu, M. (2013). Efficient monochromatic red-light-emitting PLEDs based on a series of nonconjugated Eu-polymers containing a neutral terpyridyl ligand, *J. Mater. Chem. C*, **1**, pp. 4885–4901.

29. Bunzli, J. C. G. (2010). Lanthanide luminescence for biomedical analyses and imaging, *Chem. Rev.*, **110**, pp. 2729–2755.

30. Shao, G., Yu, H., Zhang, N., He, Y., Feng, K., Yang, X., Cao, R., and Gong, M. (2014). Synthesis and photophysical properties of europium(III)-β-diketonate complexes applied in LEDs, *Phys. Chem. Chem. Phys.*, **16**, pp. 695–702.

31. Pourjavid, M. R., Norouzi, P., Ganjali, M. R., Nemati, A., Zamani, H. A., and Javaheri, M. (2009). Separation and determination of medium lanthanides: a new experiment with use of ion-exchange separation and fast Fourier transform continuous cyclic voltammetry, *Int. J. Electrochem. Sci.*, **4**, pp. 1650–1671.

32. Ganjali, M. R., Gupta, V. K., Faridbod, F., and Norouzi, P. (2016). *Lanthanides Series Determination by Various Analytical Methods* (Elsevier Science), p. 448.

33. Weissman, S. I. (1942). Intramolecular energy transfer the fluorescence of complexes of europium, *J. Chem. Phys.*, **10**, pp. 214.

34. Melby, L. R., Rose, N. J., Abramson, E., and Caris, J. C. (1964). Synthesis and fluorescence of some trivalent lanthanide complexes, *J. Am. Chem. Soc.*, **86**, pp. 5117–5125.

35. Bernanose, A. (1955). Electroluminescence of organic compounds, *Br. J. Appl. Phys.*, **6**, p. S54.

36. de Bettencourt-Dias, A. (2007). Lanthanide-based emitting materials in light-emitting diodes, *Dalton Trans.*, pp. 2229–2241.

37. Lempicki, A., and Samelson, H. (1963). Optical maser action in europium benzoylacetonatc, *Phys. Lett.*, **4**, pp. 133–135.

38. Pope, M., Kallamann, H. P., and Magnante, P. (1963). Electroluminescence in organic crystals, *J. Chem. Phys.*, **38**, pp. 2042–2043.

39. Helfrich, W., and Schneider, W. G. (1965). Recombination radiation in anthracene crystals, *Phys. Rev. Lett.*, **14**, pp. 229–231.

40. Ferre, N., Filatov, M., and Huix-Rotllant, M. (2016). *Density-Functional Methods for Excited States* (Springer International, Switzerland), Topics in Current Chemistry, **368**, p. 481.

41. Crosby, G. A. (1983). Structure, bonding, and excited states of coordination complexes, *J. Chem. Educ.*, **60**, p. 791.

42. Yin, R., Tomalia, D. A., Qin, D., and Dunham, J. (2003). Convergent self-branching polymerization, US 6632889 B1.

43. Tomalia, D. A., Baker, H., Dewald, J. R., Hall, M., Kallos, G., Martin, S., Roeck, J., Ryder, J., and Smith, P. (1985). A new class of polymers: starburst-dendritic macromolecules, *J. Polym.*, **17**, pp. 117–132.

44. Tang, C. W., and VanSlyke, S. A. (1987). Organic electroluminescent diodes, *Appl. Phys. Lett.*, **51**, p. 913.

45. VanSlyke, S. A., and Tang, C. W. (1985). Organic electroluminescent devices having improved power conversion efficiencies, US Patent No. 4539507.

46. Tang, C. W., Chen, C. H., and Goswami, R. (1988). Electroluminescent device with modified thin film luminescent zone, US Patent No. 4769292.

47. Burroughes, J. H., Bradley, D. D. C., Brown, A. R., Marks, R. N., Mackay, K., Friend, R. H., Burns, P. L., and Holmes, A. B. (1900). Light-emitting diodes based on conjugated polymers, *Nature*, **347**, pp. 539–541.

48. Friend, R., Burroughes, J., and Bradley, D. (1993). Electroluminescent devices, US Patent No. 5247190

49. Crosby, G. A., Whan, R. E., and Alire, R. M. (1961). Intramolecular energy transfer in rare earth chelates. Role of the triplet state, *J. Chem. Phys.*, **34**, p. 743.

50. Kido, J., Kimura, M., and Nagai, K. (1995). Multilayer white light-emitting organic electroluminescent device, *Science*, **267**, p. 1132.

51. Kido, J., Hongawa, K., Okuyama, K., and Nagai, K. (1994). White light-emitting organic electroluminescent devices using the poly(N-vinylcarbazole) emitter layer doped with three fluorescent dyes, *Appl. Phys. Lett.*, **64**, p. 815.

52. Liu, W. T., Wang, P. H., Lee, H. F., Huang, Y. C., and Huang, W. Y. (2013). A tunable emission prepared by proton-induced fluorescent color change materials for a potential application in pleds luminescence and display materials, devices, and processing, *ECS J. Solid State Sci. Technol.*, **2**, pp. R142–R147.

53. Ma, B., Djurovich, P. I., Garon, S., Alleyne, B., and Thompson, M. E. (2006). Platinum binuclear complexes as phosphorescent dopants for monochromatic and white OLEDs, *Adv. Funct. Mater.*, **16**, pp. 2438–2446.

54. Baek, H., Cho, H., and Lee, C. (2007). Multilayer white organic light emitting diode with optimum emitting layer sequence, *Conference on Lasers and Electro-Optics - Pacific Rim*, Seoul, pp. 1–2.

55. Kido, J., Kimura, M., and Nagai, K. (1995). Multilayer white light-emitting organic electroluminescent device, *Science*, **267**, p. 1332.

56. Kido, J., Ikeda, W., Kimura, M., and Nagai, K. (1996). White-light-emitting organic electroluminescent device using lanthanide complexes, *Jpn. J. Appl. Phys.*, **35**, p. L394.

57. Bera, D., Qian, L., Tseng, T. K., and Holloway, P. H. (2010). Quantum dots and their multimodal applications: a review, *Materials*, **3**, pp. 2260–2345.

58. Huang, J., Pfeiffer, M., Werner, A., Blochwitz, J., Leo, K., and Liu, S. (2002). Low-voltage organic electroluminescent devices using pin structures, *Appl. Phys. Lett.*, **80**, pp. 139–141.

59. Sun, M., Xin, H., Wang, K. Z., Zhang, Y. A., Jin, L. P., and Huang, C. H. (2003). Bright and monochromic red light-emitting electroluminescence devices based on a new multifunctional europium ternary complex, *Chem. Commun.*, pp. 702–703.

60. Binnemans, K. (2015). Interpretation of europium(III) spectra, *Coord. Chem. Rev.*, **295**, pp. 1–45.

61. New, E. J., Parker, D., Smith, D. G., and Walton, J. W. (2010). Development of responsive lanthanide probes for cellular applications, *Curr. Opin. Chem. Biol.*, **14**, pp. 238–246.

62. Hammell, J., Buttarazzi, L., Huang, C. H., and Morrow, J. R. (2011). Eu(III) complexes as anion-responsive luminescent sensors and paramagnetic chemical exchange saturation transfer agents, *Inorg. Chem.*, **50**, pp. 4857–4867.

63. Yang, C., Fu, L. M., Wang, Y., Zhang, J. P., Wong, W. T., Ai, X. C., Qiao, Y. F., Zou, B. S., and Gui, L. L. (2004). A highly luminescent europium complex showing visible-light-sensitized red emission: direct observation of the singlet pathway, *Angew. Chem. Int. Ed.*, **43**, pp. 5010–5013.

64. Latva, M., Takalo, H., Mukkala, V. M., Matachescu, C., Rodriguez-Ubis, J. C., and Kankare, J. (1997). Correlation between the lowest triplet state energy level of the ligand and lanthanide(III) luminescence quantum yield, *J. Lumin.*, **75**, pp. 149–169.

65. Xu, H., Sun, Q., An, Z., Wei, Y., and Liu, X. (2015). Electroluminescence from europium(III) complexes, *Coord. Chem. Rev.* **293–294**, pp. 228–249.

66. Dexter, D. L. (1953). A theory of sensitized luminescence in solids, *J. Chem. Phys.*, **21**, pp. 836–850.

67. Crosby, G. A., Whan, R. E., and Alire, R. M. (1961). Intramolecular energy transfer in rare earth chelates. Role of the triplet state, *J. Chem. Phys.*, **34**, pp. 743–748.

68. Baek, N. M., Kim, Y. H., Lee, D. H., and Seo, K. D., and Kim, H. K. (2009). Effect of coordination environment on the photophysical properties of luminescent europium(iii) complexes, *Bull. Korean Chem. Soc.*, **30**, p. 1153.

69. Armelao, L., Quici, S., Barigelletti, F., Accorsi, G., Bottaro, G., Cavazzini, M., and Tondello, E. (2010). Design of luminescent lanthanide complexes: from molecules to highly efficient photo-emitting materials, *Coord. Chem. Rev.*, **254**, pp. 487–505.

70. Wong, W. Y., and Hoa, C. L. (2009). Functional metallophosphors for effective charge carrier injection/transport: new robust OLED materials with emerging applications, *J. Mater. Chem.*, **19**, pp. 4457–4482.

71. Katkova, M. A., and Bochkarev, M. N. (2010). New trends in design of electroluminescent rare earth metallo-complexes for OLEDs, *Dalton Trans.*, **39**, pp. 6599–6612.

72. Chishio, H., Hiroshi, T., Hisahiro, H., and Tadashi, K. (1992). Transient behavior of organic thin film electroluminescence, *Appl. Phys. Lett.*, **60**, p. 1220.

73. Umberto, G., Mariacecilia, P., Christelle, F., Chiara, B., William, P., and Silvia, D. (2009). Highly efficient color-tunable OLED based on poly(9,9-dioctylfluorene) doped with a novel europium complex, *J. Phys. Chem. C*, **113**, pp. 2290–2295.

74. Lee, S. S., Ko, D., Chung, C. H., and Cho, S. M. (2002). Movement of carrier recombination zone in organic light emitting devices by applied voltage, *Synth. Met.*, **128**, pp. 51–55.

75. Hany, A., Zoran, D. P., Hu, N. X., Hor, A. M., and Gu, X. (1999). Degradation mechanism of small molecule-based organic light-emitting devices, *Science*, **283**, pp. 1900–1902.

76. Koji, I., Hiromitsu, O., and Yasuhiko, S. (1998). Exciplex formation at the organic solid-state interface: yellow emission in organic light-emitting diodes using green-fluorescent tris(8-quinolinolato) aluminum and hole-transporting molecular materials with low ionization potentials, *Appl. Phys. Lett.*, **72**, p. 636.

77. Xin, Q., Li, W. L., Che, G. B., Su, W. M., Sun, X. Y., Chu, B., and Li, B. (2006). Microheaters fabricated from a conducting composite, *Appl. Phys. Lett.*, **89**, pp. 223521–223524.

78. Park, J. S., Lee, J. W., Kim, Y. M., Bae, S. J., Jang, J., Kim, J. K., and Ju, B. K. (2005). Enhancement of efficiency for white-organic light-emitting diode with a thin electron-blocking layer sensors and displays: principles, materials, and processing, *J. Electrochem. Soc.*, **152**, pp. H196–H199.

79. Wang, J., Wang, R., Yang, J., Zheng, Z., Michael D., Carducci, T. C., Peyghambarian, N., and Ghassan, E. J. (2001). First oxadiazole-functionalized terbium(iii) β-diketonate for organic electroluminescence, *J. Am. Chem. Soc.*, **123**, p. 6179.

80. Kulkarni, A. P., Tonzola, C. J., Babel, A., and Jenekhe, S. A. (2004). Electron transport materials for organic light-emitting diodes, *Chem. Mater.*, **16**, pp. 4556–4573.

81. Wagner, D., Hoffmann, S. T., Heinemeyer, U., Munster, I., Kohler, A., and Strohriegl, P. (2013). Triazine based bipolar host materials for blue phosphorescent OLEDs, *Chem. Mater.*, **25**, pp. 3758–3765.

82. Huang, J., Watanabe, T., Ueno, K., and Yang Y. (2007). Highly efficient red-emission polymer phosphorescent light-emitting diodes based on two novel tris(1-phenylisoquinolinato-C2,N)iridium(III) derivatives, *Adv. Mater.*, **19**, pp. 739–743.

83. Bhansali, U. S., Polikarpov, E., Swensen, J. S., Chen, W.-H., Jia, H., Gaspar, D. J., Gnade, B. E., Padmaperuma, A. B., and Omary M. A. (2009). High-efficiency turquoise-blue electrophosphorescence from a Pt(II)-pyridyltriazolate complex in a phosphine oxide host, *Appl. Phys. Lett.*, **95**, p. 233304.

84. Duan, J. P., Pei, S., and Cheng, C. H. (2003). New iridium complexes as highly efficient orange–red emitters in organic light-emitting diodes, *Adv. Mater.*, **15**, pp. 224–228.

85. White, R. L. (1969). Review of recent work on the magnetic and spectroscopic properties of the rare-earth orthoferrites, *J. Appl. Phys.*, **40**, p. 1061.

86. Zhiqiang, G., Lee, C. S., Bello, I., Lee, S. T., Chen, R. M., Luh, T. Y., Shi, J., and Tang, C. W. (1999). Bright-blue electroluminescence from a silyl-substituted ter-(phenylene–vinylene) derivative, *Appl. Phys. Lett.*, **74**, p. 865.

87. Wu, Y., and Zhu, W. (2013). Organic sensitizers from D–π–A to D–A–π–A: effect of the internal electron-withdrawing units on molecular absorption, energy levels and photovoltaic performances, *Chem. Soc. Rev.*, **42**, pp. 2039–2058.

88. Zeng, W., Wu, H., Zhang, C., Huang, F., Peng, J., Yang, W., and Cao, C. (2007). Polymer light-emitting diodes with cathodes printed from conducting ag paste, *Adv. Mater.*, **19**, pp. 810–814.

89. Ho, C.-L., and Wong, W.-Y. (2013). Charge and energy transfers in functional metallophosphors and metallopolyynes, *Coord. Chem. Rev.*, **257**, pp. 1614–1649.

90. Farinola, G. M., and Ragni, R. (2011). Electroluminescent materials for white organic light emitting diodes, *Chem. Soc. Rev.*, **40**, pp. 33467–33482.

91. Amitava, P. (2002). Electroluminescence properties of systematically derivatized organic chromophores containing electron donor and acceptor groups, *Chem. Mater.*, **14**, pp. 4044–4048.

92. Xu, H., Chen, R., Sun, Q., Lai, W., Su, Q., Huang, W., and Liu, X. (2014). Recent progress in metal–organic complexes for optoelectronic applications, *Chem. Soc. Rev.*, **43**, pp. 3259–3302.

93. Bran, M. F., Castellano, J. M., Perron, D., Maher, C., Conlon, D., Bousquet, P. F., George, J., Qian, X. D., and Robinson, S. P. (1997). Chromophore-

modified bis-naphthalimides: synthesis and antitumor activity of bis-dibenz[de,h]isoquinoline-1,3-diones, *J. Med. Chem.*, **40**, pp. 449–454.

94. Singh, V., and Mishra, A. K. (2015). White light emission from vegetable extracts, *Sci. Rep.*, **5**, p. 11118.

95. Jang, J. H., Kim, H. U., Lee, S. B., Lee, J., and Hwang, D. H. (2015). Phosphorescent imidazole-based iridium(III) complex with a broad full width at half maximum for solution-processed organic light-emitting diodes, *Synth. Met.*, **203**, pp. 180–186.

96. Ramya, A. R., Varughese, S., and Reddy, M. L. P. (2014). Tunable white-light emission from mixed lanthanide (Eu^{3+}, Gd^{3+}, Tb^{3+}) coordination polymers derived from 4-(dipyridin-2-yl)-aminobenzoate, *Dalton Trans.*, **43**, pp. 10940–10946.

97. Al-Rasbi, N. K., Adams, H., and Suliman, F. E. O. (2014). Synthesis, structure and tunable white-light emission of dinuclear Eu(III) Schiff base complex, *Dyes Pigm.*, **104**, pp. 83–88.

98. Turak, A. (2013). Interfacial degradation in organic optoelectronics, *RSC Adv.*, **3**, pp. 6188–6225.

99. Sun, C. Y., Wang, X. L., Zhang, X., Qin, C., Li, P., Su, Z. M., Zhu, D. X., Shan, G. G., Shao, K. Z., Wu, H., and Li, J. (2013). Efficient and tunable white-light emission of metal–organic frameworks by iridium-complex encapsulation, *Nat. Commun.*, **4**, p. 2717.

100. Jiao, M., Guo. N, Lu, W., Jia, Y., Lv, W., Zhao, Q., Shao, B., and You, H. (2013). Synthesis, structure and photoluminescence properties of europium-, terbium-, and thulium-doped $Ca_3Bi(PO_4)_3$ phosphors, *Dalton Trans.*, **42**, pp. 12395–12402.

101. Zheng, T., and Choy, C. H. W. (2008). High-efficiency blue fluorescent organic light emitting devices based on double emission layers, *J. Phys. D: Appl. Phys.*, **41**, p. 055103.

102. Divayana, Y., Sun, X. W., Chen, B. J., Lo, G. Q., Sarma, K. R., and Kwong, D. L. (2007). Bandgap engineering in Alq3 and NPB-based organic light-emitting diodes for efficient green, blue and white emission, *Solid-State Electron.*, **51**, pp. 1618–1623.

103. Deng, Z., Lee, S. T., Webb, D. P., Chan, Y. C., and Gambling, W. A. (1999). Carrier transport in thin films of organic electroluminescent materials, *Synth. Met.*, **107**, pp. 107–109.

104. Sullivan, P., Heutz, S., Schultes, S. M., and Jones, T. S. (2004). Influence of codeposition on the performance of CuPc–C60 heterojunction photovoltaic devices, *Appl. Phys. Lett.*, **84**, pp. 1210–1212.

105. Ren, X., Alleyne, B. D., Djurovich, P. I., Adachi, C., Tsyba, I., Bau, R., and Thompson, M. E. (2004). Organometallic complexes as hole-

transporting materials in organic light-emitting diodes, *Inorg. Chem.*, **43**, pp. 1697–1707.

106. Kim, J., Song, M., Seol, J., Hwang, H., and Park, C. (2005). Fabrication of red, green, and blue organic light-emitting diodes using m-MTDATA as a common hole-injection layer, *Korean J. Chem. Eng.*, **22**, pp. 643–647.

107. Jiang, W., Duan, Y., Zhao, Y., Hou, J., and Liu, S. (2005). A novel efficient blue organic light emitting structure, *Mater. Sci. Forum*, pp. 3677–3680.

108. Ma, D., Tsuboi, T., Qiu, Y., and Duan, L. (2016). Recent progress in ionic iridium(III) complexes for organic electronic devices, *Adv. Mater.*, **29**, p. 1603253.

109. Du, B., Wang, L., Wu, H., Yang, W., Zhang, Y., Liu, R., Sun, M., Peng, J., and Cao, Y. (2007). High-efficiency electrophosphorescent copolymers containing charged iridium complexes in the side chains, *Chemistry*, **13**, pp. 7432–7442.

110. Hai, N., Bo, Z., and Tang, X. Z. (2007). Significant improvement of OLED efficiency and stability by doping both HTL and ETL with different dopant in heterojunction of polymer/small-molecules, *Chin. Phys. Soc.*, **16**, p. 3.

111. Ahn, S. Il., Kim, W. K., Ryu, S. H., Kim, K. J., Lee, S. E., Kim, S. H., Park, J. C., and Choi, K. C. (2012). OLED with a controlled molecular weight of the PVK (poly(9-vinylcarbazole)) formed by a reactive ink-jet process, *Org. Electron.*, **13**, pp. 980–984.

112. Mu, H., Li, W., Jones, R., Steckl, A., and Klotzkin, D. (2007). A comparative study of electrode effects on the electrical and luminescent characteristics of Alq3/TPD OLED: improvements due to conductive polymer (PEDOT) anode, *J. Lumin.*, **126**, pp. 225–229.

113. Gao, F. G., and Bard, A. J. (2000). Solid-state organic light-emitting diodes based on tris(2,2¢-bipyridine) ruthenium(II) complexes, *J. Am. Chem. Soc.*, **122**, pp. 7426–7427.

114. Strohriegl, P., Wagnera, D., Schrögela, P., Hoffmannb, S. T., Köhlerb, A., Heinemeyerc, U., and Münster, I. (2013). Novel host materials for blue phosphorescent OLEDs, *Proc. SPIE*, **8829**, p. 12.

115. Chen, K. L., Huang, C. J., Chen, W. R., Kang, C. C., Lan, W. H., and Lee, Y. C. (2013). The investigation on color purity of blue organic light-emitting diodes (BOLED) by hole-blocking layer, *Int. J. Photoenergy*, **2013**, p. 878537.

116. Weinfurtner, K. H., Fujikawa, H., Tokito, S., and Taga, Y. (2000). Highly efficient pure blue electroluminescence from polyfluorene: influence

of the molecular weight distribution on the aggregation tendency, *Appl. Phys. Lett.*, **76**, p. 18.

117. Nie, C., Li, F., Wu, J., Zeng, Q., Ooi, P. C., Veeramalai, C. P., and Guo, T. (2015). Solution-processed flexible blue organic light emitting diodes using graphene anode, *Vacuum*, **121**, pp. 70–74.

118. Kido, J., Nagai, K., and Okamoto, Y. (1993). Organic electroluminescent devices using lanthanide complexes, *J. Alloys Compd.*, **192**, pp. 30–33.

119. Kido, J., Nagai, K., Okamoto, Y., and Skotheim, T. (1991). Electroluminescence from polysilane film doped with europium complex, *Chem. Lett.*, **20**, pp. 1267–1270.

120. Noriyuki, T., Tetsuo, T., and Shogo, S. (1994). Strongly directed emission from controlled-spontaneous-emission electroluminescent diodes with europium complex as an emitter, *Jpn. J. Appl. Phys.*, **33**, pp. L863–L866.

121. Takeshi, S., Masayuki, F., Takanori, F., Yuji, H., Kenichi, S., and Kazuhiko, K. (1995). Novel europium complex for electroluminescent devices with sharp red emission, *Jpn. J. Appl. Phys.*, **34**, pp. 1883–1887.

122. Takeshi, S., yoshitaka, N., Yuji, H., Hisakazu, T., Tatsuro, U., and Kenichi S. (2000). Design of conjugated molecular materials for optoelectronics, *J. Mater. Chem.*, **10**, pp. 157–161.

123. Adachi, C., Baldo, M. A., and Forrest, S. R. (2000). Electroluminescence mechanisms in organic light emitting devices employing a europium chelate doped in a wide energy gap bipolar conducting host, *J. Appl. Phys.*, **87**, p. 8049.

124. Yutaka, O., Hirotake, K., Takumi, S., Hiroshi, U., and Katsumi, Y. (2001). Enhancement of electroluminescence utilizing confined energy transfer for red light emission, *Thin Solid Films*, **393**, pp. 407–411.

125. Keizou, O., Mesahiro, U., Wang, Y. F., Chen, T. M., and Nakaya, T. (1998). Red organic electroluminescent devices based on novel furan-contained Eu complex as an emitting layer, *Chem. Lett.*, pp. 801–802.

126. Keizou, O., Wang, Y. F., and Tadao, N. (1998). A novel red organic electroluminescent device using Eu complex as an emitting layer, *Synth. Met.*, **97**, pp. 113–116.

127. Zhang, Y., Lei, W., Chun, L. I., Zeng, W. J., Shi, H. H., and Yong, C. (2007). Enhanced electroluminescent efficiency based on functionalized europium complexes in polymer light-emitting diodes, *Chin. Phys. Lett.*, **24**, p. 1376.

128. Uekawn, M., Miyamoto, Y., Ikeda, H., Kaifu, K., and Nakaya, T. (1997). Synthesis and luminescent properties of europium complexes, *Synth. Met.*, **91**, pp. 259–262.

129. He, P., Wang, H. H., Liu, S. G., Shi, J. X., Wang, G., and Gong, M. L. (2009). Visible-light excitable europium (III) complexes with 2, 7-positional substituted carbazole group-containing ligands, *Inorg. Chem.*, **48**, pp. 11382–11387.

130. He, P., Wang, H. H., Liu, S. G., Shi, J. X, Wang, G., and Gong, M. L. (2010). An efficient 2-linked carbazolyl β-diketonate europium (III) complex as red phosphor applied in LED, *Appl. Phys.*, **99**, pp. 757–762.

131. Liu, S. G., Pan, R. K., Zhou, X. P., Wen, X. L., Chen, Y. Z., Wang, S., and Shi, X. B. (2013). Blue-light excitable europium (III) complex based on deprotonated 1-(9-ethyl-6, 8-dimethyl-9H-carbazol-2-yl)-4, 4, 4-trifluorobutane-1, 3-dionate and 1, 10-phenanthroline, *Inorg. Chim. Acta*, **395**, pp. 119–123.

132. Liu, S. G., Su, W. Y., Pan, R. K., Zhou, X. P., Wen, X. L., Chen, Y. Z., Wang, S., and Shi, X. B. (2013). The synthesis and luminescence of europium (III) complex based on deprotonated 1-(4-ethyl-4*H*-thieno[3,2-b]indol-6-yl)-4,4,4-trifluorobutane-1,3-dionate and 1,10-phenanthroline, *Spectrochim. Acta A*, **103**, pp. 417–422.

133. Wang, H., He, P., Yan, H., Shi, J., and Gong, M. (2011). A novel europium (III)–imidazol–diketonate–phenanthroline complex as a red phosphor applied in LED, *Inorg. Chem. Commun.*, **14**, pp. 1183–1185.

134. Xiang, N. J., Leung, L. M., So, S. K., and Gong, M. L. (2006). Preparation and photoluminescence of a novel β-diketone ligand containing electro-transporting group and its europium (III) ternary complex, *Spectrochim. Acta A*, **65**, pp. 907–911.

135. Xiang, N., Xu, Y., Wang, Z., Wang, X., Leung, L. M., Wang, J., Su, Q., and Gong, M. (2008). A single red InGaN-based light-emitting diode with a europium (III) ternary complex as mono-phosphor, *Spectrochim. Acta A*, **69**, pp. 1150–1153.

136. Sun, P. P., Duan, J. P., Lih, J. J., and Cheng, C. H. (2003). Synthesis of new europium complexes and their application in electroluminescent devices, *Adv. Funct. Mater.*, **13**, pp. 683–691.

137. Sun, P. P., Duan, J. P., Shih, H. T., and Cheng, C. H. (2002). Europium complex as a highly efficient red emitter in electroluminescent devices, *Appl. Phys. Lett.*, **81**, p. 792.

138. Li, X. N., Wu, Z. J., Si, Z. J., Liang, Z., Liu, X. J., and Zhang, H. J. (2009). Effect of secondary ligands' size on energy transfer and electroluminescent

efficiencies for a series of europium (III) complexes, a density functional theory study, *Phys. Chem. Chem. Phys.*, **11**, pp. 9687–9695.

139. Sun, J., Hu, G., She, Q., Zuo, Z., and Guo, L. (2012). Nanofibers doped with a novel red-emitting Europium complex: synthesis, characterization, photophysical property and sensing activity toward molecular oxygen, *Spectrochim. Acta A*, **91**, pp. 192–197.

140. Pereira, A., Conte, G., Gallardo, H., Zucco, C., Quirino, W. G., Legnani, C, Cremona, M., and Bechtold, I. H. (2011). OLEDs based on an europium(III) complex: {Tris(thenoyltrifluoroacetonate) [1,2,5]-thiadiazolo[3,4-f] [1,10]phenanthroline}europium(III), *J. Soc. Inf. Disp.*, **19**, pp. 793–797.

141. Fang, J., You, H., Gao, J., Lu, W., and Ma, D. (2007). Ligand effect on the performance of organic light-emitting diodes based on europium complexes, *J. Lumin.*, **124** pp. 157–161.

142. Fang, J., and Ma, D. (2003). Efficient red organic light-emitting devices based on a europium complex, *Appl. Phys. Lett.*, **83**, p. 4041.

143. Zucchi, G., Murugesan, V., Tondelier, D., Aldakov, D., Jeon, T., Yang, F., Thuery, P., Ephritikhine, M., and Geffroy. (2011). Solution, solid state, and film properties of a structurally characterized highly luminescent molecular europium plastic material excitable with visible light, *Inorg. Chem.*, **50**, pp. 4851–4856.

144. Liu, Y., Chen, K., Xing, K., Wan, Y., Jiang, H., Deng, X., Zhu, M., and Zhu, W. (2013). Conjugated and nonconjugated bipolar-transporting dinuclear europium (III) complexes involving triphenylamine and oxadiazole units: synthesis, photophysical and electroluminescent properties, *Tetrahedron.*, **69**, pp. 4679–4686.

145. Okada, K., Wang, Y. F., Chen, T. M., Kitamura, M., Nakaya, T., and Inoue, H. (1999). Bright red light-emitting organic electroluminescent devices based on a novel thiophene-containing europium complex as an emitting layer, *J. Mater. Chem.*, **9**, pp. 3023–3026.

146. Xu, H., Yin, K., and Huang, W. (2007). Highly improved electroluminescence from a series of novel Eu^{III} complexes with functional single-coordinate phosphine oxide ligands: tuning the intramolecular energy transfer, morphology, and carrier injection ability of the complexes, *Chem. Eur. J.*, **13**, pp. 10281–10293.

147. Yin, K., Xu, H., Zhong, G., Ni, G., and Huang, W. (2009). Highly improved electroluminescence from double-layer devices based on a carbazole-functionalized $europium^{3+}$ complex, *Appl. Phys. A*, **95**, pp. 595–600.

148. Xu, H., Yin, K., and Huang, W. (2008). Highly improved electroluminescence from double-layer devices based on a carbazole-functionalized europium^{3+} complex, *Chem. Phys. Chem.*, **9**, pp. 1752–1760.

149. Xu, H., Yin, K., Wang, L., and Huang, W. (2008). Bright electroluminescence from a chelate phosphine oxide Eu^{III} complex with high thermal performance, *Thin Solid Films*, **51**, pp. 8487–8492.

150. Xu, H., Wei, Y., Zhao, B., and Huang, W. (2010). Influence of bidentate structure of an aryl phosphine oxide ligand on photophysical properties of its Eu^{III} complex, *J. Rare Earths*, **28**, pp. 666–670.

151. Ambili Raj, D. B., Biju, S., and Reddy, M. L. (2009). 4,4,5,5,5-pentafluoro-1-(9H-fluoren-2-yl)-1,3-pentanedione complex of Eu^{3+} with 4,5-bis(diphenylphosphino)-9,9-dimethylxanthene oxide as a promising light-conversion molecular device, *Dalton Trans.*, pp. 7519–7528.

152. Pietraszkiewicz, M., Maciejczyk, M., Samuel, I. D. W., and Zhang, S. (2013). Highly photo- and electroluminescent 1, 3-diketonate Eu (III) complexes with spiro-fluorene-xantphos dioxide ligands: synthesis and properties, *J. Mater. Chem. C*, **1**, pp. 8028–8032.

153. Wang, J., Han, C., Xie, G., Wei, Y., Xue, Q., Yan, P., and Xu, H. (2014). Solution-processible brilliantly luminescent Eu(III) complexes with host-featured phosphine oxide ligands for monochromic red-light-emitting diodes, *Chem. Eur. J.*, **20**, pp. 11137–11148.

154. Han, C., Zhu, L., Zhao, F., Zhang, Z., Wang, J., Deng, Z., Xu, H., Li, J., Ma, D., and Yan, P. (2014). Suppressing triplet state extension for highly efficient ambipolar phosphine oxide host materials in blue PHOLEDs, *Chem. Commun.*, **50**, pp. 2670–2672.

155. Han, C., Zhao, F., Zhang, Z., Zhu, L., Xu, H., Li, J., Ma, D., and Yan, P. (2013). Constructing low-triplet-energy hosts for highly efficient blue PHOLEDs: controlling charge and exciton capture in doping systems, *Chem. Mater.*, **254**, pp. 966–976.

156. Han, C., Zhang, Z., Xu, H., Xie, G., Li, J., Zhao, Y., Deng, Z., Liu, S., and Yan, P. (2013). Convergent modulation of singlet and triplet excited states of phosphine-oxide hosts through the management of molecular structure and functional-group linkages for low-voltage-driven electrophosphorescence, *Chem. Eur. J.*, **19**, pp. 141–154.

157. Silva, C. R. De, Li, F., Huang, C., and Zheng, Z. (2008). Europium β-diketonates for red-emitting electroluminescent devices, *Thin Solid Films*, **517**, pp. 957–962.

158. You, H., Fang, J., Wang, L., Zhu, X., Huang, W., and Ma, D. (2007). Efficient red organic light-emitting diodes based on a dinuclear europium complex, *Opt. Mater.*, **29**, pp. 1514–1517.

159. Wang, H., He, P., Liu, S., Shi, J., and Gong, M. (2009). A europium(III) organic ternary complex applied in fabrication of near UV-based white light-emitting diodes, *Appl. Phys. B*, **97**, pp. 481–487.

160. Wang, H., He, P, Liu, S., Shi, J., and Gong, M. (2010). Effects of annealing temperature and thickness on microstructure and properties of sol-gel derived multilayer Al-doped ZnO films, *Inorg. Chem. Commun.*, **13**, pp. 145–148.

161. Chen, Z., Ding, F., Hao, F., Guan, M., Bian, Z., Ding, B., and Huang, C. (2010). Synthesis and electroluminescent property of novel europium complexes with oxadiazole substituted 1, 10-phenanthroline and 2, 20 -bipyridine ligands, *New J. Chem.*, **34**, pp. 487–494.

162. De Silva, C. R., Maeyer, J. R., Dawson, A., and Zheng, Z. (2007). Adducts of lanthanide β-diketonates with 2, 4, 6-tri (2-pyridyl)-1, 3, 5-triazine: synthesis, structural characterization, and photoluminescence studies, *Polyhedron*, **26**, pp. 1229–1238.

163. Panayiotidou, L., Stylianou, M., Arabatzis, N., Drouza, C., Lianos, P., Stathatos, E., and Keramidas, A. D. (2013). Synthesis, crystal structure and luminescence of novel Eu^{3+}, Sm^{3+} and Gd^{3+} complexes of 1,3,5- and 1,2,4-triazines, *Polyhedron*, **52**, pp. 856–865.

164. Law, G. L., Wong, K. L., Tam, H. L, Cheah, K. W., and Wong, W. T. (2009). White OLED with a single-component europium complex, *Inorg. Chem.*, **48**, pp. 10492–10494.

165. Kido, J., Hayase, H., Hongawa, K., Nagai, K., and Okuyama, K. (1994). Bright red light-emitting organic electroluminescent devices having a europium complex as an emitter, *Appl. Phys. Lett.*, **65**, pp. 2124–2126.

166. Heil, H., Steiger, J., Schmechel, R., and von Seggern, H. (2001). Tris(dibenzoylmethane)(monophenanthroline)europium(III) based red emitting organic light emitting diodes, *Appl. Phys.*, **90**, p. 5357

167. Peng, J., Takada, N., and Minami, N. (2002). Red electroluminescence of a europium complex dispersed in poly(N-vinylcarbazole), *Thin Solid Films*, **405**, pp. 224–227.

168. O'Riordan, A., O'Connor, E., Moynihan, S., Llinares, X., van Deun, R. Fias, P. Nockemann, P., Binnemans, K., and Redmond, G. (2005). Narrow bandwidth red electroluminescence from solution-processed lanthanide-doped polymer thin films, *Thin Solid Films*, **491**, pp. 264–269.

169. Robinson, M. R., O'Regan, M. B., and Bazan, G. C. (2000). Synthesis, morphology and optoelectronic properties of tris [(N-ethylcarbazolyl) (3′, 5′-hexyloxybenzoyl) methane] (phenanthroline) europium, *Chem. Commun.*, pp. 1645–1646.

170. McGehee, M. D., Bergstedt, T., Zhang, C., Saab, A. P., O'Regan, M. B., Bazan, G. C., Sradanov, V. I., and Heeger, A. J. (1999). Narrow bandwidth luminescence from blends with energy transfer from semiconducting conjugated polymers to europium complexes, *Adv. Mater.*, **11**, pp. 1349–1354.

171. Li, S., Zhong, G., Zhu, W., Li, F., Pan, J., Huang, W., and Tian, H. (2005). Dendritic europium complex as a single dopant for white-light electroluminescent devices, *J. Mater. Chem.*, **15**, pp. 3221–3228.

172. Liang, H., and Xic, F. (2010). Fluorescent properties of a dendritic ligand europium complex incorporated in polymethylvinylsiloxane, *Spectrochim. Acta A*, **75**, pp. 1191–1194.

173. Robinson, M. R., Ostrowski, J. C., Bazan, G. C., and McGehee, M. D. (2003). Reduced operating voltages in polymer light-emitting diodes doped with rare-earth complexes, *Adv. Mater.*, **15**, pp. 1547–1551.

174. Noto, M., Irie, K., and Era, M. (2001). Efficient red electroluminescence from new europium complex, *Chem. Lett.*, **30**, pp. 320–321.

175. Liang, C. J., Hong, Z. R., Liu, X. Y., Zhao, D. X., Li, W. L., Peng, J. B., Yu, J. Q., Lee, C. S., and Lee, S. T. (2000). Organic electroluminescent devices using europium complex as an electron-transport emitting layer, *Thin Solid Films*, **359**, pp. 14–16.

176. Liang, C. J., Zhao, D., Hong, Z. R., Zhao, D. X., and Liu, X. Y. (2000). Improved performance of electroluminescent devices based on an europium complex, *Appl. Phys. Lett.*, **76**, p. 67.

177. Hong, Z., Liang, C. Li, R., Li, W., Zhao, D., Fan, D., Wang, D., Chu, B., Zang, F., Hong, L. S., and Lee, S. (2001). Rare earth complex as a high-efficiency emitter in an electroluminescent device, *Adv. Mater.*, **13**, p. 1241.

178. Liu, Y., Wang, Y., Li, C., Huang, Y., Dang, D., Zhu, M., Zhu, W., and Cao, Y. (2014). Catalpol provides protective effects against cerebral is chaemia/reperfusion injury in gerbils, *Mater. Chem. Phys.*, **143**, pp. 1265–1270.

179. Liu, Y., Li, J., Li, C., Song, J., Zhang, Y., Peng, J., Wang, X., Zhu, M., Cao, Y., and Zhu, W. (2007). Highly efficient sharp red electroluminescence from europium complex-doped poly (9, 9- dioctylfluorene) devices, *Chem. Phys. Lett.*, **433**, pp. 331–334.

180. Liu, Y., Wang, Y., Guo, H., Zhu, M., Li, C., Peng, J., Zhu, W., and Cao, Y. (2011). Synthesis and optoelectronic characterization of a monochromic

red-emitting europium(III) complex containing triphenylamine-functionalized phenanthroline, *J. Phys. Chem. C*, **115** , pp. 4209–4216.

181. Liu, Y., Wang, Y., He, J., Mei, Q., Chen, K., Cui, J., Li, C., Zhu, M., Peng, J., Zhu, W., and Cao. Y. (2012). High-efficiency red electroluminescence from europium complex containing a neutral dipyrido (3, 2-a: 2', 3'-c) phenazine ligand in PLEDs, *Org. Electron.*, **13**, pp. 1038–1043.

182. Liu, X., Xiao, G., Chi, H., Dong, Y., Zhao, H., Lei, P., Zhang, Z., Hu, Z., Wu, S., Su, Z., and Li, W. (2010). A novel fluorinated europium ternary complex for highly efficient pure red electroluminescence, *Mater. Chem. Phys.*, **123**, pp. 289–292.

183. Sun, M., Xin, H., Wang, K. Z., Zhang, Y. A., Jin, L. P., and Huang, C. H. (2003). Bright and monochromic red light-emitting electroluminescence devices based on a new multifunctional europium ternary complex, *Chem. Commun.*, pp. 702–703.

184. Li, Y. (2012). High performance europium (III) based emitters for electroluminescence: synthesis, crystal structures, photophysical and electroluminescent properties, *J. Lumin.*, **132**, pp. 2102–2108.

185. Yuan, J. B., Li, J. H., Leung, W. L., Su, S. J., Yao, J. H., and Gong, M. L. (2005). Synthesis and electroluminescence of a novel Eu(III) ternary complex, *Chem. J. Chin. Univ.*, **26**, pp. 1787–1790.

186. Tang, H., Zhang, Z., Yuan, J., Cong, C., and Zhang, K. (2009). Synthesis, photoluminescent and electroluminescent properties of a novel europium (III) complex involving both hole- and electron-transporting functional groups, *Synth. Met.*, **159**, pp. 72–77.

187. Li, L., Liu, Y., Guo, H., Wang, Y., Cao, Y., Liang, A., Tan, H., Qi, H., Zhu, M., and Zhu, W. (2010). Synthesis, optophysical and electrochemical properties of bipolar-transporting europium (III) complexes with carbazole and oxadiazole units, *Tetrahedron*, **66**, pp. 7411–7417.

188. Liu, Y., Wang, J., Wang, Y., Zhang, Z., Zhu, M., Lei, G., and Zhu, W. (2012). Bipolar-transporting dinuclear europium (III) complexes involving carbazole and oxadiazole units: synthesis, photophysical and electroluminescent properties, *Dyes Pigm.*, **95**, pp. 322–329.

189. Xin, H., Li, F. Y., Guan, M., Huang, C. H., Sun, M., Wang, K. Z., Zhang, Y. A., and Jin, L. P. (2003). Carbazole-functionalized europium complex and its high-efficiency organic electroluminescent properties, *J. Appl. Phys.*, **94**, pp. 4729–4731.

190. Xin, H., Sun, M., Wang, K. Z., Zhang, Y. A., Jin, L., and Huang, C. H. (2004). Voltage-independent pure red devices based on a carbazole-functionalized europium complex, *Chem. Phys. Lett.*, **388**, pp. 55–57.

191. Guan, M., Bian, Z. Q., Li, F. Y., Xin, H., Huang, C. H., and J. N. (2003). Bright red light-emitting electroluminescence devices based on a functionalized europium complex, *New J. Chem.*, **27**, pp. 1731–1734.

192. Bian, Z. Q., Wang, K. Z., and Jin, L. P. (2002). Syntheses, spectroscopic and crystal structural studies of novel imidazo [4, 5-f] 1, 10-phenanthroline derivatives and their Eu (III) ternary complexes with dibenzoylmethane, *Polyhedron*, **21**, pp. 313–319.

193. Bian, Z., Gao, D., Wang, K., Jin, L., and Huang, C. (2004). Pure red organic electroluminescent devices using a novel europium (III) complex as emitting layer, *Thin Solid Films*, **460**, pp. 237–241.

194. Bian, Z. Q., Gao, D. Q., Guan, M., Xin, H., Li, F. Y., and Huang, C. H. (2004). Electroluminescent properties of three ternary europium complexes with different phenanthroline derivatives. *Science in China (Series B)*, **47**, pp. 326–334

195. Lam, M. K., Kwok, K. L., Tse, S. C., So, S. K., Yuan, J. B., Leung, L. M., and Gong, M. L. (2006). *Opt. Mater.*, **28**, pp. 709–713.

196. Xue, Q., Chen, P., Lu, J., Xie, G., Hou, J., Liu, S., Zhao, Y., Zhang, L., and Li, B. (2009). Pure red organic light-emitting diode based on a europium complex, *Solid State Electron.*, **53**, pp. 397–399.

197. Zheng, Y., Cardinali, F., Armaroli, N., and Accorsi, G. (2008). Synthesis and photoluminescence properties of heteroleptic europium (III) complexes with appended carbazole units, *Eur. J. Inorg. Chem.*, pp. 2075–2080.

198. Zheng, Y., Zhou, Y., Accorsi, G., and Armaroli, N. (2008). Synthesis and photoluminescence of a dendritic europium complex with carbazole moieties, *J. Rare Earths*, **26**, pp. 173–177.

199. Wang, S., Zhang, J., Hou, Y., Du, C., and Wu, Y. (2011). 4, 5-Diaza-9, 90 -spirobifluorene functionalized europium complex with efficient photo- and electro-luminescent properties, *J. Mater. Chem.*, **21**, pp. 7559–7561.

200. Liu, Z., Wen, F., and Li, W. (2005). Study on transient distributed model of frost on heat pump evaporator, *Thin Solid Films*, **478**, pp. 265–270.

201. Shi, M., Li, F., Yi, T., Zhang, D., Hu, H., and Huang, C. (2005). Tuning the triplet energy levels of pyrazolone ligands to match the $^5D_0$ level of europium (III), *Inorg. Chem.*, **44**, pp. 8929–8936.

202. Hu, W., Matsumura, M., Wang, M., and Jin, L. (2000). Efficient red electroluminescence from devices having multilayers of a europium complex, *Appl. Phys. Lett.*, **77**, pp. 4271–4273.

203. Hu, W., Matsumura, M., Wang, M., and Jin, L. (2000). Red electroluminescence from an organic europium complex with a triphenylphosphine oxide ligand, *Jpn. Appl. Phys. Lett.*, **39**, pp. 6445–6448.

204. Teotonio, E. E. S., Brito, H. F., Cremona, M., Quirino, W. G., Legnani, C., and Felinto, M. C. F. C. (2009). Novel electroluminescent devices containing Eu^{3+}-(2-acyl-1, 3-indandionate) complexes with TPPO ligand, *Opt. Mater.*, **32**, pp. 345–349.

205. Zhu, X. H., Wang, L. H., Ru, J., Huang, W., Fang, J. F., and Ma, D.-G. (2004). An efficient electroluminescent (2, 2′-bipyridine mono N-oxide) europium (III) β-diketonate complex, *J. Mater. Chem.*, **14**, pp. 2732–2734.

206. Wang, Y.-Y., Wang, L.-H., Zhu, X.-H., Ru, J., Huang, W., Fang, J.-F., and Ma, D.-G. (2007). Efficient electroluminescent tertiary europium (III)-diketonate complex with functional 2, 2-bipyridine ligand, *Synth. Met.*, **157**, pp. 165–169.

207. You, H., Fang, J., Gao, J., and Ma, D. (2007). Improved efficiency of organic light-emitting diodes based on a europium complex by fluorescent dye, *J. Lumin.*, **122**, pp. 687–689.

208. Wang, K., Gao, L., and Huang, C. (2003). Optical properties of the highly ordered Langmuir–Blodgett film of a strongly luminescent Eu (III) complex, *J. Photochem. Photobiol. A*, **156**, pp. 39–43.

209. Liang, F., Zhou, Q., Cheng, Y., Wang, L., Ma, D., Jing, X., and Wang, F. (2003). Oxadiazole-functionalized europium (III) β-diketonate complex for efficient red electroluminescence, *Chem. Mater.*, **15**, pp.1935–1937.

210. Zhang, L., Li, T., Li, B, Lei, B., Yue, S., and Li, W. (2007). Synthesis and electroluminescent properties of a carbozole functionalized europium (III) complex, *J. Lumin.*, **126**, pp. 682–686.

211. Guan, M., Gao, L., Wang, S., Huang, C., and Wangb, K. (2007). Syntheses and electroluminescent properties of two europium ternary complexes Eu $(DBM)_3$(PBO) and Eu $(DBM)_3$(PBT), *J. Lumin.*, **127**, pp. 489–493.

212. Liu, Y., Liang, B., Xiao, D., Zhu, M., and Zhu, W. (2009). Synthesis and characteristics of a europium complex using pyridyloxadiazole derivative as a secondary ligand, *J. Alloys Compd.*, **469**, pp. 370–373

213. Gusev, A. N., Shul'gin, V. F., Meshkova, S. B., Doga, P. G., Hasegawa, M., Alek-sandrov, G. G., Eremenko, I. L., and Linert, W. (2012). Structural and photophysical studies of europium complexes containing triazole ligands, *Inorg. Chim. Acta*, **387**, pp. 321–326.

214. Gusev, A. N., Hasegawa, M., Nishchymenko, G. A., Shul'gin, V. F., Meshkova, S. B., Doga, P., and Linert, W. (2013). Ln (III) complexes of a bis (5-(pyridine-2-yl)-1, 2, 4-triazol- 3-yl) methane ligand: synthesis, structure and fluorescent properties, *Dalton Trans.*, **42**, pp. 6936–6943.

215. Gusev, A. N., Hasegawa, M., Shul'gin, V. F., Nishchymenko, G., and Linert, W. (2014). Photophysical studies on ternary mixed ligand europium complexes containing pyridyltriazolylmethane and 1, 3-diketonate ligands, *Inorg. Chim. Acta*, **414**, pp. 71–77.

216. Gusev, A. N., Shul'gin, V. F., Nishimenko, G., Hasegawa, M., and Linert, W. (2013). Photo- and electroluminescent properties europium complexes using bistriazole ligands, *Synth. Met.*, **164**, pp. 17–21.

217. Chen, L., Wang, B., Zhang, L., Zhu, D., Li, P., Su, Z., and Li, B. (2012). Red- and white-emitting organic light-emitting diodes based on trimetallic dendritic europium (III) complex: Eu_3 (DBM) $_9$(TMMB), *Solid State Electron.*, **69**, pp. 67–71.

218. Wei, F., and Yu, G. X. (2013). Study on a star-shaped europium complex: synthesis, photophysical properties and organoelectronic performances, *J. Lumin.*, **134**, pp. 710–717.

Chapter 10

An Assorted Outlook on the Versatility of Thermoluminescence Techniques

Sumedha Tamboli,[a] S. J. Dhoble[a] and B. C. Bhatt[b]
[a]*Department of Physics, R. T. M. Nagpur University, Nagpur 440033, Maharashtra, India*
[b]*Radiochemistry Division, Bhabha Atomic Research Centre, Trombay 400085, Maharashtra, India*
sumedhatamboli@yahoo.com, sjdhoble@rediffmail.com

Ionizing radiation plays a crucial role in the fields of medicine, engineering, science and technology, and research and agriculture. Controlled use of ionizing radiation, with the adoption of recommended safety procedures, has led to wide applications of ionizing radiation in various fields. Adoption of radiation safety norms has led to harnessing the power of the atom for various technological applications. One of the most important, peaceful applications of nuclear energy is in power generation. However, indiscriminate use of ionizing radiation, that is, without adhering to proper safety procedures, may lead to excessive radiation exposure of radiation workers, which may be hazardous to their health. People who receive relatively high doses of radiation could have a

Phosphors: Synthesis and Applications
Edited by Sanjay J. Dhoble, B. Deva Prasad Raju, and Vijay Singh

ISBN 978-981-4774-49-9 (Hardcover), 978-042-9460-99-9 (eBook)
www.panstanford.com

greater risk of developing cancer later in life, depending on the level of radiation exposure. Therefore, monitoring radiation exposure of occupational workers is a mandatory requirement from national as well as international regulations. A dosimeter is used for this purpose. A dosimeter is a small radiation detection device worn by an individual, used to measure doses of ionizing radiation. Radiation dosimeters/detectors are of two types, active and passive. A passive dosimeter is a device that is used to record personal exposure to radiation (and sometimes environmental exposure), whereas an active dosimeter provides a real-time instant measure of dose accrued and dose rate. Among the active detectors are gaseous detectors (ionization chambers, Geiger–Müller [GM] counters, etc.), solid-state detectors (scintillation and semiconductor detectors), etc., whereas among the passive detectors are thermoluminescent, optically stimulated luminescent, radiophotoluminescent (RPL), and photographic film-based detectors. TL materials are useful in dosimetry applications of ionizing radiations due to availability of various types of TLDs, in different physical forms and sizes, in large numbers with high sensitivity, low photon energy, and angular dependence, wide dynamic dose response, low postirradiation fading, etc. Since the 1950s, there has been extensive research on thermoluminescence (TL) and its application in radiation dosimetry. This research, which was only widely applied beginning in the 1970s, guided the development of thermoluminescent dosimeters (TLDs), which are now the most commonly used type of dosimeter around the world for personnel monitoring applications in particular. This paper is solely devoted to the stimulating aspects of TL and the related concepts.

TL is also a good research tool for study of defects, kinetics of the TL process, and determination of kinetic parameters in TL phosphors. Knowledge of kinetic parameters is helpful in assessing the stability of TL peaks in a TL phosphor. Characterizing a TL material in terms of its dose response, kinetic parameters, photoluminescence (PL), optical absorption (OA), etc., is also helpful in understanding the underlying TL mechanism. Various routes of synthesizing TLD materials have also been included as firsthand knowledge for novice researchers in this field so as to encourage them to attempt materials development work to initiate further research work in this field. A brief discussion of the possibility of using organic materials for dosimetric purpose would pave the way to the development of new organic TLD materials.

10.1 Introduction

Initially, dosimetry measurements were dominated by ionization chamber measurements and photographic film until the development of dosimetry techniques based on luminescence and other solid-state methods [1]. Since the 1950s, there has been extensive research on thermoluminescence (TL) and its application in radiation dosimetry [2–5]. All these developments in the field of TL dosimetry, in the initial years, till the mid-1980s, led to the development and characterization of many useful thermoluminescent phosphors for radiation dosimetry, such as LiF:Mg,Ti [6, 7], $Li_2B_4O_7$:Mn [8, 9], BeO [10], $Li_2B_4O_7$:Ag,Cu [11, 12], MgB_4O_7:Dy [13, 14], LiF:Mg,Cu,P [15, 16], $CaSO_4$:Dy [17], $CaSO_4$:Tm [18, 19], CaF_2:Dy [20], CaF_2:Tm [21, 22], and Mg_2SiO_4:Tb [23, 24]. The use of TL as a method for dosimetry of ionizing radiation has been established for many decades and has found many useful applications in various fields, such as personnel and environmental monitoring, medical dosimetry, accidental dosimetry, retrospective dosimetry, archaeological and geological dating, and space dosimetry [1, 25–30]. Thermoluminescent dosimeters (TLDs) are presently the most commonly used dosimeter type in many countries around the world. Due to nonreliability of photographic film-based dosimeters for long-term stability of stored information, particularly in a tropical environment, this method of personnel dosimetry has nearly been phased out and replaced by other competent luminescent methods of personnel dosimetry, such as TL.

TLD materials have continued to gain wide acknowledgement for their prominence in personal and environmental monitoring, medical dosimetry, accidental dosimetry, etc. [4, 31, 32]. One of the important applications of TL dosimetry has been in the field of medical physics for radiodiagnosis, radiotherapy, and nuclear medicine. TLDs have become popular in these fields due to their high sensitivity, miniature size, tissue equivalence, high stability to environmental conditions, low TL fading, reusability, linear dose response, and sufficient precision and accuracy. In cancer treatment

TLD materials have immense importance for their applicability in detection and measurement of charged particle dose in ion beam therapy for tumors [33]. TLD materials proved to be superior over thermally stimulated exoelectron emission (TSEE) and thermally stimulated current (TSC) because of their high sensitivity and reliability. Their production method is commercially viable and has desired characteristics for their large-scale use in field conditions. Most of these TL materials are nontoxic. There is a large number commercially available TLD materials such as LiF:Mg,Ti (TLD-100) [6, 7], LiF:Mg,Cu,P (TLD-700H) [15, 16], $CaSO_4$:Dy (TLD-900) [17], CaF_2:Dy (TLD-200) [20], Al_2O_3:C (TLD 500) [34, 35], etc. Among these materials, lithium fluoride (LiF) family materials activated with different dopants such as magnesium and titanium (Mg, Ti) and magnesium, copper, and phosphorus (Mg, Cu, P), are useful for personnel and medical dosimetry because of good sensitivity and tissue equivalence (i.e., having an effective atomic number, Z_{eff}, 8.14, which is close to that of biological tissue, 7.4) so as to provide a response, which varies only slightly with photon energy [2, 36]. LiF:Mg,Cu,P TLDs are about 25–30 times more sensitive than LiF:Mg,Ti TLDs; therefore, the former are replacing the latter for various applications, particularly for personnel monitoring and medical dosimetry. The useful dose range of LiF:Mg,Ti and LiF:Mg,Cu,P TLDs is from micrograys to 5–10 Gy; LiF:Mg,Ti shows supralinear dose response beyond this dose and saturates around 1 kGy, whereas LiF:Mg,Cu,P shows sublinear response beyond 10 Gy [36, 37]. Now a considerable amount of research is going on in synthesizing nanostructures of some highly TL sensitive materials, such as LiF:Mg, Cu,P, $CaSO_4$:Dy, Al_2O_3, and MgB_4O_7:Dy [38, 39]. These nanostructure materials have negligible fading, reasonably good sensitivity, stable sensitivity under different heat treatments, and extended dose response. Efforts are being made to improve the characteristics of TLD materials to achieve better efficiency and to extend their applications in a broader range [40]. Thus, nanophosphors could find many potential applications in the field of radiation dosimetry [41], such as low-energy β-dosimetry (as thin films on substrates), high–linear energy transfer (LET) dosimetry, and high-dose dosimetry for technological applications (e.g., sterilization of medical products and irradiation of food products).

This phenomenon seems to be more descriptive by the term "thermally stimulated emission" (TSL) than TL. In the field of research, study of the TL phenomenon plays a vital role in understanding defects in materials [42]. When the sample, which may be a semiconductor or an insulator containing defects, is irradiated by energetic ionizing radiations, energy gets stored in this sample. When this sample is heated subsequently, the stored energy is emitted in the form of light. Heating plays the role of a stimulant or a trigger for initiating the emission of light. It is usually considered that TL is caused by electron–hole recombination, one of which is released thermally from a trap. Crystals must contain some kind of imperfection such as impurities or point defects that traps an electron or a hole [43, 44]. The intensity of emitted light is plotted as a function of temperature, and the corresponding graph obtained is called a *TL glow curve*. Analysis of the glow curve provides information regarding the radiation dose absorbed and the defects present in the sample. TL is the study of defects present in a given sample; it is a sensitive tool to determine low quantity of impurities present in a sample. The shape, number, and spectral distribution of the peak depend on the nature of excitation and defects. A glow curve must be linearly dependent on the absorbed dose of radiation so that it can be useful for radiation detection as well as archaeological dating. But it is often found that the curve attains saturation at high doses and becomes superlinear at low doses. Analysis of glow curves helps to determine physical parameters such as activation energy and frequency factor [45, 46].

10.2 Chronology of Development

Robert Boyle (1663) was the first to record the TL response scientifically by heating a diamond, giving a glimmering light. Elsholtz observed TL in fluorspar. Oldenbung (1676) wrote that material received light from fire. Du Fay in 1726 observed that when sulfur is heated it emits light. He showed that TL can be reactivated by exposing a quartz crystal to light. Hence it was declared that temperature is not the cause of TL. In 1867 Becquerel also reported luminescence from fluorspar [47]. Wiedemann and Schmidt in 1895 irradiated a specimen with an electron beam in the laboratory and observed TL;

this was the first artificially obtained TL because the sample was intentionally irradiated by ionizing energy [17, 48]. Marie Curie in 1904 conveyed that some fluorites show luminescence on heating and have the capability to show re-emission when irradiated with β-rays [49]. Wick and Slattery in 1928 synthesized phosphors and measured the TL response of these phosphors after their excitation by X-rays [50]. In 1940 at the University of Wisconsin, USA, Daniel and coworkers started studying TL properties of LiF materials. Though, in between 1956 and 1960, the work was temporarily stopped due poor dosimetric properties, Cameron continued the work on this material [6]. He doped LiF with Mg and Ti, and this phosphor was patented by the Harshaw Chemical Co., USA, and known as TLD-100, TLD-600, and TLD-700, depending on the preparation from natural lithium or enriched lithium ^{6}Li, or enriched lithium ^{7}Li, respectively. Nakajima et al. prepared LiF:Mg,Cu,P, a highly sensitive TL phosphor [16]. TL models for LiF:Mg,Ti were proposed on the basis of solid-state theory and conduction band/valence band modeling of the traffic of charge carriers [51]; later ionization density theory [52] and a theory depending on the interaction of radiation with matter were introduced [53]. On the basis of these theories models such as the extended track interaction model (ETIM), the unified interaction model (UNIM), modified track structure theory (MTST), and microdosimetric target theory (MTT) were introduced, which were more or less adequate in explaining the TL phenomenon of material [54]. In 1965 Bjarngard et al. [55] made use of $CaSO_4$:Mn in radiation dosimetry because of its high sensitivity. But due to a low-temperature glow peak (~100°C), it experiences postirradiation fading at room temperature (RT). The development of rare earth (especially Dy or Tm)-doped $CaSO_4$ in 1970 by Yamashita et al. (1971) was a remarkable development in TL dosimetry, and soon after TL became an extensively used tool for dosimetric applications because of its high sensitivity and considerably high glow peak at 220°C [17]. Daniels and coworkers (1957) investigated TL properties of Al_2O_3 for its possible use in radiation dosimetry [56]. Later, Mehta and Sengupta (1977) developed Al_2O_3:Si,Ti having five times more intensity than TLD-100; the impurities Si and Ti appear to be responsible for the observed TL in this phosphor [57]. Theoretical and experimental works on TLD materials are still going on [58–60].

10.3 Mechanism of Thermoluminescence

10.3.1 Band Theory

The mechanism of TL can be explained very well on the basis of the band theory of solids. Solids are broadly classified as conductors, semiconductors, and insulators. In conductors, the valence band overlaps with the conduction band, while in the case of insulators and semiconductors, it is separated by a forbidden energy gap. At RT, the valence band is filled with electrons and the conduction band is empty. When a crystal consists of some impurities, localized energy levels get introduced in the forbidden energy gap. These localized energy levels act as trapping centers for electrons and holes. Now, to understand the mechanism of TL we consider the energy level diagram of a crystal that consists of two localized energy levels present in the forbidden energy gap, as shown in Fig. 10.1. Initially, electrons are present in the valence band when there is no irradiation [54].

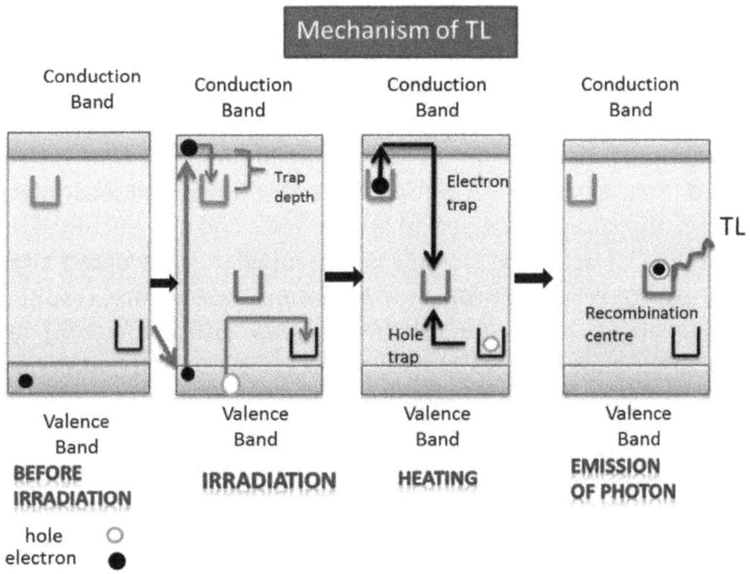

Figure 10.1 Mechanism of thermoluminescence based on band theory.

When the sample is irradiated with energetic ionizing radiations such as γ-rays, X-rays, α-particles, β-particles, neutrons, and heavy ions, electrons get excited and jump to the conduction band, and a positively charged vacancy is created in the valence band, called a hole. These electrons and holes get trapped in the electron traps and hole traps, which are present in the forbidden energy gap. The electrons will remain in the electron traps until and unless a thermal energy is given to the crystal. When the crystal is heated, electrons acquire an activation energy, which is equal to the energy difference between the electron trap and the conduction band. Electrons get released from the trap and jump to the conduction band, and from there they jump to the recombination center, where the electron–hole pair recombines to give photons, which are emitted from the crystal. Sometimes, it is radiative combination, and at other times it is nonradiative recombination. Radiative recombination gives rise to TL emission.

10.3.2 The Mobile Interstitial Model

Sagastibelza and Alvarez Rivas (1981), as an extension to the work previously undertaken by Alvarez Rivas and coworkers, proposed the mobile interstitial model for explaining TL in an alkali halide (LiF:Mg,Ti) phosphor, depending on the theory of exciton formation [61]. When phosphors are irradiated with some high-energy particles or radiation, free electrons and self-trapped holes (V_k-centers) get formed. Free electrons recombine with V_k-centers and lead to the creation of excitons. These excitons are mobile and move before their relaxation. A fraction of excitons gets trapped at some defect sites. At RT exciton relaxation causes a displacement sequence, resulting in the formation of F-centers (halide vacancy with an electron) and H-centers (halide interstitial with a hole) (Fig. 10.2). H-centers are mobile and get captured at impurity defects such as MgV or (MgV)3 in LiF:Mg,Ti, whereas F-centers are immobile. At low temperature the mobility of F-centers is low, but at high temperature F-centers also become mobile and aggregate at impurity defects such as Ti complexes in the case of LiF:Mg,Ti. When the sample is heated, H-centers get released from impurity defects and once again become mobile and, thereafter, combine with F-centers to release photons, leading to TL. Moharil traced out some anomalies and suggested the directions in which the mobile interstitial model is to be modified. He showed that although TL glow curves where successfully described by the mobile interstitial model, phototransfer thermoluminescence (PTTL) glow curves seemed difficult to be explained by this model [62].

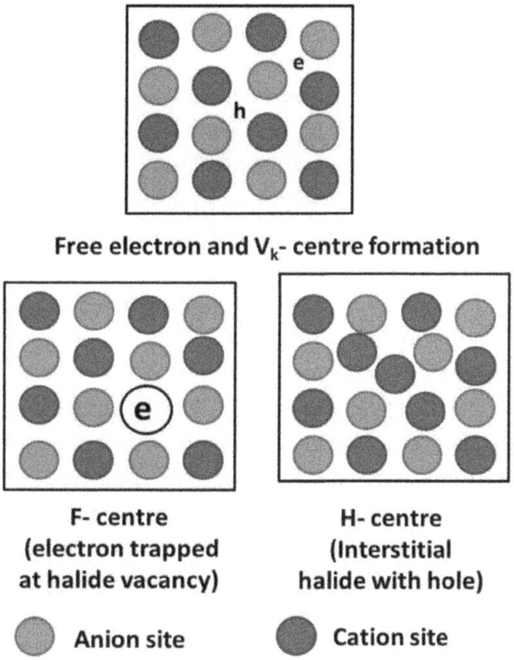

Figure 10.2 Mobile interstitial model. Formation of F-center, V_k-center, and H-center.

10.3.2.1 Phototransfer thermoluminescence and its evolution

When phosphors are irradiated with high-energy γ-rays and annealed partially, with the exception of a high-temperature glow peak, all the remaining peaks vanish. On subsequent exposure to UV light, the charge carriers from filled deep traps transfer to shallower traps, resulting in the regeneration of erased peaks. This process is called PTTL. According to the mobile interstitial model, it is assumed that during UV irradiation, F-centers migrate. Some of them recombine to give photoluminescence (PL), and others get destroyed nonradiatively. Heat generated during nonradiative recombination creates interstitials, which get trapped in shallower traps and give PTTL. Shapes of the PTTL glow curve depend on the exposure time of UV rays. Lower-temperature curves saturate easily at lower doses of UV exposure, while higher-temperature glow peaks saturate at higher doses. Saturation heights attained in the PTTL glow curve are not same as in γ-ray exposure [62].

In all alkali halides, it is observed that as the dose is increased, the glow peaks grow, saturate, and then decline, giving rise to new glow peaks at higher temperatures [63, 64]. This has been termed as "evolution of the glow peak." At a high enough dose, only a single peak occurring at high temperature can be observed. The TL process involves both traps and luminescence centers, and a decline in the intensity of the glow peak for higher doses is not necessarily due to the destruction of traps alone. This can be considered as a result of the damage to the luminescence centers as well. Thus, for higher doses (>105 rad), the luminescence centers get distorted. In such conditions, the charges released from traps corresponding to the low-temperature peaks find the distorted luminescence centers having low luminescence efficiency. Most of the charges will be retrapped in deeper traps or get destroyed by nonradiative processes. With increasing doses, the luminescence centers will get distorted to a greater extent, and they will revert to normal at higher temperatures. This gives a complete explanation of the evolution of the glow curve and the accompanying changes in the emission spectra.

10.3.3 The Track Interaction Model

The track interaction model is a simplistic model adopted for the purposes of estimating how the average distance between luminescence centers along a charged particle track compares with the average distance between tracks. It is related to high-LET radiation or heavy charge particle interaction with phosphors. The model postulates that during irradiation electrons and holes get trapped near the track of ionizing particles, and some of them activate luminescence centers, as shown in Fig. 10.3. At low doses the distance between tracks is large and therefore there is no intertrack electron–hole recombination. Recombination occurs only in the track itself, and therefore at low doses TL intensity is less. Also at low doses, the unirradiated space between the tracks is occupied by competing centers, which capture the charges and quench luminescence. At higher temperatures these charge carriers at traps get detached, giving TL on the higher-temperature side. When the phosphor is irradiated with a high dose or heavy charge particles such as neutrons, a large number of tracks is generated and the average distance between tracks becomes comparable to the distance between activators centers. Intertrack migration of charges can take place, resulting in a higher intensity of TL [65]. The track interaction model turned out to be the most successful model for explaining the supralinearity and sensitization in LiF TLD-100 [66].

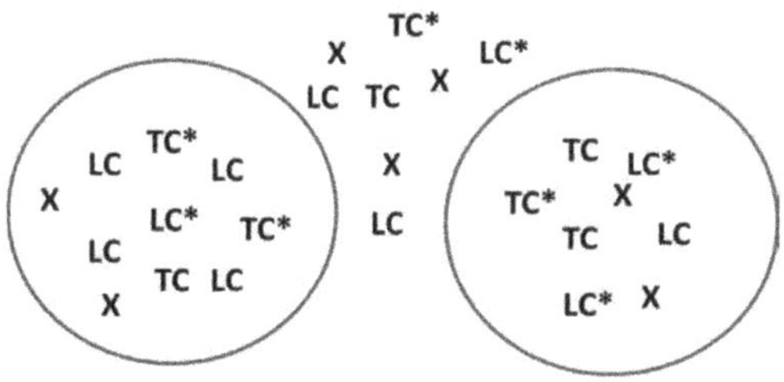

Figure 10.3 Track interaction model. The figure shows two tracks: LC is the luminescence center, TC is the trap center, X is the competing center, and LC* and TC* are centers following activation by charge carrier capture.

When a plot of the TL signal verses the radiation dose is drawn, it gives three regions, which are sublinear, linear, and superlinear. In the linear region, TL intensity increases linearly with increase in the radiation dose. In the sublinear region the rate of increase of TL intensity is less with an increase in dose, and in the superlinear region the rate of increase of TL intensity is high as compared to the irradiation dose. Superlinearity is different from supralinearity as superlinearity measures the rate of change of TL with dose, gives a correction to be made in extrapolation of linearity, it quantifies the deviation of the graph from linear behavior. Sensitization is the process of increasing TL sensitivity by irradiating phosphors with a high dose of radiation, followed by thermal annealing. This is the technique of increasing sensitivity. However, some objections have been raised against this model such as it would predict much higher sensitization than is observed experimentally [67]. Following postirradiation thermal annealing, the model is said to predict higher losses in sensitization than are actually found. Later, Moharil experimentally verified this model and found that the results can be explained very well with the help of this model [68]. According to this model, when a phosphor is exposed to ionizing radiation, an electron–hole pair gets created and some of these charges get trapped in the track of secondary radiation. On heating the phosphor, trapped charges get free and a recombination of these charges leads to TL emission. The charges can recombine in their own track or may get destroyed if they do not reach the recombination center. For higher doses, the tracks intersect with one another, and thus a larger fraction of charges recombine, owing to an increase in the recombination probability. This ultimately results in a supralinearity response for higher doses. Not all luminescent centers are used for TL; in sensitised materials luminescent centers left behind result in effectively higher intersection of tracks. Hence, the sensitization effect is observed for higher doses and the objections raised against this model do not stand in the face of experimental proof.

10.3.4 The Phenomenological Model

The phenomenological model gave the most satisfactory explanation for the supralinearity and sensitization factors of LiF TLD-100. Several models were introduced to explain supralinearity, which also postulated the presence of deep traps with early saturation. However, no deep traps were found with such properties, and the models were discarded. The phenomenological model was named so because of its inadequacy in understanding the mechanism of defect formation, photon-emitting step during TL, etc. This is a qualitative model, which is also the most satisfactory model in explaining the properties of LiF TLD-100 material. Several models attribute supralinearity to the presence of nonluminescent centers, which compete with luminescent centers. Sensitization was attributed to the removal of this competition by filling nonluminescent centers during pre-irradiation. Moharil attempted to calculate supralinearity and sensitization parameters on the basis of this model [69].

10.3.4.1 Supralinearity

Before irradiation, the sample is assumed to consist of empty traps, luminescent centers, along with a fixed number of nonluminescent centers. Irradiations fill the traps, and equal numbers of luminescent centers are formed at a linear rate. During TL readout, a fraction of released charges gets trapped by nonluminescent centers and loses energy nonradiatively. On increasing the dose rate, filled traps and luminescent centers increase but nonluminescent centers decrease. Therefore, at high doses supralinearity is observed, and a further increase in the dose leads to destroying of filled traps with a subsequent decline of TL response.

Moharil obtained an equation for the TL output at higher and lower irradiation doses [69]. Let n_0 be the rate at which traps and luminescent centers are filled. $n_{0\text{v}}$ is the number of nonluminescent centers in the sample before irradiation. α is the fraction of charges going to nonluminescent centers during readout. Then, the TL output T_0 corresponds to $(1 - \alpha)n_1d$, where d is the dose. Thus, it varies linearly with dose. For doses exceeding the limiting dose at which this limit is surmounted the TL output will be $n_1d - n_{0\text{V}}$v.

$$T_0 = (1 - \alpha)\, n_1 d, \text{ (for } d < 10^{-3} \text{ rad)}$$

$$T_0 = n_1 d - n_{0\text{V}}, \text{ (for } d > 10^{-3} \text{ rad)} \tag{10.1}$$

Assuming this model, the maximum supralinearity factor of 3.48 was obtained at a 3×10^4 rad dose and the value was in good agreement with experimental results. However, the supralinearity factor at lower doses is abrupt, and thus the modification was much needed. The new modified equation is

$$\alpha = \alpha_0 \left[1 - e^{\frac{n_{0v}}{n_{1d}}} \right] \tag{10.2}$$

and the supralinearity factor is

$$S_\text{p} = \frac{1-\alpha}{1-\alpha_0} \tag{10.3}$$

The values obtained with these equations where in best agreement with experimental values.

10.3.4.2 Sensitization

Moharil proposed a twofold mechanism for sensitization. He assumed the number of nonluminescent centers in the sensitized material, n_{0S}, to be less than n_{0V}. Also, the postirradiation thermal annealing did not destroy all the luminescence centers. If d_p is the pre-irradiation dose, then during pre-irradiation $n_1 d_p$ traps are filled, out of $n_1 d_p$ traps, $\alpha n_1 d_p$ were destroyed nonradiatively, and $(1 - \alpha)n_1 d_p$ contributed to TL. After annealing $(1 - \alpha)n_1 d_p$ luminescence centers were destroyed by recombination and $\alpha n_1 d_p$ luminescence centers were left behind. During the test dose d_t, $n_1 d_t$ traps are filled. Then luminescence centers in sensitized materials are $n_1 d_t + \alpha n_1 d_p$ and nonluminescent centers will be

$$n_{0S} = n_{0V} - \alpha n_1 d_p,$$

Hence, in sensitized materials, the fraction of charges destroyed by nonradiative recombination α_s is given by

$$\alpha_s = \alpha_0 \left[1 - e^{-\frac{n_{os}}{n_{1s}}} \right], \quad (10.4)$$

$$\alpha_s = \alpha_0 \left[1 - e^{-\frac{n_{ov - n_1 d_p \alpha}}{n_1 d_t + k\alpha n_1 d_p}} \right]. \quad (10.5)$$

The sensitization factor is

$$\frac{S}{S_0} = \frac{1 - \alpha_S}{1 - \alpha_t}, \quad (10.6)$$

where α_t is the α-factor for the test dose d_t and can be obtained from Eq. 10.2 by substituting $d = d_t$. The sensitization factor can be calculated as a function of the pre-irradiation dose using the above equations. The maximum sensitization factor was obtained for a 10^5 rad irradiation dose, while the supralinearity factor was maximum at 3×10^4 rad. It was assumed that although traps saturate with an increase in dose, the luminescence centers continue to grow up to 10^6 rad. The maximum sensitization is obtained for 10^5 rad since the number of luminescent centers increases rapidly than the number of traps filled during the exposure. This model found success in explaining a large number of experimental facts but suffered from an inadequate understanding of certain aspects of TL [52].

10.4 Preparative Methods for TL Dosimetric Materials

Many different natural minerals and synthetic inorganic compounds exhibit the phenomenon of TL. But not all the compounds satisfy the requirements to be used as TLDs. Only a few of them satisfy the requirements. For dosimetric uses, a TL material is expected to have a relatively simple glow curve with the main peak at about 200°C, high sensitivity and stability, resistance to environmental factors, independence of the radiation energy, and good linearity in the specific useful range of dose [70]. One of the most important aspects in TL material designing is the mechanism of excitation energy transformation into the output of light. The accuracy of the measurements and the threshold detection dose depend on the efficiency of energy transformation. Depending on the nature of the TL material and its composition, the role of various mechanisms of energy transfer and energy losses is different. The composition may include intrinsic defects and defects induced by impurities. Preparation methods control the structure of these defects to a large extent. Hence, it is a matter of importance to find interrelations between the various preparation techniques, structural defects, and TL properties of the material.

The final properties of a TL material are generally controlled by the preparation methods. TL materials can be synthesized in the form of powders or crystals. The use or application of the material determines whether the material should be obtained in powdered form or in crystalline form or in the form of thin films. For structural characterization, obtaining a single crystal is important. Likewise each form has some or the other advantage and application in diverse fields. Depending on the physical form required for the TL material, the preparation method is selected. Evaporation and precipitation methods are commonly used to obtain polycrystalline powder, while the Czochralski technique, zone melting, and precipitation from solutions or molten phases are used to obtain single crystals. Some of the preparation methods are described next.

10.4.1 Methods for Polycrystalline Powder Preparation

10.4.1.1 Precipitation

Precipitation is the formation of a solid in a solution during a chemical reaction. In the co-precipitation technique, required amounts of a chemical compound along with trace amounts of an activator are added to form the desired compound. When the chemical reaction occurs, the solid formed called the phase sinks to the bottom of the solution precipitate. This can occur when an insoluble substance, the precipitate, is formed in the solution due to a reaction or when a compound has supersaturated the solution. The solute phase sinks to the bottom of the solution (though it will float if it is less dense than the solvent or form a suspension). Precipitation reactions can be used for making pigments, removing salts from water, and performing qualitative chemical analysis. The precipitated compound (after appropriate washing, drying of the product) is sintered at high temperature to facilitate diffusion of the activator and formation of an appropriate number of defects in the crystallite material. Rivera reported synthesis of a $CaSO_4$,Dy phosphor by precipitation [71].

10.4.1.2 Recrystallisation from a solution

In recrystallization from a solution, an appropriate amount of a chemical compound along with the desired amount of an activator are dissolved in a solvent, and the solution thus formed is slowly evaporated to dryness to form the required TLD phosphor. In this method, the reactants are added with the desired concentration of dopants in an acid solution and then mixed homogeneously. The mixture is placed in a sealed system. It is evaporated at high temperature for a given time by carrying the acid with an air or nitrogen flow. By varying both the temperature and the gas flow, crystallization can be controlled. After the evaporation, crystals whose dimensions depend on both the initial reagents and the type and concentration of the dopants are obtained, which are then washed several times to remove the remaining acid and submitted to thermal treatment at high temperature. The powder is finally pulverized to a desired grain size. $CaSO_4$:Dy and $CaSO_4$:Tm are prepared by this method [72].

10.4.1.3 Combustion synthesis

Combustion synthesis has been proposed to prepare oxide-based phosphors of smaller-size particles [73]. This method involves a highly exothermic reaction between an organic fuel and metal salts. The method makes use of the heat energy liberated by the redox exothermic reaction at a relative low igniting temperature between metal nitrates and urea as fuel. All the precursors are taken in the nitrate form and urea is used as a reducer. Stoichiometric quantities are taken by calculating oxidizer and reducer valences. All the precursors are crushed in a mortar pestle for 1 h. The resultant paste is placed in a furnace at a temperature of 550°C–600°C. A vigorous flame is produced with a temperature of 1600°C. The resultant product formed is homogeneous and foamy.

10.4.1.4 Sol-gel technique

In the sol-gel process, a powdered ceramic sample is prepared by an alkoxide compound, which is an organometallic compound. The method proceeds through a series of chemical processes, including hydrolysis, gelation, drying, and thermal treatment. The process involves formation of a sol (a colloidal suspension of solid particles in a liquid phase) through hydrolysis and polymerization reactions and subsequent attainment of a semirigid porous gel. The final product is then obtained by removal of the solvent and residuals from the pores of the gel by aging, drying, and annealing [74].

10.4.1.5 Solid-state diffusion

Solid-state diffusion is the most extensively used method for the preparation of phosphors. In the solid-state diffusion method, the desired materials are intimately mixed, and the mixture so formed is sintered at high temperature for the formation of a compound as well as uniform dispersal of the activators in the lattice. Sometimes, flux is used to lower the melting point of the compound, thereby facilitating better incorporation of the activator. This method involves the precursors in solid form, which are mixed to form a solid solution. Heat treatment, generally ranging from 500°C to 2000°C, is required to carry out the diffusion between the cation and the anion by overcoming the lattice energy. Normally, formation of the material takes place even by crushing and then sintering the materials only once, but sometimes it is necessary to break the grain boundary formed between the surface of materials, which becomes possible by adopting a repeated crushing and sintering method [42]. This procedure is sometimes referred to as the modified solid-state diffusion method.

10.4.2 Methods for Single-Crystal Synthesis

10.4.2.1 The Czochralski technique

This is one of the major melt-growth techniques. It is widely used for growing large-size single crystals for a wide range of commercial and technological applications. One of the main advantages of the Czochralski technique is the relatively high growth rate. This method involves the growth of a single crystal from a melt of the same composition. A crystal seed is brought into contact with the surface of the melt, whose temperature is maintained slightly above the melting point. Withdrawing the seed slowly from the surface causes the melt to be pulled, and the melt solidifies in the same crystallographic orientation as the original seed. The growing crystal and the crucible with the molten usually rotate in opposite directions during extraction so as to maintain a constant temperature. Usually the process employs an inert gas (argon or xenon) at high pressure to prevent volatilization losses [11].

10.4.2.2 The Bridgman technique

The Bridgeman technique uses a fairly simple apparatus. In this technique the crucible containing the molten material is translated along the axis of a temperature gradient in a furnace. In this method a full crucible of a melt having a composition that will result in a single crystal is lowered through a baffled zone within the furnace at a slow rate. The baffles within the furnace produces a uniform temperature decrease between the two zones. The temperature set point of the upper zone is just above the melting point of the material to be grown, and the lower zone is set just below the melting point of the material. A seed is introduced at the tip of the crucible at the center of the two zones, which is the freezing point of the material. Then, as the crucible is lowered through the freezing part of the temperature zone, the rest of the crucible will freeze as a single crystal [75].

10.4.2.3 The Verneuil method

The Verneuil method (also called the flame fusion technique) is used to grow a large quantity of crystals with high melting temperature. It was originally developed for producing synthetic gemstones. A specially designed burner is required so that the crystallites will melt during the short time during which they pass through the flame front. In this technique, a hopper containing powder is adjusted close to the burner, and the powder is allowed to pass through oxygen gas. It melts while passing through the flame and crystallizes slowly on a rotating pedestal, which is adjusted to a proper height with respect to the burner. Nowadays the technique is used to grow a variety of high-quality crystals such as corundum, spinel, rutile, and strontium titanate, which are widely applied in laser devices and precision instruments, as well as in thin-film technologies as substrates [76].

10.4.2.4 Zone melting

In this method crystals are grown by slow cooling of a small molten zone. It is a method of purifying crystals, in which a narrow region of a crystal is molten, and this molten zone is moved along the crystal. The crystal is formed with a preferential orientation under these conditions. The principle is that the segregation coefficient, which is the ratio of an impurity in the solid phase to that in the liquid phase, usually is less than 1. Therefore, at the solid–liquid boundary, the impurity atoms will diffuse to the liquid region. Thus, by passing a crystal boule through a thin section of a furnace slowly, such that only a small region of the boule is molten at any time, the impurities will be segregated at the end of the crystal. Because of the lack of impurities in the leftover regions, which solidify, the boule can grow as a perfect single crystal if a seed crystal is placed at the base to initiate a chosen direction of crystal growth. When high purity is required, such as in the semiconductor industry, the impure end of the boule is cut off and the refining is repeated. In zone refining, solutes are segregated at one end of the ingot in order to purify the remainder or to concentrate the impurities. Lithium fluoride has been produced by using this method by Soifer et al. (1965) [77].

10.4.2.5 The Kyropoulos method

The Kyropoulos method was developed for growing large alkali halide crystals by the German scientist Spyro Kyropoulos. After that large sapphire crystals were also grown by using this method. In this technique, first a melt is formed and then seed introduction is done. The melt is raised through a temperature gradient to form a single crystal. The single crystal will grow from the point where the seed comes in contact with the melt [78].

10.5 Evaluation of Essential Parameters Using TL Glow Curve Analysis

A variety of analysis methods are generally used to study the TL glow curve. TL glow curves can be analyzed systematically by following a certain procedure. The three parameters E (activation energy), s (frequency factor), and b (order of kinetics) can be evaluated using several of the standard analysis methods. Samples irradiated at different doses are generally examined to carry out the analysis of different trap fillings. The results of the different analyses must be in good agreement with each other. Any disagreement between their results must be thoroughly examined and resolved. This can be done by carrying out the analysis for glow peaks measured at different heating rates, at different irradiation doses, and in the form of powdered as well as bulk samples [47].

10.5.1 Initial Rise Methods

This method is used to evaluate the activation energy of the TL glow curve. At low temperatures, the number of trapped electrons n(T) is assumed to be constant. TL intensity I_C is smaller than 15% of the maximum TL intensity. In Fig. 10.4 the graph between ln(I) versus $1/kT$ is plotted and a straight line is obtained. The slope of this line gives the activation energy. This method is applied to the initial rise portion of the curve [47].

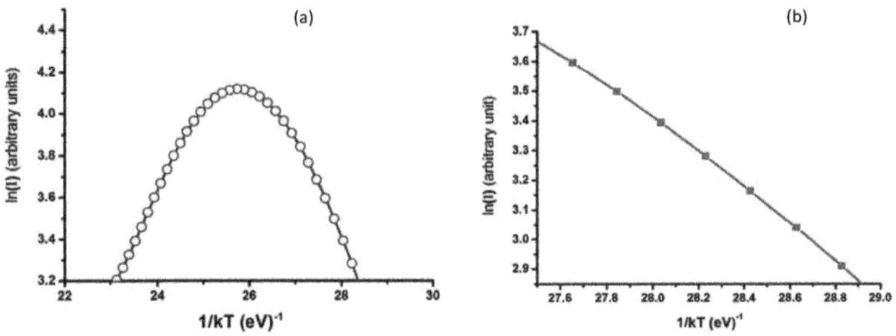

Figure 10.4 Application of the initial rise technique to the initial portion of a glow curve. Reproduced from Ref. [42] with permission of The Royal Society of Chemistry.

10.5.2 The Ilich Method

Ilich proposed a graphical method as an alternative to the initial rise method [79]. In this method, a cut-off temperature T_c, corresponding to the TL intensity I_c, is determined by drawing a tangent to the initial rise of the TL glow peak at the point I_c, as shown in Fig. 10.5. The intensity I_c is smaller than about 15% of the maximum TL intensity I_m. The intersection of the tangent to the temperature axis gives the lower temperature T_0. The activation energy is calculated using the formula

$$E = \frac{kT_C^2}{T_c - T_o},$$

where k is the Boltzmann constant.

10.5.3 Peak Shape Method or Chen's Method

This is a popular method of determining kinetic parameters such as E, b, and s by analyzing geometry of the TL glow curve. T_M is the peak temperature at the maximum, and T_1 and T_2 are, respectively, the temperatures on either side of T_M, as shown in Fig. 10.6.

Let $\tau = T_M - T_1$ be the half-width at the low-temperature side of the peak,

$\delta = T_2 - T_M$ is the half-width toward the fall-off side of the glow peak,

$\omega = T_2 - T_1$ is the total half-width, and
$\mu = \delta/\omega$ is the so-called geometrical shape or symmetry factor.

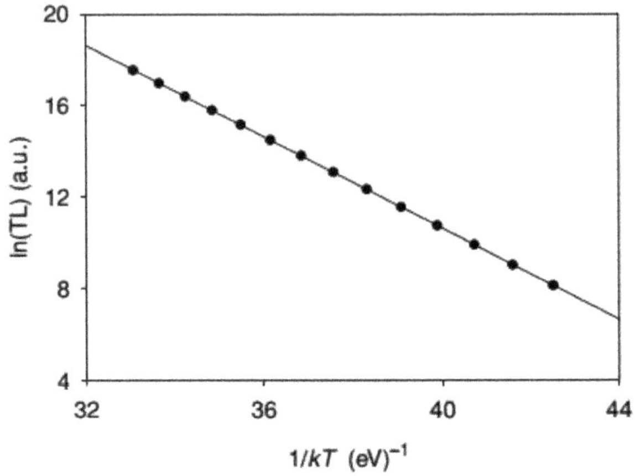

Figure 10.5 Application of the Ilich method to a TL glow curve. Reproduced from Ref. [42] with permission of The Royal Society of Chemistry.

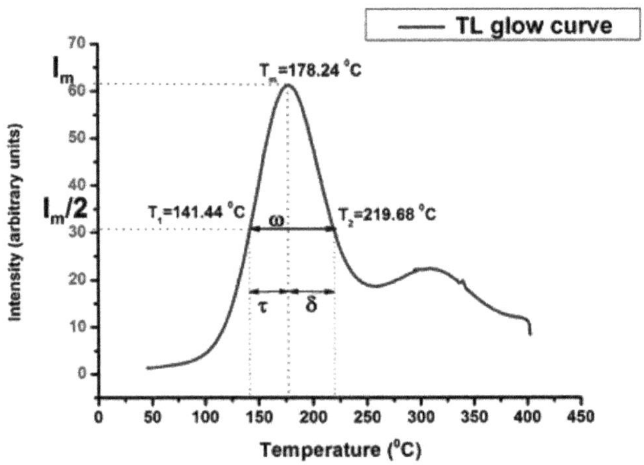

Figure 10.6 Geometrical shape quantities τ, δ, and ω. Reproduced from Ref. [42] with permission of The Royal Society of Chemistry.

Chen gave a generalized equation for activation energy:

$$E = c_\alpha \left(\frac{kT_M^2}{\alpha} \right) - b_\alpha (2kT_M) \tag{10.7}$$

The values of c_α and b_α are

$$c_\tau = 1.510 + 3.0(\mu - 0.42),\ b_\tau = 1.58 + 4.2(\mu - 0.42),$$
$$c_\delta = 0.976 + 7.3(\mu - 0.42),\ b_\delta = 0,$$
$$c_\omega = 2.52 + 10.2(\mu - 0.42),\ b_\omega = 1, \tag{10.8}$$

with $\mu = 0.42$ for the case of first-order TL glow peaks and $\mu = 0.52$ for the case of second-order peaks.

10.5.4 The Isothermal Decay Method

In this method, decay of luminescence is recorded with respect to time at constant temperature. We choose three temperatures on the rising side of the TL glow peak. The sample is irradiated with a suitable dose of X-rays and heated at one of the chosen temperatures, and by keeping that temperature constant, decay of luminescence is recorded. For the remaining two temperatures the procedure is repeated.

10.5.4.1 First-order kinetics

We know that for first-order kinetics the intensity of TL is given by

$$I = -\frac{dn}{dt},$$

and the solution for the isothermal case is

$$I(\mathrm{t}) = n_0 s e^{-(E/kT)} e^{\left[-ste^{-(E/kT)}\right]}. \tag{10.9}$$

Taking the log on both the sides of Eq. 10.9 yields

$$\ln[I(t)] = -\left[se^{-\left(\frac{E}{kT}\right)}\right]t + \left[\ln(n_0 s) - \frac{E}{kT}\right], \tag{10.10}$$

and we plot the graph of $\ln[I(t)]$ versus time, which is a straight line whose slope is given by Eq. 10.11

$$m = se^{\left(-\frac{E}{kT}\right)}. \tag{10.11}$$

When this plot does not give a straight line it can be understood that the glow peak does not obey first-order kinetics. Then we think of general-order kinetics. Again taking the log on both sides of Eq. 10.11 we get

$$\ln[m(T)] = -\frac{E}{kT} + \ln(s). \qquad (10.12)$$

Again $\ln(m)$ is plotted against time at different temperatures, which again is a straight line. The slope of this line is $m' = -E/k$ and the intercept is $\ln(s)$. Hence we get activation energy E and frequency factor s.

10.5.4.2 General-order kinetics

The TL intensity for general-order kinetics is given by

$$I = s' n^b e^{-\frac{E}{kT}},$$

where $s' = s/n_0^{b-1}$ is the pre-exponential factor. The solution of this equation for the isothermal decay case is

$$\left(\frac{I}{I_0}\right)^{\frac{b-1}{b}} - 1 = se^{\left(-\frac{E}{kT}\right)}(b-1)t. \qquad (10.13)$$

We plot the graph of $\left[\left(\frac{I}{I_0}\right)^{\frac{b-1}{b}} - 1\right]$ versus time, and it gives a straight line with slope

$$m = (b-1)e^{\left(-\frac{E}{kT}\right)}. \qquad (10.14)$$

Taking the log of Eq. 10.14 we get

$$\ln[m(T)] = -\frac{E}{kT} + \ln[s(b-1)]. \qquad (10.15)$$

The graph of $\ln[m(T)]$ is plotted against $1/T$, which is again a straight line with slope $m' = -E/k$. Determination of m at different temperatures will yield the frequency factor

$$s = \frac{(m_2)^{\frac{T_2}{T_2 - T_1}}}{(b-1)(m_1)^{\frac{T_2}{T_2 - T_1}}}. \qquad (10.16)$$

Here m_1 and m_2 are the slopes of the lines at temperatures T_1 and T_2. In this way the isothermal decay method can be used for analysing the TL glow curve and determining the trapping parameter.

10.5.5 The Variable Heating Rate Method

The maximum temperature T_M of a TL glow peak shifts to higher values as we change the linear heating rate β.

For two different heating rates β_1 and β_2 the activation energy will be given by Eq. 10.17 as

$$E = k\frac{T_{M_1}T_{M_2}}{T_{M_1} - T_{M_2}}\ln\left[\frac{\beta_1}{\beta_2}\left(\frac{T_{M_2}}{T_{M_1}}\right)^2\right]. \tag{10.17}$$

Hoogenstraaten obtained a linear relation by using differing heating rates:

$$\ln\left(\frac{T_M^2}{\beta}\right) = \frac{E}{kT_M} + \ln\frac{E}{sk} \tag{10.18}$$

The plot of $\ln\left(\frac{T_M^2}{\beta}\right)$ versus $1/T_M$ will give a straight line with slope E. Chen and Winer chose the method based on approximation in the integral in the general-order equation and obtained the following Eq. 10.19:

$$\ln\left[I_M^{b-1}\left(\frac{T_M^2}{\beta}\right)^2\right] = \frac{E}{kT_M} + c \tag{10.19}$$

This method is again used to find b [64]. By putting different values of b we obtained different sets of graphs. That value of b will be considered for which linearity is found.

10.6 Applications

The applications of TL stretch across various fields such as radiation dosimetry, geological dating, archeological dating, and solid-state physics [2, 15, 54]. Among various uses the most important application of the TL phenomenon is in radiation dosimetry. Radiation dosimetry is divided into several types; each type is described next.

10.6.1 Important Areas of Radiation Dosimetry

As the demand for the development of radiation technologies such as high-energy accelerators for research (Large Hadron Collider), hadron therapy of cancer, and fusion–fission power facilities (International Thermonuclear Experimental Reactor) are growing, the need of TLDs is also increasing. Professionals need accurate estimates of absorption of energy from radiation fields to obtain desired results. Measurements of these quantities form the basis of radiation dosimetry, and systems used for this purpose are referred to as dosimeters. TL materials have a superior position in radiation dosimetry for over the past four decades. Radiation dosimetry by using TLD materials can be divided into three main areas: (i) personnel monitoring, (ii) environmental dosimetry, and (iii) clinical dosimetry.

10.6.1.1 Personnel monitoring

It is essential to detect the radiation dose absorbed by individuals who are working in a radiative environment in order to take care of their health and prevent them from life-threatening effects of radiation. Excessive absorption of energetic radiations by individuals may lead to serious diseases such as cancer and genetic disorders for upcoming generations, so scientists working in laboratories carry TL badges with them to find the amount of radiation dose absorbed. There are maximum permissible dose (MPD) values permitted for the workers, above which workers are restricted to absorb radiation. Tissue-equivalent dosimeters are used for personal monitoring. The most commonly used TLD for personnel dosimetry is LiF:Mg,Ti (TLD-100) [80].

10.6.1.2 Environmental dosimetry

Environmental monitoring is becoming important as the pollution in the world is increasing day by day. Radiation released from nuclear fuel reprocessing, low-level waste disposal, nuclear power station accidents, and activities connected with the nuclear power industry needs to be estimated. In many countries TLD systems are placed near the radiation environment. TLD materials used for environmental dosimetry need to be have long-term stability, and a low dose of radiation (10 mGy) can be detected. Tissue equivalence of the TLD material is not necessary. For this purpose, $CaSO_4$:Dy, $CaSO_4$:Tm, CaF_2:Dy (TLD-200), and LiF:Mg,Cu,P phosphors are used [10, 15, 81, 82].

10.6.1.3 Neutron dosimetry

Fast neutrons are used in medical and health physics. They have more effect on biological tissues than X-rays and γ-rays and, hence, are used in curing diseases. Though for the detection of X-rays, γ-radiation, and β-radiation, many TLDs are readily available, for fast neutron dosimetry, TLD materials faces disadvantages of low cross section of the constituent elements to most of the nuclear reactions at high neutron energies. Hence neutron detection is a challenging task in TL dosimetry. For applying TLDs to be used in neutron detection, they need to be irradiated by keeping them in contact with hydrogen-rich organic materials used as proton radiators. Activation of some of the constituent elements of the material using threshold nuclear reactions, use of high-LET-sensitive TL peaks (250°C TL peak in CaF_2:Tm), and use of the Albedo technique are necessary. The Albedo technique is the most commonly used TLD technique for personnel monitoring in neutron fields; in this technique fast neutrons are moderated and backscattered from the body. These scattered thermal neutrons are detected by thermal neutron-sensitive TL detectors. A pair of TLDs is used to detect thermal and γ-neutrons [83]; among these one TLD should be sensitive to thermal neutrons and γ-rays (^{6}LiF), and the other should be sensitive to only γ-rays (^{7}LiF). The net dose due to neutrons (thermal, intermediate, and fast) can be estimated together with a γ-ray component. By mixing ^{6}LiF, TL materials could be made neutron sensitive and used for neutron dosimetry [84].

10.6.1.4 Charged particle dosimetry

Heavy charged particles have an inverse dose depth profile due to which they act as an ideal tool in the treatment of tumor cells, which are deeply situated in the living organism. Also, heavy charged particles are replacing conventional sources such as γ-rays because they have high relative biological effectiveness [85]. But in the course of the treatment of tumor cells with heavy charged particles, it is essential to measure the absorbed dose with great precession.

This has initiated investigations of TLDs to be used in charge particle dosimetry. However, TLD materials exhibit some disadvantages such as early saturation against high doses of a charged particle beam. But using nano-TLD materials this limitation can be removed. The nanomaterial forms of $CaSO_4$:Dy, LiF:Mg,Cu,P, $K_2Ca_2(SO_4)_3$:Eu, and $Ba_{0.97}Ca_{0.03}SO_4$:Eu possess good properties to be used for charge particle dosimetry [39].

10.6.1.5 Dosimetry in medical applications of radiation

As the use of radiation for curing diseases is growing, dosimetry of radiation is also gaining importance in the field of medicine. Medical applications such as radiation therapy, diagnostic radiology, and radiotherapy-mailed dosimetry make use of TLDs. A well-known TLD for medical applications is LiF:Mg,Ti. But nowadays LiF:Mg,Cu,P has considerably attracted researchers [15]. In the point of view of TLD applications medical dosimetry is important as improper use of TL materials in medical applications may lead to unacceptable bias and large uncertainties in the dose estimation.

10.6.1.6 Retrospective dosimetry

When accidents cause exposure to radiation with no dosimeters available in the area, the radiation dose can be detected from the material present in the immediate environment or the person in the vicinity of the incident/accident. This is called retrospective dosimetry. The atomic bomb explosions in Hiroshima and Nagasaki in Japan led to exposure of dangerous radiation in the environment. The dose was determined by using TL. Ceramic resistors, screens of cell phones, and other electronic components of personal electronic devices have shown interesting and promising results for their use in accidental dosimetry [2].

10.6.1.7 High-dose dosimetry

High-dose ionizing radiations are generally used in sterilization of medical and pharmaceutical products, food and flower preservation, treatment of electrical cables, and treatment of different materials [86]. In radiological accidents, uncontrolled quantity of radioactive material is released, which results in serious injuries and sometimes leads to death. In radiological installations and industries humans can become victims of such accidents. Dosimetry of such high doses of radiation is always mandatory. The $CaSO_4$:Dy TLD-100 dosimeter was investigated for high-dose radiation measurement, and it shows early saturation for a dosimetric peak (200°C) for 10^3 Gy of radiation exposure. Nanosized phosphors fulfil all the requirements, so they can be used for high-dose measurements; however, microsized TLDs encounter problems of early saturation at high doses, while the high-temperature peak (400°C) response is linear [87]. The deep traps present in the detector are responsible for this high-temperature peak when material is exposed to a high dose of irradiation. The Al_2O_3:C compound possess deep traps and has potential use in high radiation dosimetry. But these dosimeters were not having satisfactory performance, and the situation led to the development of nanoparticles of $CaSO_4$:Dy, LiF:Mg,Cu,P, and $K_3Na(SO_4)_2$:Eu [36, 88, 89].

10.6.1.8 Space dosimetry

Space contains a complex mix of charged particles: high-energy protons, electrons, and ions with an energy range over a wide area with changing density. Astronauts working in space are exposed to a radiation level that is a hundred times higher than the radiation on earth. Hence it is essential to determine the radiation dose absorbed by astronauts. No single TL detector is sufficient to be used across the full spectrum of radiations; combinations of dosimeters will be required. Currently this is done by passive TLDs. TLDs have proven remarkably effective in the assessment of doses in terrestrial applications and have been an important tool in space dosimetry for several decades. TLDs such as ^{6}Li (TLD 600), Al_2O_3, and polyethylene have proved to be useful in dosimetry of slow neutrons in space [2, 24].

10.6.1.9 Ultraviolet dosimetry

Ultraviolet (UV) radiations comprisenonionizing electromagnetic radiations having a wavelength from 180 nm to 400 nm. UV radiations have both disease-curing as well as disease-forming effects on humans, plants, and animals. Excess exposure to UV radiations can result in skin cancer, diseases related to the eyes, and burning of the skin. In agriculture, plant growth can be affected due to UV radiation

[90]. UV radiation has beneficial effects in the production of vitamin D. It is used for plant growth. In phototherapy treatment narrow-band UVB (313 nm) and UVA (315–400 nm) are used for treatment of psoriasis, vititligo, dermatitis, etc. It is necessary to optimize UV exposure to humans to allow consistent radiation exposure. The International Commission on Non-ionizing Radiation Protection (ICNRP) provided guidelines on the limit for UV exposure. Exposure to UVA irradiation should not exceed 1 mW/cm^2 for a time greater than 103 s and for UVC and UVB irradiation 1 J/cm^2 for a time less than 103 s. TL and PTTL of $CaSO_4$:Dy,CaF_2 and Al2O3:Si,Ti phosphors make them useful as UV radiation dosimeters.

10.6.1.10 Measurement of external radiation exposure in high-background-radiation areas

Naturally occurring background radiation is the main source of exposure for most people. This radiation occurs due to radioactive elements present in the earth's crust, such as uranium, thorium, potassium, and their decay products. This radiation also occurs from building materials. The dose range of this radiation typically varies from 1.5 to 3.5 mSv. A high level of background radiation is found in Kerala and Madras, states of India; here people are exposed to 15 mSv radiation dose per year. Also, in Brazil, Iran, and Sudan a radiation dose of 50 mSv/year is present. It is not possible to avoid this radiation completely, but we can detect and measure the radiation dose. By limiting the time of exposure experienced by people, the risk can be reduced. The determination of background radiation is possible by TL detectors. Rao et al. used $CaSO_4$:Tm and successfully estimated the dose of radiation present in villages of India [91].

10.6.2 Applications of Thermoluminescence to Archaeology

Dating of ancient pottery samples is one of the applications of TL. It can be used for dating before 30,000 years, and the minimum year will be 50 years. This is a quick and easy method of determining forgery. TL is taken from a quartz grain, which is collected from pottery. Kiln firing is the basis for the event that was started in the long past;

it will be considered as a "TL clock" for archeological dating. It is considered that TL from the time of crystallization is erased due to kiln firing. After kiln firing, pottery undergoes internal radiation due to radioactive elements present in clay, that is, uranium, thorium, and potassium. Also it absorbs external radiation [28, 92].

10.6.3 Prospecting of Radioactive Uranium-Bearing Ores

TL is one of the important tools for exploring radioactive uranium-containing ores. In this technique a quartz specimen is isolated from a field sample, and it is characterized by the TL technique. This sample is irradiated at various doses, and its TL is calculated. This facilitates the determination of the natural radiation received by the sample. The information of total radiation received by the material and its natural radioactivity give an estimate of the status of uranium as well as other radioactive sources present in the ore [93]. The TL technique can be applied for locating oil and gas reservoirs. In this method TL of naturally occurring rocks, minerals, or combinations of minerals is taken to determine the closeness of a sample to a hydrocarbon occurrence or the potential for hydrocarbons to occur in the environment of the sample. Such TL studies can be carried out on whole-rock samples, quartz, feldspars, carbonates, clays, micas, apatite, zircon, and other common or accessory rock-forming minerals.

10.6.4 Long Persistent Luminescent Phosphors

Long persistent luminescent phosphors (LLPs) are phosphors that emit UV, visible, and near-infrared (near-IR) light for minutes, hours, or days, even after excitation stops. These phosphors find applications in watch dials, defense, and street lighting. Phosphorescence shown by these phosphors is due to forbidden transitions of electrons. Trapping centers play a crucial role in photoenergy storage in LLPs. TL is widely used for studying trapping centers formed and their nature in LLPs. The TL glow curve for an LLP is obtained and by the initial rise method and the trap depth is determined [94].

10.6.5 Authenticity Testing of a Ceramic Piece of Art/Pottery

TL tests could be made on all kinds of fired pottery, such as earthenware, stoneware, and true porcelain. Even antique bronzes could be dated this way, provided they are made with a clay core, which is sometimes the case. In this, a small amount of sample is taken from the pottery and its TL spectrum is estimated. The TL test tells us whether the sample is new or antique; it measures the time of the object formed. Hence by using the TL technique authentication of a ceramic piece of pottery can be done.

10.6.6 TL of Nanophosphors

In nanophosphors, TL characteristics vary drastically compared to microphosphors. Grain size affects significantly the TL of phosphors, which was studied by Salah et al. in the nanosized and microsize $CaSO_4$:Dy commercial phosphor [95]. They found that in the case of microsize $CaSO_4$:Dy, TL saturates at a 10^3 Gy dose but not in nanosized $CaSO_4$:Dy and shows a TL response up to 10^4 Gy dose of radiation. Similarly, the nanosized LiF:Mg,Cu,P phosphor shows sensitivity for a wide dose from 0.1 to 10 Gy. Nowadays heavy ion beams are used in radiotherapy treatment. The dose of this heavy ion beam is needed to be calculated during treatment [85]. Microphosphors of commercially used TLDs saturate at a low-energy dose; hence they cannot be used for determination of such heavy ion doses. Nanocrystalline phosphors have the advantage of showing TL at high doses and can be used for the dosimetry of heavy ion beams.

10.6.7 Thermoluminescence Emission from Photosynthetic Systems

TL measurements are useful for the study of photosystem II (PS II) electron transport in intact leaves, in algal and cyanobacterial cells, and in isolated membrane complexes. The TL of photosynthetic systems was first observed by Arnold and Sherwood [96]. Photochemically separated charge pairs were stabilized on electron carriers by activation energy barriers that limit back reaction, that is, charge recombination, at physiological temperatures. TL originates from PS II via thermally stimulated delayed light emitted by singlet excited chlorophylls, generated by the recombination of the $S_2Q_A^-$, $S_2Q_B^-$, and $S_3Q_B^-$ charge pairs [97]. Besides radiative recombination, the nonradiative charge recombination pathways play an important role in TL emission. The rate of recombination can be made negligibly small by cooling the sample, before or immediately after an illumination, to temperatures where thermally stimulated reversal of charge separation becomes negligible. Then, progressive warming increases the recombination rate, thereby revealing the different types of charge pairs as successive TL bands [98].

10.7 Recent Trends in TLDs

Phosphors that are used for TL dosimetry are known as TLDs. Commonly used TLD materials are lithium fluoride, calcium fluoride, lithium borate, and calcium sulfate usually manufactured in the form of chips, pellets, small rods, or powder, which can then be encapsulated before irradiation. The most common applications of TLD materials are in personal dosimetry, environmental monitoring, and medical and reactor dosimetry. TLDs have applications in detection of γ-rays, X-rays, neutrons, and heavy ion beams. Several properties have to be considered for the choice of TLD materials for different uses:

- Materials must contain a number of traps and should give maximum efficiency of emission by radiative recombination.
- The glow curve must be simple to avoid complexities in determining various parameters. The peak should be well resolved in the case of a complex glow curve.
- It must have good storage stability at a large range of temperature and time.
- The TL response should be linear over a wide range of radiation doses.
- Fading in the material must be less.
- The material should be reusable so that after a TL readout, it can be exposed to radiations and the radiation dose can be measured.
- The phosphor should be highly sensitive.

These are the general requirements that need to be satisfied by the phosphor to be used in radiation dosimetry.

10.7.1 Sulfates

Sulfates are highly sensitive TL phosphors with a sensitivity four times more than LiF:Mg,Ti, having many properties such as inertness to environmental conditions, high storage, low cost, and a linear response over a wide range of doses of radiations. Yamashita et al. studied the properties of $CaSO_4$:Dy and $CaSO_4$:Eu and showed that sulfates form a good family for TL dosimetric applications. $CaSO_4$:Tm exhibits optically stimulated luminescence (OSL) under both IR and blue-light stimulation centered at 830 and 470 nm wavelengths, respectively. A new composite based on $CaSO_4$, using terbium as a dopant and ytterbium as a co-dopant ($CaSO_4$:Tb,Yb), was developed using a production route based on the Yamashita method for employment as a TLD. Its peak centered at 240°C, and although it was not the most intense, it could be used as a dosimetric peak due to its good sensitivity, reproducibility, and acceptable fading. Sulfates are non-tissue-equivalent phosphors, so they are not useful for clinical dosimetry; still sulfate materials are investigated and research is going on to improve their characteristics [99, 100]. Here we have discussed some recently developed sulfate phosphors.

10.7.1.1 $CaMg_3(SO_4)_4:Dy^{3+}$

This is a mixed sulfate phosphor developed by the acid distillation method by Kore et al. [101]. All the precursors, $CaSO_4$, $MgSO_4$, and a stoichiometric amount of Dy_2O_3, were dissolved in 15 mL of hot sulfuric acid. The mixture was heated at approximately 300°C for 20 h, and during heating, the highly active acid vapors were condensed using a water-cooled condenser assembly so as to prevent any spontaneous reactivity. The samples were cooled and then repeatedly washed with distilled water, followed by heating at 80°C to obtain small water-insoluble crystals. This phosphor shows 3.5 times higher sensitivity than the commercially available $CaSO_4:Dy^{3+}$ for 0.2 mol%, as shown in Fig. 10.7. The phosphor was exposed to different doses of ^{60}Co, ^{137}Cs, and C^{5+} radiation. Nearly the same type of TL response was observed for different types of radiation. The linearity response of TL glow curves for different types of radiation sources is shown in Fig. 10.8. This phosphor finds good application in carbon ion beam dosimetry along with γ-ray dosimetry over a wide range of exposures.

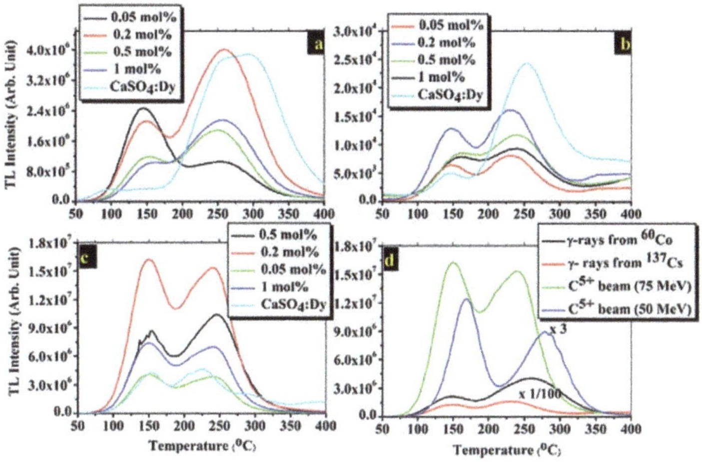

Figure 10.7 TL glow curves of the CMS phosphor irradiated with (a) γ-rays from a ^{60}Co source at a 15 Gy dose, (b) γ-rays from a ^{137}Cs source, and (c) C^{5+} ion beam at 75 MeV with a 216 kGy dose. (d) comparison between TL glow curves of the CMS phosphor C^{5+}. Reproduced from Ref. [26] with permission of The Royal Society of Chemistry.

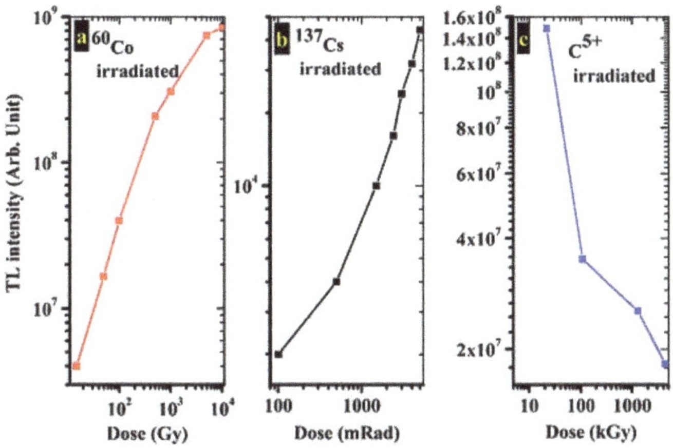

Figure 10.8 TL response curves of the CMS phosphor irradiated by γ-rays from (a) ^{60}Co, (b) ^{137}Cs, and (c) C^{5+} ion beams. Reproduced from Ref. [26] with permission of The Royal Society of Chemistry.

10.7.1.2 $K_2Ca_2(SO_4)_3:Eu^{2+},Ce^{3+}$

Salah et al. study of Ce with Eu in $K_2Ca_2(SO_4)_3$ resulted in increased TL sensitivity by 1.7 times that of the $K_2Ca_2(SO_4)_3:Eu^{2+}$ phosphor [102]. He prepared this co-doped phosphor by the solid-state route: first, K_2SO_4, chloride salts of Ce and Eu were dissolved in water and mixed homogeneously, and then they were heated to evaporate the water. This obtained powder was then ground with $CaSO_4$ in the ratio of 1:2. Heat treatment was given to the material at 1000°C for one day and it was cooled slowly. This phosphor was quenched at 700°C. The obtained powder was characterized by the X-ray diffraction technique and the formation of compound was confirmed. The dopant concentration was optimized to 0.05 mole%. This phosphor was irradiated by γ-ray source Cs^{137} with a dose rate of 1 Gy/s and a TL glow curve was obtained. It was found that due to the energy transfer mechanism from Ce to Eu, the TL intensity of $K_2Ca_2(SO_4)_3$:Eu,Ce increased by 1.7 times that of $K_2Ca_2(SO_4)_3$:Eu, as can be seen from Fig. 10.9. Also it is seven times more than $CaSO_4$:Dy. The glow peak is found at 160°C.

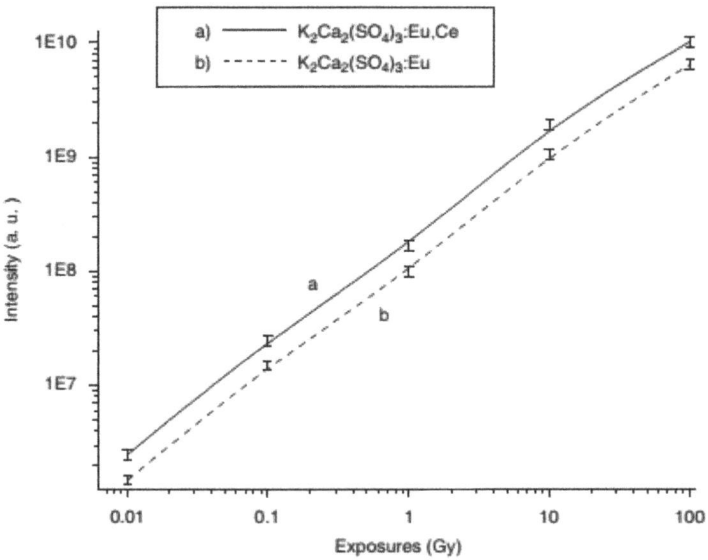

Figure 10.9 TL response of $K_2Ca_2(SO_4)_3$:Eu,Ce powder to Cs^{137}-rays and its comparison with $K_2Ca_2(SO_4)_3$:Eu. Reprinted from Ref. [102], Copyright (2006), with permission from Elsevier.

10.7.2 Aluminates

Aluminum oxide is one of the former materials studied for possible use as a dosimeter because chromium substitution for some atoms of aluminum in Al_2O_3 changes sapphire into ruby, which exhibits TL properties [103]. The alkaline earth aluminates are an important class of phosphorescence material because of their high quantum efficiency in the visible region, long persistence of phosphorescence, color purity and good stability, and good thermal and radiation resistance [104–106]. Aluminum oxide is a popular wide-band-gap dielectric material, which has several polymorphic phases of Al_2O_3 with different lattice structures and ordering. α-Al_2O_3 or corundum (with band gap $E_g = 9.4$ eV) is the most stable alumina phase; it is the most widely used for practical applications and comprehensively studied. It is used in optics and optoelectronics, in pure form as optical windows for the UV–IR range, and doped with Cr and Ti metal ions—as lasing material. Anion-defective α-Al_2O_3:C is a well-known dosimetric material [107, 108]. This material has also found application in industrial fields such as petrol chemistry, ceramic industry, pharmacy, and others. Some of the rare earth aluminates are discussed next.

10.7.2.1 $SrAl_4O_7:Dy^{3+}$

Choubey et al. synthesized 15% molar concentration $SrAl_4O_7:Dy^{3+}$ by the combustion method, wherein urea was used as a reducer and $Sr(NO_3)_2$ and $Al(NO_3)_3$ were used as nitrate precursors [109]. Dy_2O_3 was also converted into nitrate form by dissolving it in concentrated nitric acid. The synthesized material was subjected to UV absorption and TL studies. Absorption spectra displayed an absorption edge at 242 nm and the band gap was found to be 5.11 eV. For TL characterization the sample was irradiated to different doses and TL was taken. It was found that TL intensity strongly depends on doses of γ-radiation. A linear response was observed upto 2.36 kGy, suggesting that Dy(15%)-doped $SrAl_4O_7$ could be used for TL dosimetry upto 2.36 kGy. The TL glow curve is shown in Fig. 10.10. The maximum TL intensity corresponds to 172.84°C, 173.4°C, 170.6°C, 170.2°C, and 171.3°C for a dose value of 49.17 Gy, 147.5 Gy, 295 Gy, 885 Gy, and 2360 Gy, respectively.

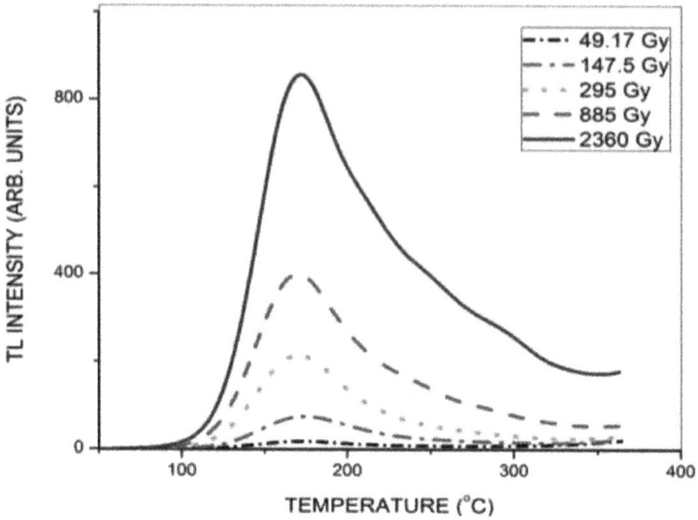

Figure 10.10 TL glow curve of $SrAl_4O_7$:Dy(15%) irradiated with different doses of γ-radiation. Reproduced with permission from Ref. [109], Copyright 2014, VBRI Press.

10.7.2.2 $CaAl_2O_4$:RE

Madhukumar et al. synthesized and studied TL properties of this phosphor [110]. The phosphor was synthesized by the solid-state synthesis method. $CaCO_3$ and Al_2O_3 and oxides of different rare earths (Ce, Dy, Sm) were taken as precursors. The precursors were mixed in a wetted medium and then calcinated at 900°C for 4 h. Again heat treatment was given for 7 h at temperature 1250°C. The phosphor was irradiated by ^{60}Co γ-rays, and it was found that 0.25% of Ce gave maximum intensity and had a linear response up to 4 kGy and a single peak was observed at 295°C.The phosphor could be used for high-temperature dosimetry.

10.7.2.3 $BaAl_2O_4$

$BaAl_2O_4$:Eu^{3+} is a long persistent phosphor studied by many researchers. Kore et al. synthesized $BaAl_2O_4$ by combustion synthesis. Precursors where taken in nitrate form, and urea was used as a reducer. Eu^{3+} was the dopant. The TL study was done by irradiating the sample with γ-radiation from ^{60}Co and a glow curve was recorded. This study is found to be useful in understanding the mechanism of TL [111].

10.7.3 Borates

Borates are excellent dosimeters with an effective atomic number close to human tissue (Z_{eff} = 7.42). Borates show higher sensitivity than TLD-100. Borates are low-cost materials and easy to produce. Schulman et al. (1967) were the first to study the TL properties of borates and Furetta et al. (2001) worked on $Li_2B_4O_7$:Cu,In and Li_2BO_7:Cu [112]. Borates like $K_2B_4O_7$, $LiSr_4(BO_3)_3$:Ce, Tb-doped $Ba_2Ca(BO_3)_2$, Tb-doped $Zn(BO_2)_2$, $Sr_2B_5O_9Cl$, MgB_4O_7:Dy,Na, and SrB_4O_7:Dy were also investigated [13, 113–115].

10.7.3.1 $Li_2B_4O_7$:Cu,In, $Li_2B_4O_7$:Cu, and $Li_2B_4O_7$:Mn

At energies such as 20 keV to 100 keV tissue equivalence is more important, since at these energies photoelectric interaction is proportional to the third power of the atomic number. Thus in this energy region materials show overresponse if they have a high atomic number compared to the atomic number of the composition of human tissue, so for the measurement of the dose in tissue, TL material should be tissue equivalent (Z_{eff} = 7.42). Though there are few dosimetric materials that are tissue equivalent, Lithium borate-based TL materials are tissue equivalent with an effective atomic number of 7.3 and hence are good dosimetric detectors for clinical purpose. First, the $Li_2B_4O_7$:Mn phosphor having low sensitivity was reported, but doping by Cu resulted in high sensitivity. A typical glow curve for $Li_2B_4O_7$:Mn consists of a single peak centered at 185°C when heated at 10°C/s, but the glow peak position has been reported to move to the (Mn) impurity level [116]. Another report has shown that TLD-800 has a general-order main glow peak with activation energy Ea = 0.58 eV at 170°C at a linear heating rate of 1°C/s [117]. However, this phosphor suffers from a drawback due to its existence only in powder form. To overcome this difficulty the Institute of Nuclear Sciences, Vinca, prepared $Li_2B_4O_7$:Cu,In and $Li_2B_4O_7$:Cu in the form of sintered pellets. Presently $Li_2B_4O_7$:Cu,In is a promising tissue-equivalent and high-sensitivity TL detector with a sensitivity close to that of LiF:Mg,Ti (TLD-100).

10.7.3.2 $LiCaBO_3$

$LiCaBO_3$ was synthesized by a high-temperature solid-state reaction by Jiang et al. by doping it with different rare earths [118]. It is difficult to prepare borate because it may get converted to a crystalline or glassy phase. Jiang reported that when Tm and Tb were used as the dopants the TL glow curve was complex and hence it was not usable for the dosimetry. The crystalline phase is necessary so as to get higher efficiency of the luminescence. Anishia et al. [119] prepared different activators doped with polycrystalline $LiCaBO_3$ phosphor powder by a simple solid-state sintering technique. It was found that rare earth–doped phosphors show good properties and silver codoping improves its properties. $LiCaBO_3$ was doped with Tm^{3+} and it was found that it shows good TL characteristics, with the main peak at 196°C. Two major well-separated useful dosimetric peaks are observed at 230°C and 430°C, along with a low-temperature peak in $LiCaBO_3:Tm^{3+}$. The TL sensitivity of these peaks to γ-radiation is about 8 times that of TLD-100 and a linear response up to 10^3 Gy. Figure 10.11 shows the TL glow of $LiCaBO_3:Tm^{3+}$.

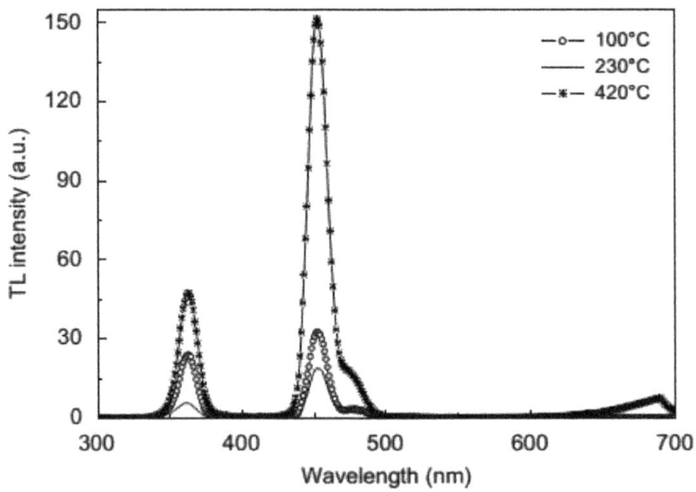

Figure 10.11 TL emissions in γ-irradiated $LiCaBO_3:Tm^+$ phosphor at different glow peak temperatures. Reprinted from Ref. [118], Copyright (2008), with permission from Elsevier.

10.7.4 Phosphates

Lightfoot et al. (1991) showed $LiCaPO_4$ to be isostructural with the sulfate $LiNaSO_4$ [120]. Though $ABPO_4$ materials were found to be efficient hosts, luminescence studies in $LiCaPO_4$ are few.

10.7.4.1 $LiCaPO_4$

Recently, More et al. observed efficient TL in phase-pure $LiCaPO_4$:Eu prepared by following the procedure prescribed by Lightfoot et al. [120, 121]. The phosphor was found to be about 4 times more sensitive than LiF TLD-100, as shown in Fig. 10.12. The $LiCaPO_4:Eu^{2+}$ phosphor was prepared using a solid-state reaction. $CaHPO_4$ and Li_2CO_3 were taken in stoichiometric quantities and the dopant was added. Precursors were mixed and fired at 300°C for 1 h. Then the mixture was fired at 650°C for 2 h and 800°C for 3 days in an alumina crucible. It was reduced in a charcoal medium for 1 h at 800°C. It was found that sensitivity of the material was 4 times that of LiF TLD-100 and hence can be useful in TL studies.

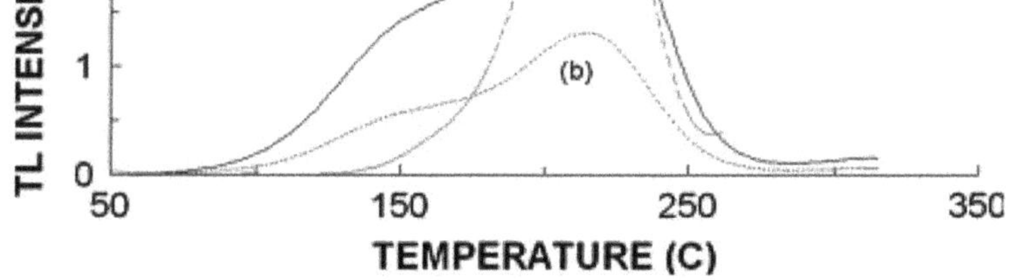

Figure 10.12 Thermoluminescence glow curves of the $LiCaPO_4$ phosphor. Reprinted from Ref. [121], Copyright (2011), with permission from Elsevier.

10.7.4.2 $NaLi_2PO_4:Eu^{3+}$

PL properties of this phosphor were studied by Shinde and Dhoble et al., while TL properties were discussed by Singh and Sahare et al. [122, 123]. This low-tissue-equivalent phosphor (Z_{eff} = 10.8) was prepared by solid-state synthesis. It has potential applications in the field of radiation dosimetry as it is a highly sensitive phosphor. The dopant concentration of 5 mol% is found to be the optimum concentration so as to get high TL sensitivity. LiOH and NaH_2PO_4 were mixed in the ratio of 2:1 in an agate mortar with ethanol, heated for 400°C for 12 h, and crushed to obtain fine particles. Again heat treatment was given for 600°C and 800°C for the same time period. Z_{eff} was determined by the formula

$$Z_{eff} = \sqrt[3]{\frac{\sum n_i X_i^4}{\sum n_i X_i}},$$

where n_i is the fractional part by weight of the whole compound occupied by the element the atomic number of which is Z_i. It was found to be 10.8 [124].

Different TL analyses were performed and the effect of annealing at different temperatures showed changes in the intensity of the TL glow curve, with a maximum intensity at 700°C. At different doses of γ-radiations TL studies were conducted for the samples. The experimental results showed that $NaLi_2PO_4$ is a highly sensitive and thermally stable TLD phosphor, and 0.5 mol% of Eu^{3+} impurity concentration gave the maximum sensitivity. The glow curve structure is simple, consisting of a main dosimetric peak at around 458 K (with two small shoulders on both sides at around 400 and 500 K), sufficiently above RT, and shows low fading. The dose response of the synthesized phosphor shows sublinear behavior up to 10 Gy of the dose and later becomes linear before it starts saturating at around 1.0 kGy, as shown in Fig. 10.13. Thus it has a wider range of doses compared to the standard commercially available phosphors, especially for high doses (e.g., TLD-700H saturates at around 100 Gy). The sensitivity was found to be high, that is, 18, 2, and 1.5 times more than the commercially used phosphors TLD-100, TLD-400, and TLD-900, respectively. A simple method of preparation, a simple glow curve structure, a wide range of dose response, low fading, and good reusability make it a suitable candidate for radiation monitoring using the TL technique.

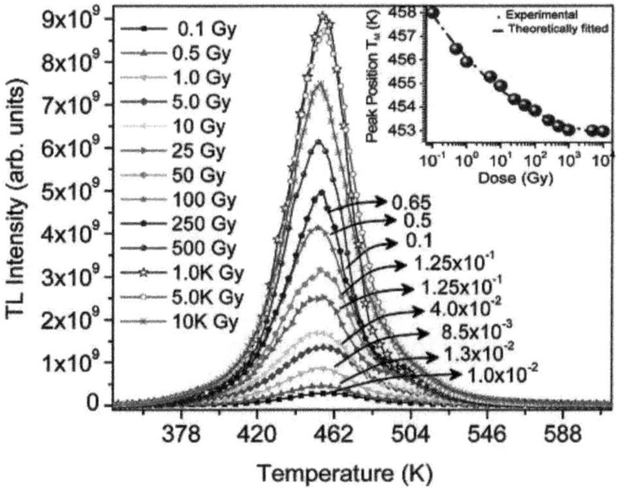

Figure 10.13 Thermoluminescence glow curves of the $NaLi_2PO_4:Eu^{3+}$ phosphor. Reprinted from Ref. [123], Copyright (2013), with permission from Elsevier.

10.8 Possibility of Using Organic Materials for TLDs

Organic materials have found a number of applications in various fields, but it is considered that they are not useful for dosimetric purposes. Though organic materials have replaced inorganic materials in various applications such as conductors, photoconductors, and organic photovoltaic materials, it is considered that no organic compound is suitable for dosimetric applications. Organic phosphors have an advantage that they are tissue equivalent, and hence, with sufficient sensitivity, hydrogen-rich organic phosphors will be useful in neutron dosimetric applications. These materials are excluded from dosimetric applications because they rarely show a glow peak above RT and virtually no TL could be recorded for irradiation at RT [125].

Still there is possibility of obtaining a good TL glow peak from organic compounds. Here some examples of organic materials are given and their TL properties discussed.

10.8.1 *n*-Alkanes

Hashimoto et al. reported a sharp glow peak at the solid-state transition point in *n*-alkanes C_{20}–C_{40} [126]. The glow peak was observed at 304–308 K, which is low, and fast fading was expected. A good signal-to-noise ratio was expected due to the sharp peak. In other organic phosphors, distribution over a wide temperature interval (full width at half maximum [FWHM] = 50 K) resulted in low intensities of TL. Thus, if a sharp peak could be obtained, as observed for C_{20}–C_{40}, and if the intensities could be enhanced by doping with dyes, then the organic phosphor will prove to be a good TLD to some extent. Moharil et al. recorded the TL of octacosane and hexacotane, which are commercially available materials [127]. These samples were exposed to γ-rays from a Co^{60} source at RT and at 268 K. Glow curves for the samples were obtained by keeping the heating rate 10 min/K. When heated in a N_2 medium, two sharp peaks were obtained, one at 322 K and another close to the melting point of the compound. The peak close to the melting point was sharp. Close to the melting point molecular oxygen can diffuse into the crystal rapidly and works as a strong quencher of detrapped electrons [128].

A weak and broad glow peak is observed around 350 K for low-temperature irradiation and reading under nitrogen, though its melting point is higher (372 K). The possible reason for the absence of a sharp peak may be the absence of solid-state transitions around the melting point of C_{60} or insufficient purity of the sample. Hashinoto et al. found that blends of alkanes having different carbon atoms exhibited broad glow peaks. Thus the use of these alkanes seems impractical for TL dosimetry.

10.8.2 Makrofol Polycarbonate Copolymer

Makrofol was annealed at 373 K for 3 h and exposed to γ-rays of 51.6 C/kg at RT. An intense peak was observed at 405 K. Vandeschumen et al. reported that doping an organic compound with dyes results in enhanced TL peak intensity [129]. But this trend was violated in the compound doped by Moharil et al. because the polycarbonate used by Vandeschumen et al. was almost amorphous and crystallinity increased during doping, while the sample used by Moharil et al.

was already crystalline and further doping resulted in degrading the crystallinity of the material. This indicates that crystallinity is a must for obtaining good TL.

10.8.3 Polyethylene

Moharil used a linear polyethylene having a molecular weight of 52,000. It was crystallized into an extended-chain crystal (ECC). At 503 K, it exhibited marked TL that shows four peaks, wherein peak P_4 is around the melting point, which is 500 times weaker than the glow peak in LiF TLD-100. Polyethylene exhibits a TL glow curve when irradiated at RT. The intensity of P_4 increases several times when ECCs are exposed to a benzene-toluene mixture. On storing it in darkness for a week and giving an exposure of γ intensity, the TL response enhances by 10 times. This intensity is 50 times less than LiF TLD-100. However, for low exposures a large amount of crystal is needed. Since there is a problem of non-uniform heating in large crystals, peak P_4 broadens artificially. These difficulties can be overcome by using better techniques such as heating by a gas jet or by doping with dyes. Problem of preparing large-size crystals can be solved by using polyethylene fibers because they become highly oriented when drawn to a high draw ratio. For heating in a nitrogen atmosphere, the TL glow curve of high-density polyethylene (HDPE) fiber drawn to a ratio of 45 shows small peaks at high temperature. Lower-temperature peaks are seen when heating is done in an air medium. This phosphor shows a TL glow peak at 431 K when it is irradiated at RT. The peak was observed only when the glow curve was recorded with nitrogen.

From these examples it is shown that well-oriented polymer fiber as well as polymer crystals exhibit good TL properties, which are significant for further studies.

10.9 Emerging Trends

Scientists nowadays are focusing their efforts to investigate TL characteristics of equipment or materials that are worn or carried by individuals in day-to-day life. This is because the risk of accidental radiation exposure due to human error or machine error will not give a second chance to use commercial dosimeters. In accidental radiation exposure, measurement of radiations should be done immediately. It is found that in such incidents glass displays and electrical components in cell phones, computers, audio-video devices, and USB cards possess the potential to be used for dosimetry purpose [130]. These electrical materials are formed on a ceramic substrate made of Al_2O_3, BeO, and SiO_2. These substrate materials are also components of commercial dosimetric materials and hence expected to show good TL emission [131]. Mrozik et al. investigated the screens of Nokia, Sony Ericsson, Samsung, and HTC touch-screen cell phones [132]. They removed the thin-film transistor (TFT) layer on each screen and observed that a good TL peak was obtained around 200°C for 500 mGy irradiation. Fiedler et al. studied the TL of chip inductors from cell phones [133]. Chip inductors were removed from old and new cell phones and irradiated with a 800 mGy dose. The TL obtained for chips shows TL peaks at 100°C, 140°C, 170°C, 270°C, and 340°C. The peak at 270°C was found to be stable, signifying the potential of chip conductors to be used for accidental dosimetry.

10.10 Conclusions

Review of the literature of TL materials reveals that a large number of materials, including several sulfates, aluminates, borates, and phosphates, satisfy the requirements needed to be a good TL phosphor. Several synthesis routes can be adopted to prepare these materials, depeding on economic feasibility and the scale of production. For industrial purposes, solid-state synthesis should be preferred over other methods due to its ability to provide large-scale production. Significant efforts have been made to devise more sensitive TL phosphors and tissue-equivalent TL materials. Different materials show a variety of TL characteristics differing from one another, which brings out a wide range of applications in environmental, personal, and clinical dosimetry. On the basis of the applications and their requirements, the necessary TL materials are developed. TL materials find a major role in radiation dosimetry, often used in the form of badges, rings, lockets, and watches. With a view to developing good and efficient TL materials, research is going on at a vast level and frequently newer and newer materials are being reported.

References

1. Bartlett, D. T. (2008). 100 years of solid state dosimetry and radiation protection dosimetry, *Radiat. Meas.*, **43**(2–6), pp. 133–138.
2. Bhatt, B. C. and Kulkarni, M. S. (2013). Thermoluminescent phosphors for radiation dosimetry, *Defect Diffus. Forum*, **347**, p. 179.
3. Bos, A. J. J. (2001). High sensitivity thermoluminescence dosimetry. *Nucl. Instrum. Methods Phys. Res., Sect. B*, **184**(1–2), pp. 3–28.
4. Kron, T. (1999). Applications of thermoluminescence dosimetry in medicine, *Radiat. Prot. Dosim.*, **85**(1), pp. 333–340.
5. Lewandowski, C. M. (2015). The effects of brief mindfulness intervention on acute pain experience: an examination of individual difference, **1**(3), p. 160.
6. Cameron, J. R. and Kenny, G. N. (1963). Thermoluminescence radiation dosimetry with lithium fluoride, *Radiat. Res.*, **19**, pp. 199–200.
7. Cameron, J. R., Suntharalingam, N. and Kenny, G. N. (1968). *Thermoluminescence Dosimetry* (University of Wisconsin Press, London).
8. Chandra, B. and Bhatt, R. C. (1981). Isothermal decay characteristics of Li2B4O7 : Mn phosphor, *Nucl. Instrum. Methods*, **184**(2–3), pp. 557–559.
9. Cooke, D. W. and Hogstrom, K. R. (1980). Thermoluminescent response of LiF and $Li_2B_4O_7$:Mn to pions, *Phys. Med. Biol.*, **25**(4), pp. 657–666.
10. Vij, D. R. and Singh, N. (1997). Thermoluminescence dosimetric properties of beryllium oxide, *J. Mater. Sci.*, **32**(11), pp. 2791–2796.
11. Rawat, N. S., Kulkarni, M. S., Tyagi, M., Ratna, P., Mishra, D. R., Singh, S. G., et al. (2012). TL and OSL studies on lithium borate single crystals doped with Cu and Ag, *J. Lumin.*, **132**(8), pp. 1969–1975.
12. Yukihara, E. G. G., Coleman, A. C. C., Bastani, S., Gustafson, T., Talghader, J, Daniels, A., et al. (2015). Particle temperature measurements in closed chamber detonations using thermoluminescence from $Li_2B_4O_7$:Ag,Cu, MgB_4O_7:Dy,Li and $CaSO_4$:Ce,Tb, *J. Lumin.*, **165**, pp. 145–152.
13. Furetta, C., Prokic, M., Salamon, R. and Kitis, G. (2000). Dosimetric characterisation of a new production of MgB_4O_7:Dy,Na thermoluminescent material, *Appl. Radiat. Isot.*, **52**(2), pp. 243–250.

14. Szabó, P. P., Pradhan, A. S. and Chandra, B. (1984). Some dosimetric parameters of MgB_4O_7:Dy sintered pellets and $CaSO_4$:Dy Teflon discs—a comparative study, *Int. J. Appl. Radiat. Isot.*, **35**(5), pp. 415-417.

15. Moscovitch, M. (1999). Personnel dosimetry using LiF:Mg,Cu,P, *Radiat. Prot. Dosim.*, **85**(1), pp. 49–56.

16. Nakajima, T., Murayama, Y., Matsuzawa, T. and Koyama, A. (1978). Development of a new highly sensitive LiF thermoluminescence dosimeter and its applications, *Nucl. Instrum. Methods*, **157**(1), pp. 155–162.

17. Yamashita, T., Nada, N., Onishi, H. and Kitamura, S. (1971). Calcium sulfate activated by thulium or dysprosium for thermoluminescence dosimetry. *Health Phys.*, **21**(2), pp. 295–300.

18. Chagas, M. A. P., Nunes, M. G., Campos, L. L. and Souza, D. N. (2010). TL properties of anhydrous CaSO4:Tm improvement, *Radiat. Meas.*, **45**(3–6), pp. 550–552.

19. Sunta, C. (1984). A review of thermoluminescence of calcium fluoride, calcium sulphate and calcium carbonate, *Radiat. Prot. Dosim.*, **8**, pp. 25–32.

20. Binder, W. and Cameron, J. R. (1969). Dosimetric properties of CaF_2: Dy, *Health Phys.*, **17**(4), pp. 613–618.

21. Danilkin, M., Lust, A., Kerikmäe, M., Seeman, V., Mändar, H. and Must, M. (2006). CaF_2:Mn extreme dosimeter: effects of Mn concentration on thermoluminescence mechanisms and properties, *Radiat. Meas.*, **41**(6), pp. 677–681.

22. Wang, H., Lin, S.-W., Weng, P.-S. and Hsu, P.-C. (1995). A study of the main glow peak for CaF_2:Mn, *Appl. Radiat. Isot.*, **46**(9), pp. 869–873.

23. Bortolin, E., Fattibene, P., Furetta, C. and Onori, S. (1993). ESR of Mg_2SiO_4:Tb TL phosphors, *Appl. Radiat. Isot.*, **44**(1–2), pp. 327–330.

24. Tawara, H., Masukawa, M., Nagamatsu, A., Kitajo, K., Kumagai, H. and Yasuda, N. (2011). Characteristics of Mg_2SiO_4:Tb (TLD-MSO-S) relevant for space radiation dosimetry, *Radiat. Meas.*, **46**(8), pp. 709–716.

25. Ionescu, C., Hoeck, V., Gruian, C. and Simon, V. (2014). Insights into the EPR characteristics of heated carbonate-rich illitic clay, *Appl. Clay Sci.*, **97–98**, pp. 138–145.

26. Kore, B. P., Dhoble, N. S., Lochab, S. P. and Dhoble, S. J. (2014). A new highly sensitive phosphor for carbon ion dosimetry, *RSC Adv.*, **4**(91), pp. 49979–49986.

27. Kunghatkar, R. G., Dhoble, S. J. and Hemne, P. S. (2016). New UVA-emitting $BaAlBO_3F_2:Eu^{2+}$ phosphor with PL and TL properties for phototherapy lamp, *Luminescence*, **31**, pp. 1503–1512.

28. Richter, D., Waiblinger, J., Rink, W. J. and Wagner, G. A. (2000). Thermoluminescence, electron spin resonance and 14C-dating of the late middle and early upper palaeolithic site of Geißenklosterle cave in Southern Germany, *J. Archaeol. Sci.*, **27**, pp. 71–89.

29. Wani, J. A., Atone, M. S. and Dhoble, S. J. (2013). The influence of different synthesis routes on thermoluminescence of $CaSO_4$:Dy,P,Ce phosphors, *Adv. Mater. Lett.*, **4**(5), pp. 363–367.

30. Zvonarev, S. V., Kortov, V. S., Shtang, T. V, Ananchenko, D. V. and Petrovykh, K. A. (2015). Effect of structural changes on luminescent and dosimetric properties of nanoscale aluminum oxide, *Appl. Radiat. Isot.*, **95**, pp. 44–47.

31. Chougaonkar, M. P., Shetty, P. G., Mayya, Y. S., Puranik, V. D., Joshi, M. L. and Kushwaha, H. S. (2008). Environmental gamma radiation monitoring around nuclear power stations in India, an Indian scenario, *J. Nucl. Sci. Technol.*, pp. 619–622.

32. Nambi, K. S. V., Mehta, N. K., Basu, A. S. and Gopalkrishanan, S. (1995). Validation of environmental TLDs used in Indian nuclear programmes, *Radiat. Prot. Dosim.*, **58**, pp. 229–231.

33. Bradley, D. A., Hugtenburg, R. P., Nisbet, A., Abdul Rahman, A. T., Issa, F., Mohd Noor, N. and Alalawi, A. (2012). Review of doped silica glass optical fibre: their TL properties and potential applications in radiation therapy dosimetry, *Appl. Radiat. Isot.*, **71** Suppl, pp. 2–11.

34. Akselrod, M. S., Kortov, V. S. and Gorelova, E. A. (1993). Preparation and properties of alpha-Al_2O_3:C, *Radiat. Prot. Dosim.*, **47**(1–4), pp. 159–164.

35. Akselrod, M. S., Kortov, V. S., Kravetsky, D. J. and Gotlib, V. I. (1990). Highly sensitive thermoluminescent anion-defect alpha-Al_2O_3:C single crystal detectors, *Radiat. Prot. Dosim.*, **33**(1–4), pp. 119–122.

36. Salah, N., Sahare, P. D. and Rupasov, A. A. (2007). Thermoluminescence of nanocrystalline LiF:Mg, Cu, P, *J. Lumin.*, **124**(2), pp. 357–364.

37. Sahare, P. D., Bakare, J. S., Dhole, S. D., Ingale, N. B. and Rupasov, A. A. (2010). Synthesis and luminescence properties of nanocrystalline LiF:Mg,Cu,P phosphor, *J. Lumin.*, **130**(2), pp. 258–265.

38. Kortov, V. S. (2010). Nanophosphors and outlooks for their use in ionizing radiation detection, *Radiat. Meas.*, **45**(3–6), pp. 512–515.

39. Salah, N. (2011). Nanocrystalline materials for the dosimetry of heavy charged particles: a review, *Radiat. Phys. Chem.*, **80**(1), pp. 1–10.

40. Lakshmanan, A. R. (1999). Photoluminescence and thermostimulated luminescence processes in rare-earth-doped $CaSO_4$ phosphors, *Prog. Mater. Sci.*, **44**, pp. 1–187.

41. Sharma, R., Bisen, D. P., Dhoble, S. J., Brahme, N. and Chandra, B. P. (2011). Mechanoluminescence and thermoluminescence of Mn doped ZnS nanocrystals, *J. Lumin.*, **131**(10), pp. 2089–2092.

42. Kore, B. P., Dhoble, N. S. and Dhoble, S. J. (2014). Synthesis and thermoluminescence characterization of $Na_6Mg(SO_4)_4$:RE (RE = Ce, Tb) phosphors, *Radiat. Meas.*, **67**, pp. 35–46.

43. Klasens, H. A., Garlick, G. F. J. and Gibson, A. F. (2002). Discussion on "the electron trap mechanism of luminescence in sulphide and silicate phosphors," *Proc. Phys. Soc.*, **61**(1), pp. 101–102.

44. Zhang, C.-X., Tang, Q., Luo, D.-L., Qiu, Z.-R., Leung, P. and Stokes, M. (2002). Investigation of the TL mechanism and defect structure in $MgSO_4$ doped with Eu and Mn, P impurities, *Radiat. Meas.*, **35**(2), pp. 161–166.

45. Kadukar, M. R., Yawalkar, P. W., Choithrani, R. and Dhoble, S. J. (2015). Mechanoluminescence, thermoluminescence, photoluminescence studies on $Ca_3Y_2Si_3O_{12}$:RE $^{3+}$ (RE $^{3+}$ = Dy $^{3+}$ and Eu^{3+}) phosphors, *Luminescence*, **30**, pp. 1219–1225.

46. Pimpalshende, D. M. and Dhoble, S. J. (2014). Evaluation of trapping parameters of γ-rays irradiated Dy^{3+} -doped $LaPO_4$ phosphors, *Luminescence*, pp. 1019–1026.

47. Pagonis, V., Kitis, G. and Furetta, C. (2006). *Numerical and Practical Exercises in Thermoluminescence* (Springer, New York).

48. Wiedemann, E. and Schmidt, G. C. (1895). Ueber Luminescenz von festen Körpern und festen Lösungen, *Ann. Phys.*, **292**(10), pp. 201–254.

49. Curie, M. (1961). *Radioactive Substances* (Greenwood Press, New York, Westpoint).

50. Wick, F. G. and Slattery, M. K. (1928). Thermoluminescence excited by x-rays, *J. Opt. Soc. Am.*, **16**(6), pp. 398–407.

51. Halperin, A. and Braner, A. A. (1960). Evaluation of thermal activation energies from glow curves, *Phys. Rev.*, **117**(2), pp. 408–415.

52. Chen, R. and Kirsh, Y. (1981). *Analysis of Thermally Stimulated Processes* (Elsevier Science Limited).

53. Horowitz, Y. and Olko, P. (2004). The effects of ionisation density on the thermoluminescence response (efficiency) of LiF:Mg,Ti and LiF:Mg,Cu,P, *Radiat. Prot. Dosim.*, **109**(4), 331–348.

54. Horowitz, Y. S., Belaish, Y. and Oster, L. (2006). Theories of TL systems: failures, successes, conflicts, trends: insights into possible future materials and techniques, *Radiat. Prot. Dosim.*, **119**(1–4), pp. 124–129.

55. Bjarngard, B. (1965). Luminescence dosimetry. In *Proceedings of The Third International Conference on Dosimetry* (Standford, California).

56. Rieke, J. K. . and Daniels, F. (1957). Thermoluminescence studies of aluminum oxide, *J. Phys. Chem.*, **61**, pp. 629–633.

57. Mehta, S. K. and Sengupta, S. (1979). Luminescence and colour centres in Al_2O_3:Si,Ti, *Nucl. Instrum. Methods*, **164**(2), pp. 349–354.

58. Dhoble, S. J., Gedam, S. C., Nagpure, I. M., Godbole, S. V., Bhide, M. K. and Moharil, S. V. (2008). Luminescence of Cu^+ in halosulphate phosphor, *J. Mater. Sci.*, **43**(9), pp. 3189–3196.

59. Gedam, S. C., Dhoble, S. J. and Moharil, S. V. (2008). Eu^{2+} and Ce^{3+} emission in sulphate based phosphors, *J. Lumin.*, **128**(1), pp. 1–6.

60. Thakre, P. S., Gedam, S. C., Dhoble, S. J. and Atram, R. G. (2011). Luminescence investigations on sulfate apatite $Na_6(SO_4)_2FCl$:RE (RE=Dy, Ce or Eu) phosphors, *J. Lumin.*, **131**(12), pp. 2683–2689.

61. Azorin, J. and Gutiérrez, A. (1990). Dosimetric characteristics of LiF: Mg, Cu, P TL phosphor prepared at ININ, Mexico, *Radiat. Prot. Dosim.*, **33**, pp. 283–286.

62. Moharil, S. V. (1984). On the mobile interstitial model for TL in LiF-TLD 100, *J. Phys. C: Solid State Phys.*, **17**, pp. 531–536.

63. Ausin, V. and Rivas, J. L. A. (2001). Thermoluminescence and annealing of F-centres in KCl gamma irradiated at room temperature, *J. Phys. C: Solid State Phys.*, **5**(1), pp. 82–96.

64. Murti, Y. V. G. S. and Murthy, K. R. N. (1974). Thermoluminescence of alkali halides irradiated at 80 K, *J. Phys. C: Solid State Phys.*, **7**(10), pp. 1918–1928.

65. Horowitz, Y. and Rosenkrantz, M. (1990). Track interaction theory for heavy charged particle TL supralinearity, *Radiat. Prot. Dosim.*, **31**, pp. 71–76.

66. Attix, F. H. (1975). Further consideration of the track–interaction model for thermoluminescence in LiF(TLD–100), *J. Appl. Phys.*, **46**(1), p. 81.

67. Lakshmanan, A. R., Bhatt, R. C. and Vohra, K. G. (1979). γ-Radiation-induced sensitization, supralinearity, and photostimulated thermoluminescence in LiF TLD-100. Correlation studies, *Phys. Status Solidi A*, **53**(2), pp. 617–625.

68. Moharil, S. V. (1983). Track interaction model for supralinearity and sensitization in LiF-TLD 100, *Phys. Status Solidi A*, **79**(2), pp. 599–605.

69. Moharil, S. V. (1983). A phenomenological model for supralinearity and sensitization in LiF-TLD 100, *Phys. Status Solidi A*, **77**(1), pp. 413–420.

70. Azorin, J. (2014). Preparation methods of thermoluminescent materials for dosimetric applications: an overview, *Appl. Radiat. Isot.*, **83**, pp. 187–191.

71. Rivera, T., Roman, J., Azorín, J., Sosa, R., Guzmán, J., Serrano, A. K., et al. (2010). Preparation of $CaSO_4$:Dy by precipitation method to gamma radiation dosimetry, *Appl. Radiat. Isot.*, **68**(4–5), pp. 623–625.

72. Das, S., Amarnath Reddy, A. and Vijaya Prakash, G. (2012). Near white light emission from K^+ ion compensated $CaSO_4$:Dy^{3+},Eu^{3+} phosphors, *Ceram. Int.*, **38**(7), pp. 5769–5773.

73. Singh, V., Gundu Rao, T. K. K. and Zhu, J.-J. J. (2008). A rapid combustion process for the preparation of $MgSrAl_{10}O_{17}$:Eu^{2+} phosphor and related luminescence and defect investigations, *J. Lumin.*, **128**(4), pp. 583–588.

74. Bajpai, N., Tiwari, A., Khan, S. A., Kher, R. S., Bramhe, N. and Dhoble, S. J. (2013). Effects of rare earth ions (Tb, Ce, Eu, Dy) on the thermoluminescence characteristics of sol-gel derived and γ-irradiated SiO_2 nanoparticles, *Luminescence*, **29**, pp. 669–673.

75. Daniel, D. J., Madhusoodanan, U., Nithya, R. and Ramasamy, P. (2014). Irradiation effect on luminescence properties of fluoroperovskite single crystal ($LiBaF_3$:Eu^{2+}), *Radiat. Phys. Chem.*, **96**, pp. 135–139.

76. Pastor, R. C. and Pastor, A. C. (1966). Crystal growth above 2200°C by the Verneuil method, *Mater. Res. Bull.*, **1**(4), pp. 275–282.

77. Soifer, L. M., Shakhnovich, M. I., Chubenko, A. I. and Blank, A. B. (1965). Vacuum ultraviolet absorption of lithium fluoride crystals grown by zone melting, *J. Appl. Spectrosc.*, **2**(1), pp. 15–18.

78. Takeuchi, N., Adachi, M. and Inabe, K. (1979). A thermoluminescent center in X-irradiated NaCl:Cu^+ single crystals, *J. Lumin.*, **18–19**, pp. 897–900.

79. Ilich, B. M. (1979). Method of determination of trap depth, *Sov. Phys. Solid State*, **21**, pp. 1880–1882.

80. Horowitz, Y., Fuks, E., Datz, H., Oster, L., Livingstone, J. and Rosenfeld, A. (2011). Mysteries of LiF TLD response following high ionization density irradiation: Glow curve shapes, dose response, the unified interaction model and modified track structure theory, *Radiat. Meas.*, **46**(12), pp. 1342–1348.

81. Weng, P. and Chen, K. (1974). Response of $CaSO_4$(Dy) phosphor to neutrons, *Nucl. Instrum. Methods*, **117**(1), pp. 89–92.

82. Zahedifar, M., Sadeghi, E. and Harooni, S. (2012). Thermoluminescence characteristics of the novel CaF_2:Dy nanoparticles prepared by using the hydrothermal method, *Nucl. Instrum. Methods Phys. Res., Sect. B*, **291**, pp. 65–72.

83. Bhadane, M. S., Mandlik, N., Patil, B. J., Dahiwale, S. S., Sature, K. R., Bhoraskar, V. N. and Dhole, S. D. (2016). $CaSO_4$:Dy microphosphor for thermal neutron dosimetry, *J. Lumin.*, **170**, pp. 226–230.

84. Furuta, Y. and Tanaka, S. (1972). Response of ^{6}LiF and ^{7}LiF thermoluminescence dosimeters to fast neutrons, *Nucl. Instrum. Methods*, **104**(2), pp. 365–374.

85. Salah, N. (2008). Carbon ions irradiation on nano- and microcrystalline $CaSO_4$:Dy, *J. Phys. D: Appl. Phys.*, **41**(15), p. 155302.

86. Lakshmanan, A. R. and Bhatt, R. C. (1979). High-level gamma-ray dosimetry using common TLD phosphors, *Phys. Med. Biol.*, **24**(6), pp. 1258–1267.

87. Hasan, F., Kitis, G. and Charalambous, S. (1985). The thermoluminescence behavior of $CaSO_4$-Dy (TLD-900) for doses up to 30 Mrad, *Nucl. Instrum. Methods Phys. Res., Sect. B*, **9**(2), pp. 218–222.

88. Rivera, T., Espinoza, A., Alvarez, R., Jime, Y., Von, S. M., Alvarez, R. and Jiménez, Y. (2012). High energy electron beams characterization using $CaSO_4$:Dy^{3+} PTFE phosphors for clinical therapy applications, *Appl. Radiat. Isot.*, **70**(7), pp. 1304–1306.

89. Sahare, P. D., Ranjan, R., Salah, N. and Lochab, S. P. (2007). $K_3Na(SO_4)_2$: Eu nanoparticles for high dose of ionizing radiation, *J. Phys. D: Appl. Phys.*, **40**(3), pp. 759–764.

90. Noh, A. M., Amin, Y. M., Mahat, R. H. and Bradley, D. A. (2001). Investigation of some commercial TLD chips/discs as UV dosimeters, *Radiat. Phys. Chem.*, **61**(3–6), pp. 497–499.

91. Rao, N. S., Parial, K., Koide, H. and Sengupta, D. (2015). Measurement of environmental external gamma radiation dose rate outside the dwellings of Southern coastal Odisha, Eastern India, **109**(3), pp. 14–17.

92. Sabtu, S. N., Mahat, R. H., Amin, Y. M., Price, D. M., Bradley, D. A. and Maah, M. J. (2015). Thermoluminescence dating analysis at the site of an ancient brick structure at Pengkalan Bujang, Malaysia, *Appl. Radiat. Isot.*, **105**, pp. 182–187.

93. Hochman, M. B. M. and Ypma, P. J. M. (1984). Thermoluminescence as a tool in uranium exploration, *J. Geochem. Explor.*, **22**(1–3), pp. 315–331.

94. Li, Y., Gecevicius, M. and Qiu, J. (2016). Long persistent phosphors—from fundamentals to applications, *Chem. Soc. Rev.*, **45**, pp. 2090–2136.

95. Salah, N. (2015). Thermoluminesence of gamma rays irradiated $CaSO_4$ nanorods doped with different elements, *Radiat. Phys. Chem.*, **106**, pp. 40–45.

96. Arnold, W. and Sherwood, H. K. (1957). Are chloroplasts semiconductors? *Proc. Natl. Acad. Sci. U. S. A.*, **43**(6217), pp. 105–114.

97. Rutherford, A. W., Govindjee, and Inoue, Y. (1984). Charge accumulation and photochemistry in leaves studied by thermoluminescence and delayed light emission, *Proc. Natl. Acad. Sci. U. S. A.*, **81**(4), pp. 1107–1111.

98. Ducruet, J. M. and Vass, I. (2009). Thermoluminescence: experimental, *Photosynth. Res.*, **101**(2–3), pp. 195–204.

99. Dhoble, S. J., Gedam, S. C., Nagpure, I. M., Godbole, S. V. and Bhide, M. K. (2007). Photoluminescence and thermoluminescence characteristics of $KXSO_4Cl$: Eu (X = Zn or Mg) halosulfate phosphor, *J. Phys. D: Appl. Phys.*, **40**(19), pp. 6039–6043.

100. Di, W. H., Willinger, M. G., Ferreira, R. A. S., Ren, X. G., Lu, S. Z. and Pinna, N. (2008). Citric acid-assisted hydrothermal synthesis of luminescent TbPO(4):Eu nanocrystals: controlled morphology and tunable emission, *J. Phys. Chem. C*, **112**(48), pp. 18815–18820.

101. Kore, B. P., Dhoble, N. S., Lochab, S. P. and Dhoble, S. J. (2015). A high sensitive phosphor for dosimetric applications, *AIP Conf. Proc.*, **1665**, p. 140029.

102. Salah, N. and Sahare, P. D. (2006). TL, PL and energy transfer in $K_2Ca_2(SO_4)_3$:Eu^{2+},Ce^{3+}, *Radiat. Meas.*, **41**(6), pp. 665–670.

103. McKeever, S. W. S. (1983). *Thermoluminescence of Solids* (Cambridge University Press, Cambridge).

104. Aitasalo, T., Dereń, P., Hölsä, J., Jungner, H., Krupa, J. C., Lastusaari, M., et al. (2003). Persistent luminescence phenomena in materials doped with rare earth ions, *J. Solid State Chem.*, **171**(1–2), pp. 114–122.

105. Meléndrez, R., Arellano-Tánori, O., Pedroza-Montero, M., Yen, W. M. and Barboza-Flores, M. (2009). Temperature dependence of persistent

luminescence in β-irradiated SrAl2O_4:Eu^{2+}, Dy^{3+} phosphor, *J. Lumin.*, **129**(7), pp. 679–685.

106. Mothudi, B. M., Ntwaeaborwa, O. M., Kumar, A., Sohn, K. and Swart, H. C. (2012). Phosphorescent and thermoluminescent properties of $SrAl_2O_4$:Eu^{2+}, Dy^{3+} phosphors prepared by solid state reaction method, *Physica B*, **407**(10), pp. 1679–1682.

107. G ksu, H. Y., Bulur, E. and Wahl, W. (1999). Beta dosimetry using thin layer a-Al_2O_3:C TL detectors, *Radiat. Prot. Dosim.*, **84**(1), pp. 451–455.

108. Pagonis, V., Chen, R. and Lawless, J. L. (2007). A quantitative kinetic model for Al_2O_3:C: TL response to ionizing radiation, *Radiat. Meas.*, **42**(2), pp. 198–204.

109. Choubey, A. K., Brahme, N., Dhoble, S. J., Bisen, D. P. and Ghormare, K. B. (2014). Thermoluminescence characterization of gamma-ray irradiated Dy^{3+} activated $SrAl_4O_7$ nanophosphor, *Adv. Mater. Lett.*, **5**(7), pp. 396–399.

110. Madhukumar, K., Babu, K. R., Prasad, K. C. A., James, J., Elias, T. S., Padmanabhan, V. and Nair, C. M. K. (2006). Thermoluminescence dosimetry of rare earth doped calcium aluminate phosphors, *Bull. Mater. Sci.*, **29**(2), pp. 119–122.

111. Kore, B. P., Dhoble, N. S. and Dhoble, S. J. (2014). Study of anomalous emission and irradiation effect on the thermoluminescence properties of barium aluminate, *J. Lumin.*, **150**, pp. 59–67.

112. Furetta, C., Prokic, M., Salamon, R., Prokic, V. and Kitis, G. (2001). Dosimetric characteristics of tissue equivalent thermoluminescent solid TL detectors based on lithium borate, *Nucl. Instrum. Methods Phys. Res., Sect. A*, **456**(3), pp. 411–417.

113. Cheng, W.-D., Zhang, H., Lin, Q.-S., Zheng, F.-K. and Chen, J.-T. (2001). Syntheses, crystal and electronic structures, and linear optics of $LiMBO_3$(M = Sr, Ba) orthoborates, *Chem. Mater.*, **13**(5), pp. 1841–1847.

114. Jiang, L. H., Zhang, Y. L., Li, C. Y., Hao, J. Q. and Su, Q. (2007). Thermoluminescence properties of Ce^{3+}-doped $LiSr_4(BO_3)_3$ phosphor, *Mater. Lett.*, **61**(29), pp. 5107–5109.

115. Manam, J. (2004). Thermally stimulated luminescence studies of undoped and doped $K_2B_4O_7$ compounds, *Nucl. Instrum. Methods Phys. Res., Sect. B*, **217**(2), pp. 314–320.

116. Lakshmanan, A. R., Chandra, B. and Bhatt, R. C. (1981). Dosimetry characteristics of thermoluminescent $Li_2B_4O_7$:Cu phosphor, *Radiat. Prot. Dosim.*, **1**(3), pp. 191–198.

117. Kafadar, V. E., Yildirim, R. G., Zebari, H. and Zebari, D. (2014). Investigation of thermoluminescence characteristics of $Li_2B_4O_7$:Mn (TLD-800). *Thermochim. Acta*, **575**, pp. 300–304.

118. Jiang, L. H., Zhang, Y. L., Li, C. Y., Pang, R., Hao, J. Q. and Su, Q. (2008). Thermoluminescence characteristics of rare-earth-doped $LiCaBO_3$ phosphor, *J. Lumin.*, **128**(12), pp. 1904–1908.

119. Anishia, S. R., Jose, M. T., Annalakshmi, O., Ponnusamy, V. and Ramasamy, V. (2010). Dosimetric properties of rare earth doped $LiCaBO_3$ thermoluminescence phosphors, *J. Lumin.*, **130**(10), pp. 1834–1840.

120. Lightfoot, P., Pienkowski, M. C., Bruce, P. G. and Abrahams, I. (1991). Synthesis and structure of $LiCaPO_4$ by combined X-ray and neutron powder diffraction, *J. Mater. Chem.*, **1**(6), p. 1061.

121. More, S. D., Wankhede, S. P., Kumar, M., Chourasiya, G. and Moharil, S. V. (2011). Synthesis and dosimetric characterization of $LiCaPO_4$:Eu phosphor, *Radiat. Meas.*, **46**(2), pp. 196–198.

122. Shinde, K. N. and Dhoble, S. J. (2013). A novel reddish-orange phosphor, $NaLi_2PO_4$:Eu^{3+}, *Luminescence*, **28**(1), pp. 93–96.

123. Singh, M., Sahare, P. D. and Kumar, P. (2013). Synthesis and dosimetry characteristics of a new high sensitivity TLD phosphor $NaLi_2PO_4$:Eu^{3+}, *Radiat. Meas.*, **59**, pp. 8–14.

124. Murty, R. C. (1965). Effective atomic numbers of heterogeneous materials, *Nature*, **207**(4995), pp. 398–399.

125. Fleming, R. J. and Hagekyriakou, J. (1984). Thermoluminescence in polymers, *Radiat. Prot. Dosim.*, **8**, p. 99.

126. Hashimoto, T., Shimada, H. and Sakai, T. (1977). Thermoluminescence from γ-ray irradiated normal paraffins, *Nature*, **268**(5617), pp. 225–226.

127. Moharil, S. V., Deshmukh, B. T. and Muthal, P. L. (1989). On the possibility of using organic phosphors in thermoluminescence dosimetry, *J. Polym. Sci., Part B: Polym. Phys.*, **27**(13), pp. 2577–2585.

128. Muthal, P. L., Deshmukh, B. T. and Moharil, S. V. (1987). Thermoluminescence in extended chain crystals of polyethylene, *Phys. Status Solidi A*, **101**(2), pp. 583–587.

129. Vanderschueren, J., Linkens, A. and Niezette, J. (1986). Effect of doping on thermoluminescence in polymers. I. 9,10-phenanthrenequinone-doped polydiancarbonate, *J. Polym. Sci. Part B: Polym. Phys.*, **24**(3), pp. 697–706.

130. Beerten, K., Woda, C. and Vanhavere, F. (2009). Thermoluminescence dosimetry of electronic components from personal objects, *Radiat. Meas.*, **44**(5–6), pp. 620–625.

131. Pradhan, A. S., Kim, J. L. and Lee, J. I. (2016). Use of ubiquitous materials for the estimation of accidental exposures, *J. Med. Phys.*, **37**(3), pp. 121–123.

132. Mrozik, A., Marczewska, B., Bilski, P. and Kłosowski, M. (2014). Investigation of thermoluminescence properties of mobile phone screen displays as dosimeters for accidental dosimetry, *Radiat. Phys. Chem.*, **104**, pp. 88–92.

133. Fiedler, I. and Woda, C. (2011). Thermoluminescence of chip inductors from mobile phones for retrospective and accident dosimetry, *Radiat. Meas.*, **46**(12), pp. 1862–1865.

Index

For Product Safety Concerns and Information please contact our EU
representative GPSR@taylorandfrancis.com
Taylor & Francis Verlag GmbH, Kaufingerstraße 24, 80331 München, Germany

www.ingramcontent.com/pod-product-compliance
Ingram Content Group UK Ltd.
Pitfield, Milton Keynes, MK11 3LW, UK
UKHW022200190926
475702UK00003B/122

* 9 7 8 9 8 1 4 7 7 4 4 9 9 *